Groundwater Modeling
by the
Finite Element Method

Water Resources Monograph 13

Groundwater Modeling by the Finite Element Method

Jonathan Istok

American Geophysical Union

Published under the aegis of the AGU Water Resources Monograph Board.

Library of Congress Cataloging-in-Publication Data

Istok, J. D.
Groundwater modeling by the finite element method
Jonathan Istok
p. cm. — (Water resources monograph ; 13)
Includes bibliographical references.
ISBN 0-87590-317-7
1. Groundwater flow—Mathematical models. 2. Finite element
method.   I. Title.   II. Series.
TC176.I79   1989
551.49'01'5118—dc20   89-18404
                                                                                      CIP

Printed in the United States of America.

To my wife and friend Joan who never let me forget that
there is more to life than "a bunch of stupid equations!"

Contents

PART THREE   APPLICATIONS

APPENDICES

            Problems   462

Appendix II   Derivation of Equations of Transient Groundwater Flow   464

            Problems   468

Appendix III Derivation of Equations of Solute Transport   469

            Problems   477

Appendix IV   Concepts from Linear Algebra used in the Finite Element Method   478

            Problems   484

Appendix V   Properties of Selected Aquifer Materials   485

            REFERENCES   487

            INDEX   493

Preface

The finite element method is now widely used to solve a variety of important problems in the field of groundwater hydrology. Thus a clear understanding of the method is essential to scientists and engineers working in this field. The goal of this book is to provide the reader with the basic skills needed to use the finite element method to solve "real-world" problems. Examples are used throughout the text to illustrate each step in the solution process.

The text is divided into three parts. In the Part 1, the basic concepts of the finite element method are presented. Chapters 2 to 6 present a step-by-step application of the finite element method to problems of groundwater flow and solute transport. Techniques for dividing an aquifer system into a suitable finite element mesh are described in Chapter 2. A number of practical "rules" are presented for locating and numbering nodes and for selecting the proper element type, size, and shape. In Chapter 3, the method of weighted residuals is used to derive the integral formulations of the equations governing steady-state and transient groundwater flow and solute transport through saturated and unsaturated porous media. The derivations are presented for one-, two-, and three-dimensional problems; the integral formulations for axisymmetric problems are derived in an exercise. The derivations are unique in that they do not require the reader to be familiar with advanced mathematics; although a basic understanding of differential and integral calculus is assumed. In Chapter 4, the important properties of element interpolation functions are discussed at length. Expressions are presented for computing the element conductance, capacitance, advection-dispersion, and sorption matrices for each element type and for assembling the element matrices into a system of linear or nonlinear equations. Procedures are also presented for modifying this system of equations for different types of boundary conditions (such as constant head or no-flow boundaries) Procedures for solving the system of equations are presented in Chapter 5. In Chapter 6, procedures are presented for using computed values of hydraulic head (or pressure head or solute concentration) to compute rates of groundwater flow and solute flux.

In Part 2, the computer implementation of the finite element method is discussed. Each chapter contains a description of one or more FORTRAN subroutines, example input data and output, and the complete source code listing. The same subroutines are also available on diskette. These subroutines, although intended for instructional purposes, contain many advanced features. Most importantly the "modular" design of these subroutines means that they form convenient "building blocks" for several different finite element computer programs.

Part 3 is concerned with applications of the material in Parts 1 and 2 to "real-world" problems. Chapter 20 discusses applications to problems of regional groundwater flow. Chapter 21 discusses solute transport with application to problems of groundwater contamination from point and diffuse sources.

Useful supplementary information is contained in the Appendices. Detailed derivations of the equations of groundwater flow and solute transport are presented in Appendices I, II, and III. A concise review of important topics from linear algebra is in Appendix IV. Typical values of physical properties for selected aquifer materials are in Appendix V.

The author would like to thank the many individuals and institutions who helped to make this book possible. Former graduate students Richard Cooper, Jeffrey Smith and Alan Rea helped with the development of the computer programs. Sang Bong Lee carefully read (and reread) early versions of the manuscript and helped me correct computational errors in the example problems. Janet Lee helped me with the computer programming (but any remaining bugs are my fault!). Joan Istok drew the example finite element meshes in Chapter 2. I also wish to thank Jing Leung, Jonathan Yap, and Elvina Lim - who typeset the entire book on a Macintosh computer. They did a terrific job!

The Oregon Agricultural Experiment Station and the U.S. Geological Survey provided financial support for this project. I also wish to thank the students in my groundwater modeling classes who taught me a lot about the finite element method while I was trying to explain it to *them*. I also wish to thank Francis Hall for his interest in this project. It provided a needed lift when my enthusiasm had almost run out.

Jonathan Istok

Department of Civil Engineering
Oregon State University
Corvallis, Oregon 97331

PART ONE

BASIC CONCEPTS

Chapter 1

INTRODUCTION

1.1   GROUNDWATER FLOW AND SOLUTE TRANSPORT MODELS

Groundwater is an important natural resource. Many agricultural, domestic, and industrial water users rely on groundwater as the sole source of low-cost, high-quality water. However, in recent years it has become apparent that many human activities can have a negative impact on both the quantity and quality of the groundwater resource. Two examples are the depletion of the groundwater resource by excessive pumping and the contamination of the groundwater resource by waste disposal and other activities. One way to objectively assess the impact of existing or proposed activities on groundwater quantity and quality is through the use of *groundwater flow and solute transport models*.

In developing a groundwater flow or solute transport model the analyst begins by preparing a *conceptual model* consisting of a list of the physical and chemical processes suspected of governing the behavior of the system being studied (e.g., groundwater seepage through soil and rock pores, laminar and turbulent water flow through large pores and rock fractures, and solute transport by advection, dispersion, and diffusion). The next step is to translate the conceptual model into mathematical terms and the result is a *mathematical model* consisting of one or more *partial differential equations* and a set of *auxillary conditions*. Solutions of the equations subject to the auxiliary conditions can be obtained by one of several methods (see below). If numerical methods are used, the collection of partial differential equations, auxilliary conditions, and numerical algorithms are referred to as a *numerical model*. If a computer program is used to implement the numerical model (as is usually done) the computer program is sometimes referred to as a *computer model* .

Existing mathematical models of groundwater flow and solute transport are necessarily greatly simplified descriptions of reality. The movement of water and solutes from the surface of the earth to the aquifer, and through the aquifer to a point of water use is an extremely complex phenomenom and many of the physical and chemical processes involved are poorly understood. It is therefore difficult to translate all of these processes into a single set of equations that apply equally well to all situations encountered in practice. Instead the usual approach has been to classify groundwater flow and solute transport problems into categories and to develop mathematical and numerical models for each category separately. In this book we will consider five such categories: (1) steady-state, saturated groundwater flow, (2) steady-state, unsaturated groundwater flow, (3) transient (or time-dependent), saturated groundwater flow, (4) transient, unsaturated groundwater flow, and (5) solute transport. The partial differential equations used in mathematical models of groundwater flow and solute transport for each problem category are:

1.   The Steady-State, Saturated Flow Equation:

$$\frac{\partial}{\partial x}\left(K_x\frac{\partial h}{\partial x}\right) + \frac{\partial}{\partial y}\left(K_y\frac{\partial h}{\partial y}\right) + \frac{\partial}{\partial z}\left(K_z\frac{\partial h}{\partial z}\right) = 0 \qquad (1.1)$$

2.   The Steady-State, Unsaturated Flow Equation:

$$\frac{\partial}{\partial x}\left(K_x(\psi)\frac{\partial \psi}{\partial x}\right) + \frac{\partial}{\partial y}\left(K_y(\psi)\frac{\partial \psi}{\partial y}\right) + \frac{\partial}{\partial z}\left(K_z(\psi)\left(\frac{\partial \psi}{\partial z} + 1\right)\right) = 0 \qquad (1.2)$$

3.   The Transient, Saturated Flow Equation:

$$\frac{\partial}{\partial x}\left(K_x\frac{\partial h}{\partial x}\right) + \frac{\partial}{\partial y}\left(K_y\frac{\partial h}{\partial y}\right) + \frac{\partial}{\partial z}\left(K_z\frac{\partial h}{\partial z}\right) = S_s\frac{\partial h}{\partial t} \qquad (1.3)$$

4.   The Transient, Unsaturated Flow Equation:

$$\frac{\partial}{\partial x}\left(K_x(\psi)\frac{\partial \psi}{\partial x}\right) + \frac{\partial}{\partial y}\left(K_y(\psi)\frac{\partial \psi}{\partial y}\right) + \frac{\partial}{\partial z}\left(K_z(\psi)\left(\frac{\partial \psi}{\partial z} + 1\right)\right) = C(\psi)\frac{\partial \psi}{\partial t} \qquad (1.4)$$

5.   The Solute Transport Equation:

$$\frac{\partial(\theta C)}{\partial t} = D_x\frac{\partial^2}{\partial x^2}(\theta C) + D_y\frac{\partial^2}{\partial y^2}(\theta C) + D_z\frac{\partial^2}{\partial z^2}(\theta C)$$

$$-\frac{\partial}{\partial x}(v_x C) - \frac{\partial}{\partial t}(\rho_b K_d C) - \lambda(\theta C + \rho_b K_d C) \qquad (1.5)$$

where h is hydraulic head, K_x, K_y, and K_z are the components of saturated hydraulic conductivity in the x, y, and z coordinate directions, t is time, ψ is pressure head, $K_x(\psi)$, $K_y(\psi)$, and $K_z(\psi)$ are the components of unsaturated hydraulic conductivity, S_s is specific storage, $C(\psi)$ is specific moisture capacity, C is solute concentration, D_x, D_y, and D_z are dispersion coefficients, θ is the volumetric water content, v_x is apparent groundwater velocity in the x coordinate direction, ρ_b is bulk density, K_d is the equilibrium distribution coefficient for a particular sorption/desorption reaction involving the solute and the porous media, and λ is the solute decay coefficient.

Equations 1.1 to 1.5 are derived in Appendices I, II, and III. These derivations should be studied carefully and the simplifying assumptions used in the derivations should always be kept in mind when using these equations to solve a particular groundwater flow or solute transport problem. Partial differential equations can also be derived for additional categories of problems including energy flow (e.g., the flow of heat in a geothermal reservoir), multiphase fluid flow (e.g., the simultaneous flow of air, water, oil, and natural gas in a petroleum reservoir), aquifer deformation (e.g., the consolidation of an aquifer due to excessive groundwater withdrawl), and more complex forms of solute transport (e.g., solute transport subject to microbial degradation). Although this book is concerned only with the application of the finite element method to the solution of equations 1.1 to 1.5, many of the same procedures also can be used to solve equations derived for other types of problems.

The mathematical model for each category of groundwater flow and solute transport problems consists of one of the partial differential equations listed above and a set of auxilliary conditions. The auxilliary conditions for equations 1.1 to 1.5 are classified as either *boundary conditions or initial conditions* (defined in sections 1.2 and 1.3, respectively). A mathematical model consisting of one or more partial differential equations and a set of boundary conditions is referred to as a *boundary value problem;* a mathematical model consisting of one or more partial differential equations, a set of boundary conditions, <u>and</u> a set of initial conditions is referred to as an *initial value problem*.

1.2   BOUNDARY VALUE PROBLEMS

Mathematical models of groundwater flow based on equations 1.1 or 1.2 are classified as boundary value problems. In boundary value problems, the analyst can specify the value of the unknown quantity or *field variable* (i.e., hydraulic head or pressure head) along portions of the aquifer boundaries. Derivatives of the field variable (i.e., rates of groundwater flow) also can be specified along portions of the aquifer boundaries (e.g. to represent groundwater recharge) or at special points within the aquifer called *point sources or sinks* (e.g. to represent groundwater withdrawl from wells). These specified values are collectively referred to as boundary conditions and when they are combined with equation 1.1 or 1.2 the result is a mathematical model that can be solved for values of the field variable at any point within the aquifer. Examples of boundary value problems and boundary conditions are in Figure 1.1. In boundary value problems, boundary conditions and computed values of the field variable do not change with time and the minimum and maximum values of the field variable always occur on the boundaries of the aquifer or at point sources or sinks.

1.3   INITIAL VALUE PROBLEMS

Mathematical models of groundwater flow and solute transport based on equations 1.3, 1.4, or 1.5 are classified as initial value problems. In initial value problems, boundary conditions, i.e., specified values of the field variable (hydraulic head, pressure head, or solute concentration) and its derivatives (rates of groundwater flow or solute flux), are specified in the same way as for boundary value problems. In addition, values of the field variable must be specified at <u>all</u> points within the aquifer at some initial time t_0 and these specified values are collectively referred to as initial conditions. When the boundary conditions and initial conditions are combined with equation 1.3, 1.4, or 1.5, the result is a mathematical model that can be solved for values of the field variable at any point in the aquifer at any time $t > t_0$. Some examples of initial value problems, boundary conditions, and initial conditions are in Figure 1.2. In initial value problems, boundary conditions and computed values of the field variable can change with time and the minimum and maximum values of the field variable at time t can occur at any point within the aquifer.

1.4   ANALYTICAL METHODS FOR SOLVING THE EQUATIONS

In general we can use two types of methods to obtain solutions to a mathematical model of groundwater flow or solute transport: *analytical methods* and *numerical methods*. When using analytical methods we seek to obtain a functional representation for the solution of the partial differential equation (e.g, a mathematical expression that gives hydraulic head as a function of position and time within the aquifer). The accuracy of analytical solutions can be very good (exact in many cases) and analytical solutions to equations 1.1 to 1.5 are widely used to study the behavior of groundwater flow and transport processes under hypothetical conditions (e.g., to determine the sensitivity of computed values of hydraulic head to values of saturated hydraulic conductivity), to

Plan View of Alluvial Aquifer

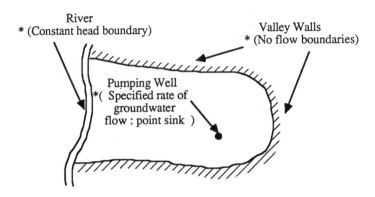

Cross-Sectional View of Earth Dam

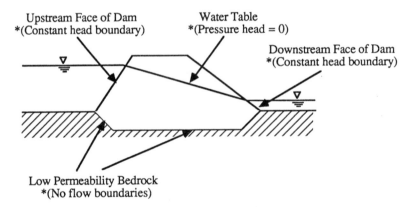

Figure 1.1   Examples of boundary value problems and boundary conditions for
              steady-state, saturated groundwater flow, * = boundary condition.

Cross-Sectional View of Aquifer

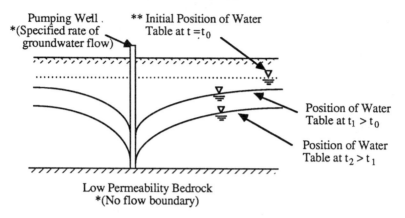

Pumping Well
*(Specified rate of groundwater flow)

** Initial Position of Water Table at t = t_0

Position of Water Table at $t_1 > t_0$

Position of Water Table at $t_2 > t_1$

Low Permeability Bedrock
*(No flow boundary)

Cross-Sectional View of Aquifer

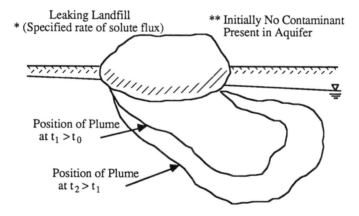

Leaking Landfill
* (Specified rate of solute flux)

** Initially No Contaminant Present in Aquifer

Position of Plume at $t_1 > t_0$

Position of Plume at $t_2 > t_1$

Figure 1.2   Examples of initial value problems, * = boundary condition, ** = initial condition.

interpret data from laboratory and field experiments (e.g., to compute values of dispersion coefficients for a soil sample in a laboratory column from the results of a miscible displacement experiment), and to verify the accuracy of solutions obtained by numerical methods (e.g., by comparing computed solute concentrations obtained using analytical and numerical methods for a wide range of apparent groundwater velocities and dispersion coefficients).

Example

Problem Statement:

One-dimensional, steady-state, groundwater flow through isotropic and homogeneous aquifer ($v_x = 0.01$ m/d; $D_x = 1$ m^2/d).

Solute transport by advection and dispersion only. No solute decay ($\lambda = 0$) or sorption of solute to porous media ($K_d = 0$).

Initially no solute is present. At time $t = 0$, solute concentration at one end of aquifer is increased instantaneously to 100 mg/l.

Compute solute concentration at $x = 100$m, $t = 500$ days

Mathematical Model:

Partial Differential Equation

$$\frac{\partial C}{\partial t} = D_x \frac{\partial^2 C}{\partial x^2} - v_x \frac{\partial C}{\partial x} \qquad \text{(see Appendix III)}$$

Boundary Conditions

$$C\,(x = 0, t > 0) = 100 \text{ mg/l}$$

Initial Conditions

$$C\,(x > 0, t = 0) = 0$$

Solution Obtained by Analytical Method (Ogata, 1970):

$$C(x,t) = \frac{C(x = 0, t > 0)}{2}\left[\text{erfc}\left(\frac{x - v_x t}{2\sqrt{D_x t}}\right) + \exp\left(\frac{v_x x}{D_x}\right)\text{erfc}\left(\frac{x + v_x t}{2\sqrt{D_x t}}\right)\right]$$

$$C(100,500) = \frac{100}{2}\left[\text{erfc}\left(\frac{100 - 0.01(500)}{2\sqrt{(1)(500)}}\right) + \exp\left(\frac{0.01(100)}{1}\right)\text{erfc}\left(\frac{100 + 0.01(500)}{2\sqrt{(1)(500)}}\right)\right]$$

$$= 50\,[\text{erfc}\,(2.124) + \exp\,(1)\,\text{erfc}\,(2.348)]$$

(values of the complementary error function, erfc( ) are tabulated in Freeze and Cherry (1979))

$$= 50 \left[0.002711 + 2.718(0.000925)\right] = 0.26 \text{ mg/l}$$

The principal limitation of analytical methods is that solutions can only be obtained by imposing severely restrictive assumptions about aquifer properties, boundary conditions, or initial conditions. For example, an assumption commonly made to obtain analytical solutions to equation 1.1 is that the aquifer is isotropic and homogeneous for hydraulic conductivity (i.e., that the components of saturated hydraulic conductivity, K_x, K_y, and K_z, are the same and do not change from point to point within the aquifer). In most field situations, however, the assumptions required to obtain solutions to groundwater flow or solute transport problems using analytical methods are not valid.

1.5   NUMERICAL METHODS FOR SOLVING THE EQUATIONS

Numerical methods do not require such restrictive assumptions. For example, it is possible to obtain numerical solutions for the case of anisotropic and nonhomogeneous aquifer properties and for problems with complex and time-dependent boundary conditions. When using numerical methods we seek a discrete approximation for the solution i.e., computed values of the field variable at a set of specified points within the aquifer at a set of specified times; the number and location of the points and the number and choice of times is determined in advance by the analyst. The accuracy of solutions obtained by numerical methods can be very good (exact in some cases) but depends on several factors including: the type of numerical method used, the complexity of the boundary and initial conditions, and the computational precision of the computer used to implement the method. In general, it is easier to obtain high-accuracy numerical solutions for steady-state groundwater flow problems than for transient groundwater flow and solute transport problems and for saturated groundwater flow problems than for unsaturated groundwater flow problems.

Several types of numerical methods have been used to solve groundwater flow and solute transport problems, the two principal ones being the *finite difference method* and the *finite element method*. Although the word "method" is singular, these terms actually refer to two rather large groups of numerical procedures.

The finite difference method was initially applied to the flow of fluids in petroleum reservoirs (Table 1.1). The method was first applied to problems of groundwater flow and solute transport in the mid-1960's. The method has a number of advantages that contribute to its continued widespread use and popularity: (1) for simple problems (e.g., one-dimensional, steady-state groundwater flow in an isotropic and homogeneous aquifer) the mathematical formulation and computer implementation are easily understood by those without advanced training in mathematics or computer programming, (2) good textbooks are available to help the beginner, (3) efficient numerical algorithms have been developed for implementing the finite difference method on computers, (4) well-documented computer programs for solving problems of groundwater flow and solute transport are widely available at little or no cost, (5) the accuracy of solutions to steady-state and transient groundwater flow problems is generally quite good, and (6) several case histories have been published that describe successful applications of the method to the solution of practical problems.

Unfortunately the finite difference method also has disadvantages: (1) the method works best for rectangular or prismatic aquifers of uniform composition; it is difficult to incorporate irregular or curved aquifer boundaries, anisotropic and heterogeneous aquifer properties, or sloping soil and rock layers into the numerical model without introducing numerous mathematical and computer programming complexities, (2) the accuracy of solutions to solute transport problems is lower than can be obtained by the finite element

Topic	References
Early Developments in Petroleum Reservoir Modeling	Bruce et al. (1953), Peaceman and Rachford (1962).
Saturated Groundwater Flow	Remson et al. (1965), Freeze and Whitherspoon (1966), Pinder and Bredehoeft (1968).
Unsaturated Groundwater Flow	Philip(1957), Aschroft et al. (1962), Freeze (1971), Brutsaert (1973).
Solute Transport	Stone and Brian (1963), Oster et al. (1970), Tanji et al. (1967), Wierenga (1977).
Application to field problems	Orlob and Woods (1967), Gambolati et al. (1973), Fleck and McDonald (1978).
Comprehensive References	Trescott and Larson (1977), Ames (1977), Mitchell and Griffiths (1980), Lapidus and Pinder (1982).
Computer Programs	Trescott et al. (1976), Konikow and Bredehoeft (1978).

Table 1.1   Selected references for the finite difference method.

method (which is now widely used in place of the finite difference method for this purpose).

The finite element method was first used to solve groundwater flow and solute transport problems in the early 1970's (Table 1.2). The method has several advantages: (1) irregular or curved aquifer boundaries, anisotropic and heterogeneous aquifer properties, and sloping soil and rock layers can be easily incorporated into the numerical model, (2) the accuracy of solutions to groundwater flow and solute transport problems is very good (exact in some cases), (3) solutions to the solute transport equation are generally more accurate than solutions obtained by the finite difference method, and (4) the finite element method lends itself to modular computer programming wherein a wide variety of types of problems can be solved using a small set of identical computer procedures.

The principal disadvantages of the finite element method for solving problems of groundwater flow and solute transport are (1) for simple problems, the finite element method requires a greater amount of mathematical and computer programming sophistication than does the finite difference method (although this disadvantage disappears for more complicated problems), (2) there are fewer well-documented computer programs and case histories available for the finite element method than for the finite difference method, and (3) there are few textbooks available to assist the beginner.

The purpose of this book is to help remove some of these disadvantages. Part 1 describes the basic principles of the finite element method as it applies to mathematical models of groundwater flow and solute transport based on equations 1.1 to 1.5. Obtaining a numerical solution to a groundwater flow or solute transport problem using the finite element method is performed in five basic steps that will be described in detail in the next five chapters. Computer implementation of each of these steps and computer programs for solving equations 1.1 to 1.5 are in Part 2. The application of the finite element method to the solution of practical groundwater flow and solute transport problems is discussed in Part 3.

Topic	References
Early Developments in Petroleum Reservoir Modeling	Price et al. (1968).
Saturated Groundwater Flow	Zienkiewicz et al. (1966), Javandel and Witherspoon (1968), Zienkiewicz and Parekh (1970), Pinder and Frind (1972).
Unsaturated Groundwater Flow	Neuman (1973), Gureghian et al. (1979), Pickens and Gillham (1980).
Solute Transport	Price et al. (1968), Guymon et al. (1970), Neuman (1973), Van Genuchten et al. (1977), Kirkner et al. (1984).
Application to field problems	Pinder (1973), Gupta and Tanji (1976), Senger and Fogg (1987).
Comprehensive References	Ziekiewicz (1971), Pinder and Gray (1977), Lapidus and Pinder (1982), Huyakorn and Pinder (1983).
Computer Programs	Neuman and Witherspoon (1970), Reeves and Duguid (1975), Segol et al. (1975), Pickens et al. (1979)

Table 1.2   Selected references for the finite element method.

NOTES AND ADDITIONAL READING

1. This text assumes the reader has a thorough understanding of the basic terminology and principles of groundwater hydrology. Readers without this background should review these subjects before proceeding. Excellent books for this purpose are Freeze and Cherry (1979), de Marsily (1986), de Wiest (1969) and Bear (1979). The reader is also assumed to have a basic knowledge of differential and integral calculus and linear algebra but no prior knowledge of numerical methods is required (a concise review of the concepts from linear algebra used in the finite element method is in Appendix IV).

2. Reviews of the historical development of groundwater flow and solute transport models are in Huyacorn and Pinder (1983) and Prickett (1975).

3. Analytical solutions to selected groundwater flow and solute transport problems are in Bear (1979), Javandel et al. (1984), and Bear and Verruijt (1987).

4. Reviews of existing computer models for solving groundwater flow and solute transport problems by the finite difference and finite element method are in Bachmat et al. (1978) and Oster (1982). These reports compare model capabilities and give references for the numerical algorithms used, user documentation, and program listings.

5. An excellent introduction to the use of the finite difference method for solving problems of groundwater flow is in Bennett (1978) which is designed as a programmed guide for self study.

6. Segerlind (1984) is an excellent introduction to the finite element method.

7. Other references for the use of finite difference and finite element methods to solve groundwater flow and solute transport problems are Remson et al. (1971) (advanced treatment of finite difference method, introduction to finite element method), Pinder and Gray (1977) (intermediate treatment of both methods), Wang and Anderson (1982) (introductory treatment of both methods, contains computer programs in FORTRAN), Huyakorn and Pinder (1983) (advanced treatment of both methods), and Bear and Verruijt (1987) (intermediate treatment of both methods, contains computer programs in BASIC).

Chapter 2

STEP 1: DISCRETIZE THE PROBLEM DOMAIN

The first step in the solution of a groundwater flow or solute transport problem by the finite element method is to *discretize* the problem domain (aquifer, soil profile, etc.). This is done by replacing the problem domain with a collection of *nodes* (*or nodal points*) and *elements* referred to as the *finite element mesh*. (Figure 2.1). Elements consist of two or more nodes joined together by line (or arc) segments. There are different element types for one-, two-, and three-dimensional problems and for problems with axisymmetry (Figure 2.2). Elements may be of any size, the size and shape of each element in the mesh can be different, and several different types of elements can be used in a single mesh. The material properties of the aquifer (e.g., hydraulic conductivity or dispersivity) must be specified for each element. The values of the material properties are usually assumed to be constant within each element but are allowed to vary from one element to the next.

The first step in the finite element method then, is to draw the finite element mesh. Although computer programs are available for this purpose, it has been the author's experience that except for very large problems (i.e., problems with more than one or two hundred nodes) or for three-dimensional problems with complex geometry, little (if any) time is saved by their use. The following procedure will be satisfactory for most problems encountered in practice. First, prepare a drawing of the problem domain to some convenient scale on a piece of graph paper. It is desirable that the drawing scale be the same in each of the coordinate directions although this is not necessary. Next, the finite element mesh is added to the original drawing or to a transparent overlay by drawing in the positions of the nodes and the element boundaries. Then, each node is assigned a node number and each element is assigned an element number (see below). As a final step, an input data file for the finite element computer program can be prepared directly from this drawing.

When preparing the finite element mesh it is important to remember that the precision of the solution obtained and the level of computational effort required to obtain a solution will be determined to a great extent by the number of nodes in the mesh. A *coarse mesh* has a smaller number of nodes and will give a lower precision than a *fine mesh*. However, the larger the number of nodes in the mesh, the greater will be the required computational effort and cost. Unfortunately, it is usually not possible to determine in advance the number of nodes required to achieve a given level of precision. In fact, the only way to determine the precision of a solution obtained by the finite element method is to repeat the calculations with a finer mesh to see if the results change significantly. For this reason, it is best to start with a coarse mesh consisting of only a few nodes. The input data for such a mesh can be prepared easily and a solution can be obtained with little computational effort. A second, finer mesh is then prepared that has a greater number of nodes in those parts of the mesh where the first solution indicates the field variable is varying rapidly or where the most precise results are required. A second solution is then obtained and compared with the first. If computed nodal values are significantly different from those obtained from the coarser mesh, the mesh is again refined and a third solution is obtained. This process is repeated until there are no significant changes in computed values of the field variable (at least in the parts of the domain of most interest). Usually no more than two or three mesh refinements are required.

<u>one-dimension</u>

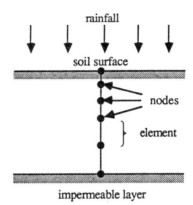

<u>two-dimensions</u>

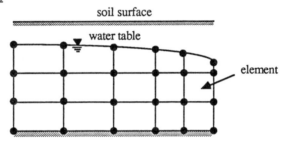

<u>three-dimensions</u>

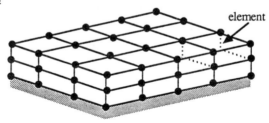

Figure 2.1   **Discretization of one-, two-, and three-dimensional problem domains.**

One-dimensional
 elements

Two-dimensional
 elements

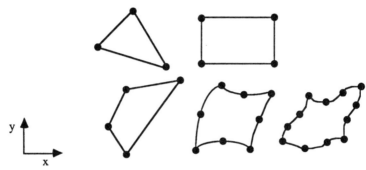

Three-dimensional
 elements

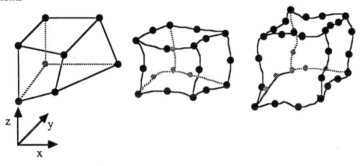

Axisymmetric
 elements

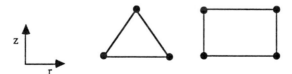

Figure 2.2   Some types of finite elements.

To prepare a finite element mesh that provides solutions with an acceptable level of precision with a reasonable amount of compuational effort requires considerable practice and for this reason, this step of the finite element method is still considered an "art" by most modelers. It helps considerably if the person drawing the mesh is familiar with groundwater flow and solute transport processes. Being able to visualize the flow or transport process is especially helpful and the use of roughly sketched flow nets is to be encouraged. It is important to remember that solutions with similar accuracy can be obtained from two meshes that appear quite different and, for this reason there is no single "correct" choice of mesh for a particular problem. The following set of "rules" describe some of the procedures used by the author to prepare a finite element mesh. These rules are by no means definitive but they should provide some initial guidance to the inexperienced modeler.

2.1 RULES FOR NODAL POINT PLACEMENT

The finite element mesh consists of several nodes (problems have been solved with as many as one million nodes but typically only a few hundred nodes are used). Each node is assigned a unique *node number* . Node numbers range from one to the number of nodes in the mesh; no "skips" in the node numbers are allowed and no two nodes can have the same node number. Each node also is assigned a set of *nodal coordinates* . These are the (x), (x,y), (x,y,z), or (r,z) coordinates of the node.

1. Place nodes along the boundaries of the problem domain, at the location of pumping wells or other point sources or sinks, and at any point where a computed value of the field variable is desired (Figure 2.3). Nodes located at points with known values of the field variable are sometimes called *Dirichlet nodes,* because they are used to represent Dirichlet boundary conditions (see section 4.5). Examples are nodes along constant-head boundaries or at points of known solute concentration (also see Chapter 20). Nodes located at points with known rates of groundwater flow or solute flux are sometimes called *Neumann nodes,* because they are used to represent Neumann boundary conditions (see section 4.5). Examples are nodes located at production and injection wells or recharge boundaries (also see Chapter 20).

2. Place nodes closest together in those parts of the problem domain where the field variable is expected to change most rapidly. This will include regions near point sources or sinks, and in any other part of the problem domain where gradients in head or solute concentration are expected to be large (Figure 2.4).

3. Place nodes along the interface between two different materials, for example along the interface between two soil or rock layers that have different hydraulic conductivities (Figure 2.5). Because material properties must be constant within an element, an interface between two different materials will also be an element boundary (see below).

4. Number the nodes to minimize the *semi-bandwidth* of the resulting system of linear equations. Minimization of the semi-bandwidth is desirable because the size of the system of linear equations created by the finite element method can be quite large (see section 4.5). When this systems of equations is operated on in matrix form, the storage capacity of many computers can be quickly exceeded. The semi-bandwidth for any mesh can be computed from: SBW = R+1, where R is the maximum difference in any two node numbers within a single element in the mesh (if the value of the field variable is specified at a node however that node is not used in the calculation of R, see section 4.4). The minimum bandwidth for a particular mesh can usually be achieved by numbering nodes across the narrow dimension of the problem domain (Figure 2.6). For problems with very complex geometry, a computer program may be required to minimize the semi-bandwidth of the matrix.

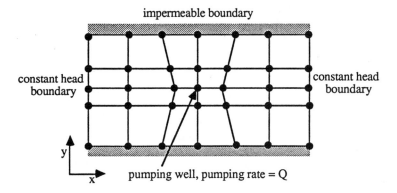

Figure 2.3   Place nodes along boundaries of problem domain and at point sources and sinks.

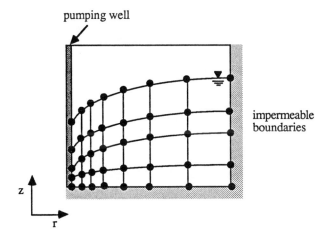

Figure 2.4   Place nodes close together where values of the field variable are expected to change rapidly.

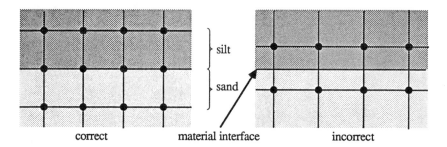

Figure 2.5   **Correct nodal placement at the interface of two different materials.**

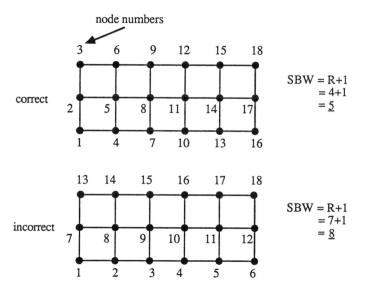

Figure 2.6   **Numbering nodes to minimize semi-bandwidth of system of equations.**

2.2   RULES FOR SELECTING ELEMENT SIZE, SHAPE, AND PLACEMENT

The size and shape of the elements in a mesh is determined primarily by the size and shape of the problem domain, the number of different types of aquifer materials, and by the number of nodes in the mesh. In problems that have a complex geometry (e.g., caused by an irregular depth to bedrock) or geologic structure (e.g., due to the presence of faults) many elements will be required. In problems with a simple geometry (e.g., shallow alluvial aquifer underlain by horizontal bedrock) fewer elements will be required. If the problem domain contains curved boundaries or interfaces different types of elements may be used than if the boundaries and interfaces consist of straight lines or planes. Elements will generally be smaller in parts of the mesh where the field variable is changing rapidly, because nodes will be placed closest together in these areas. When drawing the finite element mesh, each element is assigned a unique *element number*. In most computer programs, the element numbers begin with one and continue sequentially to the number of elements in the mesh. However, the way that element numbers are assigned will have no effect on the size or semi-bandwidth of the matrices generated during the solution process. Each element is defined using two or more nodes; the nodal coordinates define the size and shape of the element. For this reason the node numbers for each element are listed. Some convention is used to insure that node numbers for all elements of a given type in the mesh are listed in the same way (see Chapter 4). The material properties also must be specified for each element in the mesh. Because, in most cases, the material properties for several elements will be the same (e.g., all elements within a particular geologic strata) it is common to assign all elements with the same material properties to a common material set. The properties for each material set are then listed once.

1. Use the simplest type(s) of element required for a particular problem. This usually means that we use linear bar elements for one-dimensional problems, linear triangle or rectangle elements for two-dimensional problems, and linear parallelepiped elements for three-dimensional problems (see Chapter 4). However we should not hesitate to use more complex elements, especially when curved boundaries or interfaces are encountered. The biggest disadvantage in using complex elements, which can have as many as 32 nodes, is that their use can greatly increase the chance of errors occurring during the preparation of the input data.

2. The edges of adjacent elements should never overlap, nor should "gaps" appear between elements in the mesh (Figure 2.7).

3. Material properties are usually assumed to be constant within an element, but they can vary from one element to the next. Therefore no elements should overlap an interface between two different types of materials (Figure 2.5).

4. The shape of the elements can affect the accuracy of the resulting solution. In general, the use of highly distorted elements should be avoided. This is particularly important when solving transient groundwater flow or solute transport problems because the element shape influences the size of the time step required to obtain a stable solution (see Chapter 5).

5. Do not change element size abruptly; instead use a transition region to achieve a gradual change in element size (Figure 2.8).

6. Take advantage of symmetry in the problem domain to reduce the number of elements (and nodes) in the mesh (Figure 2.9). Keep in mind, however, that the boundary conditions, initial conditions, material properties and domain geometry all must display symmetry to use this approach.

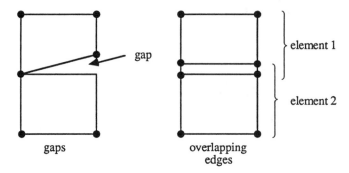

gap

element 1

element 2

gaps

overlapping
edges

Figure 2.7   Gaps and overlapping edges for adjacent elements are not permitted.

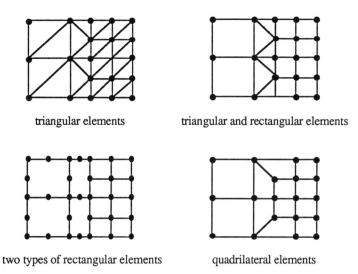

triangular elements   triangular and rectangular elements

two types of rectangular elements   quadrilateral elements

**Figure 2.8   Example transition regions for changing from a coarse mesh to a fine
mesh.**

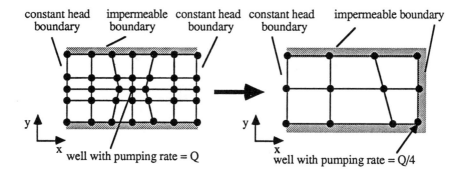

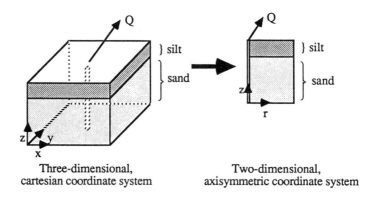

Figure 2.9   Use symmetry to reduce the number of elements and nodes in the mesh.

2.3   EXAMPLE MESHES

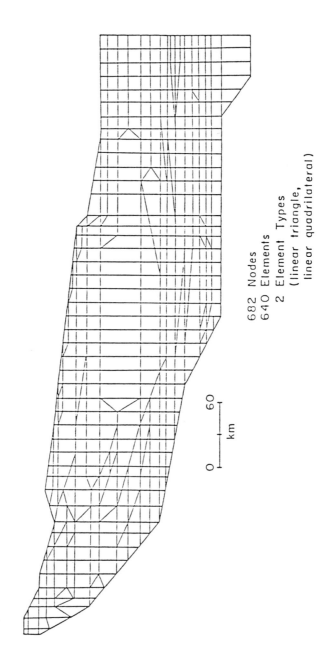

682 Nodes
640 Elements
2 Element Types
(linear triangle,
linear quadrilateral)

0   60
km

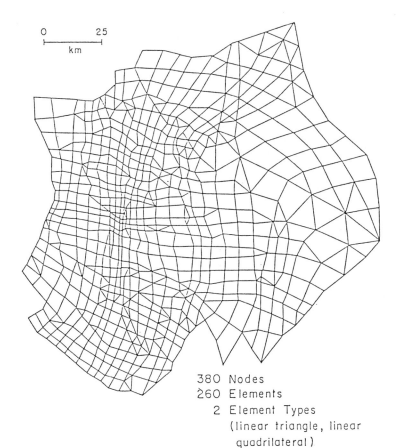

0   25
|———————|
    km

380 Nodes
260 Elements
  2 Element Types
    (linear triangle, linear
     quadrilateral)

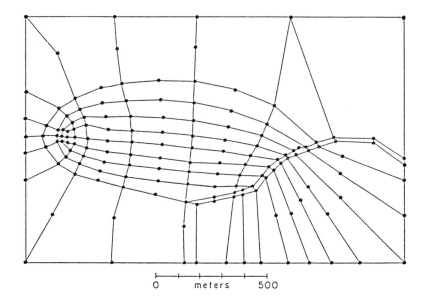

150 Nodes
 92 Elements
  3 Element Types
    (linear quadrilateral,
     quadratic quadrilateral,
     mixed linear-quadratic
     quadrilateral)

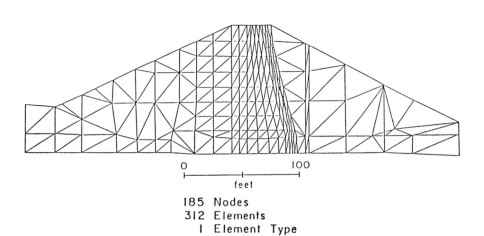

185 Nodes
312 Elements
  I Element Type
    (linear triangle).

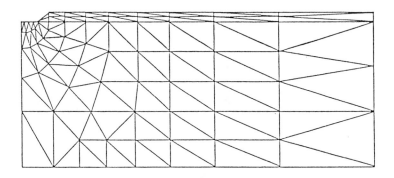

70 Nodes
124 Elements
1 Element Type
(linear triangle)

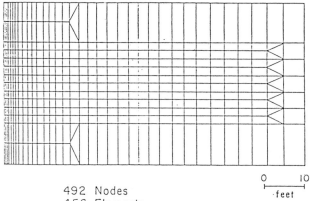

492 Nodes
456 Elements
2 Element Types
(axisymmetric linear triangle,
(axisymmetric linear rectangle)

0   10
⊢————⊣
feet

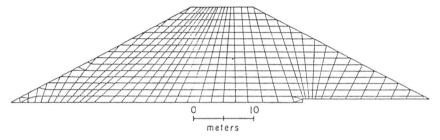

meters

450 Nodes
437 Elements
  2 Element Types
    (linear triangle,
    linear quadrilateral)

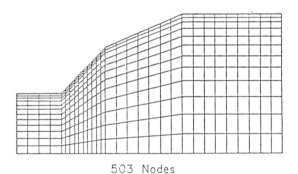

503 Nodes
360 Elements
  1 Element Type
    (linear quadrilateral)

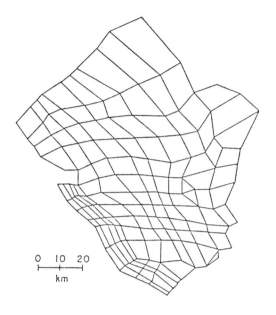

0 10 20
km

154 Nodes
130 Elements
  2 Element Types
    (linear triangle, linear
    quadrilateral)

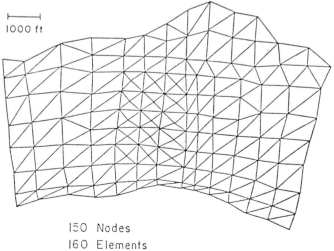

1000 ft

150 Nodes
160 Elements
  1 Element Type (Linear triangle)

Problems

For problems 1 to 5, draw a finite element mesh for the aquifer using the element types given in the problem, label node and element numbers, and compute the semi-bandwidth.

1. Plan view of alluvial aquifer

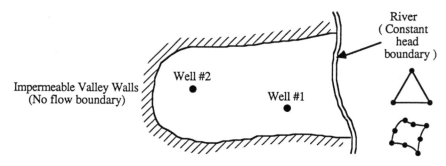

2. Plan view of sedimentary aquifer

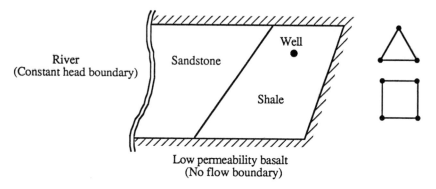

3. Cross-sectional view of sedimentary aquifer

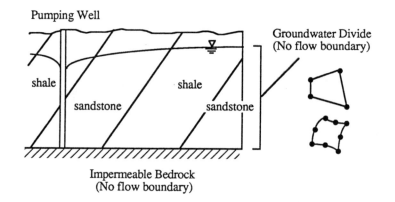

4. Plan view of alluvial aquifer

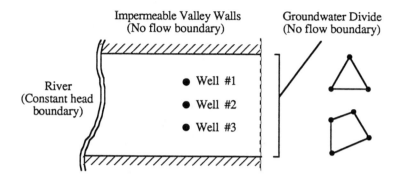

Impermeable Valley Walls
(No flow boundary)

Groundwater Divide
(No flow boundary)

River
(Constant head
boundary)

● Well #1

● Well #2

● Well #3

a) Pumping rates for all wells are equal
b) Pumping rates for all wells are not equal

5. Plan view and three cross-sections for alluvial aquifer

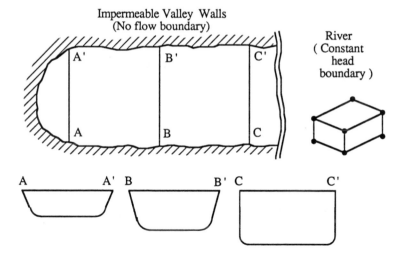

Impermeable Valley Walls
(No flow boundary)

River
(Constant
head
boundary)

6. Obtain a geologic map for an aquifer in your area. Draw a finite element mesh using a
   mixture of two-dimensional elements, label node and element numbers, and compute
   the semi-bandwidth. Speculate about appropriate boundary conditions to use with your
   mesh.

Chapter 3

STEP 2: DERIVE THE APPROXIMATING EQUATIONS

3.1 THE METHOD OF WEIGHTED RESIDUALS

The second step in the finite element method is to derive an *integral formulation* for the governing groundwater flow or solute transport equation. This integral formulation leads to a system of algebraic equations that can be solved for values of the field variable (hydraulic head, pressure head, or solute concentration) at each node in the mesh. Several methods can be used to derive the integral formulation for a particular differential equation. The *variational method* has been used to derive integral formulations for the differential equations that govern the behavior of mechanical systems e.g., in the fields of elasticity and structural mechanics. The *method of weighted residuals* is a more general approach that is widely used in groundwater flow and solute transport modeling.

In the method of weighted residuals, an *approximate solution* to the boundary or initial value problem is defined. When this approximate solution is substituted into the governing differential equation, an error or *residual* occurs at each point in the problem domain. We then force the weighted average of the residuals for each node in the finite element mesh to equal zero.

Consider a differential equation of the form

$$L(\phi(x,y,z)) - F(x,y,z) = 0 \tag{3.1}$$

where L is the differential operator, ϕ is the field variable, and F is a known function. Define an approximate solution $\hat{\phi}$ of the form

$$\hat{\phi}(x,y,z) = \sum_{i=1}^{m} N_i(x,y,z) \, \phi_i \tag{3.2}$$

where N_i are *interpolation functions*, ϕ_i are the (unknown) values of the field variable at the nodes, and m is the number of nodes in the mesh. When the approximate solution is substituted into equation 3.1 the differential equation is no longer satisfied exactly

$$L(\hat{\phi}(x,y,z)) - F(x,y,z) = R(x,y,z) \neq 0 \tag{3.3}$$

where R is the *residual* or error due to the approximate solution. The residual varies from point-to-point within the problem domain. At some points it may be large and at other points it may be small (the *sign* of the residual also can vary from point-to-point). Therefore we cannot force R to be zero at certain specified points because the residual may then become unacceptably large elsewhere in the problem domain.

In the method of weighted residuals, we force the *weighted average* of the residuals at the nodes to be equal to zero

$$\int_{\Omega} W(x,y,z)\ R(x,y,z)\ d\Omega\ =\ 0 \tag{3.4}$$

where $W(x,y,z)$ is a *weighting function* and Ω represents the problem domain. Ω will be a length in one-dimensional problems, an area in two-dimensional problems, and a volume in three-dimensional problems. Substituting equation 3.3 into equation 3.4 we have

$$\iiint_{\Omega} W(x,y,z)\left[\ L(\hat{\phi}(x,y,z)) - F(x,y,z)\ \right]\ d\Omega\ =\ 0 \tag{3.5}$$

To evaluate equation (3.5) we must specify the mathematical form of the approximate solution $\hat{\phi}$ and the weighting function W. In the finite element method $\hat{\phi}$ is defined in a piece –wise fashion over the problem domain. The value of $\hat{\phi}$, within any element e, $\hat{\phi}^{(e)}$, is given by

$$\hat{\phi}^{(e)}(x,y,z)\ =\ \sum_{i=1}^{n} N_i^{(e)}\ \phi_i \tag{3.6}$$

where $N_i^{(e)}$ are the element *interpolation functions* (one interpolation function per node), ϕ_i are the (unknown) values of the field variable at each node, and n is the number of nodes within the element. For example, the approximate solution for a one-dimensional element with two nodes i and j (Figure 3.1) can be written

$$\hat{\phi}^{(e)}(x)\ =\ N_i^{(e)}(x)\ \phi_i + N_j^{(e)}(x)\ \phi_j \tag{3.7}$$

or in matrix form

$$\hat{\phi}^{(e)}(x)\ =\ [\ N^{(e)}]\{\phi\} \tag{3.8}$$

where

$$[\ N^{(e)}]\ =\ [\ N_i^{(e)}(x)\ N_j^{(e)}(x)] \tag{3.9}$$

$$\{\phi\}\quad =\ \begin{Bmatrix} \phi_i \\ \phi_j \end{Bmatrix} \tag{3.10}$$

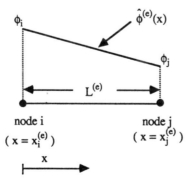

Figure 3.1   **Approximate solution for one-dimensional element with two nodes.**

For the element in Figure 3.1 the interpolation functions are *linear* functions of x

$$N_i^{(e)}(x) = \frac{x_j^{(e)} - x}{L^{(e)}} \qquad\qquad N_j^{(e)}(x) = \frac{x - x_i^{(e)}}{L^{(e)}} \tag{3.11}$$

where $x_i^{(e)}$ and $x_j^{(e)}$ are the coordinates of the nodes, and $L^{(e)}$ is the element length ($L^{(e)} = x_j^{(e)} - x_i^{(e)}$). These interpolation functions are plotted in Figure 3.2. The value of $N_i^{(e)}$ is one at node i and decreases linearly to zero at node j, while the value of $N_j^{(e)}$ is one at node j and decreases linearly to zero at node i.

At node i ($x = x_i^{(e)}$)

$$\hat{\phi}^{(e)}(x_i) = N_i^{(e)}(x_i)\,\phi_i + N_j^{(e)}(x_i)\,\phi_j$$
$$= \phi_i \tag{3.12}$$

at node j ($x = x_j^{(e)}$)

$$\hat{\phi}^{(e)}(x_j) = N_i^{(e)}(x_j)\,\phi_i + N_j^{(e)}(x_j)\,\phi_j$$
$$= \phi_j \tag{3.13}$$

and at the midpoint of the element $\left(x = \frac{x_j^{(e)} + x_i^{(e)}}{2} \right)$

$$\hat{\phi}^{(e)}(x_i) = N_i^{(e)}(x)\,\phi_i + N_j^{(e)}(x)\,\phi_j$$
$$= \frac{\phi_i + \phi_j}{2} \tag{3.14}$$

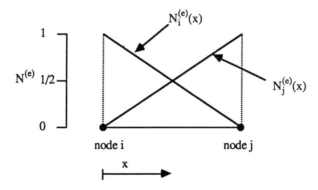

Figure 3.2   **Linear interpolation functions for one-dimensional element with two nodes.**

Several other types of interpolation functions that can be used to obtain an approximate solution for ϕ for use in solving one-, two-, and three-dimensional problems are described in Chapter 4.

In addition to the interpolation functions, the form of the weighting function W in equation 3.5 also must be specified. Several subsets of the method of weighted residuals are defined by the choice of weighting function used.

3.1.1 Subdomain Method

In the subdomain method the value of W is equal to one within a small part of the problem domain surrounding a node (the subdomain) and zero elsewhere. The size of subdomain is usually chosen to be equal to the size of the element containing the node. For a one-dimensional element the weighting function for a node is given by

$$W_i(x) = \begin{cases} 1 \text{ for } \left(x_i - \dfrac{L^{(e)}}{2} \right) \le x \le \left(x_i + \dfrac{L^{(e)}}{2} \right) \\ 0 \text{ otherwise} \end{cases} \tag{3.15}$$

where $L^{(e)}$ is the length of the element (Figure 3.3)

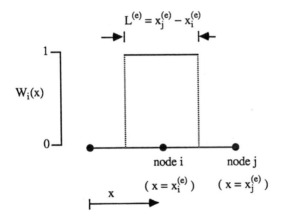

Figure 3.3   **Weighting function for node i in the subdomain method.**

3.1.2 Collocation Method

The collocation method is a special case of the subdomain method when the subdomain is chosen to be very small. For a one-dimensional element

$$W_i(x) = \delta(x_i \pm \Delta x) \tag{3.16}$$

where δ is the Dirac delta function and Δx is some small distance. This notation means that within a distance Δx of node i $W_i(x) = 1$, otherwise $W_i(x) = 0$ (Figure 3.4)

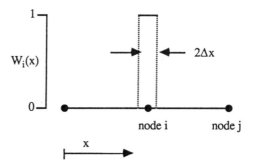

Figure 3.4   Weighting function for node i in the collocation method.

3.1.3 Galerkin's Method

In Galerkin's Method the weighting function for a node is identical to the interpolation function used to define the approximate solution $\hat{\phi}$. For the one–dimensional element with two nodes

$$W_i(x) = \frac{x_j - x}{L} \quad \text{for } x \geq x_i \tag{3.17}$$

$$W_j(x) = \frac{x - x_i}{L} \quad \text{for } x \geq x_i \tag{3.18}$$

which is plotted in Figure 3.5.

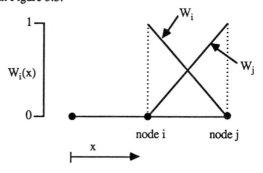

Figure 3.5   Weighting function for node i in Galerkin's Method.

Galerkin's Method is the subset of the method of weighted residuals that is most commonly used to solve groundwater flow and solute transport problems.

After specifying the form of the approximate solution and weighting function, we can evaluate the integral in equation 3.5 to obtain a system of linear equations of the form

$$[K]\{\phi\} = \{F\} \tag{3.19}$$

that can be solved for the values of the field variable at each node in the mesh. We will illustrate the entire process with an example.

3.2   A FINITE ELEMENT EXAMPLE

The column of soil in Figure 3.6 is saturated and water is flowing vertically downward at a constant rate Q. Hydraulic head is held constant at the upper and lower ends of the column and we wish to calculate the values of head at points A and B. The problem domain has been divided into a mesh with four elements and five nodes. The governing differential equation is the one-dimensional form of the steady-state, saturated groundwater flow equation derived in Appendix I

$$\frac{\partial}{\partial x}\left(K_x \frac{\partial h}{\partial x} \right) = 0 \tag{3.20}$$

where K_x is the saturated hydraulic conductivity in the x direction and h is hydraulic head.

Using the method of weighted residuals we will define an approximate solution $\hat{h}$. If this approximate solution is substituted into equation 3.20, the differential equation is no longer satisfied exactly

$$\frac{\partial}{\partial x}\left(K_x \frac{\partial \hat{h}}{\partial x} \right) = R(x) \neq 0 \tag{3.21}$$

where the residual will vary from point-to-point within the problem domain. Define the vector $\{R\}$ to be the value of residual at each node in the finite element mesh

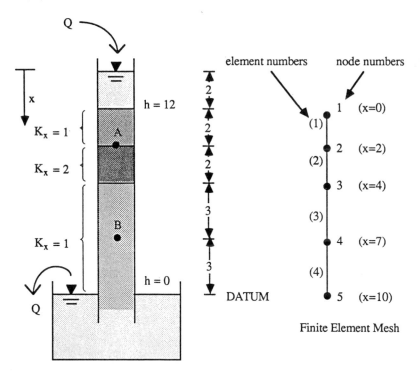

Figure 3.6   Example problem for method of weighted residuals.

$$\{R\} = \begin{Bmatrix} R(x=0) \\ R(x=2) \\ R(x=4) \\ R(x=7) \\ R(x=10) \end{Bmatrix} = \begin{Bmatrix} R_1 \\ R_2 \\ R_3 \\ R_4 \\ R_5 \end{Bmatrix} \qquad (3.22)$$

where R_1, for example is the value of the residual at node 1. The residual at any node i, R_i represents the error between the true value of hydraulic head and the approximate solution $\hat{h}$ at that node. The approximate solution at a node is determined by the values of hydraulic head at the nodes in all elements that are joined to node i, For example, elements 2 and 3 are joined to node 3. Thus the values of hydraulic head for the other nodes in these elements contribute to the residual at node 3. We can write this as

$$R_3 = R_3^{(2)} + R_3^{(3)} \qquad (3.23)$$

where the first term is the contribution of element 2 to the residual at node 3 and the second term is the contribution of element 3 to the residual at node 3. In general, we can write

$$R_i = \sum_{e=1}^{P} R_i^{(e)} \qquad (3.24)$$

where p is the number of elements that are joined to node i.

The contribution of element e to the residual at node i can be obtained from the integral formulation for that node. For the one-dimensional elements in our example

$$R_i^{(e)} = -\int_{x_i^{(e)}}^{x_j^{(e)}} N_i^{(e)} \left[K_x^{(e)} \frac{\partial^2 \hat{h}^{(e)}}{\partial x^2} \right] dx \qquad (3.25)$$

where $x_i^{(e)}$ and $x_j^{(e)}$ are the coordinates of the nodes at each end of the element, $N_i^{(e)}$ is the weighting function for node i in element e (which is identical to the interpolation function for node i in element e because we are using Galerkin's Method), and $K_x^{(e)}$ is the saturated hydraulic conductivity for the element ($K_x^{(e)}$ is assumed to be constant *within* an element but can vary from one element to the next). The equation was multiplied by a negative one for later convenience.

A similar equation can be written for the contribution of element e to the residual at any other node j joined to the element

$$R_j^{(e)} = -\int_{x_i^{(e)}}^{x_j^{(e)}} N_j^{(e)} \left[K_x^{(e)} \frac{\partial^2 \hat{h}^{(e)}}{\partial x^2} \right] dx \qquad (3.26)$$

In general, if an element has n nodes it will contribute to the residual at n nodes.

The interpolation functions for the type of elements in Figure 3.6 are in equation 3.11. From equations 3.7 and 3.11 the approximate solution $\hat{h}$ is given by

$$\hat{h}^{(e)}(x) = N_i^{(e)} h_i + N_j^{(e)} h_j$$

$$= \left(\frac{x_j^{(e)} - x}{L^{(e)}} \right) h_i + \left(\frac{x - x_i^{(e)}}{L^{(e)}} \right) h_j \qquad (3.27)$$

Because the approximate solution is a linear function of x, $\frac{\partial^2 \hat{h}}{\partial x^2}$ is not defined. The approximate solution does have a continuous first derivative, however, so we can evaluate equation 3.25 if we rewrite it in terms of $\frac{\partial \hat{h}}{\partial x}$.

Using integration by parts we can write

$$\int_{x_i^{(e)}}^{x_j^{(e)}} N_i^{(e)} \left[K_x^{(e)} \frac{\partial^2 \hat{h}^{(e)}}{\partial x^2} \right] dx = -\int_{x_i^{(e)}}^{x_j^{(e)}} K_x^{(e)} \frac{\partial N_i^{(e)}}{\partial x} \frac{\partial \hat{h}^{(e)}}{\partial x} dx + \left(N_i^{(e)} K_x^{(e)} \frac{\partial \hat{h}^{(e)}}{\partial x} \Bigg|_{x_i^{(e)}}^{x_j^{(e)}} \right) \qquad (3.28)$$

where the second term on the right-hand side of equation 3.28 represents groundwater flow across the element's surface. For elements on the exterior of the mesh this term will be used to represent specified rates of groundwater flow (Neumann boundary conditions). We will give this term the symbol $F_i^{(e)}$

$$F_i^{(e)} = \left(N_i^{(e)} K_x^{(e)} \frac{\partial \hat{h}^{(e)}}{\partial x} \Bigg|_{x_i^{(e)}}^{x_j^{(e)}} \right) \qquad (3.29)$$

$F_i^{(e)}$ will be positive if water is entering the mesh. If no flows are specified or at impermeable aquifer boundaries $F_i^{(e)}$ will be zero. For elements on the interior of the mesh, the term $F_i^{(e)}$ for adjacent elements will have opposite signs cancelling out the contribution of $F_i^{(e)}$ for the two elements for the node(s) they share. In two- or three-dimensions we have

$$F_i^{(e)} = \int_{S^{(e)}} N_i^{(e)} K_x^{(e)} \frac{\partial \hat{h}^{(e)}}{\partial x} ds \qquad (3.30)$$

where $S^{(e)}$ is the surface area of the element along the specified flow boundary (see Section 3.3).

Substituting equation 3.28 into equation 3.25 we have

$$R_i^{(e)} = -\int_{x_i^{(e)}}^{x_j^{(e)}} N_i^{(e)} \left(K_x^{(e)} \frac{\partial \hat{h}^{(e)}}{\partial x} \right) dx$$

$$= \int_{x_i^{(e)}}^{x_j^{(e)}} K_x^{(e)} \frac{\partial N_i^{(e)}}{\partial x} \frac{\partial \hat{h}^{(e)}}{\partial x} dx - \left(N_i^{(e)} K_x^{(e)} \frac{\partial \hat{h}^{(e)}}{\partial x} \Bigg|_{x_i^{(e)}}^{x_j^{(e)}} \right)^{\!\!\!0}$$

$$= \int_{x_i^{(e)}}^{x_j^{(e)}} K_x^{(e)} \frac{\partial N_i^{(e)}}{\partial x} \frac{\partial \hat{h}^{(e)}}{\partial x} dx \qquad (3.31)$$

From the definition for $\hat{h}$ (equation 3.27) we can write

$$\frac{\partial \hat{h}^{(e)}(x)}{\partial x} = \frac{\partial}{\partial x}(N_i^{(e)} h_i + N_j^{(e)} h_j) \tag{3.32}$$

From the definitions of the interpolation functions we can write

$$\frac{\partial N_i^{(e)}}{\partial x} = \frac{\partial}{\partial x}\left(\frac{x_j^{(e)} - x}{L^{(e)}} \right) = -\frac{1}{L^{(e)}} \tag{3.33}$$

$$\frac{\partial N_j^{(e)}}{\partial x} = \frac{\partial}{\partial x}\left(\frac{x - x_i^{(e)}}{L^{(e)}} \right) = \frac{1}{L^{(e)}} \tag{3.34}$$

so that

$$\frac{\partial \hat{h}^{(e)}}{\partial x} = -\frac{1}{L^{(e)}}h_i + \frac{1}{L^{(e)}}h_j$$

$$= \frac{1}{L^{(e)}}(-h_i + h_j) \tag{3.35}$$

Substituting equations 3.34 and 3.35 into equation 3.32 gives

$$R_i^{(e)} = \int_{x_i^{(e)}}^{x_j^{(e)}} K_x^{(e)} \left(-\frac{1}{L^{(e)}} \right)\left(\frac{1}{L^{(e)}} \right)(-h_i + h_j)\, dx$$

$$= -\frac{K_x^{(e)}}{L^{(e)2}}(x_j^{(e)} - x_i^{(e)})(-h_i + h_j)$$

but $x_j^{(e)} - x_i^{(e)} = L^{(e)}$ and we have

$$R_i^{(e)} = \frac{K_x^{(e)}}{L^{(e)}}(h_i - h_j) \tag{3.36a}$$

Similarly for the contribution of element e to the residual at node j

$$R_j^{(e)} = \frac{K_x^{(e)}}{L^{(e)}}(-h_i + h_j) \tag{3.36b}$$

Equations 3.36a and 3.36b can be combined and written in matrix form as

$$\left\{\begin{matrix} R_i^{(e)} \\ R_j^{(e)} \end{matrix}\right\} = \frac{K_x^{(e)}}{L^{(e)}}\begin{bmatrix} 1 & -1 \\ -1 & 1 \end{bmatrix}\left\{\begin{matrix} h_i \\ h_j \end{matrix}\right\} \tag{3.37}$$

$$= [\,K^{(e)}]\left\{\begin{matrix} h_i \\ h_j \end{matrix}\right\} \tag{3.38}$$

where

$$[\,K^{(e)}] = \frac{K_x^{(e)}}{L^{(e)}}\begin{bmatrix} 1 & -1 \\ -1 & 1 \end{bmatrix} \tag{3.39}$$

$$2\times 2$$

is called the *element conductance matrix* .

The element conductance matrix depends on the hydraulic conductivity of the aquifer material within the element ($K_x^{(e)}$), and the size ($L^{(e)}$) and shape (through the interpolation functions for the element) of the element. $[K^{(e)}]$ is always a square, symmetric matrix with a size nxn where n is the number of nodes in the element. Thus for a one-dimensional element with two nodes the size of $[K^{(e)}]$ is 2x2, for a two-dimensional element with three nodes the size of $[K^{(e)}]$ is 3x3, and so on.

We can compute the element conductance matrix for each element in the mesh in Figure 3.6 once we assign node numbers to the i th and j th nodes for each element. This is done in Figure 3.7 where the i th node for element 1 is assigned to node 1, the j th node for element 1 is assigned to node 2, the i th node for element 2 is assigned to node 2, and so on.

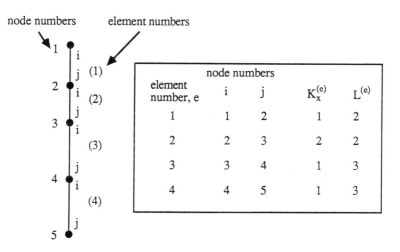

Figure 3.7   Assigning node numbers to element nodes i and j

The element conductance matrices can then be computed as follows

$$[K^{(1)}] = \frac{1}{2}\begin{bmatrix} 1 & -1 \\ -1 & 1 \end{bmatrix} = \begin{bmatrix} 1/2 & -1/2 \\ -1/2 & 1/2 \end{bmatrix}$$

$$[K^{(2)}] = \frac{2}{2}\begin{bmatrix} 1 & -1 \\ -1 & 1 \end{bmatrix} = \begin{bmatrix} 1 & -1 \\ -1 & 1 \end{bmatrix}$$

$$[K^{(3)}] = [K^{(4)}] = \frac{1}{3}\begin{bmatrix} 1 & -1 \\ -1 & 1 \end{bmatrix} = \begin{bmatrix} 1/3 & -1/3 \\ -1/3 & 1/3 \end{bmatrix}$$

We can combine the element conductance matrices to obtain a system of linear equations of the form

$$\{R\} = [K]\{h\} - \{F\} = \{0\} \tag{3.40}$$

where $\{R\}$ is the *global residual matrix* , $[K]$ is the *global conductance matrix* , $\{h\}$ is the vector of unknown hydraulic heads, and $\{F\}$ is a vector containing the specified fluxes at Neumann nodes (see section 3.3). For our example no fluxes were specified, $\{F\} = \{0\}$, and we can write

$$\{R\} = \begin{Bmatrix} R_1 \\ R_2 \\ R_3 \\ R_4 \\ R_5 \end{Bmatrix} \quad \{h\} = \begin{Bmatrix} h_1 \\ h_2 \\ h_3 \\ h_4 \\ h_5 \end{Bmatrix} \quad \{F\} = \begin{Bmatrix} 0 \\ 0 \\ 0 \\ 0 \\ 0 \end{Bmatrix} \tag{3.41}$$
$$5\times1 \qquad\qquad 5\times1 \qquad\qquad 5\times1$$

The entries of the global conductance matrix can be obtained by combining the element conductance matrices for all the elements in the mesh. An easy way to do this when the number of elements is small is to expand each element conductance matrix to the same size as the global conductance matrix. These can then be added together to form the global conductance matrix using the formula

$$[K]_{\substack{\text{global}}} = \sum_{\substack{e=1 \\ \text{expanded}}}^{m} [K^{(e)}] \tag{3.42}$$

where m is the number of elements in the mesh. For the elements in our example, the expanded form of the element conductance matrices are

$$[K^{(1)}] = \begin{bmatrix} 1/2 & -1/2 & 0 & 0 & 0 \\ -1/2 & 1/2 & 0 & 0 & 0 \\ 0 & 0 & 0 & 0 & 0 \\ 0 & 0 & 0 & 0 & 0 \\ 0 & 0 & 0 & 0 & 0 \end{bmatrix} \quad [K^{(2)}] = \begin{bmatrix} 0 & 0 & 0 & 0 & 0 \\ 0 & 1 & -1 & 0 & 0 \\ 0 & -1 & 1 & 0 & 0 \\ 0 & 0 & 0 & 0 & 0 \\ 0 & 0 & 0 & 0 & 0 \end{bmatrix}$$

$$[K^{(3)}] = \begin{bmatrix} 0 & 0 & 0 & 0 & 0 \\ 0 & 0 & 0 & 0 & 0 \\ 0 & 0 & 1/3 & -1/3 & 0 \\ 0 & 0 & -1/3 & 1/3 & 0 \\ 0 & 0 & 0 & 0 & 0 \end{bmatrix} \qquad [K^{(4)}] = \begin{bmatrix} 0 & 0 & 0 & 0 & 0 \\ 0 & 0 & 0 & 0 & 0 \\ 0 & 0 & 0 & 0 & 0 \\ 0 & 0 & 0 & 1/3 & -1/3 \\ 0 & 0 & 0 & -1/3 & 1/3 \end{bmatrix}$$

and the global conductance matrix is

$$[K] = [K^{(1)}] + [K^{(2)}] + [K^{(3)}] + [K^{(4)}]$$
global

$$= \begin{bmatrix} 1/2 & -1/2 & 0 & 0 & 0 \\ -1/2 & 1+1/2 & -1 & 0 & 0 \\ 0 & -1 & 1+1/3 & -1/3 & 0 \\ 0 & 0 & -1/3 & 1/3+1/3 & -1/3 \\ 0 & 0 & 0 & -1/3 & 1/3 \end{bmatrix}$$

$$= \begin{bmatrix} 1/2 & -1/2 & 0 & 0 & 0 \\ -1/2 & 3/2 & -1 & 0 & 0 \\ 0 & -1 & 4/3 & -1/3 & 0 \\ 0 & 0 & -1/3 & 2/3 & -1/3 \\ 0 & 0 & 0 & -1/3 & 1/3 \end{bmatrix}$$

The system of equations that result when this global conductance matrix is substituted into equation 3.40 is

$$\begin{bmatrix} 1/2 & -1/2 & 0 & 0 & 0 \\ -1/2 & 3/2 & -1 & 0 & 0 \\ 0 & -1 & 4/3 & -1/3 & 0 \\ 0 & 0 & -1/3 & 2/3 & -1/3 \\ 0 & 0 & 0 & -1/3 & 1/3 \end{bmatrix} \begin{Bmatrix} h_1 \\ h_2 \\ h_3 \\ h_4 \\ h_5 \end{Bmatrix} = \begin{Bmatrix} 0 \\ 0 \\ 0 \\ 0 \\ 0 \end{Bmatrix} \qquad (3.43)$$

But we know $h_1 = 12$ and $h_5 = 0$ (nodes 1 and 5 are sometimes called Dirichlet nodes) from the boundary conditions and we can use this information to modify equation 3.43 (the procedure is explained in section 4.5)

$$\begin{bmatrix} 3/2 & -1 & 0 \\ -1 & 4/3 & -1/3 \\ 0 & -1/3 & 2/3 \end{bmatrix} \begin{Bmatrix} h_2 \\ h_3 \\ h_4 \end{Bmatrix} = \begin{Bmatrix} 6 \\ 0 \\ 0 \end{Bmatrix} \qquad (3.44)$$

from which we obtain $h_2 = h_A = 9.33$, $h_3 = 8.0$, and $h_4 = h_B = 4.0$.

This example has illustrated each of the major steps of the finite element method. To review we first discretized the problem domain into a collection of nodes and elements (Figure 3.6). We then used the method of weighted residuals to obtain an integral formulation for the residual at each node. This integral formulation contained the differential equation written in terms of the approximate solution $\hat{h}$. Because the second derivative of approximate solution was not defined for our choice of element, we used the product rule to obtain an integral formulation for the residual at a node in terms of the first derivative of the element interpolation functions and the values of hydraulic head at the nodes. When these integrals were evaluated we obtained an expression for the element conductance matrix [K(e)]. The conductance matrix was then computed for all of the elements and, by combining these matrices, the global conductance matrix was obtained for the finite element mesh. The golbal conductance matrix is one part of a system of linear equations $[K] \{h\} = \{F\}$ where $\{F\}$ contains any specified flow rates at Neumann nodes (see Section 3.3). Finally this system of equations was modified using the known values of hydraulic head on the boundary of the mesh and then solved to obtain values of hydraulic head at the remaining nodes.

The procedure used for this example can be generalized to include two- and three-dimensional problems as well as problems of unsaturated flow, transient flow, and solute transport.

3.3   STEADY-STATE, SATURATED FLOW EQUATION

The three-dimensional form of the equation for steady-state groundwater flow through saturated porous media is written as

$$\frac{\partial}{\partial x}\left(K_x \frac{\partial h}{\partial x}\right) + \frac{\partial}{\partial y}\left(K_y \frac{\partial h}{\partial y}\right) + \frac{\partial}{\partial z}\left(K_z \frac{\partial h}{\partial z}\right) = 0 \qquad (3.45)$$

where K_x, K_y, and K_z are the saturated hydraulic conductivities of the porous media in the x, y, and z coordinate directions, and h is hydraulic head (Appendix I). As in the previous section, we will assume an approximate solution for h of the form

$$\hat{h}^{(e)} = \sum_{i=1}^{n} N_i^{(e)} h_i \qquad (3.46)$$

where $\hat{h}^{(e)}$ is the approximate solution for hydraulic head within element e, $N_i^{(e)}$ are the interpolation functions for each node within element e, n is the number of nodes within element e, and h_i are the unknown values of hydraulic head for each node within element e. When the approximate solution is substituted into equation 3.45, the differential equation is not satisfied exactly and an error or residual occurs at every point in the problem domain. The contribution of any element e to the residual at a node i to which the element is joined is

$$R_i^{(e)} = -\iiint\limits_{V^{(e)}} W_i^{(e)}(x,y,z)\left[\frac{\partial}{\partial x}\left(K_x \frac{\partial \hat{h}^{(e)}}{\partial x}\right) + \frac{\partial}{\partial y}\left(K_y \frac{\partial \hat{h}^{(e)}}{\partial y}\right) + \frac{\partial}{\partial z}\left(K_z \frac{\partial \hat{h}^{(e)}}{\partial z}\right)\right] dx\, dy\, dz$$

$$(3.47)$$

where $W_i^{(e)}$ is the weighting function for node i and the limits of the integration are chosen to represent the *volume* of element e.

In Galerkin's method we choose the weighting function for each node in the element to be equal to the interpolation function for that node, $W_i^{(e)} = N_i^{(e)}$. If we also assume that values of saturated hydraulic conductivity in the three coordinate directions are *constant within an element* (but can vary from one element to the next), equation 3.47 can be written as

$$R_i^{(e)} = -\iiint\limits_{V^{(e)}} N_i^{(e)} \left[K_x^{(e)}\frac{\partial^2 \hat{h}^{(e)}}{\partial x^2} + K_y^{(e)}\frac{\partial^2 \hat{h}^{(e)}}{\partial y^2} + K_z^{(e)}\frac{\partial^2 \hat{h}^{(e)}}{\partial z^2} \right] dx\,dy\,dz$$

(3.48)

where, for example, $K_x^{(e)}$ is the value of saturated hydraulic conductivity in the x direction within element e.

Because the second derivative of the approximate solution is not defined for most types of elements, we can use the results of equations 3.29 to 3.32 to reduce the order of the derivatives of $\hat{h}$ appearing in equation 3.48.

$$R_i^{(e)} = -\iiint\limits_{V^{(e)}} N_i^{(e)} \left[K_x^{(e)}\frac{\partial N_i^{(e)}}{\partial x}\frac{\partial \hat{h}^{(e)}}{\partial x} + K_y^{(e)}\frac{\partial N_i^{(e)}}{\partial y}\frac{\partial \hat{h}^{(e)}}{\partial y} + K_z^{(e)}\frac{\partial N_i^{(e)}}{\partial z}\frac{\partial \hat{h}^{(e)}}{\partial z} \right] dx\,dy\,dz$$
(3.49)

Equation 3.49 is the integral formulation for the three-dimensional, steady-state, saturated groundwater flow equation. If the problem domain is two-dimensional, equation 3.49 reduces to

$$R_i^{(e)} = -\iint\limits_{A^{(e)}} N_i^{(e)} \left[K_x^{(e)}\frac{\partial N_i^{(e)}}{\partial x}\frac{\partial \hat{h}^{(e)}}{\partial x} + K_y^{(e)}\frac{\partial N_i^{(e)}}{\partial y}\frac{\partial \hat{h}^{(e)}}{\partial y} \right] dx\,dy$$

(3.50)

where the limits of integration are chosen to represent the *area* of element e. If the problem domain is one-dimensional, equation 3.49 reduces to

$$R_i^{(e)} = -\int\limits_{L^{(e)}} N_i^{(e)} \left[K_x^{(e)}\frac{\partial N_i^{(e)}}{\partial x}\frac{\partial \hat{h}^{(e)}}{\partial x} \right] dx$$

(3.51)

where the limits of integration are chosen to represent the *length* of element e.

Before we can evaluate these integral equations we must first choose the type of element and interpolation functions to use. In the example in Figure 3.6, the problem domain was one-dimensional and each element had two nodes i and j. In this case the interpolation functions used were functions only of x

$$N_i^{(e)}(x) = \frac{x_j^{(e)} - x}{L^{(e)}} \quad \text{and} \quad N_j^{(e)}(x) = \frac{x - x_i^{(e)}}{L^{(e)}} \tag{3.52}$$

where $x_j^{(e)}$ and $x_i^{(e)}$ are the coordinates of the two nodes used to define the element and $L^{(e)}$ is the element length. Because each element had two nodes, it contributed to the residual at two nodes, $R_i^{(e)}$ and $R_j^{(e)}$. In the example we represented these residuals as separate integral equations

$$R_i^{(e)} = -\int_{x_i^{(e)}}^{x_j^{(e)}} N_i^{(e)} \left[K_x^{(e)} \frac{\partial N_i^{(e)}}{\partial x} \frac{\partial \hat{h}^{(e)}}{\partial x} \right] dx \tag{3.53}$$

$$R_j^{(e)} = -\int_{x_i^{(e)}}^{x_j^{(e)}} N_j^{(e)} \left[K_x^{(e)} \frac{\partial N_j^{(e)}}{\partial x} \frac{\partial \hat{h}^{(e)}}{\partial x} \right] dx \tag{3.54}$$

After evaluating these integrals the results were combined to obtain the element conductance matrix, [$K^{(e)}$]. A more direct approach is to combine equations 3.53 and 3.54 to obtain a *matrix-integral formulation* for [$K^{(e)}$]. For a one-dimensional element with two nodes [$K^{(e)}$] is given by

$$
\begin{aligned}
[K^{(e)}]_{2\times2} &= \int_{x_i^{(e)}}^{x_j^{(e)}} \begin{bmatrix} K_x^{(e)} \dfrac{\partial N_i^{(e)}}{\partial x} \dfrac{\partial N_i^{(e)}}{\partial x} & K_x^{(e)} \dfrac{\partial N_i^{(e)}}{\partial x} \dfrac{\partial N_j^{(e)}}{\partial x} \\[2ex] K_x^{(e)} \dfrac{\partial N_j^{(e)}}{\partial x} \dfrac{\partial N_i^{(e)}}{\partial x} & K_x^{(e)} \dfrac{\partial N_j^{(e)}}{\partial x} \dfrac{\partial N_j^{(e)}}{\partial x} \end{bmatrix} dx \\[4ex]
&= \int_{x_i^{(e)}}^{x_j^{(e)}} \underbrace{\begin{bmatrix} \dfrac{\partial N_i^{(e)}}{\partial x} \\[2ex] \dfrac{\partial N_j^{(e)}}{\partial x} \end{bmatrix}}_{2\times1} \underbrace{[K_x^{(e)}]}_{1\times1} \underbrace{\begin{bmatrix} \dfrac{\partial N_i^{(e)}}{\partial x} & \dfrac{\partial N_j^{(e)}}{\partial x} \end{bmatrix}}_{1\times2} dx
\end{aligned} \tag{3.55}
$$

If the one-dimensional problem had been solved using elements with 3 nodes, i, j, k, equation 3.55 would be written

$$
[K^{(e)}]_{3\times3} = \int_{x_i^{(e)}}^{x_j^{(e)}} \underbrace{\begin{bmatrix} \dfrac{\partial N_i^{(e)}}{\partial x} \\[2ex] \dfrac{\partial N_j^{(e)}}{\partial x} \\[2ex] \dfrac{\partial N_k^{(e)}}{\partial x} \end{bmatrix}}_{3\times1} \underbrace{[K_x^{(e)}]}_{1\times1} \underbrace{\begin{bmatrix} \dfrac{\partial N_i^{(e)}}{\partial x} & \dfrac{\partial N_j^{(e)}}{\partial x} & \dfrac{\partial N_k^{(e)}}{\partial x} \end{bmatrix}}_{1\times3} dx \tag{3.56}
$$

If a two-dimensional problem was being solved using elements with three nodes, i, j and k, the matrix–integral formulation for the element conductance matrix $[K^{(e)}]$ would be

$$[K^{(e)}] = \iint_{A^{(e)}} \begin{bmatrix} \dfrac{\partial N_i^{(e)}}{\partial x} & \dfrac{\partial N_i^{(e)}}{\partial y} \\[2ex] \dfrac{\partial N_j^{(e)}}{\partial x} & \dfrac{\partial N_j^{(e)}}{\partial y} \\[2ex] \dfrac{\partial N_k^{(e)}}{\partial x} & \dfrac{\partial N_k^{(e)}}{\partial y} \end{bmatrix} \begin{bmatrix} K_x^{(e)} & 0 \\[2ex] 0 & K_y^{(e)} \end{bmatrix} \begin{bmatrix} \dfrac{\partial N_i^{(e)}}{\partial x} & \dfrac{\partial N_j^{(e)}}{\partial x} & \dfrac{\partial N_k^{(e)}}{\partial x} \\[2ex] \dfrac{\partial N_i^{(e)}}{\partial y} & \dfrac{\partial N_j^{(e)}}{\partial y} & \dfrac{\partial N_k^{(e)}}{\partial y} \end{bmatrix} dx\,dy$$

$$\underset{3\times3}{} \quad \underset{A^{(e)}}{} \quad \underset{3\times2}{} \qquad \underset{2\times2}{} \qquad\qquad \underset{2\times3}{} \qquad\qquad (3.57)$$

Where $A^{(e)}$ is the area of the element e. The most general formulation for $[K^{(e)}]$ can be written for the case of a three-dimensional problem being solved using elements with n nodes.

$$\underset{n\times n}{[K^{(e)}]} = \iiint_{V^{(e)}} \begin{bmatrix} \dfrac{\partial N_1^{(e)}}{\partial x} & \dfrac{\partial N_1^{(e)}}{\partial y} & \dfrac{\partial N_1^{(e)}}{\partial z} \\[2ex] \vdots & \vdots & \vdots \\[2ex] \dfrac{\partial N_n^{(e)}}{\partial x} & \dfrac{\partial N_n^{(e)}}{\partial y} & \dfrac{\partial N_n^{(e)}}{\partial z} \end{bmatrix} \begin{bmatrix} K_x^{(e)} & 0 & 0 \\[2ex] 0 & K_y^{(e)} & 0 \\[2ex] 0 & 0 & K_z^{(e)} \end{bmatrix} \begin{bmatrix} \dfrac{\partial N_1^{(e)}}{\partial x} & \cdots & \dfrac{\partial N_n^{(e)}}{\partial x} \\[2ex] \dfrac{\partial N_1^{(e)}}{\partial y} & \cdots & \dfrac{\partial N_n^{(e)}}{\partial y} \\[2ex] \dfrac{\partial N_1^{(e)}}{\partial z} & \cdots & \dfrac{\partial N_n^{(e)}}{\partial z} \end{bmatrix} dx\,dy\,dz$$

$$\underset{V^{(e)}}{} \qquad\qquad \underset{n\times3}{} \qquad\qquad \underset{3\times3}{} \qquad\qquad \underset{3\times n}{} \qquad (3.58)$$

where $V^{(e)}$ is the volme of element e. In Chapter 4 we will learn how to evaluate equation 3.58 for several different types of elements.

If we combine equations 3.58 and equation 3.47 we can write

$$\begin{Bmatrix} R_1^{(e)} \\ \vdots \\ R_n^{(e)} \end{Bmatrix} = [K^{(e)}] \begin{Bmatrix} h_1 \\ \vdots \\ h_n \end{Bmatrix} \qquad\qquad (3.59)$$

$$\underset{n\times1}{} \qquad \underset{n\times n}{} \quad \underset{n\times1}{}$$

Equation 3.59 is written for each element in the mesh. These equations are then combined to obtain

$$\begin{Bmatrix} R_1 \\ \vdots \\ R_p \end{Bmatrix} = [K]_{global} \begin{Bmatrix} h_1 \\ \vdots \\ h_p \end{Bmatrix} \qquad\qquad (3.60)$$

$$\underset{p\times1}{} \qquad \underset{p\times p}{} \quad \underset{p\times1}{}$$

and by setting the residuals equal to zero we have

$$[K] \{h\} = \{0\} \tag{3.61}$$
global
$p \times p \ p \times 1 \qquad p \times 1$

Before we can solve this system of equations for the values of hydraulic head at the nodes, equation 3.61 must be modified to incorporate known boundary conditions. Procedures for modifying equation 3.61 for known values of hydraulic head are in section 4.5.

If flow rates are specified on the boundary of the mesh (for example to represent seepage from lakes of rivers, or recharge from the soil surface) or at points within the mesh (for example to represent groundwater withdrawl by pumping) the steady-state, saturated flow equation becomes

$$\frac{\partial}{\partial x}\left(K_x \frac{\partial h}{\partial x}\right) + \frac{\partial}{\partial y}\left(K_y \frac{\partial h}{\partial y}\right) + \frac{\partial}{\partial z}\left(K_z \frac{\partial h}{\partial z}\right) + q = 0 \tag{3.62}$$

where q is the specified flow rate. q is positive if water is flowing into the mesh and negative if water is flowing out of the mesh. The specified flow rate within element e, $q^{(e)}$, contributes to the residual at all nodes in element e. Substituting equation 3.62 into equation 3.49 gives

$$R_i^{(e)} = -\iiint\limits_{V^{(e)}} N_i^{(e)} \left[K_x^{(e)} \frac{\partial N_i^{(e)}}{\partial x} \frac{\partial \hat{h}^{(e)}}{\partial x} + K_y^{(e)} \frac{\partial N_i^{(e)}}{\partial y} \frac{\partial \hat{h}^{(e)}}{\partial y} + K_z^{(e)} \frac{\partial N_i^{(e)}}{\partial z} \frac{\partial \hat{h}^{(e)}}{\partial z} + q^{(e)} \right] dx \, dy \, dz \tag{3.63}$$

The only new term is the integral

$$\iiint\limits_{V^{(e)}} N_i^{(e)} q^{(e)} \, dx \, dy \, dz = F_i^{(e)} \tag{3.64}$$

where $F_i^{(e)}$ is the integrated specified flow rate for node i in element e. If $q^{(e)}$ represents a specified flow rate along the boundary of element e we can write (Section 3.2)

$$F_i^{(e)} = \int_{S^{(e)}} N_i^{(e)} K_x^{(e)} \frac{\partial \hat{h}^{(e)}}{\partial x} \, ds = \int_{S^{(e)}} N_i^{(e)} q \, ds \tag{3.65}$$

where $S^{(e)}$ is the surface area of element e. The evaluation of these integrals for each node in element e gives the components of the specified flow matrix for element e, $\{F^{(e)}\}$

$$\{F^{(e)}\} = \begin{Bmatrix} F_1^{(e)} \\ \vdots \\ F_n^{(e)} \end{Bmatrix} \tag{3.66}$$

Combining equation 3.66 with equation 3.59 gives

$$\begin{Bmatrix} R_1^{(e)} \\ \vdots \\ R_n^{(e)} \end{Bmatrix} = [K^{(e)}] \begin{Bmatrix} h_1 \\ \vdots \\ h_n \end{Bmatrix} - \begin{Bmatrix} F_1^{(e)} \\ \vdots \\ F_n^{(e)} \end{Bmatrix} \tag{3.67}$$

We can combine the $\{F^{(e)}\}$ for each element in the mesh to obtain the global specified flow matrix $\{F\}$

$$\{F\}_{\text{global}} = \sum_{e=1}^{m} \{F^{(e)}\}_{\text{expanded}} \tag{3.68}$$

and equation 3.61 becomes

$$\begin{array}{cc} [K] & \{h\} = \{F\} \\ \text{global} & \text{global} \\ p\times p \; p\times 1 & p\times 1 \end{array} \tag{3.69}$$

If there are no specified flow rates (i.e., no Neumann Boundary Conditions) $\{F\} = \{0\}$. The evaluation of the integrals in equations 3.64 and 3.65 and the assembly of $\{F\}$ are illustrated for a one-dimensional problem in the following example.

Example

Compute $\{F^{(e)}\}$ for each element in the mesh shown below. Assemble $\{F\}$

The node numbers for the elements are

element	node i	node j
1	1	2
2	2	3
3	3	4

For node i, element 1

$$\{F_i^{(1)}\} = \int_{S^{(1)}} N_i^{(1)} q^{(1)} \; ds$$

But $N_i = 1$ at node i and with $S^{(e)}$ equal to unity in a one–dimension problem

$$\{F_i^{(1)}\} = N_i^{(1)} q^{(1)} \int_{S^{(1)}} ds = q^{(1)} = 10m^3/d$$

Nodes 2 and 3 are not on the specified flow boundary and we can write

$$\{F_j^{(1)}\} = \{F_i^{(2)}\} = \{F_j^{(2)}\} = \{F_i^{(3)}\} = 0$$

For node j, element 3

$$\{F_j^{(3)}\} = \int_{S^{(3)}} N_j^{(3)} \, q^{(3)} \, ds = -10 m^3/d$$

and we have

$$\{F^{(1)}\} = \left\{ \begin{matrix} 10 \\ 0 \end{matrix} \right\} \begin{matrix} 1 \\ 2 \end{matrix} \qquad \{F^{(2)}\} = \left\{ \begin{matrix} 0 \\ 0 \end{matrix} \right\} \begin{matrix} 2 \\ 3 \end{matrix} \qquad \{F^{(3)}\} = \left\{ \begin{matrix} 0 \\ -10 \end{matrix} \right\} \begin{matrix} 3 \\ 4 \end{matrix}$$

and {F} is given by

$$\{F\} = \{F^{(1)}\} + \{F^{(2)}\} + \{F^{(3)}\}$$
$$\text{global} \quad \text{expanded} \quad \text{expanded} \quad \text{expanded}$$

$$= \left\{ \begin{matrix} 10 \\ 0 \\ 0 \\ 0 \end{matrix} \right\} + \left\{ \begin{matrix} 0 \\ 0 \\ 0 \\ 0 \end{matrix} \right\} + \left\{ \begin{matrix} 0 \\ 0 \\ 0 \\ -10 \end{matrix} \right\} = \left\{ \begin{matrix} 10 \\ 0 \\ 0 \\ -10 \end{matrix} \right\}$$

3.4   STEADY-STATE, UNSATURATED FLOW EQUATION

The three-dimensional form of the equation for steady-state flow through an unsaturated porous media is

$$\frac{\partial}{\partial x}\left(K_x(\psi)\frac{\partial \psi}{\partial x} \right) + \frac{\partial}{\partial y}\left(K_y(\psi)\frac{\partial \psi}{\partial y} \right) + \frac{\partial}{\partial z}\left(K_z(\psi)\left(\frac{\partial \psi}{\partial z} + 1 \right) \right) = 0 \qquad (3.70)$$

where $K_x(\psi)$, $K_y(\psi)$, and $K_z(\psi)$ are the components of unsaturated hydraulic conductivity (which are functions of the pressure head ψ) in the three coordinate directions and the z coordinate direction is assumed to be vertical (see Appendix I). The unknown quantity at each nodes is the pressure head ψ. We will assume an approximate solution for ψ, $\hat{\psi}$, of the form

$$\hat{\psi}^{(e)} = \sum_{i=1}^{n} N_i^{(e)}\psi_i \qquad (3.71)$$

Where $\hat{\psi}^{(e)}$ is the approximate solution for pressure head within element e, $N_i^{(e)}$ are the interpolation functions for each node within element e, n is the number of nodes within element e, and ψ_i are the unknown values of pressure head for each node within element e.

When the approximate solution is substituted into equation 3.70, the differential equation is not satisfied exactly and an error or residual occurs at every point in the problem domain. The contribution of any element e to the residual at a node i to which the element is joined is

$$R_i^{(e)} = -\iiint\limits_{V^{(e)}} W_i^{(e)}(x,y,z)\left[\frac{\partial}{\partial x}\left(K_x(\psi)\frac{\partial\hat\psi}{\partial x}\right) + \frac{\partial}{\partial y}\left(K_y(\psi)\frac{\partial\hat\psi}{\partial y}\right)\right.$$
$$\left. + \frac{\partial}{\partial z}\left(K_z(\psi)\left(\frac{\partial\hat\psi}{\partial z} + 1\right)\right)\right] dx\ dy\ dz \qquad (3.72)$$

Where $W_i^{(e)}$ is the element's weighting function for node i and the limits of integration are chosen to represent the volume of element e.

In Galerkin's method we choose the weighting function for each node in the element to be equal to the element's interpolation function for that node $W_i^{(e)} = N_i^{(e)}$. If we also assume that unsaturated hydraulic conductivity functions are constant within an element (but can vary from one element to the next), equation 3.72 can be written

$$R_i^{(e)} = -\iiint\limits_{V^{(e)}} N_i^{(e)}\left[K_x^{(e)}(\psi)\frac{\partial^2\hat\psi}{\partial x^2} + K_y^{(e)}(\psi)\frac{\partial^2\hat\psi}{\partial y^2} + K_z^{(e)}(\psi)\frac{\partial^2\hat\psi}{\partial z^2} + \frac{\partial K_z^{(e)}}{\partial z}(\psi)\right]dx\ dy\ dz$$
$$(3.73)$$

where, for example, $K_x^{(e)}(\psi)$ is the unsaturated hydraulic conductivity function in the x direction within element e.

Because the second derivative of the approximate solution is not defined for some types of elements, we can use the results of equations 3.28 to 3.32 to reduce the order of the derivatives of $\hat\psi$ appearing in equation 3.73

$$R_i^{(e)} = -\iiint\limits_{V^{(e)}} N_i^{(e)}\left[K_x^{(e)}(\psi)\frac{\partial N_i^{(e)}}{\partial x}\frac{\partial\hat\psi}{\partial x} + K_y^{(e)}(\psi)\frac{\partial N_i^{(e)}}{\partial y}\frac{\partial\hat\psi}{\partial y}\right.$$
$$\left. + K_z^{(e)}(\psi)\frac{\partial N_i^{(e)}}{\partial z}\frac{\partial\hat\psi}{\partial z} + \frac{\partial K_z^{(e)}}{\partial z}(\psi)\right]dx\ dy\ dz \qquad (3.74)$$

Equation 3.74 is the integral formulation for the steady-state, unsaturated flow equation.
When the porous media is relatively dry, the term $\frac{\partial K_z^{(e)}}{\partial z}(\psi)$ will be small i.e., capillary forces $\left(\frac{\partial\hat\psi}{\partial x}\right)$ are much larger than gravitational forces $\frac{\partial K_z^{(e)}}{\partial z}$. In this case the last term within the integral can be neglected in the calculation of $[K^{(e)}(\psi)]$. We will assume this is true for the remainder of this section (also see section 5.4.3). If necessary the integral can be evaluated by developing a functional form for $K_z^{(e)}(\psi)$ within element e. Of course for problems of *horizontal* flow the last two terms in the integral in equation 3.74 are always zero.

From previous work we know that we can write a matrix expression for the contribution of element e to the residuals at all nodes that join the element

$$\begin{Bmatrix} R_1^{(e)} \\ \vdots \\ R_n^{(e)} \end{Bmatrix} = \left[K^{(e)}(\psi)\right]\begin{Bmatrix} \psi_1 \\ \vdots \\ \psi_n \end{Bmatrix} \qquad (3.75)$$
$$\quad n\times 1 \qquad\qquad n\times n \qquad n\times 1$$

where the element has n nodes and $[K^{(e)}(\psi)]$ is the unsaturated form of the element conductance matrix given by

$$
[K^{(e)}(\psi)] = - \iiint\limits_{V^{(e)}}
\begin{bmatrix}
\dfrac{\partial N_1^{(e)}}{\partial x} & \dfrac{\partial N_1^{(e)}}{\partial y} & \dfrac{\partial N_1^{(e)}}{\partial z} \\[2mm]
\vdots & \vdots & \vdots \\[2mm]
\dfrac{\partial N_n^{(e)}}{\partial x} & \dfrac{\partial N_n^{(e)}}{\partial y} & \dfrac{\partial N_n^{(e)}}{\partial z}
\end{bmatrix}
\begin{bmatrix}
K_x^{(e)}(\psi) & 0 & 0 \\
0 & K_y^{(e)}(\psi) & 0 \\
0 & 0 & K_z^{(e)}(\psi)
\end{bmatrix}
\begin{bmatrix}
\dfrac{\partial N_1^{(e)}}{\partial x} & \cdots & \dfrac{\partial N_n^{(e)}}{\partial x} \\[2mm]
\dfrac{\partial N_1^{(e)}}{\partial y} & \cdots & \dfrac{\partial N_n^{(e)}}{\partial y} \\[2mm]
\dfrac{\partial N_1^{(e)}}{\partial z} & \cdots & \dfrac{\partial N_n^{(e)}}{\partial z}
\end{bmatrix}
dx\,dy\,dz
$$

$$
\phantom{[K^{(e)}(\psi)] = }\quad\; n \times 3 \qquad\qquad 3 \times 3 \qquad\qquad 3 \times n \qquad\qquad (3.76)
$$

where $V^{(e)}$ is the volume of element e. In Chapter 4 we will learn how to evaluate equation 3.76 for several different types of elements.

When we combine the element conductance matrices for all the elements of the mesh we can obtain an unsaturated form of the global conductance matrix

$$
\underset{\text{global}}{[K(\psi)]} = \sum_{\substack{e=1 \\ \text{expanded}}}^{m} \left[K^{(e)}(\psi) \right]
\tag{3.77}
$$

where there are m elements in the mesh. The dependence of the global conductance matrix on the pressure head ψ is emphasized because in the solution process we will be concerned with a system of nonlinear equations of the form

$$
\underset{\text{global}}{[K(\psi)]}
\begin{Bmatrix} \psi_1 \\ \vdots \\ \psi_p \end{Bmatrix}
=
\begin{Bmatrix} F_1 \\ \vdots \\ F_p \end{Bmatrix}
\tag{3.78}
$$

where $\psi_1 \ldots \psi_p$ are the values of pressure head at each node (there are p nodes in the mesh). For the case of unsaturated flow, $\{F\}$ will contain specified rates of groundwater flow at boundaries and at sources and sinks. If we wish to include gravitational forces, additional contributions to $\{F\}$ in equation 3.78 result from the integration of $\dfrac{\partial K_z^{(e)}}{\partial z}(\psi)$. The solution of equation 3.78 is discussed in Chapter 5.

3.5 TRANSIENT, SATURATED FLOW EQUATION

The three-dimensional form of the equation for transient groundwater flow through saturated porous media is

$$
\frac{\partial}{\partial x}\left(K_x \frac{\partial h}{\partial x}\right) + \frac{\partial}{\partial y}\left(K_y \frac{\partial h}{\partial y}\right) + \frac{\partial}{\partial z}\left(K_z \frac{\partial h}{\partial z}\right) = S_s \frac{\partial h}{\partial t}
\tag{3.79}
$$

where S_s is the specific storage of the porous media and t is time (Appendix II). The only difference between the integral formulations for steady-state and transient groundwater flow equations is the addition of the term $S_s \dfrac{\partial h}{\partial t}$. When the approximate solution for hydraulic head, $\hat{h}$ is substituted into equation 3.79, the contribution of element e to the residual at node i is

$$R_i^{(e)} = -\iiint\limits_{V^{(e)}} W_i^{(e)} \left[\frac{\partial}{\partial x}\left(K_x^{(e)}\frac{\partial \hat{h}^{(e)}}{\partial x} \right) + \frac{\partial}{\partial y}\left(K_y^{(e)}\frac{\partial \hat{h}^{(e)}}{\partial y} \right) + \frac{\partial}{\partial x}\left(K_z^{(e)}\frac{\partial \hat{h}^{(e)}}{\partial z} \right) - S_s^{(e)}\frac{\partial \hat{h}^{(e)}}{\partial t} \right] dx\, dy\, dz$$

(3.80)

where $W_i^{(e)}$ is the weighting function for node i and the limits of the integration are chosen to represent the volume of element e.

In Galerkin's method $W_i^{(e)} = N_i^{(e)}$. If we assume that values of $K_x^{(e)}$, $K_y^{(e)}$, $K_z^{(e)}$, and $S_s^{(e)}$ are constant within an element (but can vary from one element to the next), equation 3.80 can be written

$$R_i^{(e)} = -\iiint\limits_{V^{(e)}} N_i^{(e)} \left[K_x^{(e)}\frac{\partial^2 \hat{h}^{(e)}}{\partial x^2} + K_y^{(e)}\frac{\partial^2 \hat{h}^{(e)}}{\partial y^2} + K_z^{(e)}\frac{\partial^2 \hat{h}^{(e)}}{\partial z^2} - S_s^{(e)}\frac{\partial \hat{h}^{(e)}}{\partial t} \right] dx\, dy\, dz$$

$$= -\iiint\limits_{V^{(e)}} N_i^{(e)} \left[K_x^{(e)}\frac{\partial^2 \hat{h}^{(e)}}{\partial x^2} + K_y^{(e)}\frac{\partial^2 \hat{h}^{(e)}}{\partial y^2} + K_z^{(e)}\frac{\partial^2 \hat{h}^{(e)}}{\partial z^2} \right] dx\, dy\, dz$$

$$+ \iiint\limits_{V^{(e)}} N_i^{(e)} S_s^{(e)}\frac{\partial \hat{h}^{(e)}}{\partial t}\, dx\, dy\, dz$$

(3.81)

where $S_s^{(e)}$ is the specific storage for element e. We know that the first integral on the right-hand side of equation 3.81 can be written

$$\left\{ \begin{array}{c} R_1^{(e)} \\ \vdots \\ R_n^{(e)} \end{array} \right\}_K = [K^{(e)}] \left\{ \begin{array}{c} h_1 \\ \vdots \\ h_n \end{array} \right\}$$

(3.82)

where $[K^{(e)}]$ is the element conductance matrix. Similarly, the evaluation of the second integral on the right-hand side of equation 3.81 can be written

$$\left\{ \begin{array}{c} R_1^{(e)} \\ \vdots \\ R_n^{(e)} \end{array} \right\}_C = [C^{(e)}] \left\{ \begin{array}{c} \dfrac{\partial h_1}{\partial t} \\ \vdots \\ \dfrac{\partial h_n}{\partial t} \end{array} \right\}$$

(3.83)

where $[C^{(e)}]$ is called the *element capacitance matrix*.

The subscripts K and C in equations 3.82 and 3.83 are used to indicate the portion of the residual matrix represented by the first and second integrals on the right-hand side of equation 3.81.

To evaluate the second integral requires that the *time derivative* of the approximate solution be defined over the volume of the element. We can do this using interpolation functions and the values of the time derivative at the nodes, in the same manner that we defined $\hat{h}$ over the volume of the element using the interpolation functions and the values of

$\hat{h}$ at the nodes. Depending on the type of interpolation functions we use, the procedure is called either a *consistent element formulation* or a *lumped element formulation*.

Both formulations are used in practice. However, the lumped formulation is less susceptible to problems of numerical oscillation (see Chapter 5) than is the consistent formulation (also see Segerlind, 1984).

3.5.1 Consistent Element Formulation

We used interpolation functions to obtain an approximate solution for hydraulic head within an element, $\hat{h}^{(e)}$ in section 3.3. For an element with n nodes the approximate solution can be written in matrix form as

$$\hat{h}^{(e)}(x,y,z) = [\, N_1^{(e)}(x,y,z) \cdots N_n^{(e)}(x,y,z)] \begin{Bmatrix} h_1 \\ \vdots \\ h_n \end{Bmatrix} \qquad (3.84)$$

where $N_i^{(e)}$ is the interpolation function at node i and h_i is the (unknown) hydraulic head at node i. In the consistent element formulation, we use the *same* interpolation functions to define the time-derivative of the approximate solution for hydraulic head within an element, $\frac{\partial \hat{h}^{(e)}}{\partial t}$. For an element with n nodes, the time–derivative can be written in matrix form as

$$\frac{\partial \hat{h}^{(e)}}{\partial t}(x,y,z) = [\, N_1^{(e)}(x,y,z) \cdots N_n^{(e)}(x,y,z)] \begin{Bmatrix} \dfrac{\partial h_1}{\partial t} \\ \vdots \\ \dfrac{\partial h_n}{\partial t} \end{Bmatrix} \qquad (3.85)$$

where $N_i^{(e)}$ are the interpolation functions and $\frac{\partial h_i}{\partial t}$ are the (unknown) time derivatives of hydraulic head at each node.

If equation 3.85 is substituted into the second integral on the right-hand side of equation 3.81 we have

$$\iiint N_i^{(e)} S_s^{(e)} \frac{\partial \hat{h}}{\partial t}\, dx\, dy\, dz = \iiint N_i^{(e)} S_s^{(e)} \big[N_1^{(e)} \cdots N_n^{(e)} \big] \begin{Bmatrix} \dfrac{\partial h_1}{\partial t} \\ \vdots \\ \dfrac{\partial h_n^{(e)}}{\partial t} \end{Bmatrix} dx\, dy\, dz \qquad (3.86)$$

We can write equation 3.86 for each node (i = 1, 2, ..., n) in element e. This set of equations can also be written in matrix form

$$\begin{Bmatrix} R_1^{(e)} \\ \vdots \\ R_n^{(e)} \end{Bmatrix}_C = \big[C^{(e)} \big] \begin{Bmatrix} \dfrac{\partial h_1}{\partial t} \\ \vdots \\ \dfrac{\partial h_n}{\partial t} \end{Bmatrix} \qquad (3.87)$$

where

$$
\underset{n \times n}{\left[C^{(e)} \right]} = \iiint\limits_{V^{(e)}} \underset{n \times 1}{\begin{bmatrix} N_1^{(e)} \\ \vdots \\ N_n^{(e)} \end{bmatrix}} \underset{1 \times 1}{\left[S_s^{(e)} \right]} \underset{1 \times n}{\left[N_1^{(e)} \quad \cdots \quad N_n^{(e)} \right]} dx \, dy \, dz
$$

(3.88)

For two-dimensional problems equation 3.88 becomes

$$
\underset{n \times n}{\left[C^{(e)} \right]} = \iint\limits_{A^{(e)}} \underset{n \times 1}{\begin{bmatrix} N_1^{(e)} \\ \vdots \\ N_n^{(e)} \end{bmatrix}} \underset{1 \times 1}{\left[S_s^{(e)} \right]} \underset{1 \times n}{\left[N_1^{(e)} \quad \cdots \quad N_n^{(e)} \right]} dx \, dy
$$

(3.89)

where $A^{(e)}$ is the area of element e. For one–dimensional problems equation 3.88 becomes

$$
\underset{n \times n}{\left[C^{(e)} \right]} = \int\limits_{L^{(e)}} \underset{n \times 1}{\begin{bmatrix} N_1^{(e)} \\ \vdots \\ N_n^{(e)} \end{bmatrix}} \underset{1 \times 1}{\left[S_s^{(e)} \right]} \underset{1 \times n}{\left[N_1^{(e)} \quad \cdots \quad N_n^{(e)} \right]} dx
$$

(3.90)

where $L^{(e)}$ is the length of element e.

3.5.2 Lumped Element Formulation

In the lumped element formulation we also define the time-derivative of the approximate solution for hydraulic head within an element using interpolation functions and the values of the time derivative at the element's nodes. However, in this case we use *different* interpolation functions to define $\dfrac{\partial \hat{h}^{(e)}}{\partial t}$ than are used to define $\hat{h}^{(e)}$

$$
\frac{\partial \hat{h}^{(e)}}{\partial t}(x,y,z) = \left[N_i^{*(e)}(x,y,z) \quad \cdots \quad N_n^{*(e)}(x,y,z) \right] \begin{Bmatrix} \dfrac{\partial h_1}{\partial t} \\ \vdots \\ \dfrac{\partial h_n}{\partial t} \end{Bmatrix}
$$

(3.91)

where $N_i^{*(e)}$ are the interpolation functions for the time derivative of hydraulic head at each node. These interpolation functions are defined so that

$$N_i^* \, N_j^* = \begin{cases} \dfrac{1}{n} & \text{if } \; i = j \\[2mm] 0 & \text{if } \; i \neq j \end{cases} \tag{3.92}$$

where n is the number of nodes in the element. If we rewrite equation 3.88 using these interpolation functions

$$\left[C^{(e)} \right] = \underset{V^{(e)}}{\iiint} \underset{n \times 1}{\begin{bmatrix} N_1^{*(e)} \\ \vdots \\ N_n^{*(e)} \end{bmatrix}} \underset{1 \times 1}{\left[S_s^{(e)} \right]} \left[N_1^{*(e)} \; \cdots \; N_n^{*(e)} \right] dx \, dy \, dz$$

$$= S_s^{(e)} \begin{bmatrix} N_1^{*(e)}N_1^{*(e)} & \cdots & N_1^{*(e)}N_n^{(e)} \\ \vdots & \cdots & \vdots \\ N_n^{*(e)}N_1^{*(e)} & \cdots & N_n^{*(e)}N_n^{*(e)} \end{bmatrix} \underset{V^{(e)}}{\iiint} dx \, dy \, dz$$

$$\boxed{ \underset{n \times n}{\left[C^{(e)} \right]} = \underset{n \times n}{S_s^{(e)} \frac{V^{(e)}}{n} \begin{bmatrix} 1 & & \mathbf{O} \\ & \ddots & \\ \mathbf{O} & & 1 \end{bmatrix}} } \tag{3.93}$$

For example, for the case of a one-dimensional element with two nodes (n = 2) equation 3.93 becomes

$$\left[C^{(e)} \right] = S_s^{(e)} \frac{L^{(e)}}{2} \begin{bmatrix} 1 & 0 \\ 0 & 1 \end{bmatrix} \tag{3.94}$$

where $L^{(e)}$ is the length of the element. For a two-dimensional element with three nodes (n = 3), equation 3.93 becomes

$$\left[C^{(e)} \right] = S_s^{(e)} \frac{A^{(e)}}{3} \begin{bmatrix} 1 & 0 & 0 \\ 0 & 1 & 0 \\ 0 & 0 & 1 \end{bmatrix} \tag{3.95}$$

where $A^{(e)}$ is the area of element e.

3.5.3 Finite-Difference Formulation for Time-Derivative of the Approximate Solution

A *global capacitance matrix* [C] can be obtained by combining the element capacitance matrices for all the elements in the mesh in the same way that the global conductance matrix was obtained by combining the element conductance matrices in section 3.2

$$[C]_{\text{global}} = \sum_{e=1}^{m} \left[C^{(e)} \right]_{\text{expanded}} \tag{3.96}$$

The global capacitance matrix is a square, symmetric matrix with size pxp where p is the number of nodes in the mesh. By substituting the appropriate matrix formulation for each of the integrals on the right-hand side of equation 3.81, the weighted residual formulation for the transient, saturated flow equation becomes

$$\underset{\substack{\text{global} \\ \text{p×p}}}{[C]} \left\{ \begin{array}{c} \dfrac{\partial h_1}{\partial t} \\ \vdots \\ \dfrac{\partial h_p}{\partial t} \end{array} \right\} + \underset{\substack{\text{global} \\ \text{p×p}}}{[K]} \left\{ \begin{array}{c} h_1 \\ \vdots \\ h_p \end{array} \right\} = \underset{\substack{\text{global} \\ \text{p×1}}}{\{F\}} \tag{3.97}$$

If we define the two vectors $\{\dot{h}\}$ and $\{h\}$ as

$$\{\dot{h}\} = \left\{ \begin{array}{c} \dfrac{\partial h_1}{\partial t} \\ \vdots \\ \dfrac{\partial h_p}{\partial t} \end{array} \right\} \qquad \{h\} = \left\{ \begin{array}{c} h_1 \\ \vdots \\ h_p \end{array} \right\}$$

equation 3.97 can be written

$$\underset{\text{global}}{[C]} \{\dot{h}\} + \underset{\text{global}}{[K]} \{h\} = \underset{\text{global}}{\{F\}} \tag{3.98}$$

Equation 3.98 is a *system of ordinary differential equations* , whose solution provides values of h and $\dfrac{\partial h}{\partial t}$ at each node in the finite element mesh. Although several methods are available for solving this system of equations, it has become standard practice in groundwater flow and solute transport modeling to use the *finite difference method* .

From the *mean value theorem* of elementary calculus we know that we can compute the time derivative of a function h at some point ε on the interval t to t+Δt by the difference between the value of the function at the two end points of the interval (Figure 3.8)

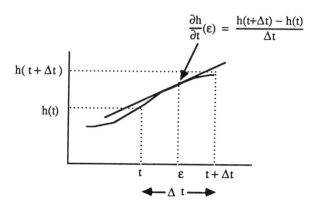

$$\frac{\partial h}{\partial t}(\varepsilon) \; = \; \frac{h(t+\Delta t) - h(t)}{\Delta t}$$

Figure 3.8   **Finite difference approximation to the time derivative for hydraulic head.**

Unfortunately the position of ε on the interval t to t+Δt is not known and different subsets of the finite difference methods have evolved based on different choices for the position of ε. From Figure 3.8

$$\frac{\partial h}{\partial t}(\varepsilon) \; = \; \frac{h(t + \Delta t) - h(t)}{\Delta t} \qquad\qquad (3.99)$$

or

$$h(\varepsilon) \; = \; h(t) + (\varepsilon - t)\frac{\partial h}{\partial t}(\varepsilon) \qquad\qquad (3.100)$$

If we define a variable ω

$$\omega \; = \; \frac{\varepsilon - t}{\Delta t} \qquad\qquad (3.101)$$

we can write

$$h(\varepsilon) \; = \; (1 - \omega)\, h(t) + \omega\, h\,(t + \Delta t) \qquad\qquad (3.102)$$

which can be extended to the vector of unknown hydraulic heads {h} and to the vector {F}

$$\{h\} \; = \; (1 - \omega)\{h\}_t + \omega\, \{h\}_{t+\Delta t} \qquad\qquad (3.103)$$

$$\{F\} \; = \; (1 - \omega)\{F\}_t + \omega\, \{F\}_{t+\Delta t} \qquad\qquad (3.104)$$

If we substitute equations 3.103 and 3.104 into equation 3.98 we have the finite difference formulation for the transient, saturated flow equation

$$([C] + \omega \, \Delta t \, [K]) \, \{h\}_{t+\Delta t}$$
$$= ([C] - (1-\omega) \, \Delta t \, [K]) \, \{h\}_t + \Delta t \, ((1 - \omega)\{F\}_t + \omega \, \{F\}_{t+\Delta t})$$
$$(3.105)$$

The solution procedure begins by specifying the initial values of $\{h\}$ (i.e., the values of head at time $t = t_0 = 0$)

$$\{h\}_{t_0} = \text{specified values}$$

Then the system of linear equations (equation 3.105) is solved to obtain values of $\{h\}$ at the end of the first time step, $\{h\}_{t_0 + \Delta t}$. We then set

$$\{h\}_t = \{h\}_{t_0+\Delta t}$$

and repeat the solution process for the next time step, and so on. Depending on the choice of ω several different subsets of the finite difference formulation are defined:

$\omega = 0 \quad \rightarrow \quad$ *Forward Difference Method*

$$[C]\{h\}_{t+\Delta t} = ([C] - \Delta t \, [K])\{h\}_t + \Delta t \, \{F\}_t \qquad (3.106)$$

$\omega = \dfrac{1}{2} \quad \rightarrow \quad$ *Central Difference* or *Crank–Nicholson Method*

$$\left([C] + \frac{\Delta t}{2}[K] \right)\{h\}_{t+\Delta t} = \left([C] - \frac{\Delta t}{2}[K] \right)\{h\}_t + \frac{\Delta t}{2}(\{F\}_t + \{F\}_{t+\Delta t}) \qquad (3.107)$$

$\omega = 1 \quad \rightarrow \quad$ *Backward Difference Method*

$$([C] + \Delta t \, [K])\{h\}_{t+\Delta t} = [C]\{h\}_t + \Delta t \, \{F\}_{t+\Delta t} \qquad (3.108)$$

3.5.4 A Finite Element Example

To illustrate the use of equation 3.105 we will again consider the column of soil from the example in Section 3.2. Initially the column is in steady-state saturated flow with a distribution of hydraulic head computed from the previous example (Figure 3.9). Then at time $t = 0$ we increase the value of hydraulic head at the upper boundary (node 1) from 12 to 20 cm. We wish to find the value of hydraulic head at each node at time $t = 1, 2, ...$ seconds.

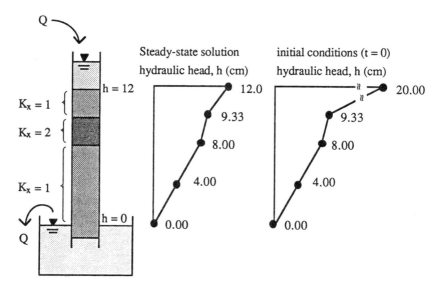

Figure 3.9   **Initial conditions for example transient, saturated flow problem.**

The governing differential equation is the one-dimensional form of equation 3.79

$$\frac{\partial}{\partial x}\left(K_x\frac{\partial h}{\partial x}\right) = S_s\frac{\partial h}{\partial t}$$

where K_x is the saturated hydraulic conductivity in the direction of flow (the x axis is directed vertically downward in this case). We will use the lumped element formulation to solve this problem. Let $S_s^{(1)} = 0.02$, $S_s^{(2)} = 0.01$, and $S_s^{(3)} = S_s^{(4)} = 0.02$. For one-dimensional elements with two nodes, the element capacitance matrices are given by equation 3.94

$$\left[C^{(1)}\right] = \frac{S_s^{(1)}L^{(1)}}{2}\begin{bmatrix}1 & 0\\0 & 1\end{bmatrix} = \frac{(0.02)(2)}{2}\begin{bmatrix}1 & 0\\0 & 1\end{bmatrix} = \begin{bmatrix}0.02 & 0\\0 & 0.02\end{bmatrix}$$

$$\left[C^{(2)}\right] = \frac{S_s^{(2)}L^{(2)}}{2}\begin{bmatrix}1 & 0\\0 & 1\end{bmatrix} = \frac{(0.01)(2)}{2}\begin{bmatrix}1 & 0\\0 & 1\end{bmatrix} = \begin{bmatrix}0.01 & 0\\0 & 0.01\end{bmatrix}$$

$$\left[C^{(3)}\right] = \left[C^{(4)}\right] = \frac{S_s^{(3)}L^{(3)}}{2}\begin{bmatrix}1 & 0\\0 & 1\end{bmatrix} = \frac{(0.02)(3)}{2}\begin{bmatrix}1 & 0\\0 & 1\end{bmatrix} = \begin{bmatrix}0.03 & 0\\0 & 0.03\end{bmatrix}$$

The global capacitance matrix is obtained by adding the expanded form of the element capacitance matrices

$$[C] = \begin{bmatrix} 0.02 & 0 & 0 & 0 & 0 \\ 0 & 0.02+0.01 & 0 & 0 & 0 \\ 0 & 0 & 0.01+0.03 & 0 & 0 \\ 0 & 0 & 0 & 0.03+0.03 & 0 \\ 0 & 0 & 0 & 0 & 0.03 \end{bmatrix}$$

$$= \begin{bmatrix} 0.02 & 0 & 0 & 0 & 0 \\ 0 & 0.03 & 0 & 0 & 0 \\ 0 & 0 & 0.04 & 0 & 0 \\ 0 & 0 & 0 & 0.06 & 0 \\ 0 & 0 & 0 & 0 & 0.03 \end{bmatrix} \tag{3.109}$$

From the previous example, the global conductance matrix is

$$[K] = \begin{bmatrix} 1/2 & -1/2 & 0 & 0 & 0 \\ -1/2 & 3/2 & -1 & 0 & 0 \\ 0 & -1 & 4/3 & -1/3 & 0 \\ 0 & 0 & -1/3 & 2/3 & -1/3 \\ 0 & 0 & 0 & -1/3 & 1/3 \end{bmatrix} \tag{3.110}$$

The initial values of hydraulic head at the nodes are

$$\{h\}_{t=0} = \begin{Bmatrix} h_1 \\ h_2 \\ h_3 \\ h_4 \\ h_5 \end{Bmatrix}_{t=0} = \begin{Bmatrix} 20.00 \\ 9.33 \\ 8.00 \\ 4.00 \\ 0.00 \end{Bmatrix} \tag{3.111}$$

We will use the backward difference formulation (equation 3.108), with a time step $\Delta t = 1$ sec. By setting $\{F\} = 0$ (no specified flow rates) the system of equations for the end of the first time step becomes

$$([C] + \Delta t\,[K])\{h\}_{t=1} = [C]\{h\}_{t=0} + \cancel{\Delta t\,\{F\}_{t=1}}^{\;0} \tag{3.112}$$

Substituting equations 3.109, 3.110, and 3.111 into equation 3.112 gives

$$\left(\begin{bmatrix} 0.02 & 0 & 0 & 0 & 0 \\ 0 & 0.03 & 0 & 0 & 0 \\ 0 & 0 & 0.04 & 0 & 0 \\ 0 & 0 & 0 & 0.06 & 0 \\ 0 & 0 & 0 & 0 & 0.03 \end{bmatrix} + (1) \begin{bmatrix} 1/2 & -1/2 & 0 & 0 & 0 \\ -1/2 & 3/2 & -1 & 0 & 0 \\ 0 & -1 & 4/3 & -1/3 & 0 \\ 0 & 0 & -1/3 & 2/3 & -1/3 \\ 0 & 0 & 0 & -1/3 & 1/3 \end{bmatrix} \right) \begin{Bmatrix} h_1 \\ h_2 \\ h_3 \\ h_4 \\ h_5 \end{Bmatrix}_{t=1}$$

$$= \begin{bmatrix} 0.02 & 0 & 0 & 0 & 0 \\ 0 & 0.03 & 0 & 0 & 0 \\ 0 & 0 & 0.04 & 0 & 0 \\ 0 & 0 & 0 & 0.06 & 0 \\ 0 & 0 & 0 & 0 & 0.03 \end{bmatrix} \begin{Bmatrix} 20.00 \\ 9.33 \\ 8.00 \\ 4.00 \\ 0.00 \end{Bmatrix}$$

which simplifies to

$$\begin{bmatrix} 0.52 & -0.50 & 0 & 0 & 0 \\ -0.50 & 1.53 & -1.00 & 0 & 0 \\ 0 & -1.00 & 1.37 & -0.33 & 0 \\ 0 & 0 & -0.33 & 0.73 & -0.33 \\ 0 & 0 & 0 & -0.33 & 0.36 \end{bmatrix} \begin{Bmatrix} h_1 \\ h_2 \\ h_3 \\ h_4 \\ h_5 \end{Bmatrix}_{t=1} = \begin{Bmatrix} 0.04 \\ 0.28 \\ 0.32 \\ 0.24 \\ 0.00 \end{Bmatrix} \qquad (3.113)$$

But $h_1 = 20$ and $h_5 = 0$ for all values of t (because the hydraulic head at the upper and lower ends of the column are held constant). Modifying equations 3.113 for these known values (see Section 4.5) gives

$$\begin{bmatrix} 1.53 & -1.00 & 0 \\ -1.00 & 1.37 & -0.33 \\ 0 & -0.33 & 0.73 \end{bmatrix} \begin{Bmatrix} h_2 \\ h_3 \\ h_4 \end{Bmatrix}_{t=1} = \begin{Bmatrix} 10.28 \\ 0.32 \\ 0.24 \end{Bmatrix}$$

which can be solved to obtain the values of hydraulic head at the end of the first time step

$$\begin{Bmatrix} h_1 \\ h_2 \\ h_3 \\ h_4 \\ h_5 \end{Bmatrix}_{t=1} = \begin{Bmatrix} 20.00 \\ 14.95 \\ 12.60 \\ 6.02 \\ 0.00 \end{Bmatrix}$$

This process is repeated for each subsequent time step until a solution is obtained for each required value of t.

3.6 TRANSIENT, UNSATURATED FLOW EQUATION

The three-dimensional form of the equation for transient groundwater flow through unsaturated porous media is written as

$$\frac{\partial}{\partial x}\left(K_x(\psi)\frac{\partial \psi}{\partial x}\right) + \frac{\partial}{\partial y}\left(K_y(\psi)\frac{\partial \psi}{\partial y}\right) + \frac{\partial}{\partial z}\left(K_z(\psi)\left(\frac{\partial \psi}{\partial z} + 1\right)\right) = C(\psi)\frac{\partial \psi}{\partial t} \qquad (3.114)$$

where $K_x(\psi)$, $K_y(\psi)$, and $K_z(\psi)$ are the unsaturated hydraulic conductivities (which are functions of the pressure head ψ) in the three coordinate directions (the z coordinate direction is assumed to be vertical), and $C(\psi)$ is the *specific moisture capacity*

$$C(\psi) = \frac{d\theta}{d\psi} \qquad (3.115)$$

where θ is the volumetric water content (Appendix II). The unknown quantity at each node is the pressure head ψ. As before we assume an approximate solution for ψ, $\hat{\psi}$ of the form

$$\hat{\psi}^{(e)} = \sum_{i=1}^{n} N_i^{(e)} \psi_i \qquad (3.116)$$

where $\hat{\psi}^{(e)}$ is the approximate solution for pressure head within element e and $N_i^{(e)}$ are the interpolation functions for each node within element e.

When the approximate solution is substituted into equation 3.114, the differential equation is not satisfied exactly and an error or residual occurs at every point in the problem domain. The contribution of any element e to the residual at node i to which the element is joined is

$$R_i^{(e)} = -\iiint\limits_{V^{(e)}} W_i^{(e)}(x,y,z) \left[\frac{\partial}{\partial x}\left(K_x(\psi)\frac{\partial \psi}{\partial x}\right) + \frac{\partial}{\partial y}\left(K_y(\psi)\frac{\partial \psi}{\partial y}\right) + \frac{\partial}{\partial z}\left(K_z(\psi)\left(\frac{\partial \psi}{\partial z} + 1\right)\right) \right.$$
$$\left. - C(\psi)\frac{\partial \psi}{\partial t} \right] dx\, dy\, dz \qquad (3.117)$$

where $W_i^{(e)}$ is the element's weighting function for node i and the limits of integration are chosen to represent the volume of element e.

In Galerkin's method we choose the weighting function for each node in the element to be equal to the element's interpolation function for that node $W_i^{(e)} = N_i^{(e)}$. If we also assume that the unsaturated hydraulic conductivity and specific moisture capacity are constant within an element (but can vary from one element to the next), and that gravitational forces are small, equation 3.117 can be written

$$R_i^{(e)} = -\iiint\limits_{V^{(e)}} N_i^{(e)}(x,y,z) \left[K_x^{(e)}(\psi)\frac{\partial^2 \hat{\psi}}{\partial x^2} + K_y^{(e)}(\psi)\frac{\partial^2 \hat{\psi}}{\partial y^2} + K_z^{(e)}(\psi)\frac{\partial^2 \hat{\psi}}{\partial z^2} - C^{(e)}(\psi)\frac{\partial \hat{\psi}}{\partial t} \right] dx\, dy\, dz$$

$$= -\iiint\limits_{V^{(e)}} N_i^{(e)} \left[K_x^{(e)}(\psi)\frac{\partial^2 \hat{\psi}}{\partial x^2} + K_y^{(e)}(\psi)\frac{\partial^2 \hat{\psi}}{\partial y^2} + K_z^{(e)}(\psi)\frac{\partial^2 \hat{\psi}}{\partial z^2} \right] dx\, dy\, dz$$

$$+ \iiint\limits_{V^{(e)}} N_i^{(e)} C^{(e)}(\psi)\frac{\partial \hat{\psi}}{\partial t}\, dx\, dy\, dz \qquad (3.118)$$

where $C^{(e)}(\psi)$ is the specific moisture capacity within element e.

We know from the results of sections 3.4 and 3.5 that the integrals in equation 3.118 can be written

$$\begin{Bmatrix} R_1^{(e)} \\ \vdots \\ R_n^{(e)} \end{Bmatrix}_K = [K^{(e)}(\psi)] \begin{Bmatrix} \psi_1 \\ \vdots \\ \psi_n \end{Bmatrix} \tag{3.119}$$

and

$$\begin{Bmatrix} R_1^{(e)} \\ \vdots \\ R_n^{(e)} \end{Bmatrix}_C = [C^{(e)}(\psi)] \begin{Bmatrix} \dfrac{\partial \psi_1}{\partial t} \\ \vdots \\ \dfrac{\partial \psi_n}{\partial t} \end{Bmatrix} \tag{3.120}$$

where $[K^{(e)}(\psi)]$ and $[C^{(e)}(\psi)]$ are the unsaturated forms of the element conductance and capacitance matrices for element e. Just as in the case of transient, saturated flow, we can use two different types interpolation functions to evaluate the integral

$$\iiint\limits_{V^{(e)}} N_i^{(e)} C^{(e)}(\psi) \frac{\partial \hat{\psi}}{\partial t} \, dx \, dy \, dz \tag{3.121}$$

and obtain the computational form for $[C^{(e)}(\psi)]$. In the *consistent element formulation* we use the *same* interpolation functions to define the time-derivative of the approximate solution for pressure head within an element, $\dfrac{\partial \hat{\psi}^{(e)}}{\partial t}$, as those used to define the approximation solution for pressure head ψ

$$\frac{\partial \hat{\psi}^{(e)}}{\partial t}(x,y,z) = [N_1^{(e)}(x,y,z) \cdots N_n^{(e)}(x,y,z)] \begin{Bmatrix} \dfrac{\partial \psi_1}{\partial t} \\ \vdots \\ \dfrac{\partial \psi_n}{\partial t} \end{Bmatrix} \tag{3.122}$$

where $N_i^{(e)}$ are the interpolation functions and $\dfrac{\partial \psi_i}{\partial t}$ are the (unknown) time derivatives of pressure head at each node within element e. For this choice of interpolation functions, we can write the unsaturated form of the element capacitance matrix as

$$[C^{(e)}(\psi)] = \iiint\limits_{V^{(e)}} \begin{bmatrix} N_1^{(e)} \\ \vdots \\ N_n^{(e)} \end{bmatrix}_{n\times 1} [C^{(e)}(\psi)]_{1\times 1} [N_1^{(e)} \cdots N_n^{(e)}]_{1\times n} \; dx \; dy \; dz$$

$$(3.123)$$

In the *lumped element formulation*, we use *different* interpolation functions to define $\dfrac{\partial\hat{\psi}^{(e)}}{\partial t}$ than are used to define $\hat{\psi}^{(e)}$

$$\frac{\partial\hat{\psi}^{(e)}}{\partial t}(x,y,z) = \left[N_1^{*(e)}(x,y,z) \cdots N_n^{*(e)}(x,y,z) \right] \begin{Bmatrix} \dfrac{\partial\psi_1}{\partial t} \\ \vdots \\ \dfrac{\partial\psi_n}{\partial t} \end{Bmatrix}$$

$$(3.124)$$

where $N_i^{*(e)}$ is the interpolation function for the time derivative at node i within element e. These interpolation functions were defined in equation 3.92 and using equation 3.93 we can immediately write

$$\begin{bmatrix} C^{(e)}(\psi) \end{bmatrix}_{n\times n} = C^{(e)}(\psi)\frac{V^{(e)}}{n} \begin{bmatrix} 1 & & O \\ & \ddots & \\ O & & 1 \end{bmatrix}_{n\times n}$$

$$(3.125)$$

where $V^{(e)}$ is the volume of the element.

The unsaturated form of the global capacitance matrix is obtained by combining the element capacitance matrices for all elements in the mesh

$$\underset{\text{global}}{[C(\psi)]} = \sum_{\substack{e=1 \\ \text{expanded}}}^{m} \left[C^{(e)}(\psi) \right]$$

$$(3.126)$$

where there are m elements in the mesh. By substituting the appropriate matrix formulations for each of the integrals on the right-hand side of equation 3.118, the weighted residual formulation for the transient, unsaturated flow equation becomes

$$\underset{\text{global}}{[C(\psi)]} \begin{Bmatrix} \dfrac{\partial\psi_1}{\partial t} \\ \vdots \\ \dfrac{\partial\psi_p}{\partial t} \end{Bmatrix} + \underset{\text{global}}{[K(\psi)]} \begin{Bmatrix} \psi_1 \\ \vdots \\ \psi_p \end{Bmatrix} = \underset{\text{global}}{\{F\}}$$

$$(3.127)$$

If we define $\{\dot{\psi}\} = \left\{\dfrac{\partial\psi}{\partial t}\right\}$ equation 3.127 can be rewritten as

$$[C(\psi)]\ \{\dot{\psi}\} + [K(\psi)]\ \{\psi\} = \{F\} \qquad (3.128)$$
$$\text{global} \qquad\qquad \text{global} \qquad\qquad \text{global}$$

Using the results of Section 3.53 we can also write the finite difference formulation for the transient, unsaturated flow equation

$$([C(\psi)] + \omega\Delta t\,[K(\psi)]\,)\{\psi\}_{t+\Delta t} = ([C(\psi)] - (1-\omega)\Delta t\,[K(\psi)]\{\psi\}_t$$
$$+\ \Delta t\,((1-\omega)\{F\}_t + \omega\{F\}_{t+\Delta t}\,) \qquad (3.129)$$

Equation 3.129 is a system of ordinary differential equations, whose solution provides values of ψ and $\dfrac{\partial\psi}{\partial t}$ at each node in the finite element mesh at each time. A modification of the finite difference method described in section 3.5.3 can be used to obtain this solution. The modified procedure will be described in Chapter 5.

3.7 SOLUTE TRANSPORT EQUATION

The three-dimensional form of the solute transport equation for *uniform* groundwater flow in the x direction is

$$\frac{\partial(\theta C)}{\partial t} = D_x\frac{\partial^2}{\partial x^2}(\theta C) + D_y\frac{\partial^2}{\partial y^2}(\theta C) + D_z\frac{\partial^2}{\partial z^2}(\theta C)$$

$$-\frac{\partial}{\partial x}(v_x C) - \frac{\partial}{\partial t}(\rho_b K_d C) - \lambda(\theta C + \rho_b K_d C) \qquad (3.130)$$

where θ is the volumetric water content of the porous media, C is solute concentration, D_x, D_y, and D_z are the dispersion coefficients of the porous media in the x, y, and z coordinate directions, v_x is the apparent groundwater velocity in the x coordinate direction, ρ_b is the bulk density of the porous media, K_d is the distribution coefficient, and λ is the solute decay constant (Appendix III).

When we solve a solute transport problem by the finite element method, the unknown quantity at each node is the solute concentration C. We begin by assuming an approximate solution for C, $\hat{C}$ of the form

$$\hat{C}^{(e)}(x,y,z) = \sum_{i=1}^{n} N_i^{(e)}\,C_i \qquad (3.131)$$

where $\hat{C}^{(e)}$ is the approximate solution for solute concentration within element e, $N_i^{(e)}$ are the interpolation functions for each node within element e, and C_i are the unknown solute concentrations for each node within element e. When the approximate solution is substituted into equation 3.130, the differential equation is not satisfied exactly and an error or residual occurs at every point in the problem domain. The contribution of element e to the residual at node i is

$$R_i^{(e)} = -\iiint\limits_{V^{(e)}} W_i^{(e)}(x,y,z)\left[D_x\frac{\partial^2}{\partial x^2}(\theta\hat{C}^{(e)}) + D_y\frac{\partial^2}{\partial y^2}(\theta\hat{C}^{(e)}) + D_z\frac{\partial^2}{\partial z^2}(\theta\hat{C}^{(e)}) - \frac{\partial}{\partial x}(v_x\hat{C}^{(e)}) \right.$$

$$\left. - \frac{\partial}{\partial t}(\rho_b K_d\hat{C}^{(e)}) - \lambda(\theta\hat{C}^{(e)} + \rho_b K_d\hat{C}^{(e)}) - \frac{\partial}{\partial t}(\theta\hat{C}^{(e)}) \right] dx\ dy\ dz$$

$$(3.132)$$

where $W_i^{(e)}$ is the element's weighting function for node i and the limits of integration are chosen to represent the volume of the element.

In Galerkin's method we choose the weighting function for each node in the element to be equal to the element's interpolation function for that node, $W_i^{(e)} = N_i^{(e)}$. If we also assume that the properties of the porous media and the apparent groundwater velocity are constant within an element (but can vary from one element to the next) equation 3.132 can be written

$$R_i^{(e)} = -\iiint\limits_{V^{(e)}} N_i^{(e)}(x,y,z)\left[D_x^{(e)}\theta^{(e)}\frac{\partial^2\hat{C}^{(e)}}{\partial x^2} + D_y^{(e)}\theta^{(e)}\frac{\partial^2\hat{C}^{(e)}}{\partial y^2} + D_z^{(e)}\theta^{(e)}\frac{\partial^2\hat{C}^{(e)}}{\partial z^2} - v_x^{(e)}\frac{\partial\hat{C}^{(e)}}{\partial x} \right.$$

$$\left. - \rho_b^{(e)}K_d^{(e)}\frac{\partial\hat{C}^{(e)}}{\partial t} - \lambda(\theta^{(e)}\hat{C}^{(e)} + \rho_b^{(e)}K_d^{(e)}\hat{C}^{(e)}) \right.$$

$$\left. - \frac{\partial}{\partial t}(\theta^{(e)}\ \hat{C}^{(e)}) \right] dx\ dy\ dz \qquad (3.133)$$

where, for example, $\theta^{(e)}$ is the volumetric water content of the porous media within element e. λ is not superscripted because it is a property of the *solute* (not the porous media) and is therefore constant from one element to the next. Because water content $\theta^{(e)}$ and the apparent groundwater velocity $v_x^{(e)}$ may or may not change with time, two separate formulations of equation 3.133 are possible.

3.7.1 Steady-State Groundwater Flow

In steady-state groundwater flow (saturated or unsaturated), the water content and apparent groundwater velocity are constant from one time step to the next. They are also constant within an element (but can vary from one element to the next). In this case equation 3.133 becomes

$$R_i^{(e)} = -\iiint\limits_{V^{(e)}} N_i^{(e)} \left[D_x^{(e)} \theta^{(e)} \frac{\partial^2 \hat{C}^{(e)}}{\partial x^2} + D_y^{(e)} \theta^{(e)} \frac{\partial^2 \hat{C}^{(e)}}{\partial y^2} + D_z^{(e)} \theta^{(e)} \frac{\partial^2 \hat{C}^{(e)}}{\partial z^2} \right] dx\, dy\, dz$$

$$+\iiint\limits_{V^{(e)}} N_i^{(e)} \left[v_x^{(e)} \frac{\partial \hat{C}^{(e)}}{\partial x} \right] dx\, dy\, dz$$

$$+\iiint\limits_{V^{(e)}} N_i^{(e)} \left[\rho_b^{(e)} K_d^{(e)} \frac{\partial \hat{C}^{(e)}}{\partial t} \right] dx\, dy\, dz$$

$$+\iiint\limits_{V^{(e)}} N_i^{(e)} \left[\lambda(\theta^{(e)} \hat{C}^{(e)} + \rho_b^{(e)} K_d^{(e)} \hat{C}^{(e)}) \right] dx\, dy\, dz$$

$$+\iiint\limits_{V^{(e)}} N_i^{(e)} \left[\theta^{(e)} \frac{\partial \hat{C}^{(e)}}{\partial t} \right] dx\, dy\, dz \qquad (3.134)$$

From our previous work with the transient groundwater flow equations we know that we can write equation 3.134 in matrix form by combining the integral expressions for each node in element e. Specifically we can write

$$\begin{Bmatrix} R_1^{(e)} \\ \vdots \\ R_n^{(e)} \end{Bmatrix} = \left[D^{(e)} \right] \begin{Bmatrix} C_1 \\ \vdots \\ C_n \end{Bmatrix} + \left[A^{(e)} \right] \begin{Bmatrix} \dfrac{\partial C_1}{\partial t} \\ \vdots \\ \dfrac{\partial C_n}{\partial t} \end{Bmatrix} \qquad (3.135)$$

where $[D^{(e)}]$ is the *element advection–dispersion matrix* and $[A^{(e)}]$ is the *element sorption matrix*. The element advection-dispersion matrix is defined as

$$[D^{(e)}]_{n \times n} = \iiint\limits_{V^{(e)}} \underbrace{\begin{bmatrix} \dfrac{\partial N_1^{(e)}}{\partial x} & \dfrac{\partial N_1^{(e)}}{\partial y} & \dfrac{\partial N_1^{(e)}}{\partial z} \\ \vdots & \vdots & \vdots \\ \dfrac{\partial N_n^{(e)}}{\partial x} & \dfrac{\partial N_n^{(e)}}{\partial y} & \dfrac{\partial N_n^{(e)}}{\partial z} \end{bmatrix}}_{n \times 3} \underbrace{\begin{bmatrix} D_x^{(e)} \theta^{(e)} & 0 & 0 \\ 0 & D_y^{(e)} \theta^{(e)} & 0 \\ 0 & 0 & D_z^{(e)} \theta^{(e)} \end{bmatrix}}_{3 \times 3} \underbrace{\begin{bmatrix} \dfrac{\partial N_1^{(e)}}{\partial x} & \cdots & \dfrac{\partial N_n^{(e)}}{\partial x} \\ \dfrac{\partial N_1^{(e)}}{\partial y} & \cdots & \dfrac{\partial N_n^{(e)}}{\partial y} \\ \dfrac{\partial N_1^{(e)}}{\partial z} & \cdots & \dfrac{\partial N_n^{(e)}}{\partial z} \end{bmatrix}}_{3 \times n} dx\, dy\, dz$$

$$+ \iiint\limits_{V^{(e)}} \underset{n\times 1}{\begin{bmatrix} N_1^{(e)} \\ \vdots \\ N_n^{(e)} \end{bmatrix}} \underset{1\times 1}{v_x^{(e)}} \underset{1\times n}{\left[\frac{\partial N_1^{(e)}}{\partial x} \cdots \frac{\partial N_n^{(e)}}{\partial x} \right]} dx\, dy\, dz$$

$$+ \iiint\limits_{V^{(e)}} \underset{n\times 1}{\begin{bmatrix} N_1^{(e)} \\ \vdots \\ N_n^{(e)} \end{bmatrix}} \underset{1\times 1}{\left[\lambda\, (\theta^{(e)} + \rho_b^{(e)} K_d^{(e)}) \right]} \underset{1\times n}{[N_1^{(e)} \cdots N_n^{(e)}]} dx\, dy\, dz \tag{3.136}$$

where $V^{(e)}$ is the volume of element e. The reader should recognize the terms in these equations as coming from the first, second and fourth integrals on the right-hand side of equation 3.134. The validity of these equations can be checked by multiplying a few of the terms and comparing the results with the integrals in equation 3.134. If the groundwater flow is not uniform (see Appendix III), equation 3.136 becomes

$$[D^{(e)}] = \iiint\limits_{V^{(e)}} \begin{bmatrix} \frac{\partial N_1^{(e)}}{\partial x} & \frac{\partial N_1^{(e)}}{\partial y} & \frac{\partial N_1^{(e)}}{\partial z} \\ \vdots & \vdots & \vdots \\ \frac{\partial N_n^{(e)}}{\partial x} & \frac{\partial N_n^{(e)}}{\partial y} & \frac{\partial N_n^{(e)}}{\partial z} \end{bmatrix} \begin{bmatrix} D_{xx}^{(e)}\theta^{(e)} & D_{xy}^{(e)}\theta^{(e)} & D_{xz}^{(e)}\theta^{(e)} \\ D_{yx}^{(e)}\theta^{(e)} & D_{yy}^{(e)}\theta^{(e)} & D_{yz}^{(e)}\theta^{(e)} \\ D_{zx}^{(e)}\theta^{(e)} & D_{zy}^{(e)}\theta^{(e)} & D_{zz}^{(e)}\theta^{(e)} \end{bmatrix} \begin{bmatrix} \frac{\partial N_1^{(e)}}{\partial x} & \cdots & \frac{\partial N_n^{(e)}}{\partial x} \\ \frac{\partial N_1^{(e)}}{\partial y} & \cdots & \frac{\partial N_n^{(e)}}{\partial y} \\ \frac{\partial N_1^{(e)}}{\partial z} & \cdots & \frac{\partial N_n^{(e)}}{\partial z} \end{bmatrix} dx\, dy\, dz$$

$$+ \iiint\limits_{V^{(e)}} \underset{n\times 3}{\begin{bmatrix} N_1^{(e)} & N_1^{(e)} & N_1^{(e)} \\ \vdots & \vdots & \vdots \\ N_n^{(e)} & N_n^{(e)} & N_n^{(e)} \end{bmatrix}} \underset{3\times 3}{\begin{bmatrix} v_x^{(e)} & 0 & 0 \\ 0 & v_y^{(e)} & 0 \\ 0 & 0 & v_z^{(e)} \end{bmatrix}} \underset{3\times n}{\begin{bmatrix} \frac{\partial N_1^{(e)}}{\partial x} & \cdots & \frac{\partial N_n^{(e)}}{\partial x} \\ \frac{\partial N_1^{(e)}}{\partial y} & \cdots & \frac{\partial N_n^{(e)}}{\partial y} \\ \frac{\partial N_1^{(e)}}{\partial z} & \cdots & \frac{\partial N_n^{(e)}}{\partial z} \end{bmatrix}}$$

$$+ \iiint\limits_{V^{(e)}} \underset{n\times 1}{\begin{bmatrix} N_1^{(e)} \\ \vdots \\ N_n^{(e)} \end{bmatrix}} \underset{1\times 1}{\left[\lambda(\rho_b^{(e)} K_d^{(e)} + \theta^{(e)}) \right]} \underset{1\times n}{[N_1^{(e)} \cdots N_n^{(e)}]} dx\, dy\, dz \tag{3.137}$$

The element sorption matrix is defined as

$$
\left[A^{(e)} \right] = \underset{V^{(e)}}{\iiint} \underset{n \times 1}{\begin{bmatrix} N_1^{(e)} \\ \vdots \\ N_n^{(e)} \end{bmatrix}} \underset{1 \times 1}{\left[\rho_b^{(e)} K_d^{(e)} + \theta^{(e)} \right]} \underset{1 \times n}{[N_1^{(e)} \cdots N_n^{(e)}]} \, dx \, dy \, dz
$$

(3.138)

if a consistent element formulation is used for the time derivative of the approximate solution $\dfrac{\partial \hat{C}}{\partial t}$. If a lumped element formulation is used for $\dfrac{\partial \hat{C}}{\partial t}$, the element sorption matrix is defined as

$$
\left[A^{(e)} \right] = (\rho_b^{(e)} K_d^{(e)} + \theta^{(e)}) \left(\frac{V^{(e)}}{n} \right) \underset{n \times n}{\begin{bmatrix} 1 & & O \\ & \ddots & \\ O & & 1 \end{bmatrix}}
$$

(3.139)

where $V^{(e)}$ is the volume of element e and n is the number of nodes within element e.

A *global advection-dispersion matrix* [D] and a *global sorption matrix* [A] can be obtained by combining the element matrices for all the elements in the mesh in the same way that the global conductance matrix was obtained by combining the element conductance matrices in Section 3.2

$$
\underset{\substack{\text{global} \\ p \times p}}{[D]} = \sum_{e=1}^{m} \underset{\substack{\text{expanded} \\ n \times n}}{\left[D^{(e)} \right]}
$$

(3.140)

$$
\underset{\substack{\text{global} \\ p \times p}}{[A]} = \sum_{e=1}^{m} \underset{\substack{\text{expanded} \\ n \times n}}{\left[A^{(e)} \right]}
$$

(3.141)

where m is the number of elements and p is the number of nodes in the mesh. The weighted residual formulation for the solute transport equation becomes

$$
\underset{\text{global}}{[D]} \begin{Bmatrix} C_1 \\ \vdots \\ C_p \end{Bmatrix} + \underset{\text{global}}{[A]} \begin{Bmatrix} \dfrac{\partial C_1}{\partial t} \\ \vdots \\ \dfrac{\partial C_p}{\partial t} \end{Bmatrix} = \underset{\text{global}}{\{F\}}
$$

(3.142)

If we define the two vectors $\{C\}$ and $\{\dot{C}\}$

$$\{C\} = \begin{Bmatrix} C_1 \\ \vdots \\ C_p \end{Bmatrix} \qquad \{\dot{C}\} = \begin{Bmatrix} \dfrac{\partial C_1}{\partial t} \\ \vdots \\ \dfrac{\partial C_p}{\partial t} \end{Bmatrix} \qquad\qquad (3.143)$$

equation 3.142 can be written as

$$[A]\{\dot{C}\} + [D]\{C\} = \{F\} \qquad\qquad (3.144)$$
$$\text{global} \qquad \text{global} \qquad \text{global}$$

Equation 3.144 is a system of ordinary differential equations, the solution of which provides values of C and $\dfrac{\partial C}{\partial t}$ at each node in the finite element mesh at each time. This equation can be solved using the finite difference method described in section 3.5.3. Using equation 3.105, we can immediately write the finite difference formulation for equation 3.144

$$([A] + \omega \Delta t\, [D]\,)\{C\}_{t+\Delta t} = ([A] - (1-\omega)\, \Delta t\, [D]\,)\{C\}_t + \Delta t\, ((1-\omega)\{F\}_t + \omega\{F\}_{t+\Delta t}\,)$$

$$(3.145)$$

The solution procedure begins by specifying the initial values of $\{C\}$

$$\{C\}_{t_0} = \text{specified values}$$

Then we solve the system of linear equations to obtain values of $\{C\}$ at the end of the first time step, $\{C\}_{t_0 + \Delta t}$. We then set

$$\{C\}_t = \{C\}_{t_0 + \Delta t}$$

in equation 3.145 and repeat the solution process for the next time step, and so on (see Chapter 5).

3.7.2 A Finite Element Example

The use of equation 3.145 is illustrated with the one-dimensional problem in Figure 3.10. Steady-state, saturated groundwater flow is occurring in a confined aquifer. Initially no solute is present. At time zero, the solute concentration along the left boundary of the aquifer is increased to 10 mg/l and remains constant thereafter. The problem domain is discretized into a mesh with five elements and six nodes. Each element has two nodes so the dispersion-advection matrix for each element is given by the one-dimensional form of equation 3.136 (n=2)

$$
\begin{bmatrix} D^{(e)} \end{bmatrix} = \int_{x_i^{(e)}}^{x_j^{(e)}} \begin{bmatrix} \dfrac{\partial N_1^{(e)}}{\partial x} \\[2mm] \dfrac{\partial N_2^{(e)}}{\partial x} \end{bmatrix} [D_x^{(e)}\theta^{(e)}] \begin{bmatrix} \dfrac{\partial N_1^{(e)}}{\partial x} & \dfrac{\partial N_2^{(e)}}{\partial x} \end{bmatrix} dx
$$

$$
\underset{2\times2}{\phantom{[D^{(e)}]}} \qquad \underset{2\times1}{} \qquad \underset{1\times1}{} \qquad \underset{1\times2}{}
$$

$$
+ \int_{x_i^{(e)}}^{x_j^{(e)}} \begin{bmatrix} N_1^{(e)} \\[2mm] N_2^{(e)} \end{bmatrix} [v_x^{(e)}] \begin{bmatrix} \dfrac{\partial N_1^{(e)}}{\partial x} & \dfrac{\partial N_2^{(e)}}{\partial x} \end{bmatrix} dx
$$

$$
\underset{2\times1}{} \quad \underset{1\times1}{} \qquad \underset{1\times2}{}
$$

$$
+ \int_{x_i^{(e)}}^{x_j^{(e)}} \begin{bmatrix} N_1^{(e)} \\[2mm] N_2^{(e)} \end{bmatrix} \left[\lambda(\theta^{(e)} + \rho_b^{(e)} K_d^{(e)}) \right] \begin{bmatrix} N_1^{(e)} & N_2^{(e)} \end{bmatrix} dx \qquad (3.146)
$$

$$
\underset{2\times1}{} \qquad\qquad \underset{1\times1}{} \qquad\qquad \underset{1\times2}{}
$$

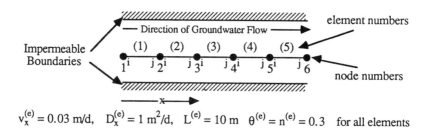

$v_x^{(e)} = 0.03$ m/d, $D_x^{(e)} = 1$ m^2/d, $L^{(e)} = 10$ m $\theta^{(e)} = n^{(e)} = 0.3$ for all elements

Figure 3.10   Example one-dimensional solute transport problem.

Now if we use the interpolation functions of the example in section 3.2 we have

$$
N_1 = \frac{x_j^{(e)} - x}{L^{(e)}} \quad , \quad \frac{\partial N_1}{\partial x} = \frac{-1}{L^{(e)}} \qquad (3.147a)
$$

$$N_2 = \frac{x - x_i^{(e)}}{L^{(e)}} \quad , \quad \frac{\partial N_2}{\partial x} = \frac{1}{L^{(e)}} \tag{3.147b}$$

for all five elements in the mesh. Since the aquifer is saturated, equation 3.146 can be divided by $\theta^{(e)}$(Appendix III). If we assume that the solute does not react with the porous media and does not decay i.e., $K_d^{(e)} = 0$ for all elements and $\lambda = 0$, and since the porous media is saturated $\theta^{(e)} = n^{(e)}$, equation 3.146 can be written

$$\begin{bmatrix} D^{(e)} \end{bmatrix}_{2\times 2} = \int_{x_i^{(e)}}^{x_j^{(e)}} \begin{bmatrix} \dfrac{-1}{L^{(e)}} \\ \dfrac{1}{L^{(e)}} \end{bmatrix} \begin{bmatrix} D_x^{(e)} \end{bmatrix} \begin{bmatrix} \dfrac{-1}{L} & \dfrac{1}{L} \end{bmatrix} dx + \int_{x_i^{(e)}}^{x_j^{(e)}} \begin{bmatrix} \dfrac{x_j^{(e)} - x}{L^{(e)}} \\ \dfrac{x - x_i^{(e)}}{L^{(e)}} \end{bmatrix} \begin{bmatrix} \dfrac{v_x^{(e)}}{\theta^{(e)}} \end{bmatrix} \begin{bmatrix} \dfrac{-1}{L^{(e)}} & \dfrac{1}{L^{(e)}} \end{bmatrix} dx$$

$$= \frac{D_x^{(e)}}{L^{(e)}} \begin{bmatrix} 1 & -1 \\ -1 & 1 \end{bmatrix} + \frac{v_x^{(e)}}{2\theta^{(e)}} \begin{bmatrix} -1 & 1 \\ -1 & 1 \end{bmatrix} \tag{3.148}$$

For the elements in Figure 3.10 these matrices are

$$\begin{bmatrix} D^{(1)} \end{bmatrix} = \frac{1}{10} \begin{bmatrix} 1 & -1 \\ -1 & 1 \end{bmatrix} + \frac{0.03}{2(0.3)} \begin{bmatrix} -1 & 1 \\ -1 & 1 \end{bmatrix}$$

$$= \frac{1}{10} \begin{bmatrix} 1 & -1 \\ -1 & 1 \end{bmatrix} + \frac{1}{20} \begin{bmatrix} -1 & 1 \\ -1 & 1 \end{bmatrix}$$

$$= \frac{1}{20} \begin{bmatrix} 1 & -1 \\ -3 & 3 \end{bmatrix} = [D^{(2)}] = [D^{(3)}] = [D^{(4)}] = [D^{(5)}]$$

In this problem we elect to use the lumped element formulation of the element sorption matrix, equation 3.139 (written here for saturated flow)

$$\begin{bmatrix} A^{(e)} \end{bmatrix} = \left(\frac{p_b^{(e)} K_d^{(e)}}{n^{(e)}} + 1 \right) \frac{L^{(e)}}{2} \begin{bmatrix} 1 & 0 \\ 0 & 1 \end{bmatrix} = \frac{L^{(e)}}{2} \begin{bmatrix} 1 & 0 \\ 0 & 1 \end{bmatrix}$$

For the elements in Figure 3.10 these matrices are

$$\begin{bmatrix} A^{(1)} \end{bmatrix} = \frac{(10)}{2} \begin{bmatrix} 1 & 0 \\ 0 & 1 \end{bmatrix}$$

$$= 5 \begin{bmatrix} 1 & 0 \\ 0 & 1 \end{bmatrix} = [A^{(2)}] = [A^{(3)}] = [A^{(4)}] = [A^{(5)}]$$

We can now assemble the global matrices [D] and [A] as follows

$$
[D] = \begin{array}{c} \text{global} \\ 6\times 6 \end{array} \begin{bmatrix}
-1/20 & -1/20 & 0 & 0 & 0 & 0 \\
-3/20 & (3+3)/20 & -1/20 & 0 & 0 & 0 \\
0 & -3/20 & (3+3)/20 & -1/20 & 0 & 0 \\
0 & 0 & -3/20 & (3+3)/20 & -1/20 & 0 \\
0 & 0 & 0 & -3/20 & (3+3)/20 & -1/20 \\
0 & 0 & 0 & 0 & -3/20 & 3/20
\end{bmatrix}
$$

$$
= \begin{bmatrix}
.05 & -.05 & 0 & 0 & 0 & 0 \\
-.15 & .30 & -.05 & 0 & 0 & 0 \\
0 & -.15 & .30 & -.05 & 0 & 0 \\
0 & 0 & -.15 & .30 & -.05 & 0 \\
0 & 0 & 0 & -.15 & .30 & -.05 \\
0 & 0 & 0 & 0 & -.15 & .05
\end{bmatrix}
$$

$$
[A] = \begin{bmatrix}
5 & 0 & 0 & 0 & 0 & 0 \\
0 & 5+5 & 0 & 0 & 0 & 0 \\
0 & 0 & 5+5 & 0 & 0 & 0 \\
0 & 0 & 0 & 5+5 & 0 & 0 \\
0 & 0 & 0 & 0 & 5+5 & 0 \\
0 & 0 & 0 & 0 & 0 & 5
\end{bmatrix}
$$

$$
= \begin{bmatrix}
5 & 0 & 0 & 0 & 0 & 0 \\
0 & 10 & 0 & 0 & 0 & 0 \\
0 & 0 & 10 & 0 & 0 & 0 \\
0 & 0 & 0 & 10 & 0 & 0 \\
0 & 0 & 0 & 0 & 10 & 0 \\
0 & 0 & 0 & 0 & 0 & 5
\end{bmatrix}
$$

We will use the backward difference form of equation 3.145 ($\omega = 1$)

$$
([A] + \Delta t\,[D]\,)\{C\}_{t+\Delta t} = [A]\{C\}_t + \Delta t\{F\}_{t+\Delta t}^{\;0}
$$

$$(3.149)$$

The solute concentrations at the nodes at time t=0 are

$$
\{C\}_{t=0} = \begin{Bmatrix} C_1 \\ C_2 \\ C_3 \\ C_4 \\ C_5 \\ C_6 \end{Bmatrix}_{t=0} = \begin{Bmatrix} 10 \\ 0 \\ 0 \\ 0 \\ 0 \\ 0 \end{Bmatrix}
$$

With a time step of 10 days ($\Delta t=10$) equation 3.149 can be solved for the solute concentrations at the end of the first time step ($t=10$)

$$([A] + \omega \Delta t\, [D])\{C\}_{t=10} = [A]\{C\}_{t=0}$$

$$\begin{bmatrix} 3.0 & -0.5 & 0 & 0 & 0 & 0 \\ -1.5 & 8.0 & -0.5 & 0 & 0 & 0 \\ 0 & -1.5 & 8.0 & -0.5 & 0 & 0 \\ 0 & 0 & -1.5 & 8.0 & -0.5 & 0 \\ 0 & 0 & 0 & -1.5 & 8.0 & -0.5 \\ 0 & 0 & 0 & 0 & -1.5 & 3.0 \end{bmatrix} \begin{Bmatrix} C_1 \\ C_2 \\ C_3 \\ C_4 \\ C_5 \\ C_6 \end{Bmatrix}_{t=0} = \begin{bmatrix} 5 & 0 & 0 & 0 & 0 & 0 \\ 0 & 10 & 0 & 0 & 0 & 0 \\ 0 & 0 & 10 & 0 & 0 & 0 \\ 0 & 0 & 0 & 10 & 0 & 0 \\ 0 & 0 & 0 & 0 & 10 & 0 \\ 0 & 0 & 0 & 0 & 0 & 5 \end{bmatrix} \begin{Bmatrix} 10 \\ 0 \\ 0 \\ 0 \\ 0 \\ 0 \end{Bmatrix} = \begin{Bmatrix} 50 \\ 0 \\ 0 \\ 0 \\ 0 \\ 0 \end{Bmatrix}$$

However, this system of equations must be modified because of the boundary condition $\{C_1\}_{t=0}=10$. Modifying this system of equations (see Section 4.5) gives

$$\begin{bmatrix} 8.0 & -0.5 & 0 & 0 & 0 \\ -1.5 & 8.0 & -0.5 & 0 & 0 \\ 0 & -1.5 & 8.0 & -0.5 & 0 \\ 0 & 0 & -1.5 & 8.0 & -0.5 \\ 0 & 0 & 0 & -1.5 & 3.0 \end{bmatrix} \begin{Bmatrix} C_2 \\ C_3 \\ C_4 \\ C_5 \\ C_6 \end{Bmatrix}_{t=10} = \begin{Bmatrix} 75.0 \\ 0 \\ 0 \\ 0 \\ 0 \end{Bmatrix}$$

which can be solved to give values of C_2 to C_6 at the end of the first time step. The solution is

$$\begin{Bmatrix} C_1 \\ C_2 \\ C_3 \\ C_4 \\ C_5 \\ C_6 \end{Bmatrix}_{t=10} = \begin{Bmatrix} 10.000 \\ 9.488 \\ 1.800 \\ 0.342 \\ 0.066 \\ 0.033 \end{Bmatrix}$$

This solution is then substituted into the right hand side of equation 3.149 and the procedure is repeated for the next time step.

3.7.3 Transient Groundwater Flow

In transient groundwater flow, the volumetric water content θ and the components of apparent groundwater velocity v_x, v_y, and v_z are functions of time t

$$\begin{aligned} \theta &= \theta(t) \\ v_x &= v_x(t) \\ v_y &= v_y(t) \\ v_z &= v_z(t) \end{aligned}$$

(3.150)

The dispersion coefficients D_x, D_y, and D_z (or D_{xx}, D_{xy}, etc) are computed using v_x, v_y, and v_z (see Appendix III) and therefore are also functions of time

$$
\begin{aligned}
D_x &= D_x(t) \\
D_y &= D_y(t) \\
D_z &= D_z(t)
\end{aligned}
\tag{3.151}
$$

if groundwater flow is uniform, or

$$
\begin{aligned}
D_{xx} &= D_{xx}(t) \\
D_{xy} &= D_{xy}(t) \\
&\vdots \\
D_{zz} &= D_{zz}(t)
\end{aligned}
\tag{3.152}
$$

if groundwater flow is not uniform.

The advection–dispersion matrix $[D^{(e)}]$ and the element sorption matrix $[A^{(e)}]$ are computed using $\theta^{(e)}$, $v_x^{(e)}$, etc., $D_x^{(e)}$, etc., and are therefore also functions of time

$$
\begin{aligned}
[D^{(e)}] &= [D^{(e)}(t)] \\
[A^{(e)}] &= [A^{(e)}(t)]
\end{aligned}
\tag{3.153}
$$

Matrix integral formulations for $[D^{(e)}(t)]$ can be obtained by substituting equations 3.150, 3.151, and 3.152 into equations 3.136 and 3.137

$[D^{(e)}(t)] =$

$$
\iiint_{V^{(e)}}
\underbrace{\begin{bmatrix}
\dfrac{\partial N_1^{(e)}}{\partial x} & \dfrac{\partial N_1^{(e)}}{\partial y} & \dfrac{\partial N_1^{(e)}}{\partial z} \\
\vdots & \vdots & \vdots \\
\dfrac{\partial N_n^{(e)}}{\partial x} & \dfrac{\partial N_n^{(e)}}{\partial y} & \dfrac{\partial N_n^{(e)}}{\partial z}
\end{bmatrix}}_{n\times3}
\underbrace{\begin{bmatrix}
D_x^{(e)}(t)\theta^{(e)}(t) & 0 & 0 \\
0 & D_y^{(e)}(t)\theta^{(e)}(t) & 0 \\
0 & 0 & D_z^{(e)}(t)\theta^{(e)}(t)
\end{bmatrix}}_{3\times3}
\underbrace{\begin{bmatrix}
\dfrac{\partial N_1^{(e)}}{\partial x} & \cdots & \dfrac{\partial N_n^{(e)}}{\partial x} \\
\dfrac{\partial N_1^{(e)}}{\partial y} & \cdots & \dfrac{\partial N_n^{(e)}}{\partial y} \\
\dfrac{\partial N_1^{(e)}}{\partial z} & \cdots & \dfrac{\partial N_n^{(e)}}{\partial z}
\end{bmatrix}}_{3\times n}
\, dx\, dy\, dz
$$

$$
+ \iiint_{V^{(e)}}
\underbrace{\begin{bmatrix} N_1^{(e)} \\ \vdots \\ N_n^{(e)} \end{bmatrix}}_{n\times1}
\underbrace{[v_x(t)]}_{1\times1}
\underbrace{\begin{bmatrix} \dfrac{\partial N_1^{(e)}}{\partial x} & \cdots & \dfrac{\partial N_n^{(e)}}{\partial x} \end{bmatrix}}_{1\times n}
\, dx\, dy\, dz
$$

$$
+ \iiint_{V^{(e)}}
\underbrace{\begin{bmatrix} N_1^{(e)} \\ \vdots \\ N_n^{(e)} \end{bmatrix}}_{n\times1}
\underbrace{[\lambda\,(\theta^{(e)}(t) + \rho_b^{(e)} K_d^{(e)})]}_{1\times1}
\underbrace{[N_1^{(e)} \cdots N_n^{(e)}]}_{1\times n}
\, dx\, dy\, dz
\tag{3.154}
$$

if groundwater flow is uniform and

$$[\,D^{(e)}(t)] =$$

$$\int\int\int_{V^{(e)}} \underbrace{\begin{bmatrix} \dfrac{\partial N_1^{(e)}}{\partial x} & \dfrac{\partial N_1^{(e)}}{\partial y} & \dfrac{\partial N_1^{(e)}}{\partial z} \\ \vdots & \vdots & \vdots \\ \dfrac{\partial N_n^{(e)}}{\partial x} & \dfrac{\partial N_n^{(e)}}{\partial y} & \dfrac{\partial N_n^{(e)}}{\partial z} \end{bmatrix}}_{n\times 3} \underbrace{\begin{bmatrix} D_{xx}^{(e)}(t)\theta^{(e)}(t) & D_{xy}^{(e)}(t)\theta^{(e)}(t) & D_{xz}^{(e)}(t)\theta^{(e)}(t) \\ D_{yx}^{(e)}(t)\theta^{(e)}(t) & D_{yy}^{(e)}(t)\theta^{(e)}(t) & D_{yz}^{(e)}(t)\theta^{(e)}(t) \\ D_{zx}^{(e)}(t)\theta^{(e)}(t) & D_{zy}^{(e)}(t)\theta^{(e)}(t) & D_{zz}^{(e)}(t)\theta^{(e)}(t) \end{bmatrix}}_{3\times 3} \underbrace{\begin{bmatrix} \dfrac{\partial N_1^{(e)}}{\partial x} & \cdots & \dfrac{\partial N_n^{(e)}}{\partial x} \\ \dfrac{\partial N_1^{(e)}}{\partial y} & \cdots & \dfrac{\partial N_n^{(e)}}{\partial y} \\ \dfrac{\partial N_1^{(e)}}{\partial z} & \cdots & \dfrac{\partial N_n^{(e)}}{\partial z} \end{bmatrix}}_{3\times n} dx\,dy\,dz$$

$$+ \int\int\int_{V^{(e)}} \underbrace{\begin{bmatrix} N_1^{(e)} & N_1^{(e)} & N_1^{(e)} \\ \vdots & \vdots & \vdots \\ N_n^{(e)} & N_n^{(e)} & N_n^{(e)} \end{bmatrix}}_{n\times 3} \underbrace{\begin{bmatrix} v_x^{(e)}(t) & 0 & 0 \\ 0 & v_y^{(e)}(t) & 0 \\ 0 & 0 & v_z^{(e)}(t) \end{bmatrix}}_{3\times 3} \underbrace{\begin{bmatrix} \dfrac{\partial N_1^{(e)}}{\partial x} & \cdots & \dfrac{\partial N_n^{(e)}}{\partial x} \\ \dfrac{\partial N_1^{(e)}}{\partial y} & \cdots & \dfrac{\partial N_n^{(e)}}{\partial y} \\ \dfrac{\partial N_1^{(e)}}{\partial z} & \cdots & \dfrac{\partial N_n^{(e)}}{\partial z} \end{bmatrix}}_{3\times n} dx\,dy\,dz$$

$$+ \int\int\int_{V^{(e)}} \underbrace{\begin{bmatrix} N_1^{(e)} \\ \vdots \\ N_n^{(e)} \end{bmatrix}}_{n\times 1} \underbrace{[\lambda\,(\theta^{(e)}(t) + \rho_b^{(e)}K_d^{(e)})]}_{1\times 1} \underbrace{[N_1^{(e)} \cdots N_n^{(e)}]}_{1\times n} dx\,dy\,dz \qquad (3.155)$$

if the groundwater flow is not uniform. The matrix integral formulation for $[A^{(e)}(t)]$ can be obtained by substituting equation 3.150 into equation 3.138

$$\left[A^{(e)}(t)\right] = \int\int\int_{V^{(e)}} \underbrace{\begin{bmatrix} N_1^{(e)} \\ \vdots \\ N_n^{(e)} \end{bmatrix}}_{n\times 1} \underbrace{[\rho_b^{(e)}K_d^{(e)} + \theta^{(e)}(t)]}_{1\times 1} \underbrace{[N_1^{(e)} \cdots N_n^{(e)}]}_{1\times n} dx\,dy\,dz \qquad (3.156)$$

The global advection-dispersion matrix and the global sorption matrix are also functions of time and the weighted residual formulation for the solute transport equation for transient groundwater flow becomes

$$\boxed{\underset{\text{global}}{[A(t)]}\{\dot{C}\} + \underset{\text{global}}{[D(t)]}\underset{\text{global}}{\{C\}} = \{F\}} \qquad (3.157)$$

The finite difference formulation for equation 3.158 is

$$
(\, [A(t + \Delta t)] + \omega \Delta t \, [D(t + \Delta t)] \,) \{C\}_{t+\Delta t}
$$
$$
= (A(t) - (1 - \omega) \, \Delta t \, [D(t)] \,) \{C\}_t + \Delta t \, ((1 - \omega)\{F\}_t + \omega\{F\}_{t+\Delta t}))
$$

$$(3.158)$$

To solve equation (3.158), we first solve the transient groundwater flow problem (either saturated or unsaturated) to obtain the values of $\theta^{(e)}(t)$, $v_x^{(e)}(t)$ etc., and $D_x^{(e)}(t)$ etc., for each element for each choice of time step used in (3.158). Then we specify the initial values of $\{C\}$

$$
\{C\}_{t_0} = \text{specified values}
$$

and compute $[D(t_0)]$, $[A(t_0)]$, $[D(t_0+\Delta t)]$ and $[A(t_0+\Delta t)]$. These are substituted into equation 3.158 which is then solved for the values of $\{C\}$ at the end of the first time step, $\{C\}_{t_0 + \Delta t}$. We then set

$$
\{C\}_t = \{C\}_{t_0 + \Delta t}
$$

compute $[D(t+\Delta t)]$ and $[A(t+\Delta t)]$, substitute these matrices into equation 3.158 and repeat the solution procedure. It should be obvious that computing each element matrix and assembling and modifying the global system of equations can be extremely time consuming.

It should be noted that <u>the procedure just described is only valid when changes in groundwater density due to changing solute concentrations in the aquifer can be assumed to be negligibly small</u> (see Appendix III). When this is not true, the groundwater flow and solute transport equations are coupled and must be solved simultaneously.

3.7.4 Saturated Groundwater Flow

The solute transport equation for saturated groundwater flow is (Appendix III)

$$
\frac{\partial C}{\partial t} = D_x \frac{\partial^2 C}{\partial x^2} + D_y \frac{\partial^2 C}{\partial y^2} + D_z \frac{\partial^2 C}{\partial z^2} - \frac{\partial}{\partial x}\left(\frac{v_x C}{n}\right)
$$
$$
- \frac{\partial}{\partial t}\left(\frac{\rho_b K_d C}{n}\right) - \lambda\left(C + \frac{\rho_b K_d C}{n}\right)
$$

$$(3.159)$$

for uniform flow and

$$
\frac{\partial C}{\partial t} = D_{xx}\frac{\partial^2 C}{\partial x^2} + D_{xy}\frac{\partial^2 C}{\partial x \partial y} + D_{xz}\frac{\partial^2 C}{\partial x \partial z} + D_{yx}\frac{\partial^2 C}{\partial y \partial x} + D_{yy}\frac{\partial^2 C}{\partial y^2} + D_{yz}\frac{\partial^2 C}{\partial y \partial z}
$$
$$
D_{zx}\frac{\partial^2 C}{\partial z \partial x} + D_{zy}\frac{\partial^2 C}{\partial z \partial y} + D_{zz}\frac{\partial^2 C}{\partial z^2} - \frac{\partial}{\partial x}\left(\frac{v_x C}{n}\right) - \frac{\partial}{\partial y}\left(\frac{v_y C}{n}\right) - \frac{\partial}{\partial z}\left(\frac{v_z C}{n}\right)
$$
$$
- \frac{\partial}{\partial t}\left(\frac{\rho_b K_d C}{n}\right) - \lambda\left(C + \frac{\rho_b K_d C}{n}\right)
$$

$$(3.160)$$

if the groundwater flow is not uniform. If the porous media is saturated $\theta = n = $ constant within an element and the element advection-dispersion matrix for steady-state groundwater flow is

$$
\underset{n \times n}{[\,D^{(e)}]} = \int\!\!\!\int\!\!\!\int_{V^{(e)}} \begin{bmatrix} \dfrac{\partial N_1^{(e)}}{\partial x} & \dfrac{\partial N_1^{(e)}}{\partial y} & \dfrac{\partial N_1^{(e)}}{\partial z} \\ \vdots & \vdots & \vdots \\ \dfrac{\partial N_n^{(e)}}{\partial x} & \dfrac{\partial N_n^{(e)}}{\partial y} & \dfrac{\partial N_n^{(e)}}{\partial z} \end{bmatrix} \begin{bmatrix} D_x^{(e)} & 0 & 0 \\ 0 & D_y^{(e)} & 0 \\ 0 & 0 & D_z^{(e)} \end{bmatrix} \begin{bmatrix} \dfrac{\partial N_1^{(e)}}{\partial x} & \cdots & \dfrac{\partial N_n^{(e)}}{\partial x} \\ \dfrac{\partial N_1^{(e)}}{\partial y} & \cdots & \dfrac{\partial N_n^{(e)}}{\partial y} \\ \dfrac{\partial N_1^{(e)}}{\partial z} & \cdots & \dfrac{\partial N_n^{(e)}}{\partial z} \end{bmatrix} dx\,dy\,dz
$$

$$
\qquad\qquad\qquad\quad V^{(e)} \qquad\qquad\qquad n \times 3 \qquad\qquad 3\times 3 \qquad\qquad\quad 3 \times n
$$

$$
+ \int\!\!\!\int\!\!\!\int_{V^{(e)}} \begin{bmatrix} N_1^{(e)} \\ \vdots \\ N_n^{(e)} \end{bmatrix} \begin{bmatrix} \dfrac{v_x^{(e)}}{n^{(e)}} \end{bmatrix} \begin{bmatrix} \dfrac{\partial N_1^{(e)}}{\partial x} & \cdots & \dfrac{\partial N_n^{(e)}}{\partial x} \end{bmatrix} dx\,dy\,dz
$$

$$
\qquad\qquad V^{(e)} \quad\; n \times 1 \quad\; 1\times 1 \qquad\qquad 1\times n
$$

$$
+ \int\!\!\!\int\!\!\!\int_{V^{(e)}} \begin{bmatrix} N_1^{(e)} \\ \vdots \\ N_n^{(e)} \end{bmatrix} \left[\lambda \left(1 + \dfrac{\rho_b^{(e)} K_d^{(e)}}{n^{(e)}} \right) \right] [N_1^{(e)} \cdots N_n^{(e)}]\, dx\,dy\,dz \qquad (3.161)
$$

$$
\qquad\qquad V^{(e)} \quad\; n \times 1 \qquad\qquad 1\times 1 \qquad\qquad\qquad 1\times n
$$

for uniform flow and

$$
\underset{}{[\,D^{(e)}]} = \int\!\!\!\int\!\!\!\int_{V^{(e)}} \begin{bmatrix} \dfrac{\partial N_1^{(e)}}{\partial x} & \dfrac{\partial N_1^{(e)}}{\partial y} & \dfrac{\partial N_1^{(e)}}{\partial z} \\ \vdots & \vdots & \vdots \\ \dfrac{\partial N_n^{(e)}}{\partial x} & \dfrac{\partial N_n^{(e)}}{\partial y} & \dfrac{\partial N_n^{(e)}}{\partial z} \end{bmatrix} \begin{bmatrix} D_{xx}^{(e)} & D_{xy}^{(e)} & D_{xz}^{(e)} \\ D_{yx}^{(e)} & D_{yy}^{(e)} & D_{yz}^{(e)} \\ D_{zx}^{(e)} & D_{zy}^{(e)} & D_{zz}^{(e)} \end{bmatrix} \begin{bmatrix} \dfrac{\partial N_1^{(e)}}{\partial x} & \cdots & \dfrac{\partial N_n^{(e)}}{\partial x} \\ \dfrac{\partial N_1^{(e)}}{\partial y} & \cdots & \dfrac{\partial N_n^{(e)}}{\partial y} \\ \dfrac{\partial N_1^{(e)}}{\partial z} & \cdots & \dfrac{\partial N_n^{(e)}}{\partial z} \end{bmatrix} dx\,dy\,dz
$$

$$
\qquad\qquad V^{(e)}
$$

$$
+ \int\!\!\!\int\!\!\!\int_{V^{(e)}} \begin{bmatrix} N_1^{(e)} & N_1^{(e)} & N_1^{(e)} \\ \vdots & \vdots & \vdots \\ N_n^{(e)} & N_n^{(e)} & N_n^{(e)} \end{bmatrix} \begin{bmatrix} \dfrac{v_x^{(e)}}{n^{(e)}} & 0 & 0 \\ 0 & \dfrac{v_y^{(e)}}{n^{(e)}} & 0 \\ 0 & 0 & \dfrac{v_z^{(e)}}{n^{(e)}} \end{bmatrix} \begin{bmatrix} \dfrac{\partial N_1^{(e)}}{\partial x} & \cdots & \dfrac{\partial N_n^{(e)}}{\partial x} \\ \dfrac{\partial N_1^{(e)}}{\partial y} & \cdots & \dfrac{\partial N_n^{(e)}}{\partial y} \\ \dfrac{\partial N_1^{(e)}}{\partial z} & \cdots & \dfrac{\partial N_n^{(e)}}{\partial z} \end{bmatrix} dx\,dy\,dz
$$

$$
\qquad\quad V^{(e)} \qquad\qquad n \times 3 \qquad\qquad\qquad 3\times 3 \qquad\qquad\qquad 3 \times n
$$

$$
+ \int\!\!\!\int\!\!\!\int_{V^{(e)}} \begin{bmatrix} N_1^{(e)} \\ \vdots \\ N_n^{(e)} \end{bmatrix} \left[\lambda \left(1 + \dfrac{\rho_b^{(e)} K_d^{(e)}}{n^{(e)}} \right) \right] [N_1^{(e)} \cdots N_n^{(e)}]\, dx\,dy\,dz \qquad (3.162)
$$

$$
\qquad\qquad V^{(e)} \quad\; n \times 1 \qquad\qquad 1\times 1 \qquad\qquad\qquad 1\times n
$$

for nonuniform flow. The element sorption matrix is

$$
\left[A^{(e)} \right] = \underset{V^{(e)}}{\int\int\int} \underset{n \times 1}{\begin{bmatrix} N_1^{(e)} \\ \vdots \\ N_n^{(e)} \end{bmatrix}} \underset{1 \times 1}{\left[1 + \frac{\rho_b^{(e)} K_d^{(e)}}{n^{(e)}} \right]} \underset{1 \times n}{[N_1^{(e)} \cdots N_n^{(e)}]} \, dx \, dy \, dz \qquad (3.163)
$$

Similar equations can be written for transient groundwater flow for a lumped formulation. The term $\left[1 + \dfrac{\rho_b^{(e)} K_d^{(e)}}{n^{(e)}} \right]$ is frequently called the *retardation factor* for the element.

NOTES AND ADDITIONAL READING

1. For problems with axisymmetry equations 3.1 to 3.5 can be written in an (r,z) coordinate system:

Steady-State, Saturated Flow Equation

$$\frac{1}{r}\frac{\partial}{\partial r}\left(K_r\, r\, \frac{\partial h}{\partial r}\right) + \frac{\partial}{\partial z}\left(K_z\frac{\partial h}{\partial z}\right) = 0 \tag{3.164}$$

Steady-State, Unsaturated Flow Equation

$$\frac{1}{r}\frac{\partial}{\partial r}\left(K_r(\psi)\, r\, \frac{\partial \psi}{\partial r}\right) + \frac{\partial}{\partial z}\left(K_z(\psi)\left(\frac{\partial \psi}{\partial z}+1\right)\right) = 0 \tag{3.165}$$

Transient, Saturated Flow Equation

$$\frac{1}{r}\frac{\partial}{\partial r}\left(K_r\, r\, \frac{\partial h}{\partial r}\right) + \frac{\partial}{\partial z}\left(K_z\frac{\partial h}{\partial z}\right) = S_s\frac{\partial h}{\partial t} \tag{3.166}$$

Transient, Unsaturated Flow Equation

$$\frac{1}{r}\frac{\partial}{\partial r}\left(K_r(\psi)\, r\, \frac{\partial \psi}{\partial r}\right) + \frac{\partial}{\partial z}\left(K_z(\psi)\left(\frac{\partial \psi}{\partial z}+1\right)\right) = C(\psi)\frac{\partial \psi}{\partial t} \tag{3.167}$$

Solute Transport Equation (Uniform Groundwater Flow)

$$\frac{\partial(\theta C)}{\partial t} = \frac{1}{r}\frac{\partial}{\partial r}\left(D_r\, r\, \frac{\partial}{\partial r}(\theta C)\right) + D_r\frac{\partial^2}{\partial z^2}(\theta C) - \frac{\partial}{\partial r}(v_r C)$$
$$- \frac{\partial}{\partial t}(\rho_b^{(e)}K_d^{(e)}C) - \lambda(\theta C + \rho_b^{(e)}K_d^{(e)}C) \tag{3.168}$$

where r is the radial coordinate direction (directed outward from the axis of symmetry) and z is the vertical coordinate direction (see, for example Figure 2.9)

2. Matrix-integral formulation for the element conductance, capacitance, advection dispersion, and sorption matrices can also be derived for equations 3.164 to 3.168. For example, the element conductance matrix for saturated flow in an axisymmetric coordinate system is

$$[K^{(e)}] = \iint_{A^{(e)}} \begin{bmatrix} \dfrac{\partial N_1^{(e)}}{\partial r} & \dfrac{\partial N_1^{(e)}}{\partial z} \\ \vdots & \vdots \\ \dfrac{\partial N_n^{(e)}}{\partial r} & \dfrac{\partial N_n^{(e)}}{\partial z} \end{bmatrix} \begin{bmatrix} K_r^{(e)} & 0 \\ 0 & K_z^{(e)} \end{bmatrix} \begin{bmatrix} \dfrac{\partial N_1^{(e)}}{\partial r} & \cdots & \dfrac{\partial N_n^{(e)}}{\partial r} \\ \dfrac{\partial N_1^{(e)}}{\partial z} & \cdots & \dfrac{\partial N_n^{(e)}}{\partial z} \end{bmatrix} 2\pi r\, dr\, dz$$

$$\phantom{[K^{(e)}] =}\quad {}_{n\times 2} \qquad\quad {}_{2\times 2} \qquad\qquad\quad {}_{2\times n} \tag{3.169}$$

Chapter 4

STEP 3: DEVELOP SYSTEM OF EQUATIONS

4.1   REQUIRED PROPERTIES OF ELEMENT INTERPOLATION FUNCTIONS

As we saw in the previous chapter, application of the method of weighted residuals to the groundwater flow and solute transport equations leads to several matrix-integral expressions (i.e., the equations for the saturated and unsaturated forms of the element conductance matrix, the saturated and unsaturated forms of the element capacitance matrix, the element dispersion-advection matrix, and the element sorption matrix) that must be evaluated for each element in the mesh. To evaluate these expressions the element interpolation functions $N_i^{(e)}$, and their derivatives $\dfrac{N_i^{(e)}}{\partial x}, \dfrac{N_i^{(e)}}{\partial y}$ and $\dfrac{N_i^{(e)}}{\partial z}$ must be known functions of the three coordinate directions x, y, and z.

Recall that the interpolation functions are used to define the approximate solution for hydraulic head (or pressure head, or solute concentration) at any point within an element. For example

$$\hat{h}^{(e)}(x, y, z) = \sum_{i=1}^{n} N_i^{(e)}(x, y, z)\, h_i \qquad (4.1)$$

where $\hat{h}^{(e)}$ is the approximate solution for hydraulic head within element e, $N_i^{(e)}$ are the interpolation functions for the nodes of element e, h_i are the unknown values of hydraulic head at the nodes of element e, and n are the number of nodes in element e. Because the interpolation functions are defined using the element's size and shape they are generally different for each element in the mesh. For example, the interpolation functions for one-dimensional elements with two nodes will be different if the lengths of the elements are different (equation 3.11). The set of interpolation functions for all elements in the mesh define an approximate solution for $\hat{h}$ (or $\hat{\psi}$ or $\hat{C}$) throughout the problem domain.

4.1.1 Continuity

The need to integrate this solution (or its derivatives) places a restriction on the types of interpolation functions that may be used: the interpolated value of $\hat{h}$ (or $\hat{\psi}$ or $\hat{C}$) *must be continuous along the boundary between adjacent elements.* That is the value of $\hat{h}$ computed at each point on the boundary between two adjacent elements must be the same regardless of which element's set of interpolation functions are used (Figure 4.1). Because the approximate solution is continuous from one element to the next, we say that $\hat{h}$ (or $\hat{\psi}$ or $\hat{C}$) is interpolated in a "piecewise continuous" manner over the problem domain. The *derivatives* of the approximate solution *do not* have to be continuous across element boundaries, however. This is so because for the integral

$$\int \frac{\partial^p \hat{h}(x)}{\partial x^p}\, dx$$

to be defined, $\hat{h}(x)$ must be continuous to the order $(p-1)$. Because all the integral equations for element matrices in Chapter 3 contain (at most) only the first derivative $(p=1)$ of $\hat{h}$ (or $\hat{\psi}$ or $\hat{C}$), $\hat{h}$ must be continuous but $\frac{\partial \hat{h}}{\partial x}$, $\frac{\partial \hat{h}}{\partial y}$, and $\frac{\partial \hat{h}}{\partial z}$ do not have to be $(p-1=0)$.

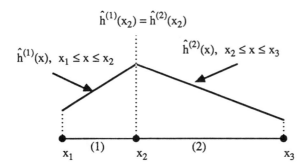

Figure 4.1   Approximate solution must be continuous along adjacent element's boundaries.

4.1.2 Convergence

When the finite element method is used to solve a groundwater flow or solute transport problem, the solution consists of the approximate value of hydraulic head (or pressure head or solute concentration) at each node. If *suitable* interpolation functions are used in the solution procedure, the *accuracy* of the approximate solution will improve as the number of nodes and elements in the mesh increases (which usually is equivalent to a decrease in the size of elements in the mesh). We say that the solution *converges* to the true solution as the number of nodes and elements in the mesh increases. Fortunately there is a simple rule that allows us to determine which types of interpolation functions possess this convergence property.

This rule has its origin in the approximate solution (equation 4.1). Consider the case of an element e that is in a portion of the problem domain where hydraulic head is constant. In this case, the value of $\hat{h}^{(e)}(x, y, z)$ is constant and should also be equal to the value of h at any node in the element, $\hat{h}^{(e)} = h_i$, $i = 1$ to n. If we call this constant value h_0 and substitute it into equation 4.1 we have

$$\hat{h}^{(e)}(x, y, z) = h_0 = \sum_{i=1}^{n} N_i^{(e)} h_i = \sum_{i=1}^{n} N_i^{(e)} h_0 \qquad (4.2)$$

which is only true if the values of all the element interpolation functions sum to one at every point within the element

$$\sum_{i=1}^{n} N_i^{(e)}(x, y, z) = 1 \quad \text{for all } (x, y, z) \text{ in } V^{(e)} \qquad (4.3)$$

Where $V^{(e)}$ is the volume of element e.   This rule insures that the elements are capable of modeling a constant head region within the mesh when such a condition exists. This rule also insures that the approximate solution converges to the true solution as the number of nodes in the mesh increases.

4.2   SUBPARAMETRIC, SUPERPARAMETRIC, AND ISOPARAMETRIC ELEMENTS

The approximate solution for hydraulic head is given by

$$\hat{h}^{(e)} = \sum_{i=1}^{n} N_i^{(e)} h_i \qquad (4.4)$$

where $N_i^{(e)}$ are the interpolation functions and h_i are the unknown values of hydraulic head at the element's nodes. It is also possible to describe the *shape* of the element using the coordinates of each node in the element and *another* set of interpolation functions for the element.   To see how this is done, let P represent an arbitrary point on the boundary of an element and let x_i, y_i, and z_i represent the coordinates of the ith node for the element (Figure 4.2). Then we can describe the position of element boundaries using the coordinates of each node and another set of interpolation functions for the element $S_i^{(e)}$

$$x(P) = \sum_{i=1}^{n} S_i^{(e)}(P) x_i \qquad (4.5a)$$

$$y(P) = \sum_{i=1}^{n} S_i^{(e)}(P) y_i \qquad (4.5b)$$

$$z(P) = \sum_{i=1}^{n} S_i^{(e)}(P) z_i \qquad (4.5c)$$

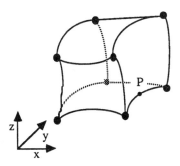

Figure 4.2   P is any point on an element boundary.

where equation 4.5a, for example, says that the x coordinate of point P is equal to the sum of the product of the interpolation function for a node evaluated at point P and the x coordinate of the node, for all nodes in the element.

Because the interpolation functions $S_i^{(e)}$ are used to define the *shape* of the element they are often called *shape functions*. Linear, quadratic, and cubic polynomials are the most common type of shape functions used in groundwater flow and solute transport modeling. For example, linear shape functions are used when the boundaries of the element can be represented by straight line segments. Quadratic shape functions are used when the boundaries of the element can be represented by quadratic curves. Similarly, linear interpolation functions are used when values of hydraulic head can be considered to vary in a linear fashion within the element. Quadratic interpolation functions are used when values of hydraulic head can be considered to vary in a (quadratic) curvilinear fashion within the element.

The order of the polynomials used for the interpolation and shape functions within an element do not have to be the same. For example, an element with straight edges (linear shape functions) can have a curvilinear variation in head (quadratic or cubic interpolation functions) (Figure 4.3). The order of polynomials used for the interpolation and shape functions are used to classify types of elements into three groups, which are illustrated for one-dimensional elements in Figure 4.4. *Subparametric elements* use polynomials for the shape functions that are a lower order than the polynomials used for the interpolation functions. In *isoparametric elements* the orders of the polynomials used for the shape and interpolation functions are the same. *Superparametric elements* use polynomials for the shape functions that are a higher order than the polynomials used for the interpolation functions.

It is important to realize that when subparametric or superparametric element types are used, not all of the nodes may have a value of hydraulic head (or pressure head or solute concentration) assigned to them. Thus in a one-dimensional, superparametric element with three nodes, quadratic shape functions, and linear interpolation functions, hydraulic head will only be computed at two of the three nodes. These nodes will also be the ones where boundary conditions are specified if the element is on the boundary of the mesh. The coordinates of all three nodes would have to be specified, however. These coordinates are used with the three shape functions to define the quadratic curve that describes the element's shape.

linear interpolation functions, linear shape functions

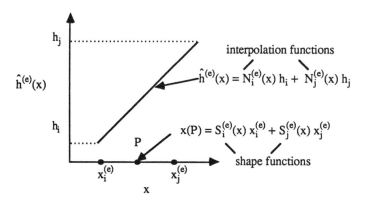

quadratic interpolation functions. linear shape functions

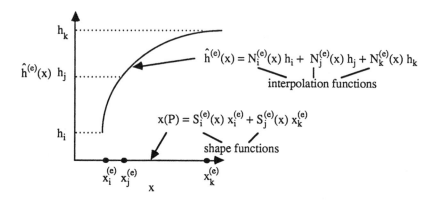

Figure 4.3   Interpolation and shape functions for two types of one-dimensional elements.

At present, isoparametric elements are used almost exclusively in groundwater flow and solute transport modeling. Because in isoparametric elements the order of the polynomials used for the shape functions and interpolation functions are identical we will refer to both types of functions as interpolation functions in the remainder of Chapter 4. The next section describes the most commonly used interpolation functions for a variety of one-, two-, and three-dimensional elements and the procedures needed to compute the element matrices.

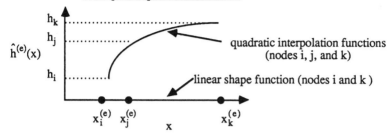

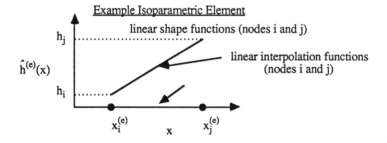

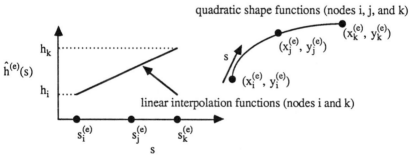

Figure 4.4   **Illustration of definitions of subparametric, isoparametric, and superparametric one-dimensional elements.**

4.3   EVALUATION OF ELEMENT MATRICES

In Chapter 3 we applied the Method of Weighted Residuals to the equations of groundwater flow and solute transport. The result was a series of matrix integral equations for each element in the mesh. These equations are listed here for reference

Element Conductance Matrix (Saturated Flow)

$$
[K^{(e)}] = \iiint\limits_{V^{(e)}}
\underbrace{\begin{bmatrix}
\dfrac{\partial N_1^{(e)}}{\partial x} & \dfrac{\partial N_1^{(e)}}{\partial y} & \dfrac{\partial N_1^{(e)}}{\partial z} \\
\vdots & \vdots & \vdots \\
\dfrac{\partial N_n^{(e)}}{\partial x} & \dfrac{\partial N_n^{(e)}}{\partial y} & \dfrac{\partial N_n^{(e)}}{\partial z}
\end{bmatrix}}_{n\times 3}
\underbrace{\begin{bmatrix}
K_x^{(e)} & 0 & 0 \\
0 & K_y^{(e)} & 0 \\
0 & 0 & K_z^{(e)}
\end{bmatrix}}_{3\times 3}
\underbrace{\begin{bmatrix}
\dfrac{\partial N_1^{(e)}}{\partial x} & \cdots & \dfrac{\partial N_n^{(e)}}{\partial x} \\
\dfrac{\partial N_1^{(e)}}{\partial y} & \cdots & \dfrac{\partial N_n^{(e)}}{\partial y} \\
\dfrac{\partial N_1^{(e)}}{\partial z} & \cdots & \dfrac{\partial N_n^{(e)}}{\partial z}
\end{bmatrix}}_{3\times n}
dx\, dy\, dz \qquad (4.6)
$$

Element Conductance Matrix (Unsaturated Flow)

$$
[K^{(e)}(\psi)] = \iiint\limits_{V^{(e)}}
\underbrace{\begin{bmatrix}
\dfrac{\partial N_1^{(e)}}{\partial x} & \dfrac{\partial N_1^{(e)}}{\partial y} & \dfrac{\partial N_1^{(e)}}{\partial z} \\
\vdots & \vdots & \vdots \\
\dfrac{\partial N_n^{(e)}}{\partial x} & \dfrac{\partial N_n^{(e)}}{\partial y} & \dfrac{\partial N_n^{(e)}}{\partial z}
\end{bmatrix}}_{n\times 3}
\underbrace{\begin{bmatrix}
K_x^{(e)}(\psi) & 0 & 0 \\
0 & K_y^{(e)}(\psi) & 0 \\
0 & 0 & K_z^{(e)}(\psi)
\end{bmatrix}}_{3\times 3}
\underbrace{\begin{bmatrix}
\dfrac{\partial N_1^{(e)}}{\partial x} & \cdots & \dfrac{\partial N_n^{(e)}}{\partial x} \\
\dfrac{\partial N_1^{(e)}}{\partial y} & \cdots & \dfrac{\partial N_n^{(e)}}{\partial y} \\
\dfrac{\partial N_1^{(e)}}{\partial z} & \cdots & \dfrac{\partial N_n^{(e)}}{\partial z}
\end{bmatrix}}_{3\times n}
dx\, dy\, dz
$$

$$(4.7)$$

Element Capacitance Matrix (Saturated Flow, Consistent Formulation)

$$
\underset{n\times n}{[C^{(e)}]} = \iiint
\underbrace{\begin{bmatrix}
N_1^{(e)} \\
\vdots \\
N_n^{(e)}
\end{bmatrix}}_{n\times 1}
\underbrace{[S_s^{(e)}]}_{1\times 1}
\underbrace{[N_1^{(e)} \cdots N_n^{(e)}]}_{1\times n}
dx\, dy\, dz \qquad (4.8)
$$

Element Capacitance Matrix (Saturated Flow, Lumped Formulation)

$$
\underset{n\times n}{[C^{(e)}]} = S_s^{(e)} \frac{V^{(e)}}{n}
\underset{n\times n}{\begin{bmatrix}
1 & & \mathbf{O} \\
& \ddots & \\
\mathbf{O} & & 1
\end{bmatrix}} \qquad (4.9)
$$

Element Capacitance Matrix (Unsaturated Flow, Consistent Formulation)

$$[\underset{n\times n}{C^{(e)}(\psi)}] = \iiint\limits_{V^{(e)}} \underset{n\times 1}{\begin{bmatrix} N_1^{(e)} \\ \vdots \\ N_n^{(e)} \end{bmatrix}} \underset{1\times 1}{[C^{(e)}(\psi)]} \underset{1\times n}{[N_1^{(e)} \cdots N_n^{(e)}]} \, dx\, dy\, dz \tag{4.10}$$

Element Capacitance Matrix (Unsaturated Flow, Lumped Formulation)

$$[\underset{n\times n}{C^{(e)}(\psi)}] = C^{(e)}(\psi)\frac{V^{(e)}}{n} \underset{n\times n}{\begin{bmatrix} 1 & & \mathbf{O} \\ & \ddots & \\ \mathbf{O} & & 1 \end{bmatrix}} \tag{4.11}$$

Element Advection-Dispersion Matrix

$$[D^{(e)}] = - \iiint\limits_{V^{(e)}} \underset{n\times n}{\begin{bmatrix} \frac{\partial N_1^{(e)}}{\partial x} & \frac{\partial N_1^{(e)}}{\partial y} & \frac{\partial N_1^{(e)}}{\partial z} \\ \vdots & \vdots & \vdots \\ \frac{\partial N_n^{(e)}}{\partial x} & \frac{\partial N_n^{(e)}}{\partial y} & \frac{\partial N_n^{(e)}}{\partial z} \end{bmatrix}} \underset{3\times 3}{\begin{bmatrix} D_{xx}^{(e)}\theta^{(e)} & D_{xy}^{(e)}\theta^{(e)} & D_{xz}^{(e)}\theta^{(e)} \\ D_{yx}^{(e)}\theta^{(e)} & D_{yy}^{(e)}\theta^{(e)} & D_{yz}^{(e)}\theta^{(e)} \\ D_{zx}^{(e)}\theta^{(e)} & D_{zy}^{(e)}\theta^{(e)} & D_{zz}^{(e)}\theta^{(e)} \end{bmatrix}} \underset{3\times n}{\begin{bmatrix} \frac{\partial N_1^{(e)}}{\partial x} & \cdots & \frac{\partial N_n^{(e)}}{\partial x} \\ \frac{\partial N_1^{(e)}}{\partial y} & \cdots & \frac{\partial N_n^{(e)}}{\partial y} \\ \frac{\partial N_1^{(e)}}{\partial z} & \cdots & \frac{\partial N_n^{(e)}}{\partial z} \end{bmatrix}} dx\, dy\, dz$$

$$+ \iiint\limits_{V^{(e)}} \underset{n\times 3}{\begin{bmatrix} N_1^{(e)} & N_1^{(e)} & N_1^{(e)} \\ \vdots & \vdots & \vdots \\ N_n^{(e)} & N_n^{(e)} & N_n^{(e)} \end{bmatrix}} \underset{3\times 3}{\begin{bmatrix} v_x^{(e)} & 0 & 0 \\ 0 & v_y^{(e)} & 0 \\ 0 & 0 & v_z^{(e)} \end{bmatrix}} \underset{3\times n}{\begin{bmatrix} \frac{\partial N_1^{(e)}}{\partial x} & \cdots & \frac{\partial N_n^{(e)}}{\partial x} \\ \frac{\partial N_1^{(e)}}{\partial y} & \cdots & \frac{\partial N_n^{(e)}}{\partial y} \\ \frac{\partial N_1^{(e)}}{\partial z} & \cdots & \frac{\partial N_n^{(e)}}{\partial z} \end{bmatrix}} dx\, dy\, dz$$

$$+ \iiint\limits_{V^{(e)}} \begin{bmatrix} N_1^{(e)} \\ \vdots \\ N_n^{(e)} \end{bmatrix} [\ \lambda(\theta^{(e)} + \rho_b^{(e)} K_d^{(e)})]\ [N_1^{(e)} \ \cdots \ N_n^{(e)}]\ dx\ dy\ dz$$

$$\qquad\qquad\quad V^{(e)} \quad\ 1\times 3 \qquad\qquad 1\times 1 \qquad\qquad 3\times n \qquad\qquad\qquad\qquad (4.12)$$

Element Sorption Matrix (Consistent Formulation)

$$[\ A^{(e)}] = \iiint\limits_{V^{(e)}} \begin{bmatrix} N_1^{(e)} \\ \vdots \\ N_n^{(e)} \end{bmatrix} [\ \rho_b^{(e)} K_d^{(e)} + \theta^{(e)}\ ]\ [\ N_1^{(e)} \ \cdots \ N_n^{(e)}]\ dx\ dy\ dz$$

$$\qquad\qquad\quad\ V^{(e)} \quad\ n\times 1 \qquad\qquad 1\times 1 \qquad\qquad 1\times n \qquad\qquad\qquad\qquad (4.13)$$

Element Sorption Matrix (Lumped Formulation)

$$[A^{(e)}] = (\rho_b^{(e)} K_d^{(e)} + \theta^{(e)}) \left(\frac{V^{(e)}}{n}\right) \begin{bmatrix} 1 & & O \\ & \ddots & \\ O & & 1 \end{bmatrix} \qquad\qquad\qquad (4.14)$$

$$\qquad\qquad\qquad\qquad\qquad\qquad\qquad n\times n$$

We can evaluate each of these matrices for any type of element once we specify the interpolation functions and their derivatives for each node in the element. When the interpolation functions have a simple form and the number of nodes is small, the integrations can be performed analytically. If the interpolation functions are complex or if the number of nodes is large, the integrations must be performed numerically.

4.3.1 Analytical Method

Certain one- and two-dimensional elements have relatively simple interpolation functions and it is possible to use analytical methods to perform the integrations required for the element matrices. The most commonly used one-dimensional element is the linear bar element used in the examples in Chapter 3 (Figure 4.5)

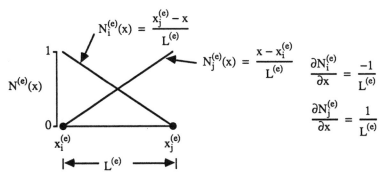

Figure 4.5   Interpolation functions and their derivatives for the linear bar element.

The element matrices for this type of element were computed in the examples in Chapter 3. The results are

$$[K^{(e)}] = \frac{K_x^{(e)}}{L^{(e)}} \begin{bmatrix} 1 & -1 \\ -1 & 1 \end{bmatrix} \qquad (4.15a)$$

2×2
saturated flow

$$[K^{(e)}(\psi)] = \frac{K_x^{(e)}(\psi)}{L^{(e)}} \begin{bmatrix} 1 & -1 \\ -1 & 1 \end{bmatrix} \qquad (4.15b)$$

2×2
unsaturated flow

$$[C^{(e)}] = \frac{S_s^{(e)} L^{(e)}}{6} \begin{bmatrix} 2 & 1 \\ 1 & 2 \end{bmatrix} \qquad (4.16a)$$

2×2
saturated flow
consistent formulation

$$[C^{(e)}] = \frac{S_s^{(e)} L^{(e)}}{2} \begin{bmatrix} 1 & 0 \\ 0 & 1 \end{bmatrix} \qquad (4.16b)$$

2×2
saturated flow
lumped formulation

$$[C^{(e)}(\psi)] = \frac{C^{(e)}(\psi) L^{(e)}}{6} \begin{bmatrix} 2 & 1 \\ 1 & 2 \end{bmatrix} \qquad (4.17a)$$

2×2
unsaturated flow
consistent formulation

$$[C^{(e)}(\psi)] = \frac{C^{(e)}(\psi) L^{(e)}}{2} \begin{bmatrix} 1 & 0 \\ 0 & 1 \end{bmatrix} \qquad (4.17b)$$

2×2
unsaturated flow
lumped formulation

$$[D^{(e)}] = \frac{D_x^{(e)} \theta^{(e)}}{L^{(e)}} \begin{bmatrix} 1 & -1 \\ -1 & 1 \end{bmatrix} + \frac{v_x^{(e)}}{2} \begin{bmatrix} -1 & 1 \\ -1 & 1 \end{bmatrix}$$

2×2

$$+ \lambda(\theta^{(e)} + \rho_b^{(e)} K_d^{(e)}) \frac{L^{(e)}}{6} \begin{bmatrix} 2 & 1 \\ 1 & 2 \end{bmatrix} \quad \text{consistent formulation} \qquad (4.18a)$$

$$\text{or} \quad + \lambda(\theta^{(e)} + \rho_b^{(e)} K_d^{(e)}) \frac{L^{(e)}}{2} \begin{bmatrix} 1 & 0 \\ 0 & 1 \end{bmatrix} \quad \text{lumped formulation} \qquad (4.18b)$$

$$[A^{(e)}] \;\; = \;\; (\rho_b^{(e)}K_d^{(e)} + \theta^{(e)}) \frac{L^{(e)}}{6} \begin{bmatrix} 2 & 1 \\ 1 & 2 \end{bmatrix} \quad \text{consistent formulation} \qquad (4.19a)$$

$$\text{or} \;\; = \;\; (\rho_b^{(e)}K_d^{(e)} + \theta^{(e)}) \frac{L^{(e)}}{2} \begin{bmatrix} 1 & 0 \\ 0 & 1 \end{bmatrix} \quad \text{lumped formulation} \qquad (4.19b)$$

A commonly used two-dimensional element is the *linear triangle* (Figure 4.6). The interpolation functions for this type of element are derived in Segerlind (1984).

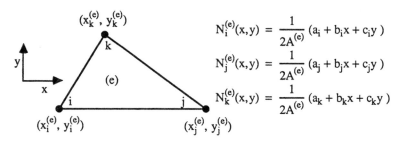

$$N_i^{(e)}(x,y) = \frac{1}{2A^{(e)}} (a_i + b_i x + c_i y)$$

$$N_j^{(e)}(x,y) = \frac{1}{2A^{(e)}} (a_j + b_j x + c_j y)$$

$$N_k^{(e)}(x,y) = \frac{1}{2A^{(e)}} (a_k + b_k x + c_k y)$$

Figure 4.6   **Interpolation functions for the linear triangle element.**

In Figure 4.6

$$a_i = x_j^{(e)} y_k^{(e)} - x_k^{(e)} y_j^{(e)} \qquad a_j = x_k^{(e)} y_i^{(e)} - x_i^{(e)} y_k^{(e)} \qquad a_k = x_i^{(e)} y_j^{(e)} - x_j^{(e)} y_i^{(e)}$$

$$b_i = y_j^{(e)} - y_k^{(e)} \qquad b_j = y_k^{(e)} - y_i^{(e)} \qquad b_k = y_i^{(e)} - y_j^{(e)}$$

$$c_i = x_k^{(e)} - x_j^{(e)} \qquad c_j = x_i^{(e)} - x_k^{(e)} \qquad c_k = x_j^{(e)} - x_i^{(e)}$$

and

$$A^{(e)} = \quad \text{Area of element}$$

$$= \frac{1}{2} \begin{vmatrix} 1 & x_i^{(e)} & y_i^{(e)} \\ 1 & x_j^{(e)} & y_j^{(e)} \\ 1 & x_k^{(e)} & y_k^{(e)} \end{vmatrix} \quad \begin{array}{l}(\text{ An equation to compute this determinant is in} \\ \text{Appendix IV Part 12c. })\end{array}$$

The derivatives of the interpolation functions are

$$\frac{\partial N_i^{(e)}}{\partial x} = \frac{b_i}{2A^{(e)}} \qquad \frac{\partial N_j^{(e)}}{\partial x} = \frac{b_j}{2A^{(e)}} \qquad \frac{\partial N_k^{(e)}}{\partial x} = \frac{b_k}{2A^{(e)}}$$

$$\frac{\partial N_i^{(e)}}{\partial y} = \frac{c_i}{2A^{(e)}} \qquad \frac{\partial N_j^{(e)}}{\partial y} = \frac{c_j}{2A^{(e)}} \qquad \frac{\partial N_k^{(e)}}{\partial y} = \frac{c_k}{2A^{(e)}}$$

The element matrices for the linear triangle element can be easily computed using an integration formula in Segerlind (1984). For a linear triangle element

$$\int_{A^{(e)}} (N_i^{(e)})^a (N_j^{(e)})^b (N_k^{(e)})^c \, dA = \frac{a!\, b!\, c!}{(a+b+c+2)!} 2A^{(e)}$$

where a, b, and c are exponents of the interpolation functions $N_i^{(e)}$, $N_j^{(e)}$, and $N_k^{(e)}$. For example consider the integral

$$\int_{A^{(e)}} N_i^{(e)} N_j^{(e)} \, dA$$

In this case a = 1, b = 1, and c = 0 and we can immediately write

$$\int_{A^{(e)}} N_i^{(e)} N_j^{(e)} \, dA = \frac{1!\, 1!\, 0!}{(1+1+0+2)!} 2A^{(e)} = \frac{A^{(e)}}{3}$$

where 0! = 1. As another example consider the integral

$$\int_{A^{(e)}} N_i^{(e)} \frac{\partial N_i^{(e)}}{\partial x} \, dA$$

In this case a = 1, b = c = 0 and we have

$$\int_{A^{(e)}} N_i^{(e)} \frac{\partial N_i^{(e)}}{\partial x} \, dA = \int_{A^{(e)}} N_i^{(e)} \frac{b_i}{2A^{(e)}} \, dA$$

$$= \frac{b_i}{2A^{(e)}} \int_{A^{(e)}} N_i^{(e)} \, dA = \frac{b_i}{2A^{(e)}} \frac{1!\, 1!\, 0!}{(1+1+0+2)!} 2A^{(e)}$$

$$= \frac{b_i}{6}$$

Example

Evaluate the matrix integral formulation for $[A^{(e)}]$ (consistent formulation) for the linear triangle element . From equation 4.13

$$[A^{(e)}] = \int_{A^{(e)}} \begin{bmatrix} N_1^{(e)} \\ N_2^{(e)} \\ N_3^{(e)} \end{bmatrix} [\rho_b^{(e)} K_d^{(e)} + \theta^{(e)}] [N_1^{(e)} \; N_2^{(e)} \; N_3^{(e)}] \, dA$$

$$= [\rho_b^{(e)}K_d^{(e)} + \theta^{(e)}] \int_{A^{(e)}} \begin{bmatrix} N_1^{(e)} \\ N_2^{(e)} \\ N_3^{(e)} \end{bmatrix} [N_1^{(e)} \; N_2^{(e)} \; N_3^{(e)}] \, dA$$

$$= [\rho_b^{(e)}K_d^{(e)} + \theta^{(e)}] \int_{A^{(e)}} \begin{bmatrix} N_1^{(e)}N_1^{(e)} & N_1^{(e)}N_2^{(e)} & N_1^{(e)}N_3^{(e)} \\ N_2^{(e)}N_1^{(e)} & N_2^{(e)}N_2^{(e)} & N_2^{(e)}N_3^{(e)} \\ N_3^{(e)}N_1^{(e)} & N_3^{(e)}N_2^{(e)} & N_3^{(e)}N_3^{(e)} \end{bmatrix} dA$$

Now

$$\int_{A^{(e)}} N_1^{(e)}N_1^{(e)} \, dA = \int_{A^{(e)}} (N_1^{(e)})^2 \, dA$$

$$= \frac{2! \, 0! \, 0!}{(2+0+0+2)!} 2A^{(e)} = \frac{A^{(e)}}{6}$$

Similarly

$$\int_{A^{(e)}} N_1^{(e)}N_2^{(e)} \, dA = \frac{1! \, 1! \, 0!}{(1+1+0+2)!} 2A^{(e)} = \frac{A^{(e)}}{12}$$

$$\int_{A^{(e)}} N_1^{(e)}N_3^{(e)} \, dA = \frac{1! \, 0! \, 1!}{(1+0+1+2)!} 2A^{(e)} = \frac{A^{(e)}}{12}$$

and so on for each term in the integral. The final result is

$$[A^{(e)}] = \frac{A^{(e)}}{12} [\rho_b^{(e)}K_d^{(e)} + \theta^{(e)}] \begin{bmatrix} 2 & 1 & 1 \\ 1 & 2 & 1 \\ 1 & 1 & 2 \end{bmatrix}$$

A similar procedure can be used to compute the other element matrices. The results are:

$$\underset{3\times3}{[K^{(e)}]} = \frac{K_x^{(e)}}{4A^{(e)}} \begin{bmatrix} b_i^2 & b_ib_j & b_ib_k \\ b_jb_i & b_j^2 & b_jb_k \\ b_kb_i & b_kb_j & b_k^2 \end{bmatrix} + \frac{K_y^{(e)}}{4A^{(e)}} \begin{bmatrix} c_i^2 & c_ic_j & c_ic_k \\ c_jc_i & c_j^2 & c_jc_k \\ c_kc_i & c_kc_j & c_k^2 \end{bmatrix} \qquad (4.20)$$

saturated flow

$$[K^{(e)}]_{3\times3} = \frac{K_x^{(e)}(\psi)}{4A^{(e)}} \begin{bmatrix} b_i^2 & b_ib_j & b_ib_k \\ b_jb_i & b_j^2 & b_jb_k \\ b_kb_i & b_kb_j & b_k^2 \end{bmatrix} + \frac{K_y^{(e)}(\psi)}{4A^{(e)}} \begin{bmatrix} c_i^2 & c_ic_j & c_ic_k \\ c_jc_i & c_j^2 & c_jc_k \\ c_kc_i & c_kc_j & c_k^2 \end{bmatrix} \quad (4.21)$$

unsaturated flow

$$[C^{(e)}]_{3\times3} = \frac{S_s^{(e)} A^{(e)}}{12} \begin{bmatrix} 2 & 1 & 1 \\ 1 & 2 & 1 \\ 1 & 1 & 2 \end{bmatrix} \quad (4.22a) \qquad [C^{(e)}]_{3\times3} = \frac{S_s^{(e)} A^{(e)}}{3} \begin{bmatrix} 1 & 0 & 0 \\ 0 & 1 & 0 \\ 0 & 0 & 1 \end{bmatrix} \quad (4.22b)$$

| saturated flow | saturated flow |
| consistent formulation | lumped formulation |

$$[C^{(e)}(\psi)]_{3\times3} = \frac{C^{(e)}(\psi) A^{(e)}}{12} \begin{bmatrix} 2 & 1 & 1 \\ 1 & 2 & 1 \\ 1 & 1 & 2 \end{bmatrix} (4.23a) \qquad [C^{(e)}(\psi)]_{3\times3} = \frac{C^{(e)}(\psi) A^{(e)}}{3} \begin{bmatrix} 1 & 0 & 0 \\ 0 & 1 & 0 \\ 0 & 0 & 1 \end{bmatrix} (4.23b)$$

| unsaturated flow | unsaturated flow |
| consistent formulation | lumped formulation |

$$[D^{(e)}]_{3\times3} = \frac{D_{xx}^{(e)} \theta^{(e)}}{4A^{(e)}} \begin{bmatrix} b_i^2 & b_ib_j & b_ib_k \\ b_jb_i & b_j^2 & b_jb_k \\ b_kb_i & b_kb_j & b_k^2 \end{bmatrix} + \frac{D_{yy}^{(e)} \theta^{(e)}}{4A^{(e)}} \begin{bmatrix} c_i^2 & c_ic_j & c_ic_k \\ c_jc_i & c_j^2 & c_jc_k \\ c_kc_i & c_kc_j & c_k^2 \end{bmatrix}$$

$$+ \frac{D_{xy}^{(e)}\theta^{(e)}}{4A^{(e)}} \begin{bmatrix} b_ic_i & b_ic_j & b_ic_k \\ b_jc_i & b_jc_j & b_jc_k \\ b_kc_i & b_kc_j & b_kc_k \end{bmatrix} + \frac{D_{yx}^{(e)}\theta^{(e)}}{4A^{(e)}} \begin{bmatrix} c_ib_i & c_ib_j & c_ib_k \\ c_jb_i & c_jb_j & c_jb_k \\ c_kb_i & c_kb_j & c_kb_k \end{bmatrix}$$

$$+ \frac{v_x^{(e)}}{6} \begin{bmatrix} b_i & b_j & b_k \\ b_i & b_j & b_k \\ b_i & b_j & b_k \end{bmatrix} + \frac{v_y^{(e)}}{6} \begin{bmatrix} c_i & c_j & c_k \\ c_i & c_j & c_k \\ c_i & c_j & c_k \end{bmatrix}$$

$$+ \ \frac{A^{(e)}}{12} \ \lambda(\theta^{(e)} + \rho_b^{(e)}K_d^{(e)}) \begin{bmatrix} 2 & 1 & 1 \\ 1 & 2 & 1 \\ 1 & 1 & 2 \end{bmatrix} \qquad \text{consistent formulation} \quad (4.24a)$$

$$\text{or} \ + \ \frac{2A^{(e)}}{3} \ \lambda(\theta^{(e)} + \rho_b^{(e)}K_d^{(e)}) \begin{bmatrix} 1 & 0 & 0 \\ 0 & 1 & 0 \\ 0 & 0 & 1 \end{bmatrix} \qquad \text{lumped formulation} \quad (4.24b)$$

$$\begin{array}{c} [\,A^{(e)}] \\ 3\times 3 \end{array} = \ \frac{A^{(e)}}{12} \ (\rho_b^{(e)}K_d^{(e)} + \theta^{(e)}) \begin{bmatrix} 2 & 1 & 1 \\ 1 & 2 & 1 \\ 1 & 1 & 2 \end{bmatrix} \qquad \text{consistent formulation} \quad (4.25a)$$

$$\text{or} \ = \ \frac{A^{(e)}}{3} \ (\rho_b^{(e)}K_d^{(e)} + \theta^{(e)}) \begin{bmatrix} 1 & 0 & 0 \\ 0 & 1 & 0 \\ 0 & 0 & 1 \end{bmatrix} \qquad \text{lumped formulation} \quad (4.25b)$$

Example

The element conductance matrix for the linear triangle element is given by

$$[K^{(e)}] = \int\int_{A^{(e)}} \begin{bmatrix} \dfrac{\partial N_1^{(e)}}{\partial x} & \dfrac{\partial N_1^{(e)}}{\partial y} \\[6pt] \dfrac{\partial N_2^{(e)}}{\partial x} & \dfrac{\partial N_2^{(e)}}{\partial y} \\[6pt] \dfrac{\partial N_3^{(e)}}{\partial x} & \dfrac{\partial N_3^{(e)}}{\partial y} \end{bmatrix} \begin{bmatrix} K_x^{(e)} & 0 \\ 0 & K_y^{(e)} \end{bmatrix} \begin{bmatrix} \dfrac{\partial N_1^{(e)}}{\partial x} & \dfrac{\partial N_2^{(e)}}{\partial x} & \dfrac{\partial N_3^{(e)}}{\partial x} \\[6pt] \dfrac{\partial N_1^{(e)}}{\partial y} & \dfrac{\partial N_2^{(e)}}{\partial y} & \dfrac{\partial N_3^{(e)}}{\partial y} \end{bmatrix} dx \ dy$$

$$= \left(\frac{1}{2A^{(e)}}\right)\left(\frac{1}{2A^{(e)}}\right) \begin{bmatrix} b_i & c_i \\ b_j & c_j \\ b_k & c_k \end{bmatrix} \begin{bmatrix} K_x^{(e)} & 0 \\ 0 & K_y^{(e)} \end{bmatrix} \begin{bmatrix} b_i & b_j & b_k \\ c_i & c_j & c_k \end{bmatrix} \int\int_{A^{(e)}} dx \ dy$$

$$= \frac{1}{4A^{(e)2}} \begin{bmatrix} b_i & c_i \\ b_j & c_j \\ b_k & c_k \end{bmatrix} \begin{bmatrix} K_x^{(e)} & 0 \\ 0 & K_y^{(e)} \end{bmatrix} \begin{bmatrix} b_i & b_j & b_k \\ c_i & c_j & c_k \end{bmatrix} A^{(e)}$$

$$= \frac{K_x^{(e)}}{4A^{(e)}} \begin{bmatrix} b_i^2 & b_ib_j & b_ib_k \\ b_jb_i & b_j^2 & b_jb_k \\ b_kb_i & b_kb_j & b_k^2 \end{bmatrix} + \frac{K_y^{(e)}}{4A^{(e)}} \begin{bmatrix} c_i^2 & c_ic_j & c_ic_k \\ c_jc_i & c_j^2 & c_jc_k \\ c_kc_i & c_kc_j & c_k^2 \end{bmatrix}$$

Example

Compute [$K^{(e)}$] for the element shown below ($K_x^{(e)} = 1$, $K_y^{(e)} = 2$)

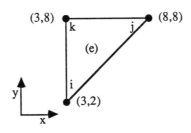

$$A^{(e)} = \frac{1}{2}\begin{vmatrix} 1 & 3 & 2 \\ 1 & 8 & 8 \\ 1 & 3 & 8 \end{vmatrix} = 15, \quad 4A^{(e)} = 60$$

$b_i = y_j^{(e)} - y_k^{(e)} = 8 - 8 = 0$ $c_i = x_k^{(e)} - x_j^{(e)} = 3 - 8 = -5$

$b_j = y_k^{(e)} - y_i^{(e)} = 8 - 2 = 6$ $c_j = x_i^{(e)} - x_k^{(e)} = 3 - 3 = 0$

$b_k = y_i^{(e)} - y_j^{(e)} = 2 - 8 = -6$ $c_k = x_j^{(e)} - x_i^{(e)} = 8 - 3 = 5$

$$[K^{(e)}] = \frac{1}{60}\begin{bmatrix} (0)(0) & (0)(6) & (0)(-6) \\ (6)(0) & (6)(6) & (6)(-6) \\ (-6)(0) & (-6)(6) & (-6)(-6) \end{bmatrix} + \frac{2}{60}\begin{bmatrix} (-5)(-5) & (-5)(0) & (-5)(5) \\ (0)(-5) & (0)(0) & (0)(5) \\ (5)(-5) & (5)(0) & (5)(5) \end{bmatrix}$$

$$= \begin{bmatrix} 5/6 & 0 & -5/6 \\ 0 & 6/10 & -6/10 \\ -5/6 & -6/10 & 43/30 \end{bmatrix}$$

Example

Compute [$D^{(e)}$] for the element in the previous example ($v_x^{(e)} = 2$, $v_y^{(e)} = 3$, $D_{xx}^{(e)} = D_{yy}^{(e)}$ $= 10$, $D_{yx}^{(e)} = D_{xy}^{(e)} = 0$, $\lambda = 0$, $K_d^{(e)} = 0$, $\theta^{(e)} = 0.3$)

$$[D^{(e)}] = \frac{10(0.3)}{60}\begin{bmatrix} 0 & 0 & 0 \\ 0 & 36 & -36 \\ 0 & -36 & 36 \end{bmatrix} + \frac{10(0.3)}{60}\begin{bmatrix} 25 & 0 & -25 \\ 0 & 0 & 0 \\ -25 & 0 & 25 \end{bmatrix}$$

$$+ \frac{2}{6}\begin{bmatrix} 0 & 6 & -6 \\ 0 & 6 & -6 \\ 0 & 6 & -6 \end{bmatrix} + \frac{3}{6}\begin{bmatrix} -5 & 0 & 5 \\ -5 & 0 & 5 \\ -5 & 0 & 5 \end{bmatrix}$$

$$= \begin{bmatrix} 0 & 0 & 0 \\ 0 & 1.8 & -1.8 \\ 0 & -1.8 & 1.8 \end{bmatrix} + \begin{bmatrix} 1.25 & 0 & -1.25 \\ 0 & 0 & 0 \\ -1.25 & 0 & 1.25 \end{bmatrix}$$

$$+ \begin{bmatrix} 0 & 2.0 & -2.0 \\ 0 & 2.0 & -2.0 \\ 0 & 2.0 & -2.0 \end{bmatrix} + \begin{bmatrix} -2.5 & 0 & 2.5 \\ -2.5 & 0 & 2.5 \\ -2.5 & 0 & 2.5 \end{bmatrix}$$

$$= \begin{bmatrix} -1.25 & 2.00 & -0.75 \\ -2.50 & 3.80 & -1.30 \\ -3.75 & 0.20 & 3.55 \end{bmatrix}$$

Another commonly used two-dimensional element is *the linear rectangle* (Figure 4.7). For this type of element, the sides of the element are required to be parallel to the x and y coordinate axes (a more flexible type of element, *the linear quadrilateral,* is described in the next section).

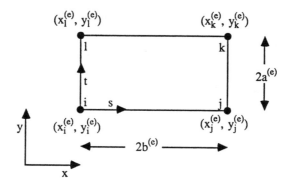

Figure 4.7   The linear rectangle element.

The interpolation functions for this type of element are derived in Segerlind (1984). Using the local (s,t) coordinate system

$$N_i^{(e)}(s,t) = \frac{(2b^{(e)} - s)(2a^{(e)} - t)}{4a^{(e)}b^{(e)}}$$

$$N_j^{(e)}(s,t) = \frac{s(2a^{(e)} - t)}{4a^{(e)}b^{(e)}}$$

$$N_k^{(e)}(s,t) = \frac{st}{4a^{(e)}b^{(e)}}$$

$$N_l^{(e)}(s,t) = \frac{(2b^{(e)} - s)t}{4a^{(e)}b^{(e)}}$$

Because the s and x axes are parallel and the t and y axes are parallel $\dfrac{N_i}{\partial x}=\dfrac{N_i}{\partial s}$ and $\dfrac{N_i}{\partial y}=\dfrac{N_i}{\partial t}$ and the derivatives of the interpolation functions are

$$\frac{\partial N_i^{(e)}}{\partial x} = \frac{t-2a^{(e)}}{4a^{(e)}b^{(e)}} \qquad\qquad \frac{\partial N_j^{(e)}}{\partial x} = \frac{2a^{(e)}-t}{4a^{(e)}b^{(e)}}$$

$$\frac{\partial N_k^{(e)}}{\partial x} = \frac{t}{4a^{(e)}b^{(e)}} \qquad\qquad \frac{\partial N_l^{(e)}}{\partial x} = \frac{-t}{4a^{(e)}b^{(e)}}$$

$$\frac{\partial N_i^{(e)}}{\partial y} = \frac{s-2b^{(e)}}{4a^{(e)}b^{(e)}} \qquad\qquad \frac{\partial N_j^{(e)}}{\partial y} = \frac{-s}{4a^{(e)}b^{(e)}}$$

$$\frac{\partial N_k^{(e)}}{\partial y} = \frac{s}{4a^{(e)}b^{(e)}} \qquad\qquad \frac{\partial N_l^{(e)}}{\partial y} = \frac{2b^{(e)}-s}{4a^{(e)}b^{(e)}}$$

The element matrices are computed in an exercise. The results are

$$\underset{\substack{4\times 4 \\ \text{saturated flow}}}{[\,K^{(e)}\,]} = \frac{K_x^{(e)}a^{(e)}}{6b^{(e)}}\begin{bmatrix} 2 & -2 & -1 & 1 \\ -2 & 2 & 1 & -1 \\ -1 & 1 & 2 & -2 \\ 1 & -1 & -2 & 2 \end{bmatrix} + \frac{K_y^{(e)}b^{(e)}}{6a^{(e)}}\begin{bmatrix} 2 & 1 & -1 & -2 \\ 1 & 2 & -2 & -1 \\ -1 & -2 & 2 & 1 \\ -2 & -1 & 1 & 2 \end{bmatrix} \qquad (4.26a)$$

$$\underset{\substack{4\times 4 \\ \text{unsaturated flow}}}{[\,K^{(e)}(\psi)\,]} = \frac{K_x^{(e)}(\psi)\,a^{(e)}}{6b^{(e)}}\begin{bmatrix} 2 & -2 & -1 & 1 \\ -2 & 2 & 1 & -1 \\ -1 & 1 & 2 & -2 \\ 1 & -1 & -2 & 2 \end{bmatrix} + \frac{K_y^{(e)}(\psi)\,b^{(e)}}{6a^{(e)}}\begin{bmatrix} 2 & 1 & -1 & -2 \\ 1 & 2 & -2 & -1 \\ -1 & -2 & 2 & 1 \\ -2 & -1 & 1 & 2 \end{bmatrix} \qquad (4.26b)$$

$$\underset{\substack{4\times 4 \\ \text{saturated flow} \\ \text{consistent formulation}}}{[\,C^{(e)}\,]} = \frac{S_s^{(e)}a^{(e)}b^{(e)}}{9}\begin{bmatrix} 4 & 2 & 1 & 2 \\ 2 & 4 & 2 & 1 \\ 1 & 2 & 4 & 2 \\ 2 & 1 & 2 & 4 \end{bmatrix} \qquad (4.27a)$$

$$\underset{\substack{4\times 4 \\ \text{saturated flow} \\ \text{lumped formulation}}}{[\,C^{(e)}\,]} = S_s^{(e)}a^{(e)}b^{(e)}\begin{bmatrix} 1 & 0 & 0 & 0 \\ 0 & 1 & 0 & 0 \\ 0 & 0 & 1 & 0 \\ 0 & 0 & 0 & 1 \end{bmatrix} \qquad (4.27b)$$

$$\underset{4 \times 4}{[\, C^{(e)}(\psi)\,]} = \frac{C^{(e)}(\psi)\, a^{(e)} b^{(e)}}{9} \begin{bmatrix} 4 & 2 & 1 & 2 \\ 2 & 4 & 2 & 1 \\ 1 & 2 & 4 & 2 \\ 2 & 1 & 2 & 4 \end{bmatrix} \tag{4.28a}$$

unsaturated flow
consistent formulation

$$\underset{4 \times 4}{[\, C^{(e)}(\psi)\,]} = C^{(e)}(\psi) a^{(e)} b^{(e)} \begin{bmatrix} 1 & 0 & 0 & 0 \\ 0 & 1 & 0 & 0 \\ 0 & 0 & 1 & 0 \\ 0 & 0 & 0 & 1 \end{bmatrix} \tag{4.28b}$$

unsaturated flow
lumped formulation

$$\underset{4 \times 4}{[\, D^{(e)}\,]} = \frac{D_{xx}^{(e)} \theta^{(e)} a^{(e)}}{6 b^{(e)}} \begin{bmatrix} 2 & -2 & -1 & 1 \\ -2 & 2 & 1 & -1 \\ -1 & 1 & 2 & -2 \\ 1 & -1 & -2 & 2 \end{bmatrix} + \frac{D_{yy}^{(e)} \theta^{(e)} b^{(e)}}{6 a^{(e)}} \begin{bmatrix} 2 & 1 & -1 & -2 \\ 1 & 2 & -2 & -1 \\ -1 & -2 & 2 & 1 \\ -2 & -1 & 1 & 2 \end{bmatrix}$$

$$+ \frac{D_{xy}^{(e)} \theta^{(e)}}{4} \begin{bmatrix} 1 & 1 & -1 & -1 \\ -1 & -1 & 1 & 1 \\ -1 & -1 & 1 & 1 \\ 1 & 1 & -1 & -1 \end{bmatrix} + \frac{D_{yx}^{(e)} \theta^{(e)}}{4} \begin{bmatrix} 1 & -1 & -1 & 1 \\ 1 & -1 & -1 & 1 \\ -1 & 1 & 1 & -1 \\ -1 & 1 & 1 & -1 \end{bmatrix}$$

$$+ \frac{v_x^{(e)} a^{(e)}}{6} \begin{bmatrix} -2 & 2 & 1 & -1 \\ -2 & 2 & 1 & -1 \\ -1 & 1 & 2 & -2 \\ -1 & 1 & 2 & -2 \end{bmatrix} + \frac{v_y^{(e)} b^{(e)}}{6} \begin{bmatrix} -2 & -1 & 1 & 2 \\ -1 & -2 & 2 & 1 \\ -1 & -2 & 2 & 1 \\ -2 & -1 & 1 & 2 \end{bmatrix}$$

$$+ \lambda(\theta^{(e)} + \rho_b^{(e)} K_d^{(e)}) \frac{a^{(e)} b^{(e)}}{9} \begin{bmatrix} 4 & 2 & 1 & 2 \\ 2 & 4 & 2 & 1 \\ 1 & 2 & 4 & 2 \\ 2 & 1 & 2 & 4 \end{bmatrix} \quad \text{consistent formulation} \tag{4.29a}$$

$$\text{or} \quad + \lambda(\theta^{(e)} + \rho_b^{(e)} K_d^{(e)}) a^{(e)} b^{(e)} \begin{bmatrix} 1 & 0 & 0 & 0 \\ 0 & 1 & 0 & 0 \\ 0 & 0 & 1 & 0 \\ 0 & 0 & 0 & 1 \end{bmatrix} \quad \text{lumped formulation} \tag{4.29b}$$

$$\underset{4 \times 4}{[\, A^{(e)}\,]} = (\rho_b^{(e)} K_d^{(e)} + \theta^{(e)}) \frac{a^{(e)} b^{(e)}}{9} \begin{bmatrix} 4 & 2 & 1 & 2 \\ 2 & 4 & 2 & 1 \\ 1 & 2 & 4 & 2 \\ 2 & 1 & 2 & 4 \end{bmatrix} \quad \text{consistent formulation} \tag{4.30a}$$

or $\quad + \quad (\rho_b^{(e)}K_d^{(e)} + \theta^{(e)})\, a^{(e)}b^{(e)} \begin{bmatrix} 1 & 0 & 0 & 0 \\ 0 & 1 & 0 & 0 \\ 0 & 0 & 1 & 0 \\ 0 & 0 & 0 & 1 \end{bmatrix} \qquad$ lumped formulation   (4.30b)

Example

The second integral in the equation for advection-dispersion matrix (equation 4.12) for the linear rectangle element is

$$\iint_{A^{(e)}} \begin{bmatrix} N_1^{(e)} & N_1^{(e)} & N_1^{(e)} \\ \vdots & \vdots & \vdots \\ N_4^{(e)} & N_4^{(e)} & N_4^{(e)} \end{bmatrix} \begin{bmatrix} v_x^{(e)} & 0 & 0 \\ 0 & v_y^{(e)} & 0 \\ 0 & 0 & v_z^{(e)} \end{bmatrix} \begin{bmatrix} \dfrac{\partial N_1^{(e)}}{\partial x} & \cdots & \dfrac{\partial N_4^{(e)}}{\partial x} \\ \dfrac{\partial N_1^{(e)}}{\partial y} & \cdots & \dfrac{\partial N_4^{(e)}}{\partial y} \end{bmatrix} dx\, dy$$

The matrix that results from the integration is

$$\frac{v_x^{(e)}a^{(e)}}{6} \begin{bmatrix} 1 & 4 & -2 & 2 \\ -2 & 2 & -1 & 1 \\ -4 & 4 & 3 & -3 \\ 5 & -2 & 2 & -2 \end{bmatrix}$$

The entry in the first row and column is obtained by evaluating the integral

$$\iint_{A^{(e)}} N_1^{(e)} v_x^{(e)} \frac{\partial N_1^{(e)}}{\partial x} dx\, dy = \int_0^{2b^{(e)}} \int_0^{2a^{(e)}} N_1^{(e)} v_x^{(e)} \frac{N_1^{(e)}}{\partial s} dt\, ds$$

$$= v_x^{(e)} \int_0^{2b^{(e)}} \int_0^{2a^{(e)}} \frac{(2b^{(e)} - s)(2a^{(e)} - t)}{4a^{(e)}b^{(e)}} \cdot \frac{(t - 2a^{(e)})}{4a^{(e)}b^{(e)}} dt\, ds$$

$$= \frac{v_x^{(e)}}{(4a^{(e)}b^{(e)})^2} \int_0^{2b^{(e)}} \int_0^{2a^{(e)}} (2b^{(e)} - s)(2a^{(e)} - t)(t - 2a^{(e)}) dt\, ds$$

$$= \frac{v_x^{(e)}}{(4a^{(e)}b^{(e)})^2} \int_0^{2b^{(e)}} \int_0^{2a^{(e)}} (s - 2b^{(e)})(2a^{(e)} - t)^2 dt\, ds$$

$$= \frac{v_x^{(e)}}{(4a^{(e)}b^{(e)})^2} \int_0^{2b^{(e)}} \int_0^{2a^{(e)}} (s - 2b^{(e)})((2a^{(e)})^2 - 4a^{(e)}t + t^2) dt\, ds$$

$$= \frac{v_x^{(e)}}{(4a^{(e)}b^{(e)})^2} \int_0^{2b^{(e)}} \left((s - 2b^{(e)}) \left((2a^{(e)})^2 t - 2a^{(e)} t^2 + \frac{t^3}{3} \right) \right) \Bigg|_0^{2a^{(e)}} ds$$

$$= -\frac{v_x^{(e)}}{(4a^{(e)}b^{(e)})^2} \int_0^{2b^{(e)}} (2b^{(e)} - s) \left((2a^{(e)})^3 - (2a^{(e)})^3 + \frac{(2a^{(e)})^3}{3} \right) ds$$

$$= -\frac{v_x^{(e)}}{(4a^{(e)}b^{(e)})^2} \int_0^{2b^{(e)}} \left(2b^{(e)} \frac{(2a^{(e)})^3}{3} - s \frac{(2a^{(e)})^3}{3} \right) ds$$

$$= -\frac{v_x^{(e)}}{(4a^{(e)}b^{(e)})^2} \left(2b^{(e)} s \frac{(2a^{(e)})^3}{3} - \frac{s^2}{2} \frac{(2a^{(e)})^3}{3} \right) \Bigg|_0^{2b^{(e)}}$$

$$= -\frac{v_x^{(e)}}{(4a^{(e)}b^{(e)})^2} \left[(2b^{(e)})(2b^{(e)}) \frac{(2a^{(e)})^3}{3} - \frac{(2b^{(e)})^2}{2} \frac{(2a^{(e)})^3}{3} \right]$$

$$= -\frac{v_x^{(e)}}{(4a^{(e)}b^{(e)})^2} (4a^{(e)}b^{(e)})^2 \frac{a^{(e)}}{3} = -\frac{v_x^{(e)} a^{(e)}}{3}$$

Example

Verify the entry in the second row and third column of equation 4.27a. From equation 4.8 we can write

$$[C^{(e)}] = \int\int_{A^{(e)}} \begin{bmatrix} N_1^{(e)} \\ \vdots \\ N_4^{(e)} \end{bmatrix} [S_s^{(e)}] [N_1^{(e)} \cdots N_4^{(e)}] \, dx \, dy$$

The integral for the entry in the second row and third column of $[C^{(e)}]$ is

$$S_s \int_0^{2b^{(e)}} \int_0^{2b^{(e)}} \frac{s}{2b^{(e)}} \left(1 - \frac{t}{2a^{(e)}} \right) \frac{st}{4a^{(e)}b^{(e)}} \, dt \, ds = \frac{S_s}{(4a^{(e)}b^{(e)})^2} \int_0^{2b^{(e)}} \int_0^{2b^{(e)}} (2a^{(e)} s - st) st \, dt \, ds$$

$$= \frac{S_s}{(4a^{(e)}b^{(e)})^2} \int_0^{2b^{(e)}} \left(a^{(e)} s^2 t^2 - \frac{s^2 t^3}{3} \right) \Bigg|_0^{2a^{(e)}} ds$$

$$= \frac{S_s}{(4a^{(e)}b^{(e)})^2} \int_0^{2b^{(e)}} \frac{4}{3} a^{3(e)} s^2 ds$$

$$= \frac{S_s}{(4a^{(e)}b^{(e)})^2} \left(\frac{4}{3} a^{3(e)} \frac{s^3}{3} \right) \Bigg|_0^{2b^{(e)}}$$

$$= \frac{S_s}{(4a^{(e)}b^{(e)})^2} \left(\frac{4}{3} \right) (a^{3(e)}) \left(\frac{8b^{3(e)}}{3} \right) = \frac{2S_s a^{(e)} b^{(e)}}{9}$$

4.3.2 Numerical Methods

In the previous section we presented equations for computing the element matrices for three simple (but commonly used) element types, the linear bar, the linear triangle, and the linear rectangle. Because the number of nodes was small and the interpolation functions were relatively simple (i.e., linear functions of x or x and y), the integrations could be performed analytically. For elements with more complex interpolation functions or a larger number of nodes, performing the required integrations analytically is awkward. Instead, the integrations are performed numerically. The numerical integration procedure is greatly simplified if the interpolation functions and their derivatives for each node are defined using a *local coordinate system* as was done for the linear rectangle element. In a local coordinate system a point within an element is assigned coordinates using a coordinate system origin *attached to the element* (e.g., the origin of the s-t coordinate system in Figure 4.7 is attached to node i). In a *global coordinate system* a point within an element is assigned coordinates using an arbitrary coordinate system origin. Interpolation functions and their derivatives can be defined using either global or local coordinates. For example, the interpolation functions for the linear bar element can be written using the global coordinate x (where the origin of the x axis can be anywhere) or the local coordinate ε (where the origin of the ε axis is at the center of the element) (Figure 4.8).

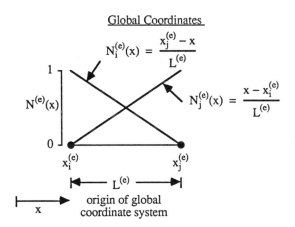

Global Coordinates

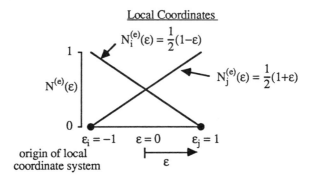

Figure 4.8 **Element interpolation functions can be defined using global or local coordinates.**

The interpolation functions defined using local coordinates have the same properties as interpolation functions defined using global coordinates namely

1. $\varepsilon_i = 1$ at $\varepsilon = \varepsilon_i$ $\varepsilon_j = 1$ at $\varepsilon = \varepsilon_j$
    $= 0$ at $\varepsilon = \varepsilon_j$ $= 0$ at $\varepsilon = \varepsilon_i$

2. $\sum_{i=1}^{n} \varepsilon_i(\varepsilon) = 1$ for all ε, $-1 \le \varepsilon \le 1$

However, when we use a local coordinate system, the integrations for the element matrices can easily be performed numerically, even when the element has curved edges. The shapes of several isoparametric elements in the local and global coordinate systems as well as the interpolation functions and their derivatives are shown in Figures 4.9 to 4.15. Derivations of the interpolation functions for these element types are in Lapidus and Pinder (1982) and Dhatt and Touzot (1984). The derivatives of the interpolation functions, and the value of the derivatives at the center of each element are also given in the figures. The notation can best be illustrated by an example. Consider the two-dimensional, linear quadrilateral element shown in Figure 4.10. The interpolation functions and their derivatives for the element's four nodes are

$$N_1 = \frac{1}{4}(1-\varepsilon)(1-\eta)$$

$$\frac{\partial N_1}{\partial \varepsilon} = -\frac{1}{4}(1-\eta) \qquad\qquad \frac{\partial N_1}{\partial \eta} = -\frac{1}{4}(1-\varepsilon)$$

$$N_2 = \frac{1}{4}(1+\varepsilon)(1-\eta)$$

$$\frac{\partial N_2}{\partial \varepsilon} = \frac{1}{4}(1-\eta) \qquad\qquad \frac{\partial N_2}{\partial \eta} = -\frac{1}{4}(1+\varepsilon)$$

$$N_3 = \frac{1}{4}(1+\varepsilon)(1+\eta)$$

$$\frac{\partial N_3}{\partial \varepsilon} = \frac{1}{4}(1+\eta) \qquad\qquad \frac{\partial N_3}{\partial \eta} = \frac{1}{4}(1+\varepsilon)$$

$$N_4 = \frac{1}{4}(1-\varepsilon)(1+\eta)$$

$$\frac{\partial N_4}{\partial \varepsilon} = -\frac{1}{4}(1+\eta) \qquad\qquad \frac{\partial N_4}{\partial \eta} = \frac{1}{4}(1-\varepsilon)$$

The choice of element type to use for a particular problem is not always clear. The calculation of the element matrices requires less computational for the linear elements than for the quadratic and cubic elements. Also curvilinear variations in head or solute concentration can usually be adequately represented by many small, linear elements. However, when gradients of head or concentration are large, quadratic or cubic elements may be preferable to linear elements because the quadratic or cubic interpolation functions can approximate curvilinear variations in the field variable with fewer nodes than linear interpolation functions. Also quadratic and cubic elements are useful when the problem domain has curved boundaries (e.g., near a well or a buried structure such as a tunnel, etc.).

In the local coordinate system, the isoparametric elements in Figures 4.9 to 4.15 have straight edges and symmetry about the ε, η, and ζ axes. However, in the global coordinate system (which is the coordinate system of the problem domain) the elements can have curved edges and asymmetric shapes. This is an important property because it means that, when we discretize the problem domain, we can better represent curved boundaries or curved interfaces between soil or rock layers. However, because the interpolation functions and their derivatives are defined in a local coordinate system while the integral formulations for the element matrices are defined in a global coordinate system, we must use a *coordinate transformation* to evaluate the integrals.

(a) linear bar element

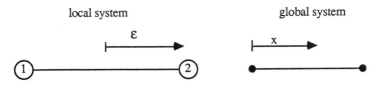

local system   global system

for nodes 1 and 2 :

$$N_i = \frac{1}{2}(1 + \varepsilon_i \varepsilon)$$

$$\frac{\partial N_i}{\partial \varepsilon} = \frac{\varepsilon_i}{2} \qquad\qquad \frac{\partial N_i}{\partial \varepsilon}\bigg|_{\varepsilon=0} = \frac{\varepsilon_i}{2}$$

where

i	1	2
ε_i	−1	1

(b) quadratic bar element

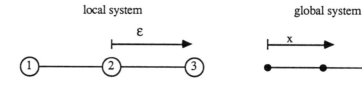

local system   global system

for nodes 1 and 3 :

$$N_i = \frac{\varepsilon_i \, \varepsilon}{2}(1 + \varepsilon_i \varepsilon)$$

$$\frac{\partial N_i}{\partial \varepsilon} = \frac{\varepsilon_i}{2}(1 + 2\varepsilon_i \varepsilon) \qquad\qquad \left. \frac{\partial N_i}{\partial \varepsilon} \right|_{\varepsilon=0} = \frac{\varepsilon_i}{2}$$

where

i	1	3
ε_i	-1	1

for node 2 :

$$N_i = (1+\varepsilon)(1 - \varepsilon)$$

$$\frac{\partial N_i}{\partial \varepsilon} = -2\varepsilon \qquad\qquad \left. \frac{\partial N_i}{\partial \varepsilon} \right|_{\varepsilon=0} = 0$$

(c) cubic bar element

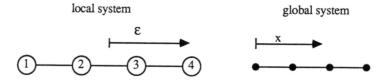

local system   global system

for nodes 1 and 4 :

$$N_i = \frac{9\varepsilon_i}{16}\left(\varepsilon + \frac{1}{3}\right)\left(\varepsilon - \frac{1}{3}\right)(\varepsilon + \varepsilon_i)$$

$$\frac{\partial N_i}{\partial \varepsilon} = \frac{9\varepsilon_i}{16}\left(3\varepsilon^2 + 2\varepsilon\varepsilon_i - \frac{1}{9}\right)$$

$$\left. \frac{\partial N_i}{\partial \varepsilon} \right|_{\varepsilon=0} = -\frac{\varepsilon_i}{16}$$

where

i	1	4
ε_i	−1	1

for nodes 2 and 3 :

$$N_i = \frac{27\varepsilon_i}{16}(\varepsilon+1)(\varepsilon-1)\left(\varepsilon-\frac{\varepsilon_i}{3}\right)$$

$$\frac{\partial N_i}{\partial\varepsilon} = \frac{27\varepsilon_i}{16}\left(3\varepsilon^2-\frac{2\varepsilon_i\,\varepsilon}{3}-1\right)$$

$$\frac{\partial N_i}{\partial\varepsilon}\bigg|_{\varepsilon=0} = -\frac{27\varepsilon_i}{16}$$

where

i	2	3
ε_i	1	−1

Figure 4.9 **Interpolation functions and their derivatives for three types of one-dimensional elements.**

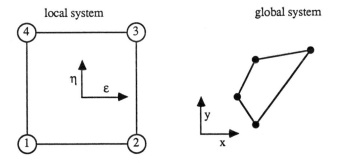

for nodes 1, 2, 3, and 4 :

$$N_i = \frac{1}{4}(1+\varepsilon_i\varepsilon)(1+\eta_i\eta)$$

$$\frac{\partial N_i}{\partial\varepsilon} = \frac{1}{4}(\varepsilon_i)(1+\eta_i\eta) \qquad \frac{\partial N_i}{\partial\varepsilon}\bigg|_{\varepsilon=0\ \eta=0} = \frac{\varepsilon_i}{4}$$

$$\frac{\partial N_i}{\partial\eta} = \frac{1}{4}(\eta_i)(1+\varepsilon_i\varepsilon) \qquad \frac{\partial N_i}{\partial\eta}\bigg|_{\varepsilon=0\ \eta=0} = \frac{\eta_i}{4}$$

where

i	1	2	3	4
ε_i	-1	1	1	-1
η_i	-1	-1	1	1

Figure 4.10   **Interpolation functions and their derivatives for linear quadrilateral elements.**

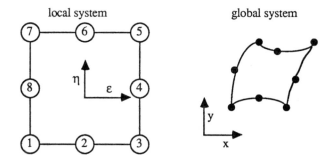

local system   global system

for nodes 1, 3, 5, and 7 :

$$N_i = \frac{1}{4}(1 + \varepsilon_i\varepsilon)(1 + \eta_i\eta)(\varepsilon_i\varepsilon + \eta_i\eta - 1)$$

$$\frac{\partial N_i}{\partial \varepsilon} = \frac{1}{4}(1 + \eta_i\eta)(2\varepsilon_i{}^2\varepsilon + \eta_i\varepsilon_i\eta) \qquad \left. \frac{\partial N_i}{\partial \varepsilon} \right|_{\varepsilon=0 \; \eta=0} = 0$$

$$\frac{\partial N_i}{\partial \eta} = \frac{1}{4}(1 + \varepsilon_i\varepsilon)(2\eta_i{}^2\eta + \eta_i\varepsilon_i\varepsilon) \qquad \left. \frac{\partial N_i}{\partial \eta} \right|_{\varepsilon=0 \; \eta=0} = 0$$

where

i	1	3	5	7
ε_i	-1	1	1	-1
η_i	-1	-1	1	1

for nodes 2 and 6 :

$$N_i = \frac{1}{2}(1 - \varepsilon^2)(1 + \eta_i\eta)$$

$$\frac{\partial N_i}{\partial \varepsilon} = -\varepsilon(1 + \eta_i\eta) \qquad \left. \frac{\partial N_i}{\partial \varepsilon} \right|_{\varepsilon=0 \; \eta=0} = 0$$

$$\frac{\partial N_i}{\partial \eta} = \frac{\eta_i}{2}(1 - \varepsilon^2) \qquad \left. \frac{\partial N_i}{\partial \eta} \right|_{\varepsilon=0 \; \eta=0} = \frac{\eta_i}{2}$$

where

i	2	6
ε_i	0	0
η_i	-1	1

for nodes 4 and 8 :

$$N_i = \frac{1}{2}(1 + \varepsilon_i\varepsilon)(1 - \eta^2)$$

$$\frac{\partial N_i}{\partial \varepsilon} = \frac{\varepsilon_i}{2}(1 - \eta^2) \qquad \frac{\partial N_i}{\partial \varepsilon}\bigg|_{\varepsilon=0\ \eta=0} = \frac{\varepsilon_i}{2}$$

$$\frac{\partial N_i}{\partial \eta} = -\eta(1 + \varepsilon_i\varepsilon) \qquad \frac{\partial N_i}{\partial \eta}\bigg|_{\varepsilon=0\ \eta=0} = 0$$

where

i	4	8
ε_i	1	-1
η_i	0	0

Figure 4.11   Interpolation functions and their derivatives for quadratic quadrilateral elements.

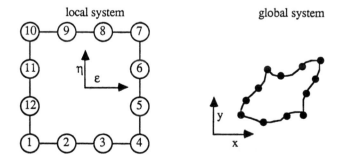

local system   global system

for nodes 1, 4, 7, and 10 :

$$N_i = \frac{1}{32}(1 + \varepsilon_i\varepsilon)(1 + \eta_i\eta)[9(\varepsilon^2 + \eta^2) - 10]$$

$$\frac{\partial N_i}{\partial \varepsilon} = \frac{1}{32}(1 + \eta_i\eta)(18\varepsilon + 27\varepsilon_i\varepsilon^2 + 9\varepsilon_i\eta^2 - 10\varepsilon_i)$$

$$\frac{\partial N_i}{\partial \eta} = \frac{1}{32}(1 + \varepsilon_i\varepsilon)(18\eta + 27\eta_i\eta^2 + 9\eta_i\varepsilon^2 - 10\eta_i)$$

$$\left.\frac{\partial N_i}{\partial \varepsilon}\right|_{\varepsilon=0\,\eta=0} = -\frac{10}{32}\varepsilon_i \qquad \left.\frac{\partial N_i}{\partial \eta}\right|_{\varepsilon=0\,\eta=0} = -\frac{10}{32}\eta_i$$

where

i	1	4	7	10
ε_i	−1	1	1	−1
η_i	−1	−1	1	1

for nodes 2, 3, 8, and 9 :

$$N_i = \frac{9}{32}(1 - \varepsilon^2)(1 + 9\varepsilon_i\varepsilon)(1 + \eta_i\eta)$$

$$\frac{\partial N_i}{\partial \varepsilon} = \frac{9}{32}(1 + \eta_i\eta)(9\varepsilon_i - 2\varepsilon - 27\varepsilon_i\varepsilon^2)$$

$$\frac{\partial N_i}{\partial \eta} = \frac{9}{32}(1 - \varepsilon^2)(\eta_i + 9\varepsilon_i\eta_i\varepsilon)$$

$$\left.\frac{\partial N_i}{\partial \varepsilon}\right|_{\varepsilon=0\,\eta=0} = \frac{81}{32}\varepsilon_i \qquad \left.\frac{\partial N_i}{\partial \eta}\right|_{\varepsilon=0\,\eta=0} = \frac{9}{32}\eta_i$$

where

i	2	3	8	9
ε_i	−1/3	1/3	1/3	−1/3
η_i	−1	−1	1	1

for nodes 5, 6, 11, and 12 :

$$N_i = \frac{9}{32}(1 - \eta^2)(1 + 9\eta_i\eta)(1 + \varepsilon_i\varepsilon)$$

$$\frac{\partial N_i}{\partial \varepsilon} = \frac{9}{32}(1 - \eta^2)(\varepsilon_i + 9\eta_i\varepsilon_i\eta)$$

$$\frac{\partial N_i}{\partial \eta} = \frac{9}{32}(1 + \varepsilon_i\varepsilon)(9\eta_i - 2\eta - 27\eta_i\eta^2)$$

$$\left.\frac{\partial N_i}{\partial \varepsilon}\right|_{\varepsilon=0\,\eta=0} = \frac{9}{32}\varepsilon_i \qquad \left.\frac{\partial N_i}{\partial \eta}\right|_{\varepsilon=0\,\eta=0} = \frac{81}{32}\eta_i$$

where

i	5	6	11	12
ϵ_i	1	1	-1	-1
η_i	$-1/3$	$1/3$	$1/3$	$-1/3$

Figure 4.12   Interpolation functions and their derivatives for cubic quadrilateral elements.

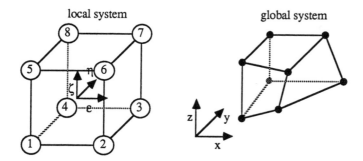

local system   global system

for nodes 1 to 8 :

$$N_i = \frac{1}{8}(1 + \epsilon_i\epsilon)(1 + \eta_i\eta)(1 + \zeta_i\zeta)$$

$$\frac{\partial N_i}{\partial \epsilon} = \frac{\epsilon_i}{8}(1 + \eta_i\eta)(1 + \zeta_i\zeta)$$

$$\frac{\partial N_i}{\partial \eta} = \frac{\eta_i}{8}(1 + \epsilon_i\epsilon)(1 + \zeta_i\zeta)$$

$$\frac{\partial N_i}{\partial \zeta} = \frac{\zeta_i}{8}(1 + \epsilon_i\epsilon)(1 + \eta_i\eta)$$

$$\left.\frac{\partial N_i}{\partial \epsilon}\right|_{\epsilon=0\ \eta=0\ \zeta=0} = \frac{\epsilon_i}{8} \qquad \left.\frac{\partial N_i}{\partial \eta}\right|_{\epsilon=0\ \eta=0\ \zeta=0} = \frac{\eta_i}{8} \qquad \left.\frac{\partial N_i}{\partial \zeta}\right|_{\epsilon=0\ \eta=0\ \zeta=0} = \frac{\zeta_i}{8}$$

where

i	1	2	3	4	5	6	7	8
ϵ_i	-1	1	1	-1	-1	1	1	-1
η_i	-1	-1	1	1	-1	-1	1	1
ζ_i	-1	-1	-1	-1	1	1	1	1

Figure 4.13   Interpolation functions and their derivatives for linear parallelepiped elements.

local system   global system

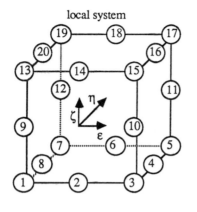 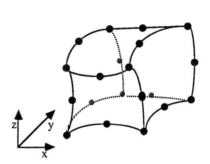

for nodes 1, 3, 5, 7, 13, 15, 17, and 19 :

$$N_i = \frac{1}{8}(1 + \varepsilon_i\varepsilon)(1 + \eta_i\eta)(1 + \zeta_i\zeta)(\varepsilon_i\varepsilon + \eta_i\eta + \zeta_i\zeta - 2)$$

$$\frac{\partial N_i}{\partial \varepsilon} = \frac{\varepsilon_i}{8}(1 + \eta_i\eta)(1 + \zeta_i\zeta)(2\varepsilon_i\varepsilon + \eta_i\eta + \zeta_i\zeta - 1)$$

$$\frac{\partial N_i}{\partial \eta} = \frac{\eta_i}{8}(1 + \varepsilon_i\varepsilon)(1 + \zeta_i\zeta)(2\eta_i\eta + \varepsilon_i\varepsilon + \zeta_i\zeta - 1)$$

$$\frac{\partial N_i}{\partial \zeta} = \frac{\zeta_i}{8}(1 + \varepsilon_i\varepsilon)(1 + \eta_i\eta)(2\zeta_i\zeta + \varepsilon_i\varepsilon + \eta_i\eta - 1)$$

$$\left.\frac{\partial N_i}{\partial \varepsilon}\right|_{\varepsilon=0\ \eta=0\ \zeta=0} = -\frac{\varepsilon_i}{8} \qquad \left.\frac{\partial N_i}{\partial \eta}\right|_{\varepsilon=0\ \eta=0\ \zeta=0} = -\frac{\eta_i}{8} \qquad \left.\frac{\partial N_i}{\partial \zeta}\right|_{\varepsilon=0\ \eta=0\ \zeta=0} = -\frac{\zeta_i}{8}$$

where

i	1	3	5	7	13	15	17	19
ε_i	−1	1	1	−1	−1	1	1	−1
η_i	−1	−1	1	1	−1	−1	1	1
ζ_i	−1	−1	−1	−1	1	1	1	1

for nodes 2, 6, 14, and 18 :

$$N_i = \frac{1}{4}(1 - \varepsilon^2)(1 + \eta_i\eta)(1 + \zeta_i\zeta)$$

$$\frac{\partial N_i}{\partial \varepsilon} = -\frac{\varepsilon}{2}(1 + \eta_i\eta)(1 + \zeta_i\zeta)$$

$$\frac{\partial N_i}{\partial \eta} = \frac{\eta_i}{4}(1 - \varepsilon^2)(1 + \zeta_i\zeta)$$

$$\frac{\partial N_i}{\partial \zeta} = \frac{\zeta_i}{4}(1 - \varepsilon^2)(1 + \eta_i\eta)$$

$$\frac{\partial N_i}{\partial \varepsilon}\bigg|_{\varepsilon=0\ \eta=0\ \zeta=0} = 0 \qquad \frac{\partial N_i}{\partial \eta}\bigg|_{\varepsilon=0\ \eta=0\ \zeta=0} = \frac{\eta_i}{4} \qquad \frac{\partial N_i}{\partial \zeta}\bigg|_{\varepsilon=0\ \eta=0\ \zeta=0} = \frac{\zeta_i}{4}$$

where

i	2	6	14	18
ε_i	0	0	0	0
η_i	−1	1	−1	1
ζ_i	−1	−1	1	1

for nodes 4, 8, 16, and 20 :

$$N_i = \frac{1}{4}(1 - \eta^2)(1 + \varepsilon_i\varepsilon)(1 + \zeta_i\zeta)$$

$$\frac{\partial N_i}{\partial \varepsilon} = \frac{\varepsilon_i}{4}(1 - \eta^2)(1 + \zeta_i\zeta)$$

$$\frac{\partial N_i}{\partial \eta} = -\frac{\eta}{2}(1 + \varepsilon_i\varepsilon)(1 + \zeta_i\zeta)$$

$$\frac{\partial N_i}{\partial \zeta} = \frac{\zeta_i}{4}(1 + \varepsilon_i\varepsilon)(1 - \eta^2)$$

$$\frac{\partial N_i}{\partial \varepsilon}\bigg|_{\varepsilon=0\ \eta=0\ \zeta=0} = \frac{\varepsilon_i}{4} \qquad \frac{\partial N_i}{\partial \eta}\bigg|_{\varepsilon=0\ \eta=0\ \zeta=0} = 0 \qquad \frac{\partial N_i}{\partial \zeta}\bigg|_{\varepsilon=0\ \eta=0\ \zeta=0} = \frac{\zeta_i}{4}$$

where

i	4	8	16	20
ε_i	1	−1	1	−1
η_i	0	0	0	0
ζ_i	−1	−1	1	1

for nodes 9, 10, 11, and 12 :

$$N_i = \frac{1}{4}(1 - \zeta^2)(1 + \varepsilon_i\varepsilon)(1 + \eta_i\eta)$$

$$\frac{\partial N_i}{\partial \varepsilon} = \frac{\varepsilon_i}{4}(1 - \zeta^2)(1 + \eta_i\eta)$$

$$\frac{\partial N_i}{\partial \eta} = \frac{\eta_i}{4}(1 - \zeta^2)(1 + \varepsilon_i \varepsilon)$$

$$\frac{\partial N_i}{\partial \zeta} = -\frac{\zeta}{2}(1 + \varepsilon_i \varepsilon)(1 + \eta_i \eta)$$

$$\left. \frac{\partial N_i}{\partial \varepsilon} \right|_{\varepsilon = 0 \; \eta = 0 \; \zeta = 0} = \frac{\varepsilon_i}{4} \qquad \left. \frac{\partial N_i}{\partial \eta} \right|_{\varepsilon = 0 \; \eta = 0 \; \zeta = 0} = \frac{\eta_i}{4} \qquad \left. \frac{\partial N_i}{\partial \zeta} \right|_{\varepsilon = 0 \; \eta = 0 \; \zeta = 0} = 0$$

where

i	9	10	11	12
ε_i	−1	1	1	−1
η_i	−1	−1	1	1
ζ_i	0	0	0	0

Figure 4.14   **Interpolation functions and their derivatives for quadratic parallelepiped elements.**

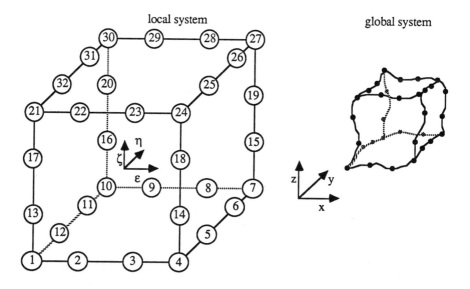

for nodes 1, 4, 7, 10, 21, 24, 27, and 30 :

$$N_i = \frac{9}{64}(1 + \varepsilon_i\varepsilon)(1 + \eta_i\eta)(1 + \zeta_i\zeta)\left[(\varepsilon^2 + \eta^2 + \zeta^2) - \frac{19}{9} \right]$$

$$\frac{\partial N_i}{\partial \varepsilon} = \frac{9}{64}(1 + \eta_i\eta)(1 + \zeta_i\zeta)\left[\varepsilon_i\left(-\frac{19}{9} + 3\varepsilon^2 + \eta^2 + \zeta^2 \right) + 2\varepsilon \right]$$

$$\frac{\partial N_i}{\partial \eta} = \frac{9}{64}(1 + \varepsilon_i\varepsilon)(1 + \zeta_i\zeta)\left[\eta_i\left(-\frac{19}{9} + \varepsilon^2 + 3\eta^2 + \zeta^2 \right) + 2\eta \right]$$

$$\frac{\partial N_i}{\partial \zeta} = \frac{9}{64}(1 + \varepsilon_i\varepsilon)(1 + \eta_i\eta)\left[\zeta_i\left(-\frac{19}{9} + \varepsilon^2 + \eta^2 + 3\zeta^2 \right) + 2\zeta \right]$$

$$\left.\frac{\partial N_i}{\partial \varepsilon}\right|_{\varepsilon=0\ \eta=0\ \zeta=0} = -\frac{19}{64}\varepsilon_i$$

$$\left.\frac{\partial N_i}{\partial \eta}\right|_{\varepsilon=0\ \eta=0\ \zeta=0} = -\frac{19}{64}\eta_i$$

$$\left.\frac{\partial N_i}{\partial \zeta}\right|_{\varepsilon=0\ \eta=0\ \zeta=0} = -\frac{19}{64}\zeta_i$$

where

i	1	4	7	10	21	24	27	30
ε_i	−1	1	1	−1	−1	1	1	−1
η_i	−1	−1	1	1	−1	−1	1	1
ζ_i	−1	−1	−1	−1	1	1	1	1

for nodes 2, 3, 8, 9, 22, 23, 28, and 29 :

$$N_i = \frac{81}{64}(1 - \varepsilon^2)\left(\frac{1}{9} + \varepsilon_i\varepsilon \right)(1 + \eta_i\eta)(1 + \zeta_i\zeta)$$

$$\frac{\partial N_i}{\partial \varepsilon} = \frac{81}{64}(1 + \eta_i\eta)(1 + \zeta_i\zeta)\left(\varepsilon_i - \frac{2}{9}\varepsilon - 3\varepsilon_i\varepsilon^2 \right)$$

$$\frac{\partial N_i}{\partial \eta} = \frac{81\eta_i}{64}(1 - \varepsilon^2)\left(\frac{1}{9} + \varepsilon_i\varepsilon \right)(1 + \zeta_i\zeta)$$

$$\frac{\partial N_i}{\partial \zeta} = \frac{81\zeta_i}{64}(1 - \varepsilon^2)\left(\frac{1}{9} + \varepsilon_i\varepsilon \right)(1 + \eta_i\eta)$$

$$\left.\frac{\partial N_i}{\partial \varepsilon}\right|_{\varepsilon=0\ \eta=0\ \zeta=0} = \frac{81}{64}\varepsilon_i$$

$$\left.\frac{\partial N_i}{\partial \eta}\right|_{\varepsilon=0\ \eta=0\ \zeta=0} = \frac{9}{64}\eta_i$$

$$\left.\frac{\partial N_i}{\partial \zeta}\right|_{\varepsilon=0\ \eta=0\ \zeta=0} = \frac{9}{64}\zeta_i$$

where

i	2	3	8	9	22	23	28	29
ε_i	−1/3	1/3	1/3	−1/3	−1/3	1/3	1/3	−1/3
η_i	−1	−1	1	1	−1	−1	1	1
ζ_i	−1	−1	−1	−1	1	1	1	1

for nodes 5, 6, 11, 12, 25, 26, 31, and 32 :

$$N_i = \frac{81}{64}(1-\eta^2)\left(\frac{1}{9}+\eta_i\eta\right)(1+\varepsilon_i\varepsilon)(1+\zeta_i\zeta)$$

$$\frac{\partial N_i}{\partial\varepsilon} = \frac{81\varepsilon_i}{64}(1-\eta^2)\left(\frac{1}{9}+\eta_i\eta\right)(1+\zeta_i\zeta)$$

$$\frac{\partial N_i}{\partial\eta} = \frac{81}{64}(1+\varepsilon_i\varepsilon)(1+\zeta_i\zeta)\left(\eta_i-\frac{2}{9}\eta-3\eta_i\eta^2\right)$$

$$\frac{\partial N_i}{\partial\zeta} = \frac{81\zeta_i}{64}(1-\eta^2)\left(\frac{1}{9}+\eta_i\eta\right)(1+\varepsilon_i\varepsilon)$$

$$\left.\frac{\partial N_i}{\partial\varepsilon}\right|_{\varepsilon=0\ \eta=0\ \zeta=0} = \frac{9}{64}\varepsilon_i$$

$$\left.\frac{\partial N_i}{\partial\eta}\right|_{\varepsilon=0\ \eta=0\ \zeta=0} = \frac{81}{64}\eta_i$$

$$\left.\frac{\partial N_i}{\partial\zeta}\right|_{\varepsilon=0\ \eta=0\ \zeta=0} = \frac{9}{64}\zeta_i$$

where

i	5	6	11	12	25	26	31	32
ε_i	1	1	−1	−1	1	1	−1	−1
η_i	−1/3	1/3	1/3	−1/3	−1/3	1/3	1/3	−1/3
ζ_i	−1	−1	−1	−1	1	1	1	1

for nodes 13, 14, 15, 16, 17, 18, 19, and 20 :

$$N_i = \frac{81}{64}(1-\zeta^2)\left(\frac{1}{9}+\zeta_i\zeta\right)(1+\varepsilon_i\varepsilon)(1+\eta_i\eta)$$

$$\frac{\partial N_i}{\partial\varepsilon} = \frac{81\varepsilon_i}{64}(1-\zeta^2)\left(\frac{1}{9}+\zeta_i\zeta\right)(1+\eta_i\eta)$$

$$\frac{\partial N_i}{\partial\eta} = \frac{81\eta_i}{64}(1-\zeta^2)\left(\frac{1}{9}+\zeta_i\zeta\right)(1+\varepsilon_i\varepsilon)$$

$$\frac{\partial N_i}{\partial\zeta} = \frac{81}{64}(1+\varepsilon_i\varepsilon)(1+\eta_i\eta)\left(\zeta_i-\frac{2}{9}\zeta-3\zeta_i\zeta^2\right)$$

$$\left.\frac{\partial N_i}{\partial \varepsilon}\right|_{\varepsilon=0\ \eta=0\ \zeta=0} = \frac{9}{64}\varepsilon_i$$

$$\left.\frac{\partial N_i}{\partial \eta}\right|_{\varepsilon=0\ \eta=0\ \zeta=0} = \frac{9}{64}\eta_i$$

$$\left.\frac{\partial N_i}{\partial \zeta}\right|_{\varepsilon=0\ \eta=0\ \zeta=0} = \frac{81}{64}\zeta_i$$

where

i	13	14	15	16	17	18	19	20
ε_i	−1	1	1	−1	−1	1	1	−1
η_i	−1	−1	1	1	−1	−1	1	1
ζ_i	−1/3	−1/3	−1/3	−1/3	1/3	1/3	1/3	1/3

Figure 4.15   Interpolation functions and their derivatives for cubic parallelepiped elements.

To illustrate the coordinate transformation process consider the one-dimensional form of the element conductance matrix for saturated flow

$$\left[K^{(e)}\right]_{2\times2} = \int_{x_i^{(e)}}^{x_j^{(e)}} \begin{bmatrix} \dfrac{\partial N_1^{(e)}}{\partial x} \\[2mm] \dfrac{\partial N_2^{(e)}}{\partial x} \end{bmatrix} \left[K_x^{(e)}\right] \begin{bmatrix} \dfrac{\partial N_1^{(e)}}{\partial x} & \dfrac{\partial N_2^{(e)}}{\partial x} \end{bmatrix} dx \qquad (4.31)$$

The derivatives of the interpolation functions are given in terms of x, but we can use a coordinate transformation of the form

$$x = f(\varepsilon)$$

to rewrite the interpolation functions and their derivatives in terms of the local coordinate ε. For the linear, bar element (Figure 4.9a)

$$N_1 = \frac{1}{2}(1-\varepsilon) \qquad\qquad N_2 = \frac{1}{2}(1+\varepsilon)$$

and

$$\frac{\partial N_1}{\partial \varepsilon} = -\frac{1}{2} \qquad\qquad \frac{\partial N_2}{\partial \varepsilon} = \frac{1}{2}$$

Using the chain rule of calculus we can immediately write

$$\frac{\partial N_1}{\partial \varepsilon} = \frac{\partial N_1}{\partial x}\frac{\partial x}{\partial \varepsilon} \qquad\qquad \frac{\partial N_2}{\partial \varepsilon} = \frac{\partial N_2}{\partial x}\frac{\partial x}{\partial \varepsilon}$$

$$= \frac{\partial N_1}{\partial x}[J] \qquad\qquad\quad = \frac{\partial N_2}{\partial x}[J]$$

The quantity $\frac{\partial x}{\partial \varepsilon}$ is called the Jacobian matrix of the coordinate transformation and, for a one-dimensional element, the Jacobian matrix [J] is a square matrix with a size of one

$$[J]_{1\times 1} = \left[\frac{\partial x}{\partial \varepsilon}\right] = \frac{\partial N_1^{(e)}}{\partial \varepsilon} x_1^{(e)} + \frac{\partial N_2^{(e)}}{\partial \varepsilon} x_2^{(e)} \tag{4.32}$$

We can also write the inverse coordinate transformation

$$\frac{\partial N_1}{\partial x} = \frac{1}{\partial x/\partial \varepsilon} \frac{\partial N_1}{\partial \varepsilon} \qquad\qquad \frac{\partial N_2}{\partial \varepsilon} = \frac{1}{\partial x/\partial \varepsilon} \frac{\partial N_2}{\partial \varepsilon}$$

$$= [J^{-1}]\frac{\partial N_1}{\partial \varepsilon} \qquad\qquad = [J^{-1}]\frac{\partial N_2}{\partial \varepsilon}$$

where $[J^{-1}]$ is the inverse of the Jacobian matrix (for a matrix of size 1×1, $[J^{-1}] = 1/[J]$, see Appendix IV).

The limits of the integration in equation 4.31 also change during the coordinate transformation (see Figure 4.8)

$$x = x_i^{(e)} \rightarrow \varepsilon = -1$$

$$x = x_j^{(e)} \rightarrow \varepsilon = 1$$

We can now rewrite equation 4.31 as

$$\left[K^{(e)}\right]_{2\times 2} = \int_{x_i^{(e)}}^{x_j^{(e)}} \begin{bmatrix} \dfrac{\partial N_1^{(e)}}{\partial x} \\[2mm] \dfrac{\partial N_2^{(e)}}{\partial x} \end{bmatrix} \left[K_x^{(e)}\right] \begin{bmatrix} \dfrac{\partial N_1^{(e)}}{\partial x} & \dfrac{\partial N_2^{(e)}}{\partial x} \end{bmatrix} dx$$

$$= \int_{-1}^{1} \begin{bmatrix} \dfrac{\partial N_1^{(e)}}{\partial \varepsilon} \\[2mm] \dfrac{\partial N_2^{(e)}}{\partial \varepsilon} \end{bmatrix} [J^{-1}]^T \underset{1\times 1}{[K_x^{(e)}]} [J^{-1}] \begin{bmatrix} \dfrac{\partial N_1^{(e)}}{\partial \varepsilon} & \dfrac{\partial N_2^{(e)}}{\partial \varepsilon} \end{bmatrix} |J|\, d\varepsilon \tag{4.33}$$

where $|J|$ is the determinant of the Jacobian matrix.

Example
   Consider the one-dimensional element in Figure 4.8 and let $x_i^{(e)} = 1$ and $x_j^{(e)} = 2$.
Then

$$[J] \quad = \frac{\partial N_1^{(e)}}{\partial \varepsilon} x_1^{(e)} + \frac{\partial N_2^{(e)}}{\partial \varepsilon} x_2^{(e)}$$

$$= \left(-\frac{1}{2}\right)(1) + \left(\frac{1}{2}\right)(2) = \frac{1}{2} = |J| \quad \text{(see Appendix IV, Part 12a)}$$

$$[J^{-1}] = 2 \qquad\qquad\qquad\qquad \text{(see Appendix IV, Part 13a)}$$

$$\begin{bmatrix} K^{(e)} \end{bmatrix}_{2\times2} = \int_{-1}^{1} \begin{bmatrix} -\dfrac{1}{2} \\ \dfrac{1}{2} \end{bmatrix} [2] \begin{bmatrix} K_x^{(e)} \end{bmatrix} [2] \begin{bmatrix} -\dfrac{1}{2} & \dfrac{1}{2} \end{bmatrix} \left(\dfrac{1}{2}\right) d\varepsilon$$

$$= \frac{K_x^{(e)}}{2} \begin{bmatrix} 1 & -1 \\ -1 & 1 \end{bmatrix} \int_{-1}^{1} d\varepsilon = K_x^{(e)} \begin{bmatrix} 1 & -1 \\ -1 & 1 \end{bmatrix} \tag{4.34}$$

which is identical to the conductance matrix obtained for the same element in the global
coordinate system, equation 4.15a ($L^{(e)} = x_j^{(e)} - x_i^{(e)} = 2 - 1 = 1$)

   For a two–dimensional element [J] and $[J^{-1}]$ are square matrices of size 2×2. For a
two-dimensional element with n nodes, the entries of [J] are given by

$$[J] = \begin{bmatrix} J_{11} & J_{12} \\ J_{21} & J_{22} \end{bmatrix} \tag{4.35}$$

where

$$J_{11} = \frac{\partial N_1}{\partial \varepsilon} x_1 + \cdots + \frac{\partial N_n}{\partial \varepsilon} x_n \ .$$

$$J_{12} = \frac{\partial N_1}{\partial \varepsilon} y_1 + \cdots + \frac{\partial N_n}{\partial \varepsilon} y_n$$

$$J_{21} = \frac{\partial N_1}{\partial \eta} x_1 + \cdots + \frac{\partial N_n}{\partial \eta} x_n$$

$$J_{22} = \frac{\partial N_1}{\partial \eta} y_1 + \cdots + \frac{\partial N_n}{\partial \eta} y_n$$

or more compactly

$$[J]_{2\times2} = \underbrace{\begin{bmatrix} \dfrac{\partial N_1}{\partial \varepsilon} & \cdots & \dfrac{\partial N_n}{\partial \varepsilon} \\ \dfrac{\partial N_1}{\partial \eta} & \cdots & \dfrac{\partial N_n}{\partial \eta} \end{bmatrix}}_{2\times n} \underbrace{\begin{bmatrix} x_1 & y_1 \\ \vdots & \vdots \\ x_n & y_n \end{bmatrix}}_{n\times 2} \tag{4.36}$$

The value of $|J|$ and the entries of $[J^{-1}]$ can be determined using the equations in Appendix IV, parts 12b and 13b.

For a three–dimensional element $[J]$ and $[J^{-1}]$ are square matrices of size 3×3. For a three-dimensional element with n nodes, the entries of $[J]$ are given by

$$[J] = \begin{bmatrix} J_{11} & J_{12} & J_{13} \\ J_{21} & J_{22} & J_{23} \\ J_{31} & J_{32} & J_{33} \end{bmatrix} \qquad (4.37)$$

where

$$J_{11} = \frac{\partial N_1}{\partial \varepsilon} x_1 + \cdots + \frac{\partial N_n}{\partial \varepsilon} x_n$$

$$J_{12} = \frac{\partial N_1}{\partial \varepsilon} y_1 + \cdots + \frac{\partial N_n}{\partial \varepsilon} y_n$$

$$J_{13} = \frac{\partial N_1}{\partial \varepsilon} z_1 + \cdots + \frac{\partial N_n}{\partial \varepsilon} z_n$$

$$J_{21} = \frac{\partial N_1}{\partial \eta} x_1 + \cdots + \frac{\partial N_n}{\partial \eta} x_n$$

$$J_{22} = \frac{\partial N_1}{\partial \eta} y_1 + \cdots + \frac{\partial N_n}{\partial \eta} y_n$$

$$J_{23} = \frac{\partial N_1}{\partial \eta} z_1 + \cdots + \frac{\partial N_n}{\partial \eta} z_n$$

$$J_{31} = \frac{\partial N_1}{\partial \zeta} x_1 + \cdots + \frac{\partial N_n}{\partial \zeta} x_n$$

$$J_{32} = \frac{\partial N_1}{\partial \zeta} y_1 + \cdots + \frac{\partial N_n}{\partial \zeta} y_n$$

$$J_{33} = \frac{\partial N_1}{\partial \zeta} z_1 + \cdots + \frac{\partial N_n}{\partial \zeta} z_n$$

or more compactly

$$\underset{3\times 3}{[J]} = \begin{bmatrix} \dfrac{\partial N_1}{\partial \varepsilon} & \cdots & \dfrac{\partial N_n}{\partial \varepsilon} \\ \dfrac{\partial N_1}{\partial \eta} & \cdots & \dfrac{\partial N_n}{\partial \eta} \\ \dfrac{\partial N_1}{\partial \zeta} & \cdots & \dfrac{\partial N_n}{\partial \zeta} \end{bmatrix} \underset{n\times 3}{\begin{bmatrix} x_1 & y_1 & z_1 \\ \vdots & \vdots & \vdots \\ x_n & y_n & z_n \end{bmatrix}} \qquad (4.38)$$

$$\underset{3\times n}{}$$

The value of $|J|$ and the entries of $[J^{-1}]$ can be determined using the equations in Appendix IV parts 12c and 13c.

In general $[J]$, $[J^{-1}]$, and $|J|$ vary from point–to–point within the element (because the interpolation function derivatives are themselves functions of ε, ε and η, or ε, η, and ζ).

Example

Compute $[J]$, $[J^{-1}]$, and $|J|$ at the point $\varepsilon = 0$, $\eta = 1$ for the element shown below

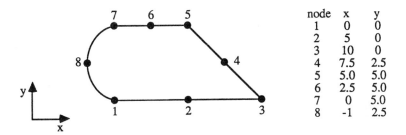

node	x	y
1	0	0
2	5	0
3	10	0
4	7.5	2.5
5	5.0	5.0
6	2.5	5.0
7	0	5.0
8	-1	2.5

The element is a quadratic quadrilateral. From Figure 4.11, the interpolation function derivatives at $\varepsilon = 0$, $\eta = 1$ are

$$\frac{\partial N_1}{\partial \varepsilon} = \frac{1}{4}(1 - 1)(2(0) + 1) = 0$$

$$\frac{\partial N_2}{\partial \varepsilon} = 0 \qquad \frac{\partial N_3}{\partial \varepsilon} = 0 \qquad \frac{\partial N_4}{\partial \varepsilon} = 0$$

$$\frac{\partial N_5}{\partial \varepsilon} = \frac{1}{2} \qquad \frac{\partial N_6}{\partial \varepsilon} = 0 \qquad \frac{\partial N_7}{\partial \varepsilon} = -\frac{1}{2}$$

$$\frac{\partial N_8}{\partial \varepsilon} = 0$$

$$\frac{\partial N_1}{\partial \eta} = \frac{1}{4}(1 - 0)(2(1) + 0) = \frac{1}{2}$$

$$\frac{\partial N_2}{\partial \eta} = -\frac{1}{2} \qquad \frac{\partial N_3}{\partial \eta} = \frac{1}{2} \qquad \frac{\partial N_4}{\partial \eta} = -1$$

$$\frac{\partial N_5}{\partial \eta} = \frac{1}{2} \qquad \frac{\partial N_6}{\partial \eta} = \frac{1}{2} \qquad \frac{\partial N_7}{\partial \eta} = \frac{1}{2}$$

$$\frac{\partial N_8}{\partial \eta} = -1$$

$$[J] = \begin{bmatrix} 0 & 0 & 0 & 0 & 1/2 & 0 & -1/2 & 0 \\ 1/2 & -1/2 & 1/2 & -1 & 1/2 & 1/2 & 1/2 & -1 \end{bmatrix} \begin{bmatrix} 0 & 0 \\ 5 & 0 \\ 10 & 0 \\ 7.5 & 2.5 \\ 5.0 & 5.0 \\ 2.5 & 5.0 \\ 0 & 5.0 \\ -1 & 2.5 \end{bmatrix}$$

$$= \begin{bmatrix} 2.50 & 0 \\ -0.25 & 2.50 \end{bmatrix}$$

$$|J| = (2.50)(2.50) - (0)(-0.25) = 6.25$$

$$[J^{-1}] = \frac{1}{6.25} \begin{bmatrix} 2.5 & 0 \\ 0.25 & 2.5 \end{bmatrix} = \begin{bmatrix} 0.40 & 0 \\ 0.04 & 0.40 \end{bmatrix}$$

The coordinate transformation can be used to rewrite the integrals in Chapter 3. These transformed integral equations can be used to compute the element matrices for each type of element shown in Figures 4.9 to 4.15.

Element Conductance Matrix (Saturated flow)

One dimensional elements:

$$\left[K^{(e)} \right]_{n \times n} = \int_{-1}^{1} \underbrace{\begin{bmatrix} \dfrac{\partial N_1^{(e)}}{\partial \varepsilon} \\ \vdots \\ \dfrac{\partial N_n^{(e)}}{\partial \varepsilon} \end{bmatrix}}_{n \times 1} \underbrace{[J^{-1}]^T}_{1 \times 1} \underbrace{\left[K_x^{(e)} \right]}_{1 \times 1} \underbrace{[J^{-1}]}_{1 \times 1} \underbrace{\begin{bmatrix} \dfrac{\partial N_1^{(e)}}{\partial \varepsilon} & \cdots & \dfrac{\partial N_n^{(e)}}{\partial \varepsilon} \end{bmatrix}}_{1 \times n} |J| \, d\varepsilon \qquad (4.39)$$

Two-dimensional elements:

$$\left[K^{(e)} \right]_{n \times n} = \int_{-1}^{1}\int_{-1}^{1} \underbrace{\begin{bmatrix} \dfrac{\partial N_1^{(e)}}{\partial \varepsilon} & \dfrac{\partial N_1^{(e)}}{\partial \eta} \\ \vdots & \vdots \\ \dfrac{\partial N_n^{(e)}}{\partial \varepsilon} & \dfrac{\partial N_n^{(e)}}{\partial \eta} \end{bmatrix}}_{n \times 2} [J^{-1}]^T \underbrace{\begin{bmatrix} K_x^{(e)} & 0 \\ 0 & K_y^{(e)} \end{bmatrix}}_{2 \times 2} [J^{-1}] \underbrace{\begin{bmatrix} \dfrac{\partial N_1^{(e)}}{\partial \varepsilon} & \cdots & \dfrac{\partial N_n^{(e)}}{\partial \varepsilon} \\ \dfrac{\partial N_1^{(e)}}{\partial \eta} & \cdots & \dfrac{\partial N_n^{(e)}}{\partial \eta} \end{bmatrix}}_{2 \times n} |J| \, d\varepsilon \, d\eta$$

$$\qquad\qquad (4.40)$$

Three-dimensional elements:

$$\left[K^{(e)}\right]_{n\times n} = \int_{-1}^{1}\int_{-1}^{1}\int_{-1}^{1} \begin{bmatrix} \dfrac{\partial N_1^{(e)}}{\partial\varepsilon} & \dfrac{\partial N_1^{(e)}}{\partial\eta} & \dfrac{\partial N_1^{(e)}}{\partial\zeta} \\ \vdots & \vdots & \vdots \\ \dfrac{\partial N_n^{(e)}}{\partial\varepsilon} & \dfrac{\partial N_n^{(e)}}{\partial\eta} & \dfrac{\partial N_n^{(e)}}{\partial\zeta} \end{bmatrix}_{n\times 3} [J^{-1}]^T \begin{bmatrix} K_x^{(e)} & 0 & 0 \\ 0 & K_y^{(e)} & 0 \\ 0 & 0 & K_z^{(e)} \end{bmatrix}_{3\times 3} [J^{-1}] \begin{bmatrix} \dfrac{\partial N_1^{(e)}}{\partial\varepsilon} & \cdots & \dfrac{\partial N_n^{(e)}}{\partial\varepsilon} \\ \dfrac{\partial N_1^{(e)}}{\partial\eta} & \cdots & \dfrac{\partial N_n^{(e)}}{\partial\eta} \\ \dfrac{\partial N_1^{(e)}}{\partial\zeta} & \cdots & \dfrac{\partial N_n^{(e)}}{\partial\zeta} \end{bmatrix}_{3\times n} |J|\, d\varepsilon\, d\eta\, d\zeta$$

$$\hspace{12cm}(4.41)$$

Element Conductance Matrix (Unsaturated flow)

One-dimensional elements:

$$\left[K^{(e)}(\psi)\right]_{n\times n} = \int_{-1}^{1} \begin{bmatrix} \dfrac{\partial N_1^{(e)}}{\partial\varepsilon} \\ \vdots \\ \dfrac{\partial N_n^{(e)}}{\partial\varepsilon} \end{bmatrix}_{n\times 1} [J^{-1}]_{1\times 1}^T \left[K_x^{(e)}(\psi)\right]_{1\times 1} [J^{-1}]_{1\times 1} \begin{bmatrix} \dfrac{\partial N_1^{(e)}}{\partial\varepsilon} & \cdots & \dfrac{\partial N_n^{(e)}}{\partial\varepsilon} \end{bmatrix}_{1\times n} |J|\, d\varepsilon \hspace{2cm} (4.42)$$

Two-dimensional elements:

$$\left[K^{(e)}(\psi)\right]_{n\times n} = \int_{-1}^{1}\int_{-1}^{1} \begin{bmatrix} \dfrac{\partial N_1^{(e)}}{\partial\varepsilon} & \dfrac{\partial N_1^{(e)}}{\partial\eta} \\ \vdots & \vdots \\ \dfrac{\partial N_n^{(e)}}{\partial\varepsilon} & \dfrac{\partial N_n^{(e)}}{\partial\eta} \end{bmatrix}_{n\times 2} [J^{-1}]_{2\times 2}^T \begin{bmatrix} K_x^{(e)}(\psi) & 0 \\ 0 & K_y^{(e)}(\psi) \end{bmatrix}_{2\times 2} [J^{-1}]_{2\times 2} \begin{bmatrix} \dfrac{\partial N_1^{(e)}}{\partial\varepsilon} & \cdots & \dfrac{\partial N_n^{(e)}}{\partial\varepsilon} \\ \dfrac{\partial N_1^{(e)}}{\partial\eta} & \cdots & \dfrac{\partial N_n^{(e)}}{\partial\eta} \end{bmatrix}_{2\times n} |J|\, d\varepsilon\, d\eta$$

$$\hspace{12cm}(4.43)$$

Three-dimensional elements:

$$\left[K^{(e)}(\psi)\right]_{n\times n} =$$

$$\int_{-1}^{1}\int_{-1}^{1}\int_{-1}^{1} \begin{bmatrix} \dfrac{\partial N_1^{(e)}}{\partial\varepsilon} & \dfrac{\partial N_1^{(e)}}{\partial\eta} & \dfrac{\partial N_1^{(e)}}{\partial\zeta} \\ \vdots & \vdots & \vdots \\ \dfrac{\partial N_n^{(e)}}{\partial\varepsilon} & \dfrac{\partial N_n^{(e)}}{\partial\eta} & \dfrac{\partial N_n^{(e)}}{\partial\zeta} \end{bmatrix}_{n\times 3} [J^{-1}]^T \begin{bmatrix} K_x^{(e)}(\psi) & 0 & 0 \\ 0 & K_y^{(e)}(\psi) & 0 \\ 0 & 0 & K_z^{(e)}(\psi) \end{bmatrix}_{3\times 3} [J^{-1}] \begin{bmatrix} \dfrac{\partial N_1^{(e)}}{\partial\varepsilon} & \cdots & \dfrac{\partial N_n^{(e)}}{\partial\varepsilon} \\ \dfrac{\partial N_1^{(e)}}{\partial\eta} & \cdots & \dfrac{\partial N_n^{(e)}}{\partial\eta} \\ \dfrac{\partial N_1^{(e)}}{\partial\zeta} & \cdots & \dfrac{\partial N_n^{(e)}}{\partial\zeta} \end{bmatrix}_{3\times n} |J|\, d\varepsilon\, d\eta\, d\zeta$$

$$\hspace{12cm}(4.44)$$

Element Capacitance Matrix (Saturated Flow, Consistent Formulation)

<u>One-dimensional elements:</u>

$$\left[\underset{n\times n}{C^{(e)}}\right] = \int_{-1}^{1} \underset{n\times 1}{\begin{bmatrix} N_1^{(e)} \\ \vdots \\ N_n^{(e)} \end{bmatrix}} \underset{1\times 1}{\left[S_s^{(e)}\right]} \underset{1\times n}{\left[N_1^{(e)} \cdots N_n^{(e)}\right]} |J| \, d\varepsilon \tag{4.45}$$

<u>Two-dimensional elements:</u>

$$\left[\underset{n\times n}{C^{(e)}}\right] = \int_{-1}^{1}\int_{-1}^{1} \underset{n\times 1}{\begin{bmatrix} N_1^{(e)} \\ \vdots \\ N_n^{(e)} \end{bmatrix}} \underset{1\times 1}{\left[S_s^{(e)}\right]} \underset{1\times n}{\left[N_1^{(e)} \cdots N_n^{(e)}\right]} |J| \, d\varepsilon \, d\eta \tag{4.46}$$

<u>Three-dimensional elements:</u>

$$\left[\underset{n\times n}{C^{(e)}}\right] = \int_{-1}^{1}\int_{-1}^{1}\int_{-1}^{1} \underset{n\times 1}{\begin{bmatrix} N_1^{(e)} \\ \vdots \\ N_n^{(e)} \end{bmatrix}} \cdot \underset{1\times 1}{\left[S_s^{(e)}\right]} \underset{1\times n}{\left[N_1^{(e)} \cdots N_n^{(e)}\right]} |J| \, d\varepsilon \, d\eta \, d\zeta \tag{4.47}$$

Element Capacitance Matrix (Saturated Flow, Lumped Formulation)

<u>One-, Two-, or Three-dimensional elements:</u>

$$\left[\underset{n\times n}{C^{(e)}}\right] = \frac{S_s^{(e)}V^{(e)}}{n} |J| \underset{n\times n}{\begin{bmatrix} 1 & & O \\ & \ddots & \\ O & & 1 \end{bmatrix}} \tag{4.48}$$

Element Capacitance Matrix (Unsaturated Flow, Consistent Formulation)

One-dimensional elements:

$$
\underset{n\times n}{[C^{(e)}(\psi)]} = \int_{-1}^{1} \underset{n\times 1}{\begin{bmatrix} N_1^{(e)} \\ \vdots \\ N_n^{(e)} \end{bmatrix}} \underset{1\times 1}{[C^{(e)}(\psi)]} \underset{1\times n}{\left[N_1^{(e)} \ \cdots \ N_n^{(e)} \right]} |J| \, d\varepsilon
\tag{4.49}
$$

Two-dimensional elements:

$$
\underset{n\times n}{[C^{(e)}(\psi)]} = \int_{-1}^{1}\int_{-1}^{1} \underset{n\times 1}{\begin{bmatrix} N_1^{(e)} \\ \vdots \\ N_n^{(e)} \end{bmatrix}} \underset{1\times 1}{[C^{(e)}(\psi)]} \underset{1\times n}{\left[N_1^{(e)} \ \cdots \ N_n^{(e)} \right]} |J| \, d\varepsilon \, d\eta
\tag{4.50}
$$

Three-dimensional elements:

$$
\underset{n\times n}{[C^{(e)}(\psi)]} = \int_{-1}^{1}\int_{-1}^{1}\int_{-1}^{1} \underset{n\times 1}{\begin{bmatrix} N_1^{(e)} \\ \vdots \\ N_n^{(e)} \end{bmatrix}} \underset{1\times 1}{[C^{(e)}(\psi)]} \underset{1\times n}{\left[N_1^{(e)} \ \cdots \ N_n^{(e)} \right]} |J| \, d\varepsilon \, d\eta \, d\zeta
\tag{4.51}
$$

Element Advection-Dispersion Matrix

One-dimensional elements:

$$
\underset{n\times n}{\left[D^{(e)} \right]} = \int_{-1}^{1} \underset{n\times 1}{\begin{bmatrix} \dfrac{\partial N_1^{(e)}}{\partial\varepsilon} \\ \vdots \\ \dfrac{\partial N_n^{(e)}}{\partial\varepsilon} \end{bmatrix}} \underset{1\times 1}{[J^{-1}]^T} \underset{1\times 1}{\left[D_x^{(e)} \theta^{(e)} \right]} \underset{1\times 1}{[J^{-1}]} \underset{1\times n}{\left[\dfrac{\partial N_1^{(e)}}{\partial\varepsilon} \ \cdots \ \dfrac{\partial N_n^{(e)}}{\partial\varepsilon} \right]} |J| \, d\varepsilon
$$

$$
+ \int_{-1}^{1} \underset{n\times 1}{\begin{bmatrix} N_1^{(e)} \\ \vdots \\ N_n^{(e)} \end{bmatrix}} \underset{1\times 1}{[v_x^{(e)}]} \underset{1\times 1}{[J^{-1}]} \underset{1\times n}{\left[\dfrac{\partial N_1^{(e)}}{\partial\varepsilon} \ \cdots \ \dfrac{\partial N_n^{(e)}}{\partial\varepsilon} \right]} |J| \, d\varepsilon
$$

$$+ \int_{-1}^{1} \begin{bmatrix} N_1^{(e)} \\ \vdots \\ N_n^{(e)} \end{bmatrix}_{n \times 1} [\lambda(\theta^{(e)} + \rho_b^{(e)} K_d^{(e)})]_{1 \times 1} \begin{bmatrix} N_1^{(e)} & \cdots & N_n^{(e)} \end{bmatrix}_{1 \times n} |J| \, d\varepsilon \qquad (4.52)$$

Two-dimensional elements:

$$\begin{bmatrix} D^{(e)} \end{bmatrix}_{n \times n} = \int_{-1}^{1} \int_{-1}^{1} \begin{bmatrix} \dfrac{\partial N_1^{(e)}}{\partial \varepsilon} & \dfrac{\partial N_1^{(e)}}{\partial \eta} \\ \vdots & \vdots \\ \dfrac{\partial N_n^{(e)}}{\partial \varepsilon} & \dfrac{\partial N_n^{(e)}}{\partial \eta} \end{bmatrix}_{n \times 2} [J^{-1}]^T_{2 \times 2} \begin{bmatrix} D_{xx}^{(e)} \theta^{(e)} & D_{xy}^{(e)} \theta^{(e)} \\ D_{yx}^{(e)} \theta^{(e)} & D_{yy}^{(e)} \theta^{(e)} \end{bmatrix}_{2 \times 2} [J^{-1}]_{2 \times 2} \begin{bmatrix} \dfrac{\partial N_1^{(e)}}{\partial \varepsilon} & \cdots & \dfrac{\partial N_n^{(e)}}{\partial \varepsilon} \\ \dfrac{\partial N_1^{(e)}}{\partial \eta} & \cdots & \dfrac{\partial N_n^{(e)}}{\partial \eta} \end{bmatrix}_{2 \times n} |J| \, d\varepsilon \, d\eta$$

$$+ \int_{-1}^{1} \int_{-1}^{1} \begin{bmatrix} N_1^{(e)} & N_1^{(e)} \\ \vdots & \vdots \\ N_n^{(e)} & N_n^{(e)} \end{bmatrix}_{n \times 2} \begin{bmatrix} v_x^{(e)} & 0 \\ 0 & v_y^{(e)} \end{bmatrix}_{2 \times 2} [J^{-1}]_{2 \times 2} \begin{bmatrix} \dfrac{\partial N_1^{(e)}}{\partial \varepsilon} & \cdots & \dfrac{\partial N_n^{(e)}}{\partial \varepsilon} \\ \dfrac{\partial N_1^{(e)}}{\partial \eta} & \cdots & \dfrac{\partial N_n^{(e)}}{\partial \eta} \end{bmatrix}_{2 \times n} |J| \, d\varepsilon \, d\eta$$

$$+ \int_{-1}^{1} \int_{-1}^{1} \begin{bmatrix} N_1^{(e)} \\ \vdots \\ N_n^{(e)} \end{bmatrix}_{n \times 1} [\lambda (\theta^{(e)} + \rho_b^{(e)} K_d^{(e)})]_{1 \times 1} [N_1^{(e)} \cdots N_n^{(e)}]_{1 \times n} |J| \, d\varepsilon \, d\eta \qquad (4.53)$$

Three-dimensional elements:

$$\begin{bmatrix} D^{(e)} \end{bmatrix}_{n \times n} =$$

$$\int_{-1}^{1} \int_{-1}^{1} \int_{-1}^{1} \begin{bmatrix} \dfrac{\partial N_1^{(e)}}{\partial \varepsilon} & \dfrac{\partial N_1^{(e)}}{\partial \eta} & \dfrac{\partial N_1^{(e)}}{\partial \zeta} \\ \vdots & \vdots & \vdots \\ \dfrac{\partial N_n^{(e)}}{\partial \varepsilon} & \dfrac{\partial N_n^{(e)}}{\partial \eta} & \dfrac{\partial N_n^{(e)}}{\partial \zeta} \end{bmatrix}_{n \times 3} [J^{-1}]^T_{3 \times 3} \begin{bmatrix} D_{xx}^{(e)} \theta^{(e)} & D_{xy}^{(e)} \theta^{(e)} & D_{xz}^{(e)} \theta^{(e)} \\ D_{yx}^{(e)} \theta^{(e)} & D_{yy}^{(e)} \theta^{(e)} & D_{yz}^{(e)} \theta^{(e)} \\ D_{zx}^{(e)} \theta^{(e)} & D_{zy}^{(e)} \theta^{(e)} & D_{zz}^{(e)} \theta^{(e)} \end{bmatrix}_{3 \times 3} [J^{-1}]_{3 \times 3} \begin{bmatrix} \dfrac{\partial N_1^{(e)}}{\partial \varepsilon} & \cdots & \dfrac{\partial N_n^{(e)}}{\partial \varepsilon} \\ \dfrac{\partial N_1^{(e)}}{\partial \eta} & \cdots & \dfrac{\partial N_n^{(e)}}{\partial \eta} \\ \dfrac{\partial N_1^{(e)}}{\partial \zeta} & \cdots & \dfrac{\partial N_n^{(e)}}{\partial \zeta} \end{bmatrix}_{3 \times n} |J| \, d\varepsilon \, d\eta \, d\zeta$$

$$
+ \int_{-1}^{1}\int_{-1}^{1}\int_{-1}^{1} \underbrace{\begin{bmatrix} N_1^{(e)} & N_1^{(e)} & N_1^{(e)} \\ \vdots & \vdots & \vdots \\ N_n^{(e)} & N_n^{(e)} & N_n^{(e)} \end{bmatrix}}_{n\times 3} \underbrace{\begin{bmatrix} v_x^{(e)} & 0 & 0 \\ 0 & v_y^{(e)} & 0 \\ 0 & 0 & v_z^{(e)} \end{bmatrix}}_{3\times 3} \underbrace{[J^{-1}]}_{3\times 3} \underbrace{\begin{bmatrix} \dfrac{\partial N_1^{(e)}}{\partial \varepsilon} & \cdots & \dfrac{\partial N_n^{(e)}}{\partial \varepsilon} \\ \dfrac{\partial N_1^{(e)}}{\partial \eta} & \cdots & \dfrac{\partial N_n^{(e)}}{\partial \eta} \\ \dfrac{\partial N_1^{(e)}}{\partial \zeta} & \cdots & \dfrac{\partial N_n^{(e)}}{\partial \zeta} \end{bmatrix}}_{3\times n} |J|\, d\varepsilon\, d\eta\, d\zeta
$$

$$
+ \int_{-1}^{1}\int_{-1}^{1}\int_{-1}^{1} \underbrace{\begin{bmatrix} N_1^{(e)} \\ \vdots \\ N_n^{(e)} \end{bmatrix}}_{n\times 1} \underbrace{[\lambda\,(\theta^{(e)} + \rho_b^{(e)}K_d^{(e)})]}_{1\times 1} \underbrace{[N_1^{(e)} \cdots N_n^{(e)}]}_{1\times n} |J|\, d\varepsilon\, d\eta\, d\zeta
$$

$$(4.54)$$

Element Sorption Matrix

One-dimensional elements:

$$
\underbrace{\left[A^{(e)}\right]}_{n\times n} = \int_{-1}^{1} \underbrace{\begin{bmatrix} N_1^{(e)} \\ \vdots \\ N_n^{(e)} \end{bmatrix}}_{n\times 1} \underbrace{[\rho_b^{(e)}K_d^{(e)} + \theta^{(e)}]}_{1\times 1} \underbrace{\left[N_1^{(e)} \cdots N_n^{(e)}\right]}_{1\times n} |J|\, d\varepsilon
$$

$$(4.55)$$

Two-dimensional elements:

$$
\underbrace{\left[A^{(e)}\right]}_{n\times n} = \int_{-1}^{1}\int_{-1}^{1} \underbrace{\begin{bmatrix} N_1^{(e)} \\ \vdots \\ N_n^{(e)} \end{bmatrix}}_{n\times 1} \underbrace{[\rho_b^{(e)}K_d^{(e)} + \theta^{(e)}]}_{1\times 1} \underbrace{[N_1^{(e)} \cdots N_n^{(e)}]}_{1\times n} |J|\, d\varepsilon\, d\eta
$$

$$(4.56)$$

Three-dimensional elements:

$$
\underbrace{\left[A^{(e)}\right]}_{n\times n} = \int_{-1}^{1}\int_{-1}^{1}\int_{-1}^{1} \underbrace{\begin{bmatrix} N_1^{(e)} \\ \vdots \\ N_n^{(e)} \end{bmatrix}}_{n\times 1} \underbrace{[\rho_b^{(e)}K_d^{(e)} + \theta^{(e)}]}_{1\times 1} \underbrace{[N_1^{(e)} \cdots N_n^{(e)}]}_{1\times n} |J|\, d\varepsilon\, d\eta\, d\zeta
$$

$$(4.57)$$

When the number of nodes for an element is greater than two, or when the order of the polynomial for the interpolation functions is greater than one, or for two-dimensional or three-dimensional elements, it is convenient to use numerical methods to evaluate the integrals in equations 4.39 to 4.57. Specifically the method of *Gauss quadrature* can be used. In this method, we obtain a numerical approximation to the integral of a function over an interval by computing the weighted sum of values of the function for specific points on the interval. In one-dimension, the equation for Gauss quadrature is

$$\int_{-1}^{1} f(\varepsilon)\, d\varepsilon = \sum_{i=1}^{N_\varepsilon} W_i(\varepsilon_i)\, f(\varepsilon_i), \qquad 0 \le W_i(\varepsilon_i) \le 1 \tag{4.58}$$

where $W_i(\varepsilon_i)$ is the weight assigned to the value of the function f at the *Gauss point* $\varepsilon = \varepsilon_i$, $-1 \le \varepsilon_i \le 1$, and N_ε is the number of Gauss points on the interval. In two-dimensions, the equation for Gauss quadrature is

$$\int_{-1}^{1}\int_{-1}^{1} f(\varepsilon,\eta)\, d\varepsilon\, d\eta = \sum_{i=1}^{N_\varepsilon}\sum_{j=1}^{N_\eta} W_i(\varepsilon_i)\, W_j(\eta_j)\, f(\varepsilon_i,\eta_j)$$
$$0 \le W_i(\varepsilon_i) \le 1$$
$$0 \le W_j(\eta_j) \le 1 \tag{4.59}$$

where $W_i(\varepsilon_i)$ and $W_i(\eta_j)$ are the weights assigned to the value of the function f at the Gauss point $(\varepsilon = \varepsilon_i, \eta = \eta_i)$ and N_ε and N_η are the number of Gauss points on the intervals $-1 \le \varepsilon_i \le 1$ and $-1 \le \eta_i \le 1$.

In three dimensions, the equation for Gauss quadrature is

$$\int_{-1}^{1}\int_{-1}^{1}\int_{-1}^{1} f(\varepsilon,\eta,\zeta)\, d\varepsilon\, d\eta\, d\zeta = \sum_{i=1}^{N_\varepsilon}\sum_{j=1}^{N_\eta}\sum_{k=1}^{N_\zeta} W_i(\varepsilon_i)\, W_j(\eta_j)\, W_k(\zeta_k)\, f(\varepsilon_i,\eta_j,\zeta_k)$$
$$0 \le W_i(\varepsilon_i) \le 1$$
$$0 \le W_j(\eta_j) \le 1$$
$$0 \le W_k(\zeta_k) \le 1 \tag{4.60}$$

where $W_i(\varepsilon_i)$, $W_j(\eta_j)$, and $W_k(\zeta_k)$ are the weights assigned to the value of function f at the Gauss point $(\varepsilon = \varepsilon_i, \eta = \eta_i, \zeta = \zeta_i)$ and N_ε, N_η and N_ζ are the number of Gauss points on the intervals $-1 \le \varepsilon_i \le 1$ and $-1 \le \eta_i \le 1$, and $-1 \le \zeta_i \le 1$.

The number and location of the Gauss points and the values of the weights are selected to achieve the greatest accuracy. If the function f is a polynomial, Gauss quadrature can provide an <u>exact</u> integration. A total of (n+1)/2 Gauss points are required to obtain an exact integration for a polynomial function of order n. If the quantity (n + 1)/2 is not a whole number it is rounded up to the next largest integer. For example, if n = 2, (n + 1)/2 = 3/2 = 1.5 which is rounded up to 2 and 2 Gauss points are required. The number and locations of the Gauss points and the values of the weights, for polynomial functions of orders 0 to 9 are in Table 4.1.

Table 4.1 Locations of Gauss points and values of weights for exact integration of a polynomial function by Gauss quadrature (after Dhatt and Touzot, 1984).

Order of polynomial	Number of Gauss points	Locations(s) of Gauss Points	Weight(s)
0 or 1	1	0	2
2 or 3	2	$1/\sqrt{3}$	1
		$-1/\sqrt{3}$	1
4 or 5	3	0	8/9
		$\sqrt{3/5}$	5/9
		$-\sqrt{3/5}$	5/9
6 or 7	4	$\sqrt{\dfrac{3-2\sqrt{6/5}}{7}}$	$\dfrac{1}{2}+\dfrac{1}{6\sqrt{6/5}}$
		$-\sqrt{\dfrac{3-2\sqrt{6/5}}{7}}$	$\dfrac{1}{2}+\dfrac{1}{6\sqrt{6/5}}$
		$\sqrt{\dfrac{3+2\sqrt{6/5}}{7}}$	$\dfrac{1}{2}-\dfrac{1}{6\sqrt{6/5}}$
		$-\sqrt{\dfrac{3+2\sqrt{6/5}}{7}}$	$\dfrac{1}{2}-\dfrac{1}{6\sqrt{6/5}}$
8 or 9	5	0	$\dfrac{128}{225}$
		$\dfrac{1}{3}\sqrt{5-4\sqrt{5/14}}$	$\dfrac{161}{450}+\dfrac{13}{180\sqrt{5/14}}$
		$-\dfrac{1}{3}\sqrt{5-4\sqrt{5/14}}$	$\dfrac{161}{450}+\dfrac{13}{180\sqrt{5/14}}$
		$\dfrac{1}{3}\sqrt{5+4\sqrt{5/14}}$	$\dfrac{161}{450}-\dfrac{13}{180\sqrt{5/14}}$
		$-\dfrac{1}{3}\sqrt{5+4\sqrt{5/14}}$	$\dfrac{161}{450}-\dfrac{13}{180\sqrt{5/14}}$

For example, consider the integral

$$\int_{-1}^{1} (\varepsilon + 3)\, d\varepsilon$$

The analytical solution is

$$\int_{-1}^{1} (\varepsilon + 3)\, d\varepsilon = \left. \frac{\varepsilon^2}{2} + 3\varepsilon \right|_{-1}^{1} = \left(\frac{1}{2} + 3 \right) - \left(\frac{1}{2} - 3 \right) = 6$$

The highest-order polynomial in the function to be integrated is 1 ($n = 1$) and the number of Gauss points required is also 1 (Table 4.1). Using equation 4.58, the numerical solution is

$$\int_{-1}^{1} (\varepsilon + 3)\, d\varepsilon = \sum_{i=1}^{1} W_i(\varepsilon_i)\, f(\varepsilon_i) = W_i(\varepsilon = \varepsilon_i)\, f(\varepsilon_i)$$

$$= W(\varepsilon = 0)\, f(\varepsilon = 0) = 2 \cdot (0 + 3) = 6$$

which is the same as the analytical solution.

For another example, consider the integral

$$\int_{-1}^{1} (\varepsilon^2 + \varepsilon)\, d\varepsilon$$

The analytical solution is

$$\int_{-1}^{1} (\varepsilon^2 + \varepsilon)\, d\varepsilon = \left. \frac{\varepsilon^3}{3} + \frac{\varepsilon^2}{2} \right|_{-1}^{1} = \left(\frac{1}{3} + \frac{1}{2} \right) - \left(-\frac{1}{3} + \frac{1}{2} \right) = \frac{2}{3}$$

The highest-order polynomial in the function to be integrated is 2 ($n = 2$) and the number of Gauss points required is also 2 (Table 4.1). Using equation 4.58, the numerical solution is

$$\int_{-1}^{1} (\varepsilon^2 + \varepsilon)\, d\varepsilon = \sum_{i=1}^{1} W_i(\varepsilon_i)\, f(\varepsilon_i)$$

$$= W\left(\varepsilon = \frac{1}{\sqrt{3}} \right) f\left(\varepsilon = \frac{1}{\sqrt{3}} \right) + W\left(\varepsilon = \frac{-1}{\sqrt{3}} \right) f\left(\varepsilon = \frac{-1}{\sqrt{3}} \right)$$

$$= 1 \cdot \left[\left(\frac{1}{\sqrt{3}} \right)^2 + \frac{1}{\sqrt{3}} \right] + 1 \cdot \left[\left(-\frac{1}{\sqrt{3}} \right)^2 - \frac{1}{\sqrt{3}} \right]$$

$$= \frac{2}{3}$$

which is the same as the analytical solution.

Example

Evaluate the integral

$$\int_{-1}^{1}\int_{-1}^{1} \eta^4\varepsilon^2 + \eta^3 \, d\varepsilon \, d\eta$$

The highest order polynomial term is quartic (η^4) and from Table 4.1 we find that three Gauss points are required in each direction for a total of nine points. They are (ε,η):

(0, 0)	($\sqrt{3/5}$, 0)	($-\sqrt{3/5}$, 0)
(0, $\sqrt{3/5}$)	($\sqrt{3/5}$, $\sqrt{3/5}$)	($-\sqrt{3/5}$, $\sqrt{3/5}$)
(0, $-\sqrt{3/5}$)	($\sqrt{3/5}$, $-\sqrt{3/5}$)	($-\sqrt{3/5}$, $-\sqrt{3/5}$)

The value of the integral can be computed using equation 4.59

$$\sum_{i=1}^{3}\sum_{j=1}^{3} W_i(\varepsilon_i) \, W_j(\eta_j)(\eta_j^4\varepsilon_i^2 + \eta_j^3)$$

$$= \left(\frac{8}{9}\right)\left(\frac{8}{9}\right)[(0)^4(0)^2 + (0)^3] + \left(\frac{8}{9}\right)\left(\frac{5}{9}\right)[(\sqrt{3/5})^4(0)^2 + (\sqrt{3/5})^3]$$

$$+ \left(\frac{8}{9}\right)\left(\frac{5}{9}\right)[(-\sqrt{3/5})^4(0)^2 + (-\sqrt{3/5})^3]$$

$$+ \left(\frac{5}{9}\right)\left(\frac{8}{9}\right)[(0)^4(\sqrt{3/5})^2 + (0)^3] + \left(\frac{5}{9}\right)\left(\frac{5}{9}\right)[(\sqrt{3/5})^4(\sqrt{3/5})^2 + (\sqrt{3/5})^3]$$

$$+ \left(\frac{5}{9}\right)\left(\frac{5}{9}\right)[(-\sqrt{3/5})^4(\sqrt{3/5})^2 + (-\sqrt{3/5})^3]$$

$$+ \left(\frac{5}{9}\right)\left(\frac{8}{9}\right)[(0)^4(-\sqrt{3/5})^2 + (0)^3] + \left(\frac{5}{9}\right)\left(\frac{5}{9}\right)[(\sqrt{3/5})^4(-\sqrt{3/5})^2 + (\sqrt{3/5})^3]$$

$$+ \left(\frac{5}{9}\right)\left(\frac{5}{9}\right)[(-\sqrt{3/5})^4(-\sqrt{3/5})^2 + (-\sqrt{3/5})^3]$$

$$= \quad 0 + 0.230 - 0.230 + 0 + 0.210 - 0.077 + 0 + 0.210 - 0.077$$
$$= \quad \underline{0.266}$$

Example

Evaluate the integral

$$\int_{-1}^{1}\int_{-1}^{1}\int_{-1}^{1} \varepsilon\eta^2\zeta^3 + \varepsilon^2\eta\zeta^2 \, d\varepsilon \, d\eta \, d\zeta$$

The highest order polynomial term is cubic (ζ^3) and from Table 4.1 we find that two Gauss points are required in each direction for a total of eight points. They are (ε,η,ξ):

$$\left(\frac{1}{\sqrt{3}}, \frac{1}{\sqrt{3}}, \frac{1}{\sqrt{3}}\right) \qquad \left(-\frac{1}{\sqrt{3}}, \frac{1}{\sqrt{3}}, \frac{1}{\sqrt{3}}\right)$$

$$\left(\frac{1}{\sqrt{3}}, \frac{1}{\sqrt{3}}, -\frac{1}{\sqrt{3}}\right) \qquad \left(-\frac{1}{\sqrt{3}}, -\frac{1}{\sqrt{3}}, \frac{1}{\sqrt{3}}\right)$$

$$\left(\frac{1}{\sqrt{3}}, -\frac{1}{\sqrt{3}}, \frac{1}{\sqrt{3}}\right) \qquad \left(-\frac{1}{\sqrt{3}}, \frac{1}{\sqrt{3}}, -\frac{1}{\sqrt{3}}\right)$$

$$\left(\frac{1}{\sqrt{3}}, -\frac{1}{\sqrt{3}}, -\frac{1}{\sqrt{3}}\right) \qquad \left(-\frac{1}{\sqrt{3}}, -\frac{1}{\sqrt{3}}, -\frac{1}{\sqrt{3}}\right)$$

The value of the integral can be computed using equation 4.60

$$\sum_{i=1}^{2}\sum_{j=1}^{2}\sum_{k=1}^{2} W_i(\varepsilon_i)\, W_j(\eta_j)\, W_k(\eta_k)\left[\varepsilon_i\eta_j^2\zeta_k^3 + \varepsilon_i^2\eta_j\zeta_k^2\right]$$

(Note: Values of weights equal 1 at all Gauss points)

$$= \left[\left(\frac{1}{\sqrt{3}}\right)\left(\frac{1}{\sqrt{3}}\right)^2\left(\frac{1}{\sqrt{3}}\right)^3 + \left(\frac{1}{\sqrt{3}}\right)^2\left(\frac{1}{\sqrt{3}}\right)\left(\frac{1}{\sqrt{3}}\right)^2\right]$$

$$+\left[\left(\frac{1}{\sqrt{3}}\right)\left(\frac{1}{\sqrt{3}}\right)^2\left(-\frac{1}{\sqrt{3}}\right)^3 + \left(\frac{1}{\sqrt{3}}\right)^2\left(\frac{1}{\sqrt{3}}\right)\left(-\frac{1}{\sqrt{3}}\right)^2\right]$$

$$+\left[\left(\frac{1}{\sqrt{3}}\right)\left(-\frac{1}{\sqrt{3}}\right)^2\left(\frac{1}{\sqrt{3}}\right)^3 + \left(\frac{1}{\sqrt{3}}\right)^2\left(-\frac{1}{\sqrt{3}}\right)\left(\frac{1}{\sqrt{3}}\right)^2\right]$$

$$+\left[\left(\frac{1}{\sqrt{3}}\right)\left(-\frac{1}{\sqrt{3}}\right)^2\left(-\frac{1}{\sqrt{3}}\right)^3 + \left(\frac{1}{\sqrt{3}}\right)^2\left(-\frac{1}{\sqrt{3}}\right)\left(-\frac{1}{\sqrt{3}}\right)^2\right]$$

$$+\left[\left(-\frac{1}{\sqrt{3}}\right)\left(\frac{1}{\sqrt{3}}\right)^2\left(\frac{1}{\sqrt{3}}\right)^3 + \left(-\frac{1}{\sqrt{3}}\right)^2\left(\frac{1}{\sqrt{3}}\right)\left(\frac{1}{\sqrt{3}}\right)^2\right]$$

$$+\left[\left(-\frac{1}{\sqrt{3}}\right)\left(-\frac{1}{\sqrt{3}}\right)^2\left(\frac{1}{\sqrt{3}}\right)^3 + \left(-\frac{1}{\sqrt{3}}\right)^2\left(-\frac{1}{\sqrt{3}}\right)\left(\frac{1}{\sqrt{3}}\right)^2\right]$$

$$+\left[\left(-\frac{1}{\sqrt{3}}\right)\left(\frac{1}{\sqrt{3}}\right)^2\left(-\frac{1}{\sqrt{3}}\right)^3 + \left(-\frac{1}{\sqrt{3}}\right)^2\left(\frac{1}{\sqrt{3}}\right)\left(-\frac{1}{\sqrt{3}}\right)^2\right]$$

$$+\left[\left(-\frac{1}{\sqrt{3}}\right)\left(-\frac{1}{\sqrt{3}}\right)^2\left(-\frac{1}{\sqrt{3}}\right)^3 + \left(-\frac{1}{\sqrt{3}}\right)^2\left(-\frac{1}{\sqrt{3}}\right)\left(-\frac{1}{\sqrt{3}}\right)^2\right]$$

$$= \quad 0.1012 + 0.0272 - 0.0272 - 0.1012 + 0.0272 - 0.1012 + 0.1012 - 0.0272$$
$$= \quad \underline{0}$$

The application of Gauss quadrature to the computation of the element matrices leads to the following equations.

Element Conductance Matrix (Saturated Flow)

One-dimensional elements:

$$
\left[K^{(e)} \right]_{n \times n} = \sum_{i=1}^{N_\varepsilon} W_i(\varepsilon_i) \left\{ \begin{bmatrix} \dfrac{\partial N_1^{(e)}(\varepsilon_i)}{\partial \varepsilon} \\ \vdots \\ \dfrac{\partial N_n^{(e)}(\varepsilon_i)}{\partial \varepsilon} \end{bmatrix}_{n \times 1} \cdot [J(\varepsilon_i)^{-1}]^T \left[K_x^{(e)} \right]_{1 \times 1} [J(\varepsilon_i)^{-1}] \begin{bmatrix} \dfrac{\partial N_1^{(e)}(\varepsilon_i)}{\partial \varepsilon} & \cdots & \dfrac{\partial N_n^{(e)}(\varepsilon_i)}{\partial \varepsilon} \end{bmatrix}_{1 \times n} |J(\varepsilon_i)| \right\}
$$

$$\text{(4.61)}$$

Two-dimensional elements:

$$
\left[K^{(e)} \right]_{n \times n} = \sum_{i=1}^{N_\varepsilon} \sum_{j=1}^{N_\eta} W_i(\varepsilon_i) W_j(\eta_j) \left\{ \begin{bmatrix} \dfrac{\partial N_1^{(e)}(\varepsilon_i,\eta_i)}{\partial \varepsilon} & \dfrac{\partial N_1^{(e)}(\varepsilon_i,\eta_i)}{\partial \eta} \\ \vdots & \vdots \\ \dfrac{\partial N_n^{(e)}(\varepsilon_i,\eta_i)}{\partial \varepsilon} & \dfrac{\partial N_n^{(e)}(\varepsilon_i,\eta_i)}{\partial \eta} \end{bmatrix}_{n \times 2} [J(\varepsilon_i,\eta_i)^{-1}]^T \begin{bmatrix} K_x^{(e)} & 0 \\ 0 & K_y^{(e)} \end{bmatrix}_{2 \times 2} \right.
$$

$$
\left. [J(\varepsilon_i,\eta_i)]^{-1}_{2 \times 2} \begin{bmatrix} \dfrac{\partial N_1^{(e)}(\varepsilon_i,\eta_i)}{\partial \varepsilon} & \cdots & \dfrac{\partial N_n^{(e)}(\varepsilon_i,\eta_i)}{\partial \varepsilon} \\ \dfrac{\partial N_1^{(e)}(\varepsilon_i,\eta_i)}{\partial \eta} & \cdots & \dfrac{\partial N_n^{(e)}(\varepsilon_i,\eta_i)}{\partial \eta} \end{bmatrix}_{2 \times n} |J(\varepsilon_i,\eta_i)| \right\} \quad \text{(4.62)}
$$

Three-dimensional elements:

$$
\left[K^{(e)} \right]_{n \times n} = \sum_{i=1}^{N_\varepsilon} \sum_{j=1}^{N_\eta} \sum_{k=1}^{N_\zeta} W_i(\varepsilon_i) W_j(\eta_j) W_k(\zeta_k) \left\{ \begin{bmatrix} \dfrac{\partial N_1^{(e)}(\varepsilon_i,\eta_j,h_\zeta)}{\partial \varepsilon} & \dfrac{\partial N_1^{(e)}(\varepsilon_i,\eta_j,h_\zeta)}{\partial \eta} & \dfrac{\partial N_1^{(e)}(\varepsilon_i,\eta_j,h_\zeta)}{\partial \zeta} \\ \vdots & \vdots & \vdots \\ \dfrac{\partial N_n^{(e)}(\varepsilon_i,\eta_j,h_\zeta)}{\partial \varepsilon} & \dfrac{\partial N_n^{(e)}(\varepsilon_i,\eta_j,h_\zeta)}{\partial \eta} & \dfrac{\partial N_n^{(e)}(\varepsilon_i,\eta_j,h_\zeta)}{\partial \zeta} \end{bmatrix}_{n \times 3} \right.
$$

$$[J(\varepsilon_i,\eta_j,\zeta_k)^{-1}]^T \underset{3\times3}{\begin{bmatrix} K_x^{(e)} & 0 & 0 \\ 0 & K_y^{(e)} & 0 \\ 0 & 0 & K_z^{(e)} \end{bmatrix}} \underset{3\times3}{[J(\varepsilon_i,\eta_j,\zeta_k)^{-1}]} \left. \begin{bmatrix} \dfrac{\partial N_1^{(e)}(\varepsilon_i,\eta_j,\zeta_k)}{\partial\varepsilon} & \cdots & \dfrac{\partial N_n^{(e)}(\varepsilon_i,\eta_j,\zeta_k)}{\partial\varepsilon} \\[2mm] \dfrac{\partial N_1^{(e)}(\varepsilon_i,\eta_j,\zeta_k)}{\partial\eta} & \cdots & \dfrac{\partial N_n^{(e)}(\varepsilon_i,\eta_j,\zeta_k)}{\partial\eta} \\[2mm] \dfrac{\partial N_1^{(e)}(\varepsilon_i,\eta_j,\zeta_k)}{\partial\zeta} & \cdots & \dfrac{\partial N_n^{(e)}(\varepsilon_i,\eta_j,\zeta_k)}{\partial\zeta} \end{bmatrix} \left| J(\varepsilon_i,\eta_j,\zeta_k)\right| \right\}$$

$$\underset{3\times n}{} \tag{4.63}$$

Element Conductance Matrix (Unsaturated Flow)

These equations can be obtained by substituting $K_x^{(e)}(\psi)$, $K_y^{(e)}(\psi)$, and $K_z^{(e)}(\psi)$ for $K_x^{(e)}$, $K_y^{(e)}$, and $K_z^{(e)}$ in equations 4.61, 4.62, and 4.63.

Element Capacitance Matrix (Saturated Flow, Consistent Formulation)

<u>One-dimensional elements:</u>

$$\underset{n\times n}{[C^{(e)}]} = \sum_{i=1}^{N_\varepsilon} W_i(\varepsilon_i) \left\{ \underset{n\times1}{\begin{bmatrix} N_1^{(e)}(\varepsilon_i) \\ \vdots \\ N_n^{(e)}(\varepsilon_i) \end{bmatrix}} \underset{1\times1}{[S_s^{(e)}]} \underset{1\times n}{[N_1^{(e)}(\varepsilon_i) \cdots N_n^{(e)}(\varepsilon_i)]} \left| J(\varepsilon_i)\right| \right\} \tag{4.64}$$

<u>Two-dimensional elements:</u>

$$\underset{n\times n}{[C^{(e)}]} = \sum_{i=1}^{N_\varepsilon}\sum_{j=1}^{N_\eta} W_i(\varepsilon_i) W_j(\eta_j) \left\{ \underset{n\times1}{\begin{bmatrix} N_1^{(e)}(\varepsilon_i,\eta_j) \\ \vdots \\ N_n^{(e)}(\varepsilon_i,\eta_j) \end{bmatrix}} \underset{1\times1}{[S_s^{(e)}]} \underset{1\times n}{[N_1^{(e)}(\varepsilon_i,\eta_j) \cdots N_n^{(e)}(\varepsilon_i,\eta_j)]} \left| J(\varepsilon_i,\eta_j)\right| \right\} \tag{4.65}$$

Three-dimensional elements:

$$[C^{(e)}] = \sum_{i=1}^{N_\epsilon} \sum_{j=1}^{N_\eta} \sum_{k=1}^{N_\zeta} W_i(\epsilon_i)\, W_j(\eta_j)\, W_k(\zeta_k) \left\{ \begin{bmatrix} N_1^{(e)}(\epsilon_i,\eta_j,\zeta_k) \\ \vdots \\ N_n^{(e)}(\epsilon_i,\eta_j,\zeta_k) \end{bmatrix}_{1\times 1} [S_s^{(e)}] \right.$$

$$\underset{n\times 1}{}$$

$$\left. \underset{1\times n}{[N_1^{(e)}(\epsilon_i,\eta_j,\zeta_k) \cdots N_n^{(e)}(\epsilon_i,\eta_j,\zeta_k)]}\, |\, J(\epsilon_i,\eta_j,\zeta_k)| \right\}$$

(4.66)

Element Capacitance Matrix (Saturated Flow, Lumped Formulation)

No numerical integration is required; use equation 4.9.

Element Capacitance Matrix (Unsaturated Flow, Consistent Formulation)

These equations can be obtained by substituting $C^{(e)}(\psi)$ for $S_s^{(e)}$ in equations 4.64, 4.65, and 4.66.

Element Capacitance Matrix (Unsaturated Flow, Lumped Formulation)

No numerical integration is required; use equation 4.11.

Element Advection-Dispersion Matrix

One-dimensional elements:

$$\underset{n\times n}{[D^{(e)}]} = \sum_{i=1}^{N_\epsilon} W_i(\epsilon_i) \left\{ \begin{bmatrix} \dfrac{\partial N_1^{(e)}(\epsilon_i)}{\partial \epsilon} \\ \vdots \\ \dfrac{\partial N_n^{(e)}(\epsilon_i)}{\partial \epsilon} \end{bmatrix}_{n\times 1} \underset{1\times 1}{[J(\epsilon_i)^{-1}]^T} \underset{1\times 1}{[D_x^{(e)}\theta^{(e)}]} \underset{1\times 1}{[J(\epsilon_i)^{-1}]} \underset{1\times n}{\left[\dfrac{\partial N_1^{(e)}(\epsilon_i)}{\partial \epsilon} \cdots \dfrac{\partial N_n^{(e)}(\epsilon_i)}{\partial \epsilon} \right]} |J(\epsilon_i)| \right\}$$

$$+ \sum_{i=1}^{N_\epsilon} W_i(\epsilon_i) \left\{ \begin{bmatrix} N_1^{(e)}(\epsilon_i) \\ \vdots \\ N_n^{(e)}(\epsilon_i) \end{bmatrix}_{n\times 1} \underset{1\times 1}{[v_x^{(e)}]} \underset{1\times 1}{[J(\epsilon_i^{-1})]^T} \underset{1\times n}{\left[\dfrac{\partial N_1^{(e)}(\epsilon_i)}{\partial \epsilon} \cdots \dfrac{\partial N_n^{(e)}(\epsilon_i)}{\partial \epsilon} \right]} |J(\epsilon_i)| \right\}$$

$$+ \sum_{i=1}^{N_\epsilon} W_i(\epsilon_i) \left\{ \begin{bmatrix} N_1^{(e)}(\epsilon_i) \\ \vdots \\ N_n^{(e)}(\epsilon_i) \end{bmatrix}_{n\times 1} \underset{1\times 1}{[\lambda(\theta^{(e)} + \rho_b^{(e)} K_d^{(e)})]} \underset{1\times n}{[N_1^{(e)}(\epsilon_i) \cdots N_n^{(e)}(\epsilon_i)]} |J(\epsilon_i)| \right\}$$

(4.67)

Two-dimensional elements:

$$
\underset{n \times n}{[D^{(e)}]} = \sum_{i=1}^{N_\varepsilon} \sum_{j=1}^{N_\eta} W_i(\varepsilon_i)\, W_i(\eta_j) \left\{ \underset{n \times 2}{\begin{bmatrix} \dfrac{\partial N_1^{(e)}}{\partial \varepsilon}(\varepsilon_i,\eta_j) & \dfrac{\partial N_1^{(e)}}{\partial \eta}(\varepsilon_i,\eta_j) \\ \vdots & \vdots \\ \dfrac{\partial N_n^{(e)}}{\partial \varepsilon}(\varepsilon_i,\eta_j) & \dfrac{\partial N_n^{(e)}}{\partial \eta}(\varepsilon_i,\eta_j) \end{bmatrix}} \underset{2 \times 2}{[J(\varepsilon_i,\eta_j)^{-1}]^T} \underset{2 \times 2}{\begin{bmatrix} D_{xx}^{(e)}\theta^{(e)} & D_{xy}^{(e)}\theta^{(e)} \\ D_{yx}^{(e)}\theta^{(e)} & D_{yy}^{(e)}\theta^{(e)} \end{bmatrix}} \right.
$$

$$
\underset{2 \times 2}{[J(\varepsilon_i,\eta_j)^{-1}]} \underset{2 \times n}{\begin{bmatrix} \dfrac{\partial N_1^{(e)}}{\partial \varepsilon}(\varepsilon_i,\eta_j) & \cdots & \dfrac{\partial N_n^{(e)}}{\partial \varepsilon}(\varepsilon_i,\eta_j) \\ \dfrac{\partial N_1^{(e)}}{\partial \eta}(\varepsilon_i,\eta_j) & \cdots & \dfrac{\partial N_n^{(e)}}{\partial \eta}(\varepsilon_i,\eta_j) \end{bmatrix}} \; |J(\varepsilon_i,\eta_j)| \Bigg\}
$$

$$
+ \sum_{i=1}^{N_\varepsilon} \sum_{j=1}^{N_\eta} W_i(\varepsilon_i)\, W_j(\eta_j) \left\{ \underset{n \times 2}{\begin{bmatrix} N_1^{(e)}(\varepsilon_i,\eta_j) & N_1^{(e)}(\varepsilon_i,\eta_j) \\ \vdots & \vdots \\ N_n^{(e)}(\varepsilon_i,\eta_j) & N_n^{(e)}(\varepsilon_i,\eta_j) \end{bmatrix}} \underset{2 \times 2}{\begin{bmatrix} v_x^{(e)} & 0 \\ 0 & v_y^{(e)} \end{bmatrix}} \underset{2 \times 2}{[J(\varepsilon_i,\eta_j)^{-1}]} \right.
$$

$$
\underset{2 \times n}{\begin{bmatrix} \dfrac{\partial N_1^{(e)}}{\partial \varepsilon}(\varepsilon_i,\eta_j) & \cdots & \dfrac{\partial N_n^{(e)}}{\partial \varepsilon}(\varepsilon_i,\eta_j) \\ \dfrac{\partial N_1^{(e)}}{\partial \eta}(\varepsilon_i,\eta_j) & \cdots & \dfrac{\partial N_n^{(e)}}{\partial \eta}(\varepsilon_i,\eta_j) \end{bmatrix}} \; |J(\varepsilon_i,\eta_j)| \Bigg\}
$$

$$
+ \sum_{i=1}^{N_\varepsilon} \sum_{j=1}^{N_\eta} W_i(\varepsilon_i)\, W_j(\eta_j) \left\{ \underset{n \times 1}{\begin{bmatrix} N_1^{(e)}(\varepsilon_i,\eta_j) \\ \vdots \\ N_n^{(e)}(\varepsilon_i,\eta_j) \end{bmatrix}} \underset{1 \times 1}{[\lambda(\theta^{(e)} + \rho_b^{(e)} K_d^{(e)})]} \right.
$$

$$
\underset{1 \times n}{\left[N_1^{(e)}(\varepsilon_i,\eta_j) \; \cdots \; N_n^{(e)}(\varepsilon_i,\eta_j) \right]} \; |J(\varepsilon_i,\eta_j)| \Bigg\} \tag{4.68}
$$

Three-dimensional elements:

$$[D^{(e)}] =$$

$$\underset{n \times n}{} \sum_{i=1}^{N_\varepsilon} \sum_{j=1}^{N_\eta} \sum_{k=1}^{N_\zeta} W_i(\varepsilon_i) \, W_j(\eta_j) \, W_k(\zeta_k) \underset{n \times 3}{\begin{bmatrix} \dfrac{\partial N_1^{(e)}}{\partial \varepsilon}(\varepsilon_i,\eta_j,\zeta_k) & \dfrac{\partial N_1^{(e)}}{\partial \eta}(\varepsilon_i,\eta_j,\zeta_k) & \dfrac{\partial N_1^{(e)}}{\partial \zeta}(\varepsilon_i,\eta_j,\zeta_k) \\ \vdots & \vdots & \vdots \\ \dfrac{\partial N_n^{(e)}}{\partial \varepsilon}(\varepsilon_i,\eta_j,\zeta_k) & \dfrac{\partial N_n^{(e)}}{\partial \eta}(\varepsilon_i,\eta_j,\zeta_k) & \dfrac{\partial N_n^{(e)}}{\partial \zeta}(\varepsilon_i,\eta_j,\zeta_k) \end{bmatrix}} \underset{3 \times 3}{[J(\varepsilon_i,\eta_j,\zeta_k)^{-1}]^T}$$

$$\underset{3 \times 3}{\begin{bmatrix} D_{xx}^{(e)}\theta^{(e)} & D_{xy}^{(e)}\theta^{(e)} & D_{xz}^{(e)}\theta^{(e)} \\ D_{yx}^{(e)}\theta^{(e)} & D_{yy}^{(e)}\theta^{(e)} & D_{yz}^{(e)}\theta^{(e)} \\ D_{zx}^{(e)}\theta^{(e)} & D_{zy}^{(e)}\theta^{(e)} & D_{zz}^{(e)}\theta^{(e)} \end{bmatrix}} \underset{3 \times 3}{[J(\varepsilon_i,\eta_j,\zeta_k)^{-1}]} \underset{n \times 3}{\begin{bmatrix} \dfrac{\partial N_1^{(e)}}{\partial \varepsilon}(\varepsilon_i,\eta_j,\zeta_k) \cdots \dfrac{\partial N_n^{(e)}}{\partial \varepsilon}(\varepsilon_i,\eta_j,\zeta_k) \\ \dfrac{\partial N_1^{(e)}}{\partial \eta}(\varepsilon_i,\eta_j,\zeta_k) \cdots \dfrac{\partial N_n^{(e)}}{\partial \eta}(\varepsilon_i,\eta_j,\zeta_k) \\ \dfrac{\partial N_1^{(e)}}{\partial \zeta}(\varepsilon_i,\eta_j,\zeta_k) \cdots \dfrac{\partial N_n^{(e)}}{\partial \zeta}(\varepsilon_i,\eta_j,\zeta_k) \end{bmatrix}} |J(\varepsilon_i,\eta_j,\zeta_k)| \Bigg\}$$

$$+ \sum_{i=1}^{N_\varepsilon} \sum_{j=1}^{N_\eta} \sum_{k=1}^{N_\zeta} W_i(\varepsilon_i) \, W_j(\eta_j) \, W_k(\zeta_k) \left\{ \underset{n \times 3}{\begin{bmatrix} N_1^{(e)}(\varepsilon_i,\eta_j,\zeta_k) & N_1^{(e)}(\varepsilon_i,\eta_j,\zeta_k) & N_1^{(e)}(\varepsilon_i,\eta_j,\zeta_k) \\ \vdots & \vdots & \vdots \\ N_n^{(e)}(\varepsilon_i,\eta_j,\zeta_k) & N_n^{(e)}(\varepsilon_i,\eta_j,\zeta_k) & N_n^{(e)}(\varepsilon_i,\eta_j,\zeta_k) \end{bmatrix}} \right.$$

$$\underset{3 \times 3}{\begin{bmatrix} v_x^{(e)} & 0 & 0 \\ 0 & v_y^{(e)} & 0 \\ 0 & 0 & v_z^{(e)} \end{bmatrix}} \underset{3 \times 3}{[J(\varepsilon_i,\eta_j,\zeta_k)^{-1}]} \underset{3 \times n}{\begin{bmatrix} \dfrac{\partial N_1^{(e)}}{\partial \varepsilon}(\varepsilon_i,\eta_j,\zeta_k) \cdots \dfrac{\partial N_n^{(e)}}{\partial \varepsilon}(\varepsilon_i,\eta_j,\zeta_k) \\ \dfrac{\partial N_1^{(e)}}{\partial \eta}(\varepsilon_i,\eta_j,\zeta_k) \cdots \dfrac{\partial N_n^{(e)}}{\partial \eta}(\varepsilon_i,\eta_j,\zeta_k) \\ \dfrac{\partial N_1^{(e)}}{\partial \zeta}(\varepsilon_i,\eta_j,\zeta_k) \cdots \dfrac{\partial N_n^{(e)}}{\partial \zeta}(\varepsilon_i,\eta_j,\zeta_k) \end{bmatrix}} |J(\varepsilon_i,\eta_j,\zeta_k)| \Bigg\}$$

$$+ \sum_{i=1}^{N_\varepsilon} \sum_{j=1}^{N_\eta} \sum_{k=1}^{N_\zeta} W_i(\varepsilon_i) \, W_j(\eta_j) \, W_k(\zeta_k) \left\{ \underset{n \times 1}{\begin{bmatrix} N_1^{(e)}(\varepsilon_i,\eta_j,\zeta_k) \\ \vdots \\ N_n^{(e)}(\varepsilon_i,\eta_j,\zeta_k) \end{bmatrix}} \underset{1 \times 1}{[\lambda(\theta^{(e)} + \rho_b^{(e)} K_d^{(e)})]} \right.$$

$$\underset{1 \times n}{\left[N_1^{(e)}(\varepsilon_i,\eta_j,\zeta_k) \cdots N_n^{(e)}(\varepsilon_i,\eta_j,\zeta_k) \right]} |J(\varepsilon_i,\eta_j,\zeta_k)| \Bigg\} \qquad (4.69)$$

Element Sorption Matrix

One-dimensional elements:

$$[A^{(e)}]_{n \times n} = \sum_{i=1}^{N_\varepsilon} W_i(\varepsilon_i) \left\{ \begin{bmatrix} N_1^{(e)}(\varepsilon_i) \\ \vdots \\ N_n^{(e)}(\varepsilon_i) \end{bmatrix}_{n \times 1} [\rho_b^{(e)} K_d^{(e)} + \theta^{(e)}]_{1 \times 1} \begin{bmatrix} N_1^{(e)}(\varepsilon_i) \cdots N_n^{(e)}(\varepsilon_i) \end{bmatrix}_{1 \times n} |J(\varepsilon_i)| \right\}$$

(4.70)

Two-dimensional elements:

$$[A^{(e)}]_{n \times n} = \sum_{i=1}^{N_\varepsilon} \sum_{j=1}^{N_\eta} W_i(\varepsilon_i)\, W_j(\eta_j) \left\{ \begin{bmatrix} N_1^{(e)}(\varepsilon_i) \\ \vdots \\ N_n^{(e)}(\varepsilon_i) \end{bmatrix}_{n \times 1} [\rho_b^{(e)} K_d^{(e)} + \theta^{(e)}]_{1 \times 1} \right.$$
$$\left. \begin{bmatrix} N_1^{(e)}(\varepsilon_i,\eta_j) \cdots N_n^{(e)}(\varepsilon_i,\eta_j) \end{bmatrix}_{1 \times n} |J(\varepsilon_i,\eta_j)| \right\}$$

(4.71)

Three-dimensional elements:

$$[A^{(e)}]_{n \times n} = \sum_{i=1}^{N_\varepsilon} \sum_{j=1}^{N_\eta} \sum_{k=1}^{N_\zeta} W_i(\varepsilon_i)\, W_j(\eta_j)\, W_k(\zeta_k) \left\{ \begin{bmatrix} N_1^{(e)}(\varepsilon_i,\eta_j,\zeta_k) \\ \vdots \\ N_n^{(e)}(\varepsilon_i,\eta_j,\zeta_k) \end{bmatrix}_{n \times 1} [\rho_b^{(e)} K_d^{(e)} + \theta^{(e)}]_{1 \times 1} \right.$$
$$\left. \begin{bmatrix} N_1^{(e)}(\varepsilon_i,\eta_j,\zeta_k) \cdots N_n^{(e)}(\varepsilon_i,\eta_j,\zeta_k) \end{bmatrix}_{1 \times n} |J(\varepsilon_i,\eta_j,\zeta_k)| \right\}$$

(4.72)

The order of polynomial for the element matrices for the element types in Figures 4.9 to 4.15 are in Table 4.2

Table 4.2 Highest order of polynomial appearing in integral equations for the element matrices for the element types in Figures 4.9 to 4.15.

	$[K^{(e)}]$ or $[K^{(e)}(\psi)]$	$[C^{(e)}]$ or $[C^{(e)}(\psi)]$ (Consistent Formulation)	$[D^{(e)}]$	$[A^{(e)}]$	Figure
One-dimensional Elements					
linear	0	2	2	2	4.9(a)
quadratic	2	4	4	4	4.9(b)
cubic	4	6	6	6	4.9(c)
Two-dimensional Elements					
linear	2	2	2	2	4.10
quadratic	4	4	4	4	4.11
cubic	6	6	6	6	4.12
Three-dimensional Elements					
linear	2	2	2	2	4.13
quadratic	4	4	4	4	4.14
cubic	6	6	6	6	4.15

Example

Determine the number of Gauss points to use in equation 4.69 for the element advection–dispersion matrix $[D^{(e)}]$ for the cubic parallelepiped element (Figure 4.15).

From equation 4.69, $[D^{(e)}]$ consists of three terms containing the dispersion coefficients, the apparent groundwater velocities, and the solute decay and distribution coefficients. The first term contains products of the derivatives of the interpolation functions and terms such as $\dfrac{\partial N_i}{\partial \varepsilon}\dfrac{\partial N_i}{\partial \varepsilon}$ are polynomials of order six (e.g. terms such as $\varepsilon^3 \cdot \varepsilon^3$ or $\eta^3 \cdot \eta^3$) and can be integrated using 4-point Gauss quadrature, $N_\varepsilon = N_\eta = N_\zeta = 4$ (Table 4.1). The second term contains products of the interpolation functions and their derivatives. Terms such as $N_i \dfrac{\partial N_i}{\partial \varepsilon}$ are polynomials of order six and can also be integrated using 4–point Gauss quadrature. The third term in equation 4.69 contains products of the interpolation functions and terms such as $N_i N_i$ are polynomials of order six and can be integrated using 4-point Gauss quadrature, $N_\varepsilon = N_\eta = N_\zeta = 5$ (Table 4.1). The locations of the 64 Gauss points are $(0,0,0)$, $\left(0,0,\sqrt{\dfrac{3-2\sqrt{6/5}}{7}}\right)$, $\left(0,0,-\sqrt{\dfrac{3-2\sqrt{6/5}}{7}}\right)$ ... $\left(-\sqrt{\dfrac{3+2\sqrt{6/5}}{7}},-\sqrt{\dfrac{3+2\sqrt{6/5}}{7}},\right.$ $\left.-\sqrt{\dfrac{3+2\sqrt{6/5}}{7}}\right)$.

Example

Compute $[K^{(e)}]$ for the linear, quadrilateral element shown below ($K_x^{(e)} = 1$, $K_y^{(e)} = 2$).

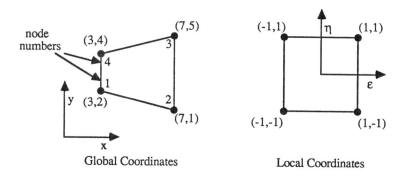

Global Coordinates   Local Coordinates

Using equation 4.62 and the derivatives in Figure 4.10 we can write

$$
\begin{bmatrix} K^{(e)} \\ 4\times 4 \end{bmatrix} = \sum_{i=1}^{N_\varepsilon} \sum_{j=1}^{N_\eta} W_i(\varepsilon_i) W_j(\eta_j) \left\{ \begin{bmatrix} -\frac{1}{4}(1-\eta) & -\frac{1}{4}(1-\varepsilon) \\ \frac{1}{4}(1-\eta) & -\frac{1}{4}(1+\varepsilon) \\ \frac{1}{4}(1+\eta) & \frac{1}{4}(1+\varepsilon) \\ -\frac{1}{4}(1+\eta) & \frac{1}{4}(1-\varepsilon) \end{bmatrix} [J(\varepsilon_i\eta_j)^{-1}]^T \begin{bmatrix} 1 & 0 \\ 0 & 2 \end{bmatrix} [J(\varepsilon_i,\eta_j)^{-1}] \right.
$$

$$
\underset{4\times 2}{} \qquad \underset{2\times 2}{} \qquad \underset{2\times 2}{} \qquad \underset{2\times 2}{}
$$

$$
\left. \begin{bmatrix} -\frac{1}{4}(1-\eta) & \frac{1}{4}(1-\eta) & \frac{1}{4}(1+\eta) & -\frac{1}{4}(1+\eta) \\ -\frac{1}{4}(1-\varepsilon) & -\frac{1}{4}(1+\varepsilon) & \frac{1}{4}(1+\varepsilon) & \frac{1}{4}(1-\varepsilon) \end{bmatrix} |J(\varepsilon_i,\eta_j)| \right\}
$$

$$
\underset{2\times 4}{}
$$

The highest-order polynomial in this equation is quadratic (n = 2). For example,

$$
\frac{1}{4}(1 - \eta) \cdot \frac{1}{4}(1 + \eta) = \frac{1}{16}(1 + \eta^2)
$$

Referring to Table 4.1, two Gauss points are required in each direction (i.e. $N_\varepsilon = N_\eta = 2$). The locations of these Gauss points are (ε, η): $(-1/\sqrt{3}, -1/\sqrt{3})$, $(-1/\sqrt{3}, 1/\sqrt{3})$, $(1/\sqrt{3}, 1/\sqrt{3})$ and $(1/\sqrt{3}, -1/\sqrt{3})$. At the first Gauss point $(\varepsilon_1 = -1/\sqrt{3}, \eta_1 = -1/\sqrt{3})$ the Jacobian matrix (equation 4.36) is

$$
[J] = \begin{bmatrix} -\frac{1}{4}\left(1+\frac{1}{\sqrt{3}}\right) & \frac{1}{4}\left(1+\frac{1}{\sqrt{3}}\right) & \frac{1}{4}\left(1-\frac{1}{\sqrt{3}}\right) & -\frac{1}{4}\left(1-\frac{1}{\sqrt{3}}\right) \\ -\frac{1}{4}\left(1+\frac{1}{\sqrt{3}}\right) & -\frac{1}{4}\left(1-\frac{1}{\sqrt{3}}\right) & \frac{1}{4}\left(1-\frac{1}{\sqrt{3}}\right) & \frac{1}{4}\left(1+\frac{1}{\sqrt{3}}\right) \end{bmatrix} \begin{bmatrix} 3 & 2 \\ 7 & 1 \\ 7 & 5 \\ 3 & 4 \end{bmatrix}
$$

$$
= \begin{bmatrix} -0.3943 & 0.3943 & 0.1057 & -0.1057 \\ -0.3943 & -0.1057 & 0.1057 & 0.3943 \end{bmatrix} \begin{bmatrix} 3 & 2 \\ 7 & 1 \\ 7 & 5 \\ 3 & 4 \end{bmatrix}
$$

$$
= \begin{bmatrix} 2.0000 & -0.2887 \\ 0.0000 & 1.2113 \end{bmatrix}
$$

$$
|J| = (2.000)(1.2113) - (0.000)(-0.2887) = 2.4228
$$

and

$$
[J^{-1}] = \begin{bmatrix} 0.5000 & 0.1192 \\ 0.0000 & 0.8255 \end{bmatrix}
$$

The contribution of the first Gauss point to the element conductance matrix is

$$[K^{(e)}(\varepsilon_1,\eta_1)] = (1)(1)\begin{bmatrix} -0.3943 & -0.3943 \\ 0.3943 & -0.1057 \\ 0.1057 & 0.1057 \\ -0.1057 & 0.3943 \end{bmatrix}\begin{bmatrix} 0.5000 & 0.0000 \\ 0.1192 & 0.8255 \end{bmatrix}\begin{bmatrix} 1 & 0 \\ 0 & 2 \end{bmatrix}\begin{bmatrix} 0.5000 & 0.1192 \\ 0.0000 & 0.8255 \end{bmatrix}$$

$$\times \begin{bmatrix} -0.3943 & 0.3943 & 0.1057 & -0.1057 \\ -0.3943 & -0.1057 & 0.1057 & 0.3943 \end{bmatrix}(2.4226)$$

$$= \begin{bmatrix} 0.6578 & 0.0285 & -0.1763 & -0.5099 \\ 0.0285 & 0.1194 & -0.0076 & -0.1403 \\ -0.1763 & -0.0076 & 0.0473 & 0.1367 \\ -0.5099 & -0.1403 & 0.1367 & 0.5135 \end{bmatrix}$$

At the second Gauss point ($\varepsilon_1 = -1/\sqrt{3}$, $\eta_2 = 1/\sqrt{3}$), the Jacobian matrix is

$$[J] = \begin{bmatrix} -\frac{1}{4}\left(1-\frac{1}{\sqrt{3}}\right) & \frac{1}{4}\left(1-\frac{1}{\sqrt{3}}\right) & \frac{1}{4}\left(1+\frac{1}{\sqrt{3}}\right) & -\frac{1}{4}\left(1+\frac{1}{\sqrt{3}}\right) \\ -\frac{1}{4}\left(1+\frac{1}{\sqrt{3}}\right) & -\frac{1}{4}\left(1-\frac{1}{\sqrt{3}}\right) & \frac{1}{4}\left(1-\frac{1}{\sqrt{3}}\right) & \frac{1}{4}\left(1+\frac{1}{\sqrt{3}}\right) \end{bmatrix}\begin{bmatrix} 3 & 2 \\ 7 & 1 \\ 7 & 5 \\ 3 & 4 \end{bmatrix}$$

$$= \begin{bmatrix} -0.1057 & 0.1057 & 0.3943 & -0.3943 \\ -0.3943 & -0.1057 & 0.1057 & 0.3943 \end{bmatrix}\begin{bmatrix} 3 & 2 \\ 7 & 1 \\ 7 & 5 \\ 3 & 4 \end{bmatrix}$$

$$= \begin{bmatrix} 2.0000 & 0.2886 \\ 0.0000 & 1.2114 \end{bmatrix}$$

$$|J| = 2.4228 \qquad \text{and} \qquad [J^{-1}] = \begin{bmatrix} 0.5000 & -0.1192 \\ 0.0000 & 0.8255 \end{bmatrix}$$

The contribution of the second Gauss point to the element conductance matrix is

$$[K^{(e)}(\varepsilon_1,\eta_2)] = (1)(1)\begin{bmatrix} -0.1057 & -0.3943 \\ 0.1057 & -0.1057 \\ 0.3943 & 0.1057 \\ -0.3943 & 0.3943 \end{bmatrix}\begin{bmatrix} 0.5000 & 0.0000 \\ -0.1192 & 0.8255 \end{bmatrix}\begin{bmatrix} 1 & 0 \\ 0 & 2 \end{bmatrix}$$

$$\times \begin{bmatrix} 0.5000 & -0.1192 \\ 0.0000 & 0.8255 \end{bmatrix}\begin{bmatrix} -0.1057 & 0.1057 & 0.3943 & -0.3943 \\ -0.3943 & -0.1057 & 0.1057 & 0.3943 \end{bmatrix}$$

$$\times \quad (2.4224)$$

$$= \begin{bmatrix} 0.5135 & 0.1367 & -0.1403 & -0.5099 \\ 0.1367 & 0.0473 & -0.0076 & -0.1763 \\ -0.1403 & -0.0076 & 0.1194 & 0.0285 \\ -0.5099 & -0.1763 & 0.0285 & 0.6578 \end{bmatrix}$$

At the third Gauss point ($\varepsilon_2 = 1/\sqrt{3}$, $\eta_1 = -1/\sqrt{3}$) we have

$$[J] = \begin{bmatrix} 2.0000 & 0.2886 \\ 0.0000 & 1.7886 \end{bmatrix} \qquad |J| = 3.5772$$

$$[J^{-1}] = \begin{bmatrix} 0.5000 & 0.0807 \\ 0.0000 & 0.5591 \end{bmatrix}$$

and the contribution of the third Gauss point to the element conductance matrix is

$$[K^{(e)}(\varepsilon_2, \eta_1)] = \begin{bmatrix} 0.1763 & -0.0284 & -0.1555 & 0.0076 \\ -0.0284 & 0.4455 & -0.2976 & -0.1194 \\ -0.1555 & -0.2976 & 0.3733 & 0.0798 \\ 0.0076 & -0.1194 & 0.0798 & 0.0320 \end{bmatrix}$$

At the fourth Gauss point $(\varepsilon_2 = 1/\sqrt{3}, \eta_2 = 1/\sqrt{3})$ we have

$$[J] = \begin{bmatrix} 2.0000 & 0.2886 \\ 0.0000 & 1.7886 \end{bmatrix} \qquad |J| = 3.5772$$

$$[J^{-1}] = \begin{bmatrix} 0.5000 & -0.0807 \\ 0.0000 & 0.5591 \end{bmatrix}$$

and the contribution of the fourth Gauss point to the element conductance matrix is

$$[K(\varepsilon_2, \eta_2)] = \begin{bmatrix} 0.0318 & 0.0793 & -0.1188 & 0.0076 \\ 0.0793 & 0.3713 & -0.2960 & -0.1546 \\ -0.1188 & -0.2960 & 0.4430 & -0.0283 \\ 0.0076 & -0.1546 & -0.0283 & 0.1753 \end{bmatrix}$$

The element conductance matrix is the sum of the contributions at the four Gauss points

$$[K^{(e)}] = [K^{(e)}(\varepsilon_1, \eta_1)] + [K^{(e)}(\varepsilon_1, \eta_2)] + [K^{(e)}(\varepsilon_2, \eta_1)] + [(K^{(e)}(\varepsilon_2, \eta_2)]$$

$$= \begin{bmatrix} 1.37952 & 0.21649 & -0.59149 & -1.00452 \\ 0.21649 & 0.98551 & -0.61051 & -0.59149 \\ -0.59149 & -0.61051 & 0.98551 & 0.21649 \\ -1.00452 & -0.59149 & 0.21649 & 1.37952 \end{bmatrix}$$

4.4   ASSEMBLING THE GLOBAL SYSTEM OF EQUATIONS

After the element matrices have been computed, they must be combined to obtain the global matrices needed to solve for unknown heads or solute concentrations at the nodes. This process is called *assembling the global system of equations*. The assembly process can be written

$$
[M] = \sum_{e=1}^{m} [M^{(e)}] \tag{4.73}
$$
global

where [M] is any global matrix, e is the element number, $[M^{(e)}]$ is an element matrix, and m
global
is the number of elements in the mesh. For example the global conductance matrix for a mesh can be obtained by combining the element conductance matrices for each element in the mesh

$$
[K] = \sum_{e=1}^{m} [K^{(e)}] \tag{4.74a}
$$
global

Similarly for the other types of global matrices

$$
\{F\} = \sum_{e=1}^{m} \{F^{(e)}\} \tag{4.74b}
$$
global

$$
[K(\psi)] = \sum_{e=1}^{m} [K^{(e)}(\psi)] \tag{4.74c}
$$
global

$$
[C] = \sum_{e=1}^{m} [C^{(e)}] \tag{4.74d}
$$
global

$$
[C(\psi)] = \sum_{e=1}^{m} [C^{(e)}(\psi)] \tag{4.74e}
$$
global

$$
[D] = \sum_{e=1}^{m} [D^{(e)}] \tag{4.74f}
$$
global

$$
[A] = \sum_{e=1}^{m} [A^{(e)}] \tag{4.74g}
$$
global

The summations in equation 4.74 can be performed by direct matrix addition only if the element matrices are first expanded to the same size as the global matrices. This is done by adding rows and columns of zeros to the element matrices.

For example, the element conductance matrix for element 1 in the example problem in section 3.2 was

$$[K^{(1)}] = \begin{bmatrix} \dfrac{1}{2} & -\dfrac{1}{2} \\ -\dfrac{1}{2} & \dfrac{1}{2} \end{bmatrix} \begin{matrix} 1 \\ 2 \end{matrix} \quad \text{element node numbers} \tag{4.75}$$

where the numbers above the columns and to the right of the rows of the element conductance matrix are the node numbers for the nodes within element 1. Element 1 had two nodes ($n = 2$) and the size of $[K^{(1)}]$ is 2×2. The finite element mesh had 5 nodes and therefore the size of the global conductance matrix $[K]_{\text{global}}$ is 5×5. We can expand the conductance matrix for element 1 to this size by adding zeros to the rows and columns containing node 3, 4, and 5.

$$[K^{(1)}]_{\text{expanded}} = \begin{bmatrix} \dfrac{1}{2} & -\dfrac{1}{2} & 0 & 0 & 0 \\ -\dfrac{1}{2} & \dfrac{1}{2} & 0 & 0 & 0 \\ 0 & 0 & 0 & 0 & 0 \\ 0 & 0 & 0 & 0 & 0 \\ 0 & 0 & 0 & 0 & 0 \end{bmatrix} \begin{matrix} 1 \\ 2 \\ 3 \\ 4 \\ 5 \end{matrix} \quad \text{mesh node numbers} \tag{4.76}$$

Let $K_{ij}^{(e)}$ refer to the entry in the ith row and j^{th} column of the expanded element conductance matrix. $K_{11}^{(1)} = 1/2$ is nonzero because element 1 contains node number 1. Similarly for $K_{22}^{(1)}$. $K_{12}^{(1)} = K_{21}^{(1)} = -1/2$ are nonzero because element 1 contains both nodes 1 and 2. $K_{13}^{(1)} = K_{14}^{(e)} = K_{15}^{(e)} = K_{23}^{(e)} = \cdots K_{55}^{(e)} = 0$ because element 1 does not contain nodes 3, 4 or 5.

We can also write the expanded form of the element conductance matrix for the other elements in the mesh (elements 2, 3, and 4)

$$[K^{(2)}]_{\text{expanded}} = \begin{bmatrix} 0 & 0 & 0 & 0 & 0 \\ 0 & 1 & -1 & 0 & 0 \\ 0 & -1 & 1 & 0 & 0 \\ 0 & 0 & 0 & 0 & 0 \\ 0 & 0 & 0 & 0 & 0 \end{bmatrix} \begin{matrix} 1 \\ 2 \\ 3 \\ 4 \\ 5 \end{matrix} \tag{4.77}$$

$$[K^{(3)}]_{\text{expanded}} = \begin{array}{cc} & \begin{array}{ccccc} 1 & 2 & 3 & 4 & 5 \end{array} \\ \begin{bmatrix} 0 & 0 & 0 & 0 & 0 \\ 0 & 0 & 0 & 0 & 0 \\ 0 & 0 & 1/3 & -1/3 & 0 \\ 0 & 0 & -1/3 & 1/3 & 0 \\ 0 & 0 & 0 & 0 & 0 \end{bmatrix} & \begin{array}{c} 1 \\ 2 \\ 3 \\ 4 \\ 5 \end{array} \end{array} \tag{4.78}$$

$$[K^{(4)}]_{\text{expanded}} = \begin{array}{cc} & \begin{array}{ccccc} 1 & 2 & 3 & 4 & 5 \end{array} \\ \begin{bmatrix} 0 & 0 & 0 & 0 & 0 \\ 0 & 0 & 0 & 0 & 0 \\ 0 & 0 & 0 & 0 & 0 \\ 0 & 0 & 0 & 1/3 & -1/3 \\ 0 & 0 & 0 & -1/3 & 1/3 \end{bmatrix} & \begin{array}{c} 1 \\ 2 \\ 3 \\ 4 \\ 5 \end{array} \end{array} \tag{4.79}$$

The global conductance matrix for the mesh can then be assembled by direct matrix addition of the expanded matrices

$$[K]_{\text{global}} = [K^{(1)}]_{\text{expanded}} + [K^{(2)}]_{\text{expanded}} + [K^{(3)}]_{\text{expanded}} + [K^{(4)}]_{\text{expanded}}$$

$$= \begin{array}{cc} & \begin{array}{ccccc} 1 & 2 & 3 & 4 & 5 \end{array} \\ \begin{bmatrix} 1/2 & -1/2 & 0 & 0 & 0 \\ -1/2 & 3/2 & -1 & 0 & 0 \\ 0 & -1 & 4/3 & -1/3 & 0 \\ 0 & 0 & -1/3 & 2/3 & -1/3 \\ 0 & 0 & 0 & -1/3 & 1/3 \end{bmatrix} & \begin{array}{c} 1 \\ 2 \\ 3 \\ 4 \\ 5 \end{array} \end{array} \tag{4.80}$$

For this problem [K] is symmetric and has a semi–bandwidth of two (SBW = 2).
global

Consider the same mesh with a different choice of node numbers (Figure 4.16)

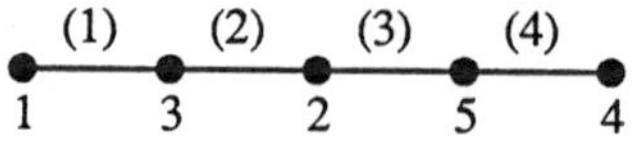

Figure 4.16   Finite element mesh with a different choice of node numbers.

The expanded form of the element conductance matrices are

$$[K^{(1)}]_{\text{expanded}} = \begin{array}{cc} & \begin{array}{ccccc} 1 & 2 & 3 & 4 & 5 \end{array} \\ \begin{bmatrix} 1/2 & 0 & -1/2 & 0 & 0 \\ 0 & 0 & 0 & 0 & 0 \\ -1/2 & 0 & 1/2 & 0 & 0 \\ 0 & 0 & 0 & 0 & 0 \\ 0 & 0 & 0 & 0 & 0 \end{bmatrix} & \begin{array}{c} 1 \\ 2 \\ 3 \\ 4 \\ 5 \end{array} \end{array} \tag{4.81}$$

$$[K^{(2)}] = \begin{array}{c} \\ \text{expanded} \end{array} \begin{bmatrix} 0 & 0 & 0 & 0 & 0 \\ 0 & 1 & -1 & 0 & 0 \\ 0 & -1 & 1 & 0 & 0 \\ 0 & 0 & 0 & 0 & 0 \\ 0 & 0 & 0 & 0 & 0 \end{bmatrix} \begin{array}{c} 1 \\ 2 \\ 3 \\ 4 \\ 5 \end{array} \qquad (4.82)$$

$$[K^{(3)}] = \begin{array}{c} \\ \text{expanded} \end{array} \begin{bmatrix} 0 & 0 & 0 & 0 & 0 \\ 0 & 1/3 & 0 & 0 & -1/3 \\ 0 & 0 & 0 & 0 & 0 \\ 0 & 0 & 0 & 0 & 0 \\ 0 & -1/3 & 0 & 0 & 1/3 \end{bmatrix} \begin{array}{c} 1 \\ 2 \\ 3 \\ 4 \\ 5 \end{array} \qquad (4.83)$$

$$[K^{(4)}] = \begin{array}{c} \\ \text{expanded} \end{array} \begin{bmatrix} 0 & 0 & 0 & 0 & 0 \\ 0 & 0 & 0 & 0 & 0 \\ 0 & 0 & 0 & 0 & 0 \\ 0 & 0 & 0 & 1/3 & -1/3 \\ 0 & 0 & 0 & -1/3 & 1/3 \end{bmatrix} \begin{array}{c} 1 \\ 2 \\ 3 \\ 4 \\ 5 \end{array} \qquad (4.84)$$

and the global conductance matrix for the mesh is

$$[K]_{\text{global}} = [K^{(1)}]_{\text{expanded}} + [K^{(2)}]_{\text{expanded}} + [K^{(3)}]_{\text{expanded}} + [K^{(4)}]_{\text{expanded}}$$

$$= \begin{bmatrix} 1/2 & 0 & -1/2 & 0 & 0 \\ 0 & 4/3 & -1 & 0 & -1/3 \\ -1/2 & -1 & 3/2 & 0 & 0 \\ 0 & 0 & 0 & 1/3 & -1/3 \\ 0 & -1/3 & 0 & -1/3 & 2/3 \end{bmatrix}$$

[K] is still symmetric (and will be symmetric for any choice of node numbers) but the semi-global
bandwidth has increased to 3. Using the formula from Chapter 2, SBW = R + 1 where R is the maximum difference in node numbers for any element in the mesh. For the mesh in Figure 4.16 R = 3 (computed in element 3) and SBW = 3 + 1 = 4.

The effect of choice of node numbers on the entries of the global conductance matrix and its semi-bandwidth for this mesh are shown in Figure 4.17.

Assembling the global matrices by direct matrix addition requires that each element matrix first be expanded to size p x p, where p is the number of nodes in the mesh. Although this procedure helps us visualize the assembly process it is inconvenient in that we must manipulate a large number of zero entries in each expanded matrix. In computer programs, direct matrix addition is also wasteful of computer memory because each expanded element matrix is the same size as the global matrices.

A more computationally efficient procedure is to assemble the global matrices by component addition. In this procedure the element matrices are not expanded to the size of the global matrices by adding rows and columns of zeros. Instead, the element node numbers are used to assign entries in the element matrices to their proper position in the global matrices. The procedure can be used for one-, two-, or three-dimensional problems and for meshes containing several different element types. In the examples that follow, the element number for each term in the global matrix is shown in parentheses as a superscript.

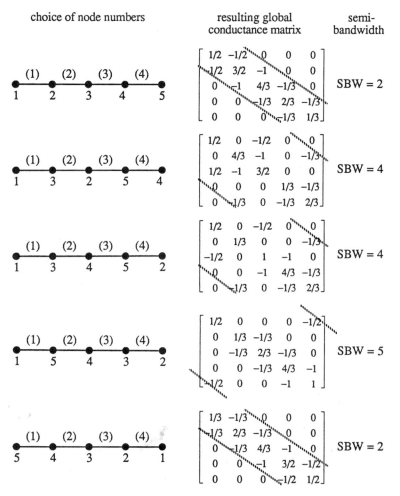

Figure 4.17   Effect of choice of node numbers on entries and semi-bandwidth of global conductance matrix.

Example

Compute $[K^{(e)}]$ and $[C^{(e)}]$ (consistent formulation) for the elements in the mesh shown below. Assemble $[K]_{global}$ and $[C]_{global}$

$$L^{(1)} = 4 \qquad L^{(2)} = 6 \qquad L^{(3)} = 5 \qquad L^{(4)} = 4$$
$$K_x^{(1)} = 2 \qquad K_x^{(2)} = 1 \qquad K_x^{(3)} = 2 \qquad K_x^{(4)} = 4$$
$$S_s^{(1)} = 1 \qquad S_s^{(2)} = 1 \qquad S_s^{(3)} = 6 \qquad S_s^{(4)} = 3$$

Using equations 4.15a and 4.16a

$$[K^{(1)}] = \begin{bmatrix} 1/2 & -1/2 \\ -1/2 & 1/2 \end{bmatrix} \begin{matrix} 1 \\ 3 \end{matrix} \quad [K^{(2)}] = \begin{bmatrix} 1/6 & -1/6 \\ -1/6 & 1/6 \end{bmatrix} \begin{matrix} 3 \\ 2 \end{matrix} \quad [K^{(3)}] = \begin{bmatrix} 2/5 & -2/5 \\ -2/5 & 2/5 \end{bmatrix} \begin{matrix} 2 \\ 4 \end{matrix}$$

$$[K^{(4)}] = \begin{bmatrix} 1 & -1 \\ -1 & 1 \end{bmatrix} \begin{matrix} 4 \\ 5 \end{matrix}$$

$$[C^{(1)}] = \begin{bmatrix} 4/3 & 2/3 \\ 2/3 & 4/3 \end{bmatrix} \begin{matrix} 1 \\ 3 \end{matrix} \quad [C^{(2)}] = \begin{bmatrix} 2 & 1 \\ 1 & 2 \end{bmatrix} \begin{matrix} 3 \\ 2 \end{matrix} \quad [C^{(3)}] = \begin{bmatrix} 10 & 5 \\ 5 & 10 \end{bmatrix} \begin{matrix} 2 \\ 4 \end{matrix}$$

$$[C^{(4)}] = \begin{bmatrix} 4 & 2 \\ 2 & 4 \end{bmatrix} \begin{matrix} 4 \\ 5 \end{matrix}$$

$$[K]_{global} = \begin{bmatrix} 1/2^{(1)} & 0 & -1/2^{(1)} & 0 & 0 \\ 0 & 1/6^{(2)}+2/5^{(3)} & -1/6^{(2)} & -2/5^{(3)} & 0 \\ -1/2^{(1)} & -1/6^{(2)} & 1/2^{(1)}+1/6^{(2)} & 0 & 0 \\ 0 & -2/5^{(3)} & 0 & 2/5^{(3)}+1^{(4)} & -1^{(4)} \\ 0 & 0 & 0 & -1^{(4)} & 1^{(4)} \end{bmatrix} \begin{matrix} 1 \\ 2 \\ 3 \\ 4 \\ 5 \end{matrix}$$

$$= \begin{bmatrix} 1/2 & 0 & -1/2 & 0 & 0 \\ 0 & 17/30 & -1/6 & -2/5 & 0 \\ -1/2 & -1/6 & 2/3 & 0 & 0 \\ 0 & -2/5 & 0 & 7/5 & -1 \\ 0 & 0 & 0 & -1 & 1 \end{bmatrix}$$

$$\begin{array}{ccccc} 1 & 2 & 3 & 4 & 5 \end{array}$$

$$[C]_{global} = \begin{bmatrix} 4/3^{(1)} & 0 & 2/3^{(1)} & 0 & 0 \\ 0 & 2^{(2)}+10^{(3)} & 1^{(2)} & 5^{(3)} & 0 \\ 0 & 1^{(2)} & 2^{(2)} & 0 & 0 \\ 0 & 5^{(3)} & 0 & 10^{(3)}+4^{(4)} & 2^{(4)} \\ 0 & 0 & 0 & 2^{(4)} & 4^{(4)} \end{bmatrix} \begin{array}{c} 1 \\ 2 \\ 3 \\ 4 \\ 5 \end{array}$$

$$= \begin{bmatrix} 4/3 & 0 & 2/3 & 0 & 0 \\ 0 & 12 & 1 & 5 & 0 \\ 0 & 1 & 2 & 0 & 0 \\ 0 & 5 & 0 & 14 & 2 \\ 0 & 0 & 0 & 2 & 4 \end{bmatrix}$$

Example

Compute $[K^{(e)}]$ and $[C^{(e)}]$ (lumped formulation) for the elements in the mesh shown below. Assemble $[K]_{global}$ and $[C]_{global}$.

$$K_x^{(1)} = K_y^{(1)} = K_x^{(2)} = K_y^{(2)} = 2$$
$$K_x^{(3)} = K_y^{(3)} = K_x^{(4)} = K_y^{(4)} = 4$$
$$S_s^{(1)} = S_s^{(2)} = S_s^{(3)} = S_s^{(4)} = 3$$

Using equations 4.20 and 4.22b with

element	node numbers		
	i	j	k
1	1	3	2
2	2	3	4
3	3	5	4
4	6	4	5

we have

$$[K^{(1)}] = \frac{2}{2}\begin{bmatrix} (-1)^2 & (-1)(1) & (-1)(0) \\ (1)(-1) & (1)^2 & (1)(0) \\ (0)(-1) & (0)(1) & (0)(0) \end{bmatrix} + \frac{2}{2}\begin{bmatrix} (-1)^2 & (-1)(0) & (-1)(1) \\ (0)(-1) & (0)^2 & (0)(1) \\ (1)(-1) & (1)(0) & (1)^2 \end{bmatrix}$$

$$= \begin{array}{ccc} 1 & 3 & 2 \end{array} \atop \begin{bmatrix} 2 & -1 & -1 \\ -1 & 1 & 0 \\ -1 & 0 & 1 \end{bmatrix} \begin{array}{c} 1 \\ 3 \\ 2 \end{array}$$

$$[K^{(2)}] = \frac{2}{2}\begin{bmatrix} (-1)^2 & (-1)(0) & (-1)(1) \\ (0)(-1) & (0)^2 & (0)(1) \\ (1)(-1) & (1)(0) & (1)^2 \end{bmatrix} + \frac{2}{2}\begin{bmatrix} (0)^2 & (0)(-1) & (0)(1) \\ (-1)(0) & (-1)^2 & (-1)(1) \\ (1)(0) & (1)(-1) & (1)^2 \end{bmatrix}$$

$$= \begin{array}{ccc} 2 & 3 & 4 \end{array} \atop \begin{bmatrix} 1 & 0 & -1 \\ 0 & 1 & -1 \\ -1 & -1 & 2 \end{bmatrix} \begin{array}{c} 2 \\ 3 \\ 4 \end{array}$$

$$[K^{(3)}] = \frac{4}{3}\begin{bmatrix} (-1)^2 & (-1)(1) & (-1)(0) \\ (1)(-1) & (1)^2 & (1)(0) \\ (0)(-1) & (0)(1) & (0)^2 \end{bmatrix} + \frac{4}{3}\begin{bmatrix} (-1.5)^2 & (-1.5)(0) & (-1.5)(1.5) \\ (0)(-1.5) & (0)^2 & (0)(1.5) \\ (1.5)(-1.5) & (1.5)(0) & (1.5)(1.5) \end{bmatrix}$$

$$= \begin{array}{ccc} 3 & 5 & 4 \end{array} \atop \begin{bmatrix} 4.3 & -1.3 & -3.0 \\ -1.3 & 1.3 & 0.0 \\ -3.0 & 0.0 & 3.0 \end{bmatrix} \begin{array}{c} 3 \\ 5 \\ 4 \end{array}$$

$$[K^{(4)}] = \frac{4}{3}\begin{bmatrix} (1)^2 & (1)(-1) & (1)(0) \\ (-1)(1) & (-1)^2 & (-1)(0) \\ (0)(1) & (0)(-1) & (0)^2 \end{bmatrix} + \frac{4}{3}\begin{bmatrix} (1.5)^2 & (1.5)(0) & (1.5)(-1.5) \\ (0)(1.5) & (0)^2 & (0)(-1.5) \\ (-1.5)(1.5) & (-1.5)(0) & (-1.5)^2 \end{bmatrix}$$

$$= \begin{array}{ccc} 6 & 4 & 5 \end{array}$$
$$= \begin{bmatrix} 4.3 & -1.3 & -3.0 \\ -1.3 & 1.3 & 0.0 \\ -3.0 & 0.0 & 3.0 \end{bmatrix} \begin{array}{c} 6 \\ 4 \\ 5 \end{array}$$

$$\begin{array}{ccccccc} & 1 & 2 & 3 & 4 & 5 & 6 \end{array}$$

$$[K]_{global} = \begin{bmatrix} 2^{(1)} & -1^{(1)} & -1^{(2)} & 0 & 0 & 0 \\ -1^{(1)} & 1^{(1)}+1^{(2)} & 0^{(1)}+0^{(2)} & -1^{(2)} & 0 & 0 \\ -1^{(1)} & 0^{(1)}+0^{(2)} & 1^{(1)}+1^{(2)}+4.3^{(3)} & -1^{(2)}-3^{(3)} & -1.3^{(3)} & 0 \\ 0 & -1^{(2)} & -1^{(2)}-3^{(3)} & 2^{(2)}+3^{(3)}+1.3^{(4)} & 0^{(3)}+0^{(4)} & -1.3^{(4)} \\ 0 & 0 & -1.3^{(3)} & 0^{(3)}+0^{(4)} & 1.3^{(3)}+3^{(4)} & -3^{(4)} \\ 0 & 0 & 0 & -1.3^{(4)} & -3^{(4)} & 4.3^{(4)} \end{bmatrix} \begin{array}{c} 1 \\ 2 \\ 3 \\ 4 \\ 5 \\ 6 \end{array}$$

$$= \begin{bmatrix} 2 & -1 & -1 & 0 & 0 & 0 \\ -1 & 2 & 0 & -1 & 0 & 0 \\ -1 & 0 & 6.3 & -4 & -1.3 & 0 \\ 0 & -1 & -4 & 6.3 & 0 & -1.3 \\ 0 & 0 & -1.3 & 0 & 4.3 & -3 \\ 0 & 0 & 0 & -1.3 & -3 & 4.3 \end{bmatrix}$$

$$[C^{(1)}] = \begin{array}{ccc} 1 & 3 & 2 \end{array}$$
$$[C^{(1)}] = \begin{bmatrix} 0.5 & 0 & 0 \\ 0 & 0.5 & 0 \\ 0 & 0 & 0.5 \end{bmatrix} \begin{array}{c} 1 \\ 3 \\ 2 \end{array}$$

$$[C^{(2)}] = \begin{array}{ccc} 2 & 3 & 4 \end{array}$$
$$[C^{(2)}] = \begin{bmatrix} 0.5 & 0 & 0 \\ 0 & 0.5 & 0 \\ 0 & 0 & 0.5 \end{bmatrix} \begin{array}{c} 2 \\ 3 \\ 4 \end{array}$$

$$[C^{(3)}] = \begin{array}{ccc} 3 & 5 & 4 \end{array}$$
$$[C^{(3)}] = \begin{bmatrix} 0.75 & 0 & 0 \\ 0 & 0.75 & 0 \\ 0 & 0 & 0.75 \end{bmatrix} \begin{array}{c} 3 \\ 5 \\ 4 \end{array}$$

$$[C^{(4)}] = \begin{array}{ccc} 6 & 4 & 5 \end{array}$$
$$[C^{(4)}] = \begin{bmatrix} 0.75 & 0 & 0 \\ 0 & 0.75 & 0 \\ 0 & 0 & 0.75 \end{bmatrix} \begin{array}{c} 6 \\ 4 \\ 5 \end{array}$$

$$[C]_{global} = \begin{bmatrix} 0.5^{(1)} & 0 & 0 & 0 & 0 & 0 \\ 0 & 0.5^{(1)}+0.5^{(2)} & 0 & 0 & 0 & 0 \\ 0 & 0 & 0.5^{(1)}+0.5^{(2)}+0.75^{(3)} & 0 & 0 & 0 \\ 0 & 0 & 0 & 0.5^{(2)}+0.75^{(3)}+0.75^{(4)} & 0 & 0 \\ 0 & 0 & 0 & 0 & 0.75^{(3)}+0.75^{(4)} & 0 \\ 0 & 0 & 0 & 0 & 0 & 0.75^{(4)} \end{bmatrix} \begin{matrix} 1 \\ 2 \\ 3 \\ 4 \\ 5 \\ 6 \end{matrix}$$

$$= \begin{bmatrix} 0.5 & 0 & 0 & 0 & 0 & 0 \\ 0 & 1.0 & 0 & 0 & 0 & 0 \\ 0 & 0 & 1.75 & 0 & 0 & 0 \\ 0 & 0 & 0 & 2.0 & 0 & 0 \\ 0 & 0 & 0 & 0 & 1.5 & 0 \\ 0 & 0 & 0 & 0 & 0 & 0.75 \end{bmatrix}$$

Example

Compute $[C^{(e)}]$ (consistent formulation) for the elements in the mesh shown below.

Assemble $[C]_{global}$.

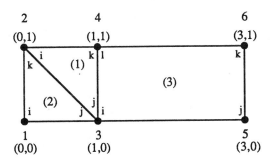

$$S_s^{(1)} = 3, \quad S_s^{(2)} = 6, \quad S_s^{(3)} = 6$$

Using equations 4.22a and 4.27a with

element	node numbers i	j	k	l
1	1	3	2	
2	2	3	4	
3	3	5	6	4

$$[C^{(1)}] = \frac{(3)(0.5)}{12}\begin{bmatrix} 2 & 1 & 1 \\ 1 & 2 & 1 \\ 1 & 1 & 2 \end{bmatrix} = \frac{1}{8}\begin{matrix} & 1 & 3 & 4 \\ \begin{matrix} 1 \\ 3 \\ 4 \end{matrix} & \left[\begin{matrix} 2 & 1 & 1 \\ 1 & 2 & 1 \\ 1 & 1 & 2 \end{matrix}\right] \end{matrix}$$

$$[C^{(2)}] = \frac{(6)(0.5)}{12}\begin{bmatrix} 2 & 1 & 1 \\ 1 & 2 & 1 \\ 1 & 1 & 2 \end{bmatrix} = \frac{1}{4}\begin{matrix} & 2 & 3 & 4 \\ \begin{matrix} 2 \\ 3 \\ 4 \end{matrix} & \left[\begin{matrix} 2 & 1 & 1 \\ 1 & 2 & 1 \\ 1 & 1 & 2 \end{matrix}\right] \end{matrix}$$

$$[C^{(3)}] = \frac{(6)(1)(2)}{9}\begin{bmatrix} 4 & 2 & 1 & 2 \\ 2 & 4 & 2 & 1 \\ 1 & 2 & 4 & 2 \\ 2 & 1 & 2 & 4 \end{bmatrix} = \frac{4}{3}\begin{matrix} & 3 & 5 & 6 & 4 \\ \begin{matrix} 3 \\ 5 \\ 6 \\ 4 \end{matrix} & \left[\begin{matrix} 4 & 2 & 1 & 2 \\ 2 & 4 & 2 & 1 \\ 1 & 2 & 4 & 2 \\ 2 & 1 & 2 & 4 \end{matrix}\right] \end{matrix}$$

$[C]$ global $=$

	1	2	3	4	5	6	
	$1/4^{(1)}$	0	$1/8^{(1)}$	$1/8^{(1)}$	0	0	1
	0	$1/2^{(2)}$	$1/4^{(2)}$	$1/4^{(2)}$	0	0	2
	$1/8^{(1)}$	$1/4^{(2)}$	$1/4^{(1)}+1/2^{(2)}+16/3^{(3)}$	$1/8^{(1)}+1/4^{(2)}+8/3^{(3)}$	$8/3^{(3)}$	$4/3^{(3)}$	3
	$1/8^{(1)}$	$1/4^{(2)}$	$1/8^{(1)}+1/4^{(2)}+8/3^{(3)}$	$1/4^{(1)}+1/2^{(2)}+16/3^{(3)}$	$4/3^{(3)}$	$8/3^{(3)}$	4
	0	0	$8/3^{(3)}$	$4/3^{(3)}$	$16/3^{(3)}$	$8/3^{(3)}$	5
	0	0	$4/3^{(3)}$	$8/3^{(3)}$	$8/3^{(3)}$	$16/3^{(3)}$	6

$$= \begin{bmatrix} 0.25 & 0 & 0.125 & 0.125 & 0 & 0 \\ 0 & 0.5 & 0.250 & 0.250 & 0 & 0 \\ 0.125 & 0.25 & 6.607 & 3.042 & 2.667 & 1.333 \\ 0.125 & 0.25 & 3.042 & 6.083 & 1.333 & 2.667 \\ 0 & 0 & 2.667 & 1.333 & 5.333 & 2.667 \\ 0 & 0 & 1.333 & 2.667 & 2.667 & 5.333 \end{bmatrix}$$

4.5 MODIFICATION OF GLOBAL SYSTEM OF EQUATIONS TO INCORPORATE BOUNDARY CONDITIONS

4.5.1 Dirichlet Boundary Conditions

In most problems the value of the field variable (hydraulic head, pressure head, or solute concentration) is specified at one or more nodes, sometimes called Dirichlet nodes. These specified values constitute the Dirichlet boundary conditions needed to solve the governing differential equation of groundwater flow or solute transport. When Dirichlet boundary conditions are specified, the global system of equations must be modified before a solution can be obtained. The modification procedure reduces the size of the global system of equations to a size (p-d) x (p-d) where p is the number of nodes in the mesh and d is the number of nodes with specified values of the field variable. For example consider the system of equations from the example problem in Section 3.2.

$$\begin{bmatrix} 1/2 & -1/2 & 0 & 0 & 0 \\ -1/2 & 3/2 & -1 & 0 & 0 \\ 0 & -1 & 4/3 & -1/3 & 0 \\ 0 & 0 & -1/3 & 2/3 & -1/3 \\ 0 & 0 & 0 & -1/3 & 1/3 \end{bmatrix} \begin{Bmatrix} h_1 \\ h_2 \\ h_3 \\ h_4 \\ h_5 \end{Bmatrix} = \begin{Bmatrix} 0 \\ 0 \\ 0 \\ 0 \\ 0 \end{Bmatrix} \qquad (4.85)$$

In this example the values of hydraulic head were specified at nodes 1 and 5, $h_1 = 12$ and $h_5 = 0$. This means that the first and fifth rows of equation 4.85 are not needed and can be crossed out :

$$\begin{bmatrix} \cancel{1/2} & \cancel{-1/2} & \cancel{0} & \cancel{0} & \cancel{0} \\ -1/2 & 3/2 & -1 & 0 & 0 \\ 0 & -1 & 4/3 & -1/3 & 0 \\ 0 & 0 & -1/3 & 2/3 & -1/3 \\ \cancel{0} & \cancel{0} & \cancel{0} & \cancel{-1/3} & \cancel{1/3} \end{bmatrix} \begin{Bmatrix} h_1 = 12 \\ h_2 \\ h_3 \\ h_4 \\ h_5 = 0 \end{Bmatrix} = \begin{Bmatrix} \cancel{0} \\ 0 \\ 0 \\ 0 \\ \cancel{0} \end{Bmatrix}$$

We then modify the remaining equations to eliminate columns 1 and 5. For row 2 we have

$$-\frac{1}{2}h_1 + \frac{3}{2}h_2 - h_3 = 0$$

or

$$\frac{3}{2}h_2 - h_3 = \frac{1}{2}h_1 = \frac{1}{2}(12) = 6$$

and for row 4 we have

$$-\frac{1}{3}h_3 + \frac{2}{3}h_4 - \frac{1}{3}h_5 = 0$$

but $h_5 = 0$, so row 4 becomes

$$-\frac{1}{3}h_3 + \frac{2}{3}h_4 = 0$$

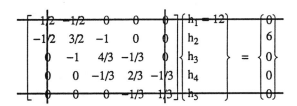

and the modified system of equations becomes

$$\begin{bmatrix} 3/2 & -1 & 0 \\ -1 & 4/3 & -1/3 \\ 0 & -1/3 & 2/3 \end{bmatrix} \begin{Bmatrix} h_2 \\ h_3 \\ h_4 \end{Bmatrix} = \begin{Bmatrix} 6 \\ 0 \\ 0 \end{Bmatrix}$$

which can be solved to give $h_2 = 9.33$, $h_3 = 8.0$, and $h_4 = 4.0$.

Example

Modify the following system of equations if $\psi_1 = 10$ and $\psi_4 = 5$. Solve for ψ_2, ψ_3, and ψ_5

$$\begin{bmatrix} 1/2 & 0 & -1/2 & 0 & 0 \\ 0 & 17/30 & -1/6 & 0 & -2/5 \\ -1/2 & -1/6 & 2/3 & 0 & 0 \\ 0 & 0 & 0 & 1 & -1 \\ 0 & -2/5 & 0 & -1 & 7/5 \end{bmatrix} \begin{Bmatrix} \psi_1 \\ \psi_2 \\ \psi_3 \\ \psi_4 \\ \psi_5 \end{Bmatrix} = \begin{Bmatrix} 0 \\ 0 \\ 0 \\ 0 \\ 0 \end{Bmatrix}$$

Crossing out rows 1 and 4 and columns 1 and 4

For row 2 we have

$$(0)(10) + (17/30)\psi_2 - (1/6)\psi_3 + 0(5) - (2/5)\psi_5 = 0$$

or

$$(17/30)\psi_2 - (1/6)\psi_3 - (2/5)\psi_5 = 0$$

For row 3 :

$$-(1/2)(10) - (1/6)\psi_2 + (2/3)\psi_3 + (0)(5) - (0)\psi_5 = 0$$

or

$$-(1/6)\psi_2 + (2/3)\psi_3 + (0)\psi_5 = 5$$

For row 5 :

$$(0)(10) - (2/5)\psi_2 + (0)\psi_3 - (1)(5) + (7/5)\psi_5 = 0$$

or

$$-(2/5)\psi_2 + (0)\psi_3 + (7/5)\psi_5 = 5$$

and the modified system of equations becomes

$$\begin{bmatrix} 17/30 & -1/6 & -2/5 \\ -1/6 & 2/3 & 0 \\ -2/5 & 0 & 7/5 \end{bmatrix} \begin{Bmatrix} \psi_2 \\ \psi_3 \\ \psi_5 \end{Bmatrix} = \begin{Bmatrix} 0 \\ 5 \\ 5 \end{Bmatrix}$$

from which we obtain $\psi_2 = 6.52$, $\psi_3 = 9.13$, and $\psi_5 = 5.44$.

4.5.2 Neumann Boundary Conditions

Rates of groundwater flow or solute flux can be specified at one or more nodes, sometimes called Neumann nodes. These specified values constitute Neumann boundary conditions and can be used, for example, to represent specified rates of groundwater or solute recharge at the soil surface, or the injection or withdrawl of groundwater or solutes at wells. When Neumann boundary conditions are specified, some entries in the global $\{F\}$ matrix are nonzero. The equation for computing the contribution of element e to the global $\{F\}$ matrix at node i $\{F_i^{(e)}\}$ for the steady-state saturated flow equation was given in Section 3.3 as

$$\{F_i^{(e)}\} = \iiint\limits_{V^{(e)}} N_i^{(e)} q^{(e)} \, dx \, dy \, dz \tag{4.86}$$

where $q^{(e)}$ is the specified groundwater flow rate within element e (positive if groundwater is flowing into the element). This equation can also be used to compute $\{F_i^{(e)}\}$ for unsaturated and transient groundwater flow. The corresponding equation for $\{F_i^{(e)}\}$ in the solute transport equation is

$$\{F_i^{(e)}\} = \iiint\limits_{V^{(e)}} N_i^{(e)} q_c^{(e)} \, dx \, dy \, dz \tag{4.87}$$

where $q_c^{(e)}$ is the specified solute flux within element e.

There are two common situations encountered in practice, The first is the *point source* or *point sink* representing, for example, a well as seen in a plan (map) view of an aquifer. If the point source or sink is located at a node, equations 4.86 and 4.87 become

$$\{F_i^{(e)}\} = q \tag{4.88}$$

and

$$\{F_i^{(e)}\} = q_c \tag{4.89}$$

for the node where the point source or point sink is located and zero at all other nodes in the element. For example if a point sink of -10m³/d is located at node j in the element in Figure 4.18 we have $\{F_j^{(e)}\} = -10$ and $\{F_i^{(e)}\} = \{F_k^{(e)}\} = 0$. The specified flux matrix for the element is $\{F^{(e)}\} = \begin{Bmatrix} 0 \\ -10 \\ 0 \end{Bmatrix}$

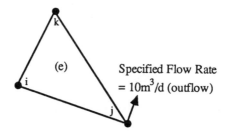

Figure 4.18   Point sink located at a node.

We must be careful not to count the contribution from a point source or sink more than once. For example if a point source of 20 m^3/d is located at node 3 in the mesh in Figure 4.19 it makes no difference whether we consider the point source to be located at node j in element 1, node k in element 2, node j in element 3, or node i in element 4. But we can only count the contribution of 20 m^3/d to node 4 once.

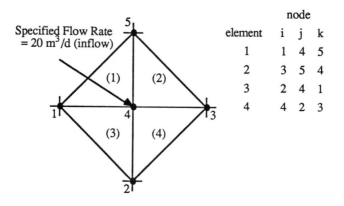

Figure 4.19   Point source located at a node.

If we assign the point source to node k in element 2 the specified flow matrices for the elements are

$$\{F^{(1)}\} = \begin{Bmatrix} 0 \\ 0 \\ 0 \end{Bmatrix} \begin{matrix} 1 \\ 4 \\ 5 \end{matrix} \qquad \{F^{(2)}\} = \begin{Bmatrix} 0 \\ 0 \\ 20 \end{Bmatrix} \begin{matrix} 3 \\ 5 \\ 4 \end{matrix} \qquad \{F^{(3)}\} = \begin{Bmatrix} 0 \\ 0 \\ 0 \end{Bmatrix} \begin{matrix} 2 \\ 4 \\ 1 \end{matrix} \qquad \{F^{(4)}\} = \begin{Bmatrix} 0 \\ 0 \\ 0 \end{Bmatrix} \begin{matrix} 4 \\ 2 \\ 3 \end{matrix}$$

and the global {F} matrix becomes

$$\{F\}_{\text{global}} = \begin{bmatrix} 0^{(1)} + 0^{(3)} \\ 0^{(3)} + 0^{(3)} \\ 0^{(2)} + 0^{(3)} \\ 0^{(1)} + 20^{(2)} + 0^{(3)} + 0^{(4)} \\ 0^{(1)} + 0^{(2)} \end{bmatrix} = \begin{Bmatrix} 0 \\ 0 \\ 0 \\ 20 \\ 0 \end{Bmatrix}$$

where the superscripts indicate each element's contribution to $\{F\}$. Another approach is to divide the point source at node 4 among the four elements joining that node. In this case the element matrices are

$$\{F^{(1)}\} = \begin{Bmatrix} 0 \\ 20/4 \\ 0 \end{Bmatrix} \begin{matrix} 1 \\ 4 \\ 5 \end{matrix} \quad \{F^{(2)}\} = \begin{Bmatrix} 0 \\ 0 \\ 20/4 \end{Bmatrix} \begin{matrix} 3 \\ 5 \\ 4 \end{matrix} \quad \{F^{(3)}\} = \begin{Bmatrix} 0 \\ 20/4 \\ 0 \end{Bmatrix} \begin{matrix} 2 \\ 4 \\ 1 \end{matrix} \quad \{F^{(4)}\} = \begin{Bmatrix} 20/4 \\ 0 \\ 0 \end{Bmatrix} \begin{matrix} 4 \\ 2 \\ 3 \end{matrix}$$

and the global $\{F\}$ matrix becomes

$$\{F\}_{\text{global}} = \begin{bmatrix} 0^{(1)} + 0^{(3)} \\ 0^{(3)} + 0^{(4)} \\ 0^{(2)} + 0^{(4)} \\ 20/4^{(1)} + 20/4^{(2)} + 20/4^{(3)} + 20/4^{(4)} \\ 0^{(1)} + 0^{(2)} \end{bmatrix} = \begin{Bmatrix} 0 \\ 0 \\ 0 \\ 20 \\ 0 \end{Bmatrix}$$

If the source or sink is not located at a node, the specified flow or flux is divided among the nodes of the *element that contains the point source or sink*. The rule is

$$\begin{Bmatrix} F_1^{(e)} \\ \vdots \\ F_n^{(e)} \end{Bmatrix} = q^{(e)} \begin{Bmatrix} N_1^{(e)}(x_0,y_0,z_0) \\ \vdots \\ N_n^{(e)}(x_0,y_0,z_0) \end{Bmatrix} \tag{4.90}$$

for specified rates of groundwater flow, and

$$\begin{Bmatrix} F_1^{(e)} \\ \vdots \\ F_n^{(e)} \end{Bmatrix} = q_c^{(e)} \begin{bmatrix} N_1^{(e)}(x_0,y_0,z_0) \\ \vdots \\ N_n^{(e)}(x_0,y_0,z_0) \end{bmatrix} \tag{4.91}$$

for specified rates of solute flux where n is the number of nodes in element e and the coordinates of the point source or sink are (x_0,y_0,z_0). The application of equations 4.90 and 4.91 is illustrated by the following examples.

Example

Calculate $\{F^{(e)}\}$ for the element shown below $(q = -25\text{m}^3/\text{d})$.

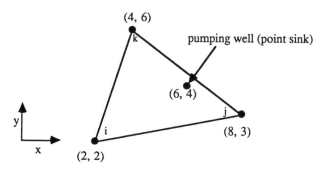

$$2A^{(e)} = \begin{vmatrix} 1 & 2 & 2 \\ 1 & 8 & 3 \\ 1 & 4 & 6 \end{vmatrix} = 22$$

$a_i = (8)(6) - (4)(3) = 36$
$b_i = 3 - 6 = -3$
$c_i = 4 - 8 = -4$

$a_j = (4)(2) - (2)(6) = -4$
$b_j = 6 - 2 = 4$
$c_j = 2 - 4 = -2$

$a_k = (2)(3) - (8)(2) = -10$
$b_k = 2 - 3 = -1$
$c_k = 8 - 2 = 6$

and

$N_i^{(e)}(6,4) = \dfrac{1}{22}(36 - 3(6) - 4(4)) = 2/22$

$N_j^{(e)}(6,4) = \dfrac{1}{22}(-4 + 4(6) - 2(4)) = 12/22$

$N_k^{(e)}(6,4) = \dfrac{1}{22}(-10 - 1(6) + 6(4)) = 8/22$

(Note that the interpolation functions sum to one at the point which is a useful check on the calculations)

$$\{F^{(e)}\} = -25 \begin{Bmatrix} 2/22 \\ 12/22 \\ 8/22 \end{Bmatrix} = \begin{Bmatrix} -2.273 \\ -13.636 \\ -9.091 \end{Bmatrix}$$

Example

Calculate $\{F^{(e)}\}$ for the two elements shown below ($q_c = 10$ kg/d)

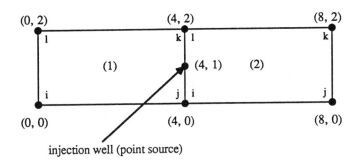

injection well (point source)

The point source can be assigned to either element (or it can be divided between the two elements) We will arbitrarily assign it to element 2. For element 2, $N_i^{(2)}(4,1) = N_l^{(2)}(4,1) = 1/2$ and $N_j^{(2)}(4,1) = N_k^{(2)}(4,1) = 0$ and we have

$$\{F^{(1)}\} = \begin{Bmatrix} 0 \\ 0 \\ 0 \\ 0 \end{Bmatrix} \qquad\qquad \{F^{(2)}\} = 10 \begin{Bmatrix} 1/2 \\ 0 \\ 0 \\ 1/2 \end{Bmatrix} = \begin{Bmatrix} 5 \\ 0 \\ 0 \\ 5 \end{Bmatrix}$$

In the case of a *distributed source or sink* the rate of groundwater or solute flow is specified for a portion of the length, surface area, or volume of an element. In a one-dimensional problem we may wish to specify a flow rate along the length of the element (Figure 4.20a). In this case $\{F_i^{(e)}\}$ is given by

$$\{F_i^{(e)}\} = \int_{L^{(e)}} N_i^{(e)} q \, dx \tag{4.92}$$

for groundwater flow problems or

$$\{F_i^{(e)}\} = \int_{L^{:(e)}} N_i^{(e)} q_c \, dx \tag{4.93}$$

for solute transport problems, where $L^{(e)}$ is the length of the element. The specified rates q and q_c can be functions of x.

Example

Compute $\{F^{(e)}\}$ for the element shown

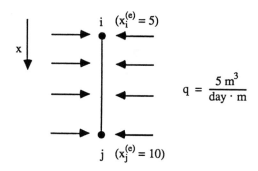

The interpolation functions for the linear bar element are in Figure 4.9. For node i

$$\{F_i^{(e)}\} = \int_{L^{(e)}} \frac{x_j^{(e)} - x}{L^{(e)}} q\, dx \qquad = q \int_{x_i^{(e)}}^{x_j^{(e)}} \frac{x_j^{(e)} - x}{L^{(e)}} dx$$

$$= q \left(\frac{x_j^{(e)} x}{L^{(e)}} - \frac{x^2}{2L^{(e)}} \right) \Bigg|_{x_i^{(e)}}^{x_j^{(e)}} \qquad = q \frac{L^{(e)}}{2}$$

$$= \frac{5m^3}{dm} \left(\frac{10 - 5}{2} \right) m \qquad = \frac{12.5m^3}{d} \qquad = F_j^{(e)}$$

and

$$\{F^{(e)}\} = \begin{Bmatrix} 12.5 \\ 12.5 \end{Bmatrix}$$

For the linear bar element, the distributed source or sink is divided equally between the two nodes of the element (as long as q or q_c are constant along the length of the element). For quadratic or cubic bar elements, the division is different (Figure 4.21). The formulas can be confirmed by integrating the interpolation functions along the length of the element.

quadratic bar element

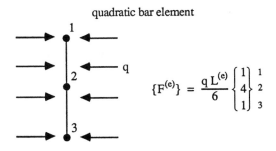

$$\{F^{(e)}\} \;=\; \frac{q\,L^{(e)}}{6}\begin{Bmatrix} 1 \\ 4 \\ 1 \end{Bmatrix}\begin{matrix} 1 \\ 2 \\ 3 \end{matrix}$$

cubic bar element

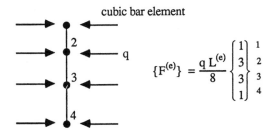

$$\{F^{(e)}\} \;=\; \frac{q\,L^{(e)}}{8}\begin{Bmatrix} 1 \\ 3 \\ 3 \\ 1 \end{Bmatrix}\begin{matrix} 1 \\ 2 \\ 3 \\ 4 \end{matrix}$$

Figure 4.21   **Division of distributed source or sink to nodes in two-types of one-dimensional elements.**

In two- or three-dimensional elements we may wish to specify the rate of groundwater flow or solute flux along one or more element boundaries or surfaces.

In this case $\{F_i^{(e)}\}$ is given by

$$\{F_i^{(e)}\} \;=\; \int_{S^{(e)}} N_i^{(e)} q \, ds \tag{4.94}$$

for groundwater flow problems or

$$\{F_i^{(e)}\} \;=\; \int_{S^{(e)}} N_i^{(e)} q_c \, ds \tag{4.95}$$

where $S^{(e)}$ is the portion of the boundary or surface of element e for which q is specified. Similar formulas can be developed for sources or sinks that are distributed throughout an element volume but these are rarely used in groundwater flow or solute transport modeling.

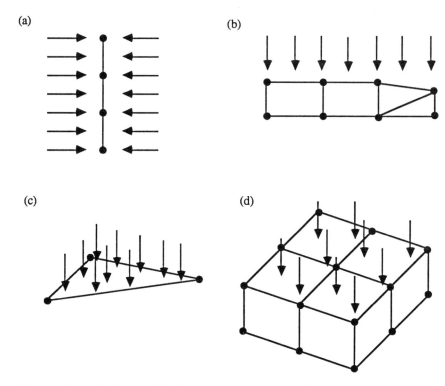

(a)

(b)

(c)

(d)

Figure 4.20   Typical applications of distributed source/sinks for one-, two-, and three-dimensional elements.

Example

Compute $\{F^{(e)}\}$ for the elements shown below

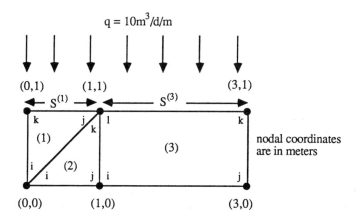

$q = 10 \text{m}^3/\text{d/m}$

(0,1)   (1,1)   (3,1)

$\leftarrow S^{(1)} \rightarrow$ $\leftarrow \quad S^{(3)} \quad \rightarrow$

nodal coordinates are in meters

(0,0)   (1,0)   (3,0)

For element 1,

$$F_i^{(1)} = 0$$

$$F_j^{(1)} = \int_0^1 N_j^{(1)}(x,y)\, q^{(1)}\, ds = \int_0^1 \frac{1}{2A^{(e)}}(a_j + b_j x + c_j y)\, q^{(1)}\, ds$$

but $A^{(1)} = 0.5$, $a_j = 0$, $b_j = 1$, $c_j = 0$, and $q^{(1)} = 10$

$$F_j^{(1)} = \int_0^1 \frac{1}{0.5}(x)(10)\, ds = 5m^3/d = F_k^{(1)}$$

and

$$\{F^{(1)}\} = \begin{Bmatrix} 0 \\ 5 \\ 5 \end{Bmatrix}$$

For element 2,

$$\{F^{(2)}\} = \begin{Bmatrix} 0 \\ 0 \\ 0 \end{Bmatrix} \qquad \text{(why ?)}$$

For element 3,

$$F_i^{(3)} = F_j^{(3)} = 0$$

$$F_k^{(3)} = \int_0^2 N_k^{(3)}(s,t)\, q^{(3)}\, ds$$

$$= \int_0^2 \left(\frac{st}{4a^{(e)}b^{(e)}} \right) 10\, ds$$

but $2a^{(e)} = 1$, $2b^{(e)} = 2$, and $t = 1$ along the boundary and we have

$$F_k^{(3)} = \int_0^2 \left(\frac{s}{2} \right) 10\, ds = 10m^3/d = F_l^{(3)}$$

and

$$\{F^{(3)}\} = \begin{Bmatrix} 0 \\ 0 \\ 10 \\ 10 \end{Bmatrix}$$

In the previous example the element interpolation functions were linear and the distributed source was divided equally between the two nodes of the element on the boundary $S^{(e)}$. For quadratic or cubic quadrilateral elements the division would be different (Figure 4.22)

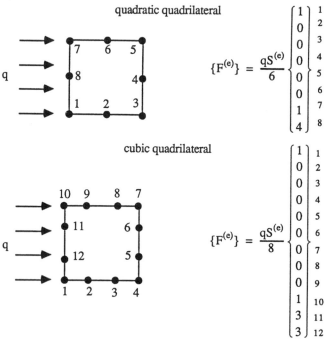

quadratic quadrilateral

$$\{F^{(e)}\} = \frac{qS^{(e)}}{6} \begin{Bmatrix} 1 \\ 0 \\ 0 \\ 0 \\ 0 \\ 0 \\ 1 \\ 4 \end{Bmatrix} \begin{matrix} 1 \\ 2 \\ 3 \\ 4 \\ 5 \\ 6 \\ 7 \\ 8 \end{matrix}$$

cubic quadrilateral

$$\{F^{(e)}\} = \frac{qS^{(e)}}{8} \begin{Bmatrix} 1 \\ 0 \\ 0 \\ 0 \\ 0 \\ 0 \\ 0 \\ 0 \\ 0 \\ 1 \\ 3 \\ 3 \end{Bmatrix} \begin{matrix} 1 \\ 2 \\ 3 \\ 4 \\ 5 \\ 6 \\ 7 \\ 8 \\ 9 \\ 10 \\ 11 \\ 12 \end{matrix}$$

Figure 4.22   Division of distributed source or sink to nodes in two-types of two-dimensional elements.

Example

Compute $\{F^{(e)}\}$ for the elements shown below

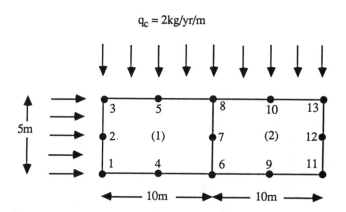

$q_c = 2\text{kg/yr/m}$

5m

10m   10m

For element 1

$$F_1^{(1)} = \frac{(5)(2)}{6} = 1.7 \text{ kg/yr}$$

$$F_2^{(1)} = \frac{(5)(2)(4)}{6} = 6.7 \text{ kg/yr}$$

$$F_3^{(1)} = \frac{(5)(2)}{6} + \frac{(10)(2)}{6} = 5.0 \text{ kg/yr}$$

$$F_5^{(1)} = \frac{(10)(2)(4)}{6} = 13.3 \text{ kg/yr}$$

$$F_8^{(1)} = \frac{(10)(2)}{6} + \frac{(10)(2)}{6} = 6.7 \text{ kg/yr}$$

$$F_{10}^{(2)} = \frac{(10)(2)(4)}{6} = 13.3 \text{ kg/yr}$$

$$F_{13}^{(2)} = \frac{(10)(2)}{6} = 3.3 \text{ kg/yr}$$

and

$$\{F\} = \begin{Bmatrix} 1.7 \\ 6.7 \\ 5.0 \\ 0 \\ 13.3 \\ 0 \\ 0 \\ 6.7 \\ 0 \\ 13.3 \\ 0 \\ 0 \\ 3.3 \end{Bmatrix} \begin{matrix} 1 \\ 2 \\ 3 \\ 4 \\ 5 \\ 6 \\ 7 \\ 8 \\ 9 \\ 10 \\ 11 \\ 12 \\ 13 \end{matrix}$$

NOTES AND ADDITIONAL READING

1. For problems with axisymmetry, the element conductance matrix is given by equation 3.169. This equation is valid for the linear triangle and rectangle elements (Figure 4.6 and 4.7) and for the quadrilateral elements (Figures 4.10 to 4.12). However, the interpolation functions and their derivatives must be written using an axisymmetric (r,z) coordinate system. For example, consider the axisymmetric, linear triangle element. The interpolation functions are in Figure 4.23 (Segerlind, 1984).

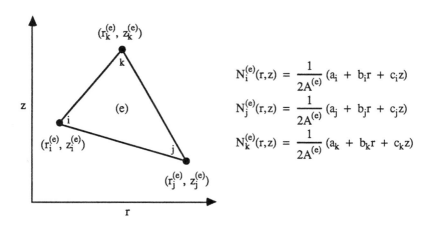

$$N_i^{(e)}(r,z) = \frac{1}{2A^{(e)}} (a_i + b_i r + c_i z)$$

$$N_j^{(e)}(r,z) = \frac{1}{2A^{(e)}} (a_j + b_j r + c_j z)$$

$$N_k^{(e)}(r,z) = \frac{1}{2A^{(e)}} (a_k + b_k r + c_k z)$$

Figure 4.23   Interpolation functions for the axisymmetric, linear triangle element.

The derivatives of the interpolation functions are

$$\frac{\partial N_i^{(e)}}{\partial r} = \frac{b_i}{2A^{(e)}} \qquad \frac{\partial N_j^{(e)}}{\partial r} = \frac{b_j}{2A^{(e)}} \qquad \frac{\partial N_k^{(e)}}{\partial r} = \frac{b_k}{2A^{(e)}}$$

$$\frac{\partial N_i^{(e)}}{\partial z} = \frac{c_i}{2A^{(e)}} \qquad \frac{\partial N_j^{(e)}}{\partial z} = \frac{c_j}{2A^{(e)}} \qquad \frac{\partial N_k^{(e)}}{\partial z} = \frac{c_k}{2A^{(e)}}$$

where $A^{(e)}$ = Area of element

$$= \frac{1}{2} \begin{vmatrix} 1 & r_i^{(e)} & z_i^{(e)} \\ 1 & r_j^{(e)} & z_j^{(e)} \\ 1 & r_k^{(e)} & z_k^{(e)} \end{vmatrix}$$

$$a_i = r_j^{(e)} z_k^{(e)} - r_k^{(e)} z_j^{(e)} \qquad a_j = r_k^{(e)} z_i^{(e)} - r_i^{(e)} z_k^{(e)} \qquad a_k = r_i^{(e)} z_j^{(e)} - r_j^{(e)} z_i^{(e)}$$

$$b_i = z_j^{(e)} - z_k^{(e)} \qquad b_j = z_k^{(e)} - z_i^{(e)} \qquad b_k = z_i^{(e)} - z_j^{(e)}$$

$$c_i = r_k^{(e)} - r_j^{(e)} \qquad c_j = r_i^{(e)} - r_k^{(e)} \qquad c_k = r_j^{(e)} - r_i^{(e)}$$

The element matrices are easily computed because of the relation

$$\iint\limits_{A^{(e)}} 2\pi r \, dr \, dz = 2\pi \bar{r}^{(e)} A^{(e)}$$

where $\bar{r}^{(e)}$ is the r coordinate of the centroid of $A^{(e)}$

$$\bar{r}^{(e)} = \frac{r_i^{(e)} + r_j^{(e)} + r_k^{(e)}}{3}$$

The results are

$$\underset{3 \times 3}{[K^{(e)}]} = \frac{2\pi \bar{r}^{(e)} K_r^{(e)}}{4A^{(e)}} \begin{bmatrix} b_i^2 & b_i b_j & b_i b_k \\ b_j b_i & b_j^2 & b_j b_k \\ b_k b_i & b_k b_j & b_k^2 \end{bmatrix} + \frac{2\pi \bar{r}^{(e)} K_z^{(e)}}{4A^{(e)}} \begin{bmatrix} c_i^2 & c_i c_j & c_i c_k \\ c_j c_i & c_j^2 & c_j c_k \\ c_k c_i & c_k c_j & c_k^2 \end{bmatrix} \qquad (4.96)$$

Problems

1. Verify that the approximate solution for $\hat{C}$ is continuous along the boundary joining the pairs of elements shown below

a)

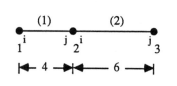

c)

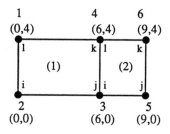

b)

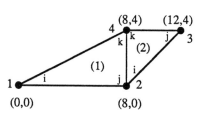

2. Verify that the interpolation functions for the elements shown below sum to one at an arbitrary point (x) or (x, y)

a)

c)

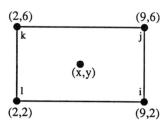

b)

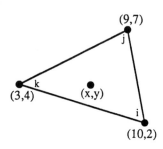

3. Calculate $[K^{(e)}]$, $[C^{(e)}]$, $[D^{(e)}]$, and $[A^{(e)}]$ (consistent formulation) for the elements in problem 2 if $K_x = K_y = 2$, $D_{xx} = D_{yy} = 1$, $D_{xy} = D_{yx} = 2$, $D_{yy} = 1$, $v_x = 10$, $v_y = 1$, $S_s = 1$, $\theta = n = 0.4$, $\rho_b = 1.1$, $K_d = 5$, and $\lambda = 0$.

4. Derive equation 4.18a by substituting the interpolation functions and interpolation function derivatives for the linear bar element into equation 4.12 (remember that, for one-dimensional elements the integrations are performed over the length of the element).

5 Derive equation 4.20 or 4.21 by substituting the interpolation function derivatives for the linear triangle element into equation 4.6 or 4.7 (remember that, for two-dimensional elements, the integrations for $[K^{(e)}]$ and $[K^{(e)}(\psi)]$ are performed over the area of the element).

6. Derive the following equations for the linear triangle element

a) Equation 4.22a   c) Equation 4.25a
b) Equation 4.24b

7. Plot N_i, $\dfrac{\partial N_i}{\partial \varepsilon}$, and $\dfrac{\partial N_i}{\partial \eta}$, $i = 1$ to 12 for the cubic quadrilateral element in Figure 4.12 along the line $\varepsilon = 0$ from $\eta = -1$ to $\eta = 1$ and along the line $\eta = 0$ from $\varepsilon = -1$ to $\varepsilon = 1$.

8. Given the interpolation functions for the linear rectangle element (Figure 4.8) verify the entries in the first row and second column of the element matrices in

a) Equation 4.26   c) Equation 4.29
b) Equation 4.27   b) Equation 4.30

9. Repeat problem 5 for the third row and second column of the element matrices.

10. Verify that the interpolation functions for the following element types sum to one at an arbitrary point (ε_0), (ε_0, η_0), or $(\varepsilon_0, \eta_0, \zeta_0)$.

a) Linear bar (Figure 4.9a).
b) Quadratic bar (Figure 4.9b).
c) Linear quadrilateral (Figure 4.10).
d) Linear parallelepiped (Figure 4.13).

11. Compute [J], [J^{-1}], and |J| at the center ($\varepsilon = \eta = 0$) of the elements shown below

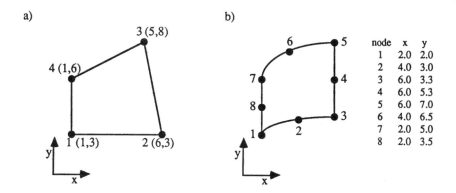

a)

3 (5,8)

4 (1,6)

1 (1,3)   2 (6,3)

b)

node	x	y
1	2.0	2.0
2	4.0	3.0
3	6.0	3.3
4	6.0	5.3
5	6.0	7.0
6	4.0	6.5
7	2.0	5.0
8	2.0	3.5

c)

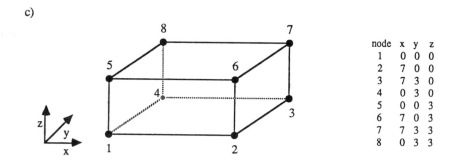

node	x	y	z
1	0	0	0
2	7	0	0
3	7	3	0
4	0	3	0
5	0	0	3
6	7	0	3
7	7	3	3
8	0	3	3

12. Perform the following integrations analytically and numerically using Gauss quadrature

a) $\displaystyle\int_{-1}^{1} \varepsilon^3 + 3\varepsilon 2 + 9\varepsilon\, d\varepsilon$

b) $\displaystyle\int_{-1}^{1} 8\varepsilon^4 + \varepsilon^2\, d\varepsilon$

c) $\displaystyle\int_{-1}^{1}\int_{-1}^{1} \varepsilon^2 + 2\varepsilon\eta + \eta^2\, d\varepsilon\, d\eta$

d) $\displaystyle\int_{-1}^{1}\int_{-1}^{1} \varepsilon^4 + 2\varepsilon^2\eta^3 + \eta^2\, d\varepsilon\, d\eta$

e) $\displaystyle\int_{-1}^{1}\int_{-1}^{1}\int_{-1}^{1} \varepsilon\eta\zeta\, d\varepsilon\, d\eta\, d\zeta$

f) $\displaystyle\int_{-1}^{1}\int_{-1}^{1}\int_{-1}^{1} \varepsilon^3\eta + \varepsilon\eta\zeta^2 + 3\zeta^2\, d\varepsilon\, d\eta\, d\zeta$

13. Compute the contribution of the first Gauss point $\left(\frac{1}{\sqrt{3}}, \frac{1}{\sqrt{3}}\right)$ to the element conductance matrix $[K^{(e)}]$ for the linear quadrilateral elements shown below if $K_x = 1$ and $K_y = 2$ for all elements

a)

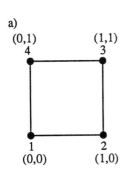

b)

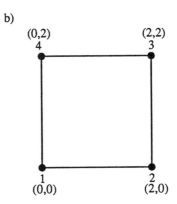

c)

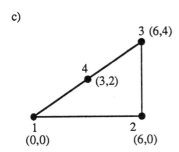

d)

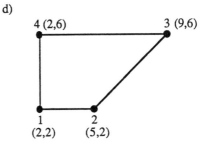

14. Repeat problem 13 except compute the contribution of the first Gauss point to the element capacitance matrix $[C^{(e)}]$ (consistent formulation). Let $S_s = 5$ for all elements.

15. Repeat problem 13 for the element in problem 11b.

16. Repeat problem 14 for the element in problem 11b.

17. Repeat problem 13 for the element in problem 11c.

18. Repeat problem 14 for the element in problem 11c.

19. Given the following element conductance matrices assemble $[K]$ for the mesh shown below global

$$[K^{(1)}] = \frac{1}{2}\begin{bmatrix} 1 & -1 & 0 \\ -1 & 2 & -1 \\ 0 & -1 & 1 \end{bmatrix}$$

$$[K^{(3)}] = \frac{1}{2}\begin{bmatrix} 1 & -1 & 0 \\ -1 & 2 & -1 \\ 0 & -1 & 1 \end{bmatrix}$$

$$[K^{(2)}] = \frac{1}{6}\begin{bmatrix} 4 & -1 & -2 & -1 \\ -1 & 4 & -1 & -2 \\ -2 & -1 & 4 & -1 \\ -1 & -2 & -1 & 4 \end{bmatrix}$$

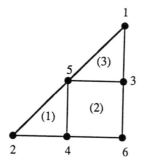

element	node numbers			
	i	j	k	l
1	1	2	4	--
2	2	3	5	4
3	4	6	5	--

20. Repeat problem 19 if the nodes are numbered as shown below

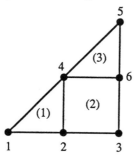

21. Write the system of equations

$$[K] \{h\} = \{F\}$$
$$\text{global} \qquad \text{global}$$

for problem 19 with

$$\{F\} = \begin{Bmatrix} 175 \\ 175 \\ 175 \\ 0 \\ 0 \\ 0 \end{Bmatrix}$$
$$\text{global}$$

Given that $h_6 = 0$, modify the system of equations and solve for $h_1, h_2, h_3, h_4,$ and h_5.

22. Repeat problem 21 if $h_1 = h_2 = h_3 = 0$.

23. Solve the following one-dimensional, steady-state groundwater flow problem if $h_1 = 10$ m and $h_5 = 0$ m. Plot $h(x)$, $0 \le x \le 19$ m

$(x = 0)$ (1)   (2)   (3)   (4) $(x = 19)$

  1   2   3   4   5

     $L^{(1)} = 4$ $L^{(2)} = 6$ $L^{(3)} = 5$ $L^{(4)} = 4$ m

     $K_x^{(1)} = 2$ $K_x^{(2)} = 1$ $K_x^{(3)} = 2$ $K_x^{(4)} = 4$ m/d

24. Repeat problem 23 if $K_x^{(1)} = K_x^{(2)} = K_x^{(3)} = K_x^{(4)} = 2$ m/d.

25. Given the following element conductance matrices assemble [K] for the mesh show
                                                              global
    below. Modify the system of equations and solve for h at each node if $h_1 = h_2 = 10$
    and $h_3 = h_6 = h_8 = 0$, $\{F\} = \{0\}$.

$$[K^{(1)}] = \frac{1}{2}\begin{bmatrix} 1 & -1 & 0 \\ -1 & 2 & -1 \\ 0 & -1 & 1 \end{bmatrix} \qquad [K^{(2)}] = \frac{1}{6}\begin{bmatrix} 4 & -1 & -2 & -1 \\ -1 & 4 & -1 & -2 \\ -2 & -1 & 4 & -1 \\ -1 & -2 & -1 & 4 \end{bmatrix}$$

$$[K^{(3)}] = \begin{bmatrix} 1.45 & -1.74 & 0.72 & -0.36 & 0.64 & -0.92 & 0.53 & -0.31 \\ -1.74 & 3.82 & -1.74 & 0 & -0.92 & 1.51 & -0.92 & 0 \\ 0.74 & -1.74 & 1.44 & -0.31 & 0.53 & -0.92 & 0.64 & -0.36 \\ -0.36 & 0 & -0.31 & 1.96 & -0.31 & 0 & -0.36 & -0.62 \\ 0.64 & -0.92 & 0.53 & -0.31 & 1.44 & -1.74 & 0.72 & -0.36 \\ -0.92 & 1.51 & -0.92 & 0 & -1.74 & 3.82 & -1.74 & 0 \\ 0.53 & -0.92 & 0.64 & -0.36 & 0.72 & -1.74 & 1.44 & -0.31 \\ -0.31 & 0 & -0.36 & -0.62 & -0.36 & 0 & -0.31 & 1.96 \end{bmatrix}$$

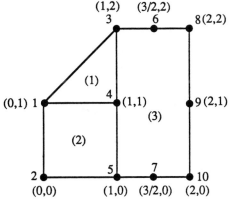

element	node numbers			
	i	j	k	l
1	1	4	3	-
2	2	5	4	1

element	node numbers							
	1	2	3	4	5	6	7	8
3	5	7	10	9	8	6	3	4

$$K_x^{(e)} = K_y^{(e)} = 1 \text{ for all elements}$$

26. Given the following element conductance matrices assemble [K] for the mesh show
global
below. Modify the system of equations and solve for h at each node if $h_5 = 0$ and $\{F\}^T$ = {10,10,10,10,0,0,0,0,-10,-10,-10,-10}.

$$[K^{(1)}] = \begin{bmatrix} 0.64 & 0.17 & 0.00 & 0.15 & 0.22 & -0.17 & -0.22 & -0.36 \\ 0.17 & 0.64 & 0.15 & 0.00 & -0.17 & -0.22 & -0.36 & -0.21 \\ 0.00 & 0.15 & 0.51 & 0.11 & -0.10 & -0.13 & -0.32 & -0.22 \\ 0.15 & 0.00 & 0.11 & 0.51 & -0.13 & -0.10 & -0.22 & -0.32 \\ -0.22 & -0.17 & -0.10 & -0.13 & 0.51 & 0.15 & -0.05 & 0.01 \\ -0.17 & -0.22 & -0.13 & -0.10 & 0.15 & 0.51 & 0.01 & -0.05 \\ -0.21 & -0.36 & -0.32 & -0.22 & -0.05 & 0.01 & 0.84 & 0.32 \\ -0.36 & -0.21 & -0.22 & -0.32 & 0.01 & -0.05 & 0.32 & 0.84 \end{bmatrix}$$

$$[K^{(2)}] = \begin{bmatrix} 0.84 & 0.32 & -0.05 & 0.05 & -0.32 & -0.22 & -0.21 & -0.36 \\ 0.32 & 0.84 & 0.05 & -0.05 & -0.22 & -0.32 & -0.36 & -0.21 \\ -0.05 & 0.05 & 0.51 & 0.15 & -0.10 & -0.13 & -0.22 & -0.17 \\ 0.05 & -0.05 & 0.15 & 0.51 & -0.13 & -0.10 & -0.17 & -0.22 \\ -0.32 & -0.22 & -0.10 & -0.13 & 0.51 & 0.11 & 0.00 & 0.15 \\ -0.22 & -0.32 & -0.13 & -0.10 & 0.11 & 0.51 & 0.15 & 0.00 \\ -0.21 & -0.36 & -0.22 & -0.17 & 0.00 & 0.15 & 0.64 & 0.17 \\ -0.36 & -0.21 & -0.17 & -0.22 & 0.15 & 0.00 & 0.17 & 0.64 \end{bmatrix}$$

element	node numbers							
i = 1	2	3	4	5	6	7	8
1	2	3	7	8	1	4	6	5
2	8	7	11	12	5	6	10	9

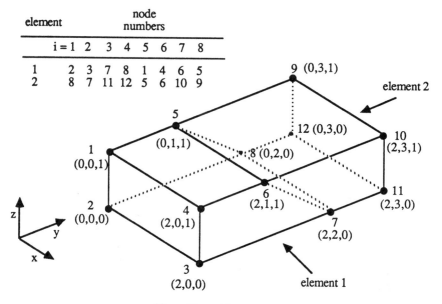

$K_x^{(e)} = K_y^{(e)} = K_z^{(e)} = 1$ for all elements

27. Assume that the solution to problem 23 represents initial conditions. At time $t = 0$ h_1 is increased to 15 m and held constant thereafter. Solve for $h_2 \cdots h_5$ at times $t = 1, 2,$ and 3 d. Use a time step of 1d. $S_s^{(1)} = 1$, $S_s^{(2)} = 1$, $S_s^{(3)} = 6$, and $S_s^{(4)} = 3 \ m^{-1}$.

28. Compute $\{F^{(e)}\}$ for the elements shown below (nodal coordinates in meters)

a)   b)

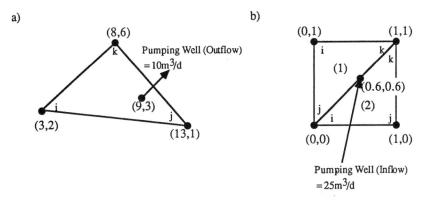

c)   d)

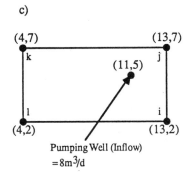
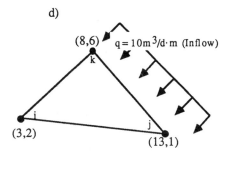

e)   f)

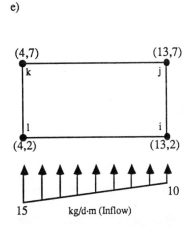
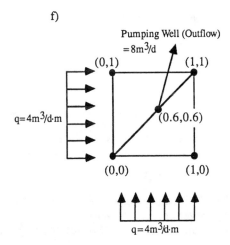

Chapter 5

STEP 4: SOLVE SYSTEM OF EQUATIONS

As we saw in Chapter 3 the application of Galerkin's method to the equations of groundwater flow and solute transport results in systems of equations that can be written in matrix form:

1. Steady-State, Saturated Flow Equation

$$[K] \{h\} = \{F\} \tag{5.1}$$

2. Steady-State, Unsaturated Flow Equation

$$[K(\psi)]\{\psi\} = \{F\} \tag{5.2}$$

3. Transient, Saturated Flow Equation

$$([C] + \omega\Delta t\, [K]\,)\{h\}_{t+\Delta t} = ([C] - (1-\omega)\, \Delta t\, [K]\,)\{h\}_t$$
$$+ \Delta t\, (\, (1-\omega)\{F\}_t + \omega\{F\}_{t+\Delta t}\,) \tag{5.3}$$

4. Transient, Unsaturated Flow Equation

$$(\, [C(\psi)] + \omega\Delta t\, [K(\psi)]\,)\{\psi\}_{t+\Delta t} = (\, [C(\psi)] - (1-\omega)\Delta t\, [K(\psi)]\,)\{\psi\}_t$$
$$+ \Delta t\, (\, (1-\omega)\{F\}_t + \omega\{F\}_{t+\Delta t}\,) \tag{5.4}$$

5. Solute Transport Equation

$$([A] + \omega\Delta t\, [D]\,)\{C\}_{t+\Delta t} = ([A] - (1-\omega)\Delta t\, [D]\,)\{C\}_t$$
$$+ \Delta t\, (\, (1-\omega)\{F\}_t + \omega\{F\}_{t+\Delta t}\,) \tag{5.5}$$

This chapter describes methods that can be used to solve these systems of equations to obtain values of hydraulic head, pressure head, or solute concentration at each node in the mesh (and for each time-step in the case of transient flow or solute transport problems). Equations 5.1, 5.3, and 5.5 are systems of *linear* equations of the form

$$[M]\{X\} = \{B\} \tag{5.6}$$

where $[M]$ is matrix of known coefficients m_{ij}, $\{X\}$ is a vector of the unknowns x_i, and $\{B\}$ is a vector of known values b_i. Equation 5.6 is a system of linear equations because none of the coefficients in $[M]$ are a function of the unknowns $\{X\}$. Equations 5.2 and 5.4

are systems of *nonlinear* equations of the form

$$[M(X)]\{X\} = \{B\} \qquad (5.7)$$

where $[M(X)]$ is a matrix of known coefficients with entries $m_{ij}(x)$, $\{X\}$ is a vector of the unknowns x_i, and $\{B\}$ is a vector of known values b_i. Equation 5.7 is nonlinear because some or all of the coefficients in $[M(X)]$ are functions of the unknowns $\{X\}$. We will first describe a method that can be used to solve equation 5.6.

5.1 PROCEDURE FOR SOLVING SYSTEM OF LINEAR EQUATIONS

Equation 5.6 can be written in an expanded form as follows

$$\begin{bmatrix} m_{11} & m_{12} & \cdots & m_{1n} \\ m_{21} & m_{22} & \cdots & m_{2n} \\ \vdots & \vdots & \cdots & \vdots \\ m_{n1} & m_{n2} & \cdots & m_{nn} \end{bmatrix} \begin{Bmatrix} x_1 \\ x_2 \\ \vdots \\ x_n \end{Bmatrix} = \begin{Bmatrix} b_1 \\ b_2 \\ \vdots \\ b_n \end{Bmatrix} \qquad (5.8)$$

There are several different numerical methods that can be used to solve equation 5.8. In selecting a method for use in solving the equations of groundwater flow and solute transport the following criteria should be considered:

1. $[M]$ is diagonally dominant (i.e., for any row the entry on the main diagonal is larger than the other entries in the row), banded, and sparse (i.e., contains many zero entries).
2. $[M]$ may or may not be symmetrical (see Appendix IV).
3. For each matrix $[M]$ we may wish to solve equation 5.8 for several different right-hand side vectors $\{B\}$.

One choice of method that meets these criteria and which has been widely used for this purpose is the Choleski method (Cook, 1981). We will first describe the Choleski method for the case when $[M]$ is stored in full matrix storage mode because the notation is simpler. We will then describe a modification of the Choleski method for the case when $[M]$ is stored in vector storage mode. Vector storage reduces the sizes of the arrays that must be stored and manipulated when the Choleski method is implemented in a computer program (see Chapter 13).

5.1.1 Choleski Method for Nonsymmetric Matrix in Full Matrix Storage

The Choleski method is a direct method for solving a system of linear algebraic equations which makes use of the fact that any square matrix $[M]$ can be expressed as the product of a lower triangular matrix $[L]$ and an upper triangular matrix $[U]$

$$[M] = [L][U] \qquad (5.9a)$$

or

$$\begin{bmatrix} m_{11} & m_{12} & \cdots & m_{1n} \\ m_{21} & m_{22} & \cdots & m_{2n} \\ \vdots & \vdots & \cdots & \vdots \\ m_{n1} & m_{n2} & \cdots & m_{nn} \end{bmatrix} = \begin{bmatrix} l_{11} & 0 & \cdots & \cdots & 0 \\ l_{21} & l_{22} & 0 & \cdots & 0 \\ \vdots & \vdots & \vdots & & \vdots \\ l_{n1} & l_{n2} & l_{n3} & \cdots & l_{nn} \end{bmatrix} \begin{bmatrix} 1 & u_{12} & u_{13} & \cdots & u_{1n} \\ 0 & 1 & u_{23} & \cdots & u_{2n} \\ 0 & 0 & 1 & \cdots & u_{3n} \\ \vdots & \vdots & \vdots & & \vdots \\ 0 & 0 & 0 & \cdots & 1 \end{bmatrix} \qquad (5.9b)$$

We say that matrix [M] is *decomposed* or *factored* into the product of two triangular matrices and this step of the Choleski method is sometimes referred to as the *triangular decomposition* of [M]. The entries of [L] and [U] are given by

$$l_{ij} = m_{ij} - \sum_{k=1}^{j-1} l_{ik} u_{kj}, \quad i \geq j \tag{5.10a}$$

$$l_{ij} = 0, \quad\quad\quad\quad\quad i < j \tag{5.10b}$$

$$u_{ij} = \frac{m_{ij} - \sum_{k=1}^{i-1} l_{ik} u_{kj}}{l_{ii}}, \quad i < j \tag{5.10c}$$

$$u_{ij} = 1, \quad i = j \tag{5.10d}$$

$$u_{ij} = 0, \quad i > j \tag{5.10e}$$

Example

Perform triangular decomposition on the following matrix

$$[M] = \begin{bmatrix} m_{11} & m_{12} & m_{13} & m_{14} \\ m_{21} & m_{22} & m_{23} & m_{24} \\ m_{31} & m_{32} & m_{33} & m_{34} \\ m_{41} & m_{42} & m_{43} & m_{44} \end{bmatrix} = \begin{bmatrix} 2 & 1 & -1 & 1 \\ -1 & 4 & 1 & -1 \\ -2 & -1 & 4 & 1 \\ -1 & -2 & 1 & 2 \end{bmatrix}$$

$l_{11} = m_{11} = 2$

$l_{12} = l_{13} = l_{14} = 0$

$u_{11} = 1$

$u_{12} = m_{12}/l_{11} = 1/2 = 0.5$

$u_{13} = m_{13}/l_{11} = -1/2 = -0.5$

$u_{14} = m_{14}/l_{11} = 1/2 = 0.5$

$l_{21} = m_{21} = -1$

$l_{22} = m_{22} - l_{21}u_{12} = 4 - (-1)(0.5) = 4.5$

$l_{23} = l_{24} = 0$

$u_{21} = 0, \quad u_{22} = 1$

$u_{23} = (m_{23} - l_{21}u_{13})/l_{22} = (1 - (-1)(-0.5))/4.5 = 0.11$

$u_{24} = (m_{24} - l_{21}u_{14})/l_{22} = (-1 - (-1)(0.5))/4.5 = -0.11$

$l_{31} = m_{31} = -2$

$l_{32} = m_{32} - l_{31}u_{12} = -1 - (-2)(0.5) = 0.0$

$l_{33} = m_{33} - l_{31}u_{13} - l_{32}u_{23} = 4 - (-2)(-0.5) - (0)(0.11) = 3.0$

$l_{34} = 0$

$u_{31} = u_{32} = 0, \quad u_{33} = 1$

$u_{34} = (m_{34} - l_{31}u_{14} - l_{32}u_{24})/l_{33} = (1-(-2)(0.5) - (0)(-0.11)/3.0 = 0.67$

$l_{41} = m_{41} = -1$

$l_{42} = m_{42} - l_{41}u_{12} = -2 -(-1)(0.5) = -1.5$

$l_{43} = m_{43} - l_{41}u_{13} - l_{42}u_{23} = 1 - (-1)(-0.5) - (-1.5)(0.11) = 0.67$

$l_{44} = m_{44} - l_{41}u_{14} - l_{42}u_{24} - l_{43}u_{34}$

$\quad = 2 - (-1)(0.5) - (-1.5)(-0.11) - (0.67)(0.67) = 1.89$

$u_{41} = u_{42} = u_{43} = 0, \quad u_{44} = 1$

$$[L] = \begin{bmatrix} 2.0 & 0.0 & 0.0 & 0.0 \\ -1.0 & 4.5 & 0.0 & 0.0 \\ -2.0 & 0.0 & 3.0 & 0.0 \\ -1.0 & -1.5 & 0.67 & 1.89 \end{bmatrix} \quad [U] = \begin{bmatrix} 1.0 & 0.5 & -0.5 & 0.5 \\ 0.0 & 1.0 & 0.11 & -0.11 \\ 0.0 & 0.0 & 1.0 & 0.67 \\ 0.0 & 0.0 & 0.0 & 1.0 \end{bmatrix}$$

and it is easy to verify that $[M]= [L][U]$.

Once $[M]$ has been decomposed into lower and upper triangular matrices the solution of the system of equations for any choice of $\{B\}$ is very simple. Because $[M] = [L][U]$ we can write

$$[M]\{X\} = \{B\} = [L][U]\{X\} = \{B\} \tag{5.11}$$

If we define a vector $\{Z\}$ as

$$[U]\{X\} = \{Z\} \tag{5.12}$$

we can write

$$[L]\{Z\} = \{B\} \tag{5.13a}$$

$$\begin{aligned} l_{11}z_1 &= b_1 \\ l_{21}z_1 + l_{22}z_2 &= b_2 \\ \vdots \quad &\quad \vdots \quad \vdots \\ l_{n1}z_1 + l_{n2}z_2 + \cdots + l_{nn}z_n &= b_n \end{aligned} \tag{5.13b}$$

which can be solved for the values in $\{Z\}$ using

$$z_i = \frac{\left(b_i - \sum_{k=1}^{i-1} l_{ik} z_k\right)}{l_{ii}}, \qquad i = 1 \text{ to } n \tag{5.14}$$

After we have computed the values in $\{Z\}$ we can then solve for the values in $\{X\}$ using equation 5.12 (this step is sometimes called "backward substitution")

$$x_1 + u_{12}x_2 + u_{13}x_3 + \cdots + u_{1n}x_n = z_1$$
$$x_2 + u_{23}x_3 + \cdots + u_{2n}x_n = z_2$$
$$x_3 + \cdots + u_{3n}x_n = z_3 \qquad (5.15)$$
$$\vdots \qquad \vdots$$
$$x_n = z_n$$

The solution is given by

$$x_{n+1-i} = z_{n+1-i} - \sum_{k=1}^{i-1} u_{n+1-i,n+1-k}\, x_{n+1-k}, \qquad i = 1 \text{ to } n \qquad (5.16)$$

Example

Solve the following system of equations

$$\begin{bmatrix} 2 & 1 & -1 & 1 \\ -1 & 4 & 1 & -1 \\ -2 & -1 & 4 & 1 \\ -1 & -2 & 1 & 2 \end{bmatrix} \begin{Bmatrix} x_1 \\ x_2 \\ x_3 \\ x_4 \end{Bmatrix} = \begin{Bmatrix} 1 \\ 0 \\ 0 \\ 0 \end{Bmatrix}$$

Triangular decomposition of [M] was performed in the previous example and using equation 5.13 we have

$$\begin{bmatrix} 2.0 & 0 & 0 & 0 \\ -1.0 & 4.5 & 0 & 0 \\ -2.0 & 0.0 & 3.0 & 0 \\ -1.0 & -1.5 & 0.67 & 1.89 \end{bmatrix} \begin{Bmatrix} z_1 \\ z_2 \\ z_3 \\ z_4 \end{Bmatrix} = \begin{Bmatrix} 1 \\ 0 \\ 0 \\ 0 \end{Bmatrix}$$

$$z_1 = b_1 / l_{11} = 1/2 = 0.50$$
$$z_2 = (b_2 - l_{21}z_1) / l_{22} = (0-(-1.0)(0.5)) / 4.5 = 0.11$$
$$z_3 = (b_3 - l_{31}z_1 - l_{32}z_2) / l_{33} = (0 - (-2.0)(0.5) - (0)(0.11) / 3 = 0.33$$
$$z_4 = (b_4 - l_{41}z_1 - l_{42}z_2 - l_{43}z_3) / l_{44}$$
$$= (0-(-1.0)(0.5) - (-1.5)(0.11) - (0.67)(0.33)) / 1.89 = 0.24$$

In the next step we have

$$\begin{bmatrix} 1.0 & 0.5 & -0.5 & 0.5 \\ 0 & 1.0 & 0.11 & -0.11 \\ 0 & 0 & 1.0 & 0.67 \\ 0 & 0 & 0 & 1.0 \end{bmatrix} \begin{Bmatrix} x_1 \\ x_2 \\ x_3 \\ x_4 \end{Bmatrix} = \begin{Bmatrix} 0.50 \\ 0.11 \\ 0.33 \\ 0.24 \end{Bmatrix}$$

and

$$x_4 = z_4 = 0.24$$
$$x_3 = z_3 - u_{34}x_4 = 0.33 - (0.67)(0.24) = 0.17$$
$$x_2 = z_2 - u_{24}x_4 - u_{23}x_3$$
$$= 0.11 - (-0.11)(0.24) - (0.11)(0.17) = 0.12$$
$$x_1 = z_1 - u_{14}x_4 - u_{13}x_3 - u_{12}x_2$$
$$= 0.50 - (0.5)(0.24) - (-0.5)(0.17) - (0.50)(0.12) = 0.41$$

$$\{X\} = \begin{Bmatrix} 0.41 \\ 0.12 \\ 0.17 \\ 0.24 \end{Bmatrix}$$

5.1.2 Choleski Method for Symmetric Matrix in Full Matrix Storage

If [M] is a symmetric matrix, [M] can be decomposed into the product of an upper triangular matrix and its transpose

$$[M] = [U]^T[U] \tag{5.17}$$

where the entries of [U] are given by

$$u_{ij} = \left(m_{ij} - \sum_{k=1}^{i-1} u_{ki}^2 \right)^{1/2}, \qquad i = j \tag{5.18a}$$

$$u_{ij} = \frac{m_{ij} - \sum_{k=1}^{i-1} u_{ki}\, u_{kj}}{u_{ii}}, \qquad i < j \tag{5.18b}$$

$$u_{ij} = 0, \qquad\qquad\qquad i > j \tag{5.18c}$$

Example

Perform triangular decomposition on the following matrix

$$[M] = \begin{bmatrix} m_{11} & m_{12} & m_{13} & m_{14} \\ m_{21} & m_{22} & m_{23} & m_{24} \\ m_{31} & m_{32} & m_{33} & m_{34} \\ m_{41} & m_{42} & m_{43} & m_{44} \end{bmatrix} = \begin{bmatrix} 2 & -1 & 2 & 1 \\ -1 & 6 & -1 & 1 \\ 2 & -1 & 6 & -1 \\ 1 & 1 & -1 & 2 \end{bmatrix}$$

$$u_{11} = (m_{11})^{1/2} = (2)^{1/2} = 1.414$$
$$u_{12} = m_{12}/u_{11} = -1/1.414 = -0.707$$
$$u_{13} = m_{13}/u_{11} = 2/1.414 = 1.414$$
$$u_{14} = m_{14}/u_{11} = 1/1.414 = 0.707$$

$$u_{21} = 0$$
$$u_{22} = (m_{22} - u_{12}^2)^{1/2} = (6 - (-0.707)^2)^{1/2} = 2.345$$
$$u_{23} = (m_{23} - u_{12}u_{13})/u_{22} = (-1 - (-0.707)(1.414))/2.345 = 0.000$$
$$u_{24} = (m_{24} - u_{12}u_{14})/u_{22} = (1 - (-0.707)(0.707))/2.345 = 0.640$$

$$u_{31} = u_{32} = 0$$
$$u_{33} = (m_{33} - u_{13}^2 - u_{23}^2)^{1/2} = (6 - (1.414)^2 - (0)^2)^{1/2} = 2.000$$
$$u_{34} = (m_{34} - u_{13}u_{14} - u_{23}u_{24})/u_{33}$$
$$= (-1 (1.4145)(0.707) - (0.000)(0.640))/2.000 = -1.000$$

$$u_{41} = u_{42} = u_{43} = 0$$
$$u_{44} = (m_{44} - u_{14}^2 - u_{24}^2 - u_{34}^2)^{1/2}$$
$$= (2 - (0.707)^2 - (0.640)^2 - (-1.000)^2)^{1/2} = 0.301$$

$$[U] = \begin{bmatrix} 1.414 & -0.707 & 1.414 & 0.707 \\ 0 & 2.345 & 0.000 & 0.640 \\ 0 & 0 & 2.000 & -1.000 \\ 0 & 0 & 0 & 0.301 \end{bmatrix}$$

$$[U]^T = \begin{bmatrix} 1.414 & 0 & 0 & 0 \\ -0.707 & 2.345 & 0 & 0 \\ 1.414 & 0.000 & 2.000 & 0 \\ 0.707 & 0.640 & -1.000 & 0.301 \end{bmatrix}$$

and it is easy to verify that $[M] = [U]^T[U]$.

Once the symmetric matrix $[M]$ has been decomposed the solution of the system of equations for any choice of $\{B\}$ is very simple. Because $[M] = [U]^T[U]$ we can write

$$[M]\{X\} = \{B\} = [U]^T[U]\{X\} = \{B\} \tag{5.19}$$

Then by defining a vector $\{Z\}$ as

$$[U]\{X\} = \{Z\} \tag{5.20}$$

we can write

$$[U]^T\{Z\} = \{B\} \tag{5.21}$$

If the entries in [U] are

$$[U] = \begin{bmatrix} u_{11} & u_{12} & \cdots & \cdots & u_{1n} \\ 0 & u_{22} & \cdots & \cdots & u_{2n} \\ 0 & 0 & u_{33} & \cdots & u_{3n} \\ \vdots & \vdots & \vdots & & \vdots \\ 0 & 0 & 0 & \cdots & u_{nn} \end{bmatrix} \tag{5.22}$$

the entries in $[U]^T$ are

$$[U^T] = \begin{bmatrix} u_{11}^T & 0 & 0 & \cdots & 0 \\ u_{21}^T & u_{22}^T & 0 & \cdots & 0 \\ \vdots & \vdots & u_{33}^T & \cdots & 0 \\ \vdots & \vdots & \vdots & & \vdots \\ u_{n1}^T & u_{n2}^T & u_{n3}^T & \cdots & u_{nn}^T \end{bmatrix} = \begin{bmatrix} u_{11} & 0 & 0 & \cdots & 0 \\ u_{12} & u_{22} & 0 & \cdots & 0 \\ \vdots & \vdots & u_{33} & \cdots & 0 \\ \vdots & \vdots & \vdots & & \vdots \\ u_{1n} & u_{2n} & u_{3n} & \cdots & u_{nn} \end{bmatrix} \tag{5.23}$$

and equation 5.21 can be written

$$\begin{aligned} u_{11}z_1 & & = b_1 \\ u_{12}z_1 + u_{22}z_2 & & = b_2 \\ \vdots \quad \vdots & & \vdots \ \vdots \\ u_{1n}z_1 + u_{2n}z_3 + \cdots + u_{nn}z_n & & = b_3 \end{aligned} \tag{5.24}$$

which can be solved for the values in {Z} using

$$z_i = \frac{\left(b_i - \displaystyle\sum_{k=1}^{i-1} u_{ki}\, z_k \right)}{u_{ii}}, \qquad i = 1 \text{ to } n \tag{5.25}$$

After we have computed the values in {Z} we can then solve for the values in {X} using equation 5.20

$$\begin{aligned} u_{11}x_1 + u_{12}x_2 + u_{13}x_3 + \cdots + u_{1n}x_n &= z_1 \\ u_{22}x_2 + u_{23}x_3 + \cdots + u_{2n}x_n &= z_2 \\ u_{33}x_3 + \cdots + u_{3n}x_n &= z_3 \\ \vdots \qquad \vdots \\ u_{nn}x_n &= z_n \end{aligned} \tag{5.26}$$

The solution is given by

$$x_{n+1-i} = \frac{\left(z_{n+1-i} - \displaystyle\sum_{k=1}^{i-1} u_{n+1-i,n+1-k} \, x_{n+1-k} \right)}{u_{n+1-i,n+1-i}} \qquad (5.27)$$

Example

Solve the following system of equations

$$\begin{bmatrix} 2 & -1 & 2 & 1 \\ -1 & 6 & -1 & 1 \\ 2 & -1 & 6 & -1 \\ 1 & 1 & -1 & 2 \end{bmatrix} \begin{bmatrix} x_1 \\ x_2 \\ x_3 \\ x_4 \end{bmatrix} = \begin{Bmatrix} 1 \\ 0 \\ 0 \\ 0 \end{Bmatrix}$$

Triangular decomposition of [M] was performed in the previous example and using equation 5.24 we have

$$\begin{bmatrix} 1.414 & 0 & 0 & 0 \\ -0.707 & 2.345 & 0 & 0 \\ 1.414 & 0.000 & 2.000 & 0 \\ 0.707 & 0.640 & -1.000 & 0.301 \end{bmatrix} \begin{bmatrix} z_1 \\ z_2 \\ z_3 \\ z_4 \end{bmatrix} = \begin{Bmatrix} 1 \\ 0 \\ 0 \\ 0 \end{Bmatrix}$$

$z_1 = b_1 / u_{11} = 1 / 1.414 = 0.707$

$z_2 = (b_2 - u_{12}z_1) / u_{22} = (0 - (-0.707)(0.707))/2.3456 = 0.213$

$z_3 = (b_3 - u_{13}z_1 - u_{23}z_2) / u_{33}$

$\quad = (0 - (1.414)(0.707) - (0.000)(0.213))/2.000 = -0.5000$

$z_4 = (b_4 - u_{14}z_1 - u_{24}z_2 - u_{34}z_3)u_{44}$

$\quad = (0 - (0.707)(0.707) - (0.640)(0.213) - (-1.000)(-0.500))/0.301 = -3.775$

In the next step we have

$$\begin{bmatrix} 1.414 & -0.707 & 1.414 & 0.707 \\ 0 & 2.345 & 0.000 & 0.640 \\ 0 & 0 & 2.000 & -1.000 \\ 0 & 0 & 0 & 0.301 \end{bmatrix} \begin{bmatrix} x_1 \\ x_2 \\ x_3 \\ x_4 \end{bmatrix} = \begin{Bmatrix} 0.707 \\ 0.213 \\ -0.500 \\ -3.775 \end{Bmatrix}$$

and

$$x_4 = z_4 / u_{44} = -3.775/0.301 = -12.54$$

$$x_3 = (z_3 - u_{34}x_4) / u_{33} = (-0.500 - (-1.000)(-12.54))/2.000 = -6.52$$

$$x_2 = (z_2 - u_{24}x_4 - u_{23}x_3) / u_{22}$$
$$= (0.213 - (0.640)(-12.54) - (0.000)(-6.52))/2.345 = 3.51$$

$$x_1 = (z_1 - u_{14}x_4 - u_{13}x_3 - u_{12}x_2) / u_{11}$$
$$= (0.707 - (0.707)(-12.54) - (1.414)(-6.52) - (-0.707)(3.51))/1.414$$
$$= 15.05$$

$$\{X\} = \begin{Bmatrix} 15.05 \\ 3.51 \\ -6.52 \\ -12.54 \end{Bmatrix}$$

5.1.3 Choleski Method for Nonsymmetric Matrix in Vector Storage

We have seen that the application of the finite element method to the equations of groundwater flow and solute transport results in systems of equations that are banded. Large savings in computer storage can be achieved by assembling and solving the system of equations in *vector storage*. In vector storage only the entries within the band are stored; the entries outside the band (which are all zero) are discarded. Consider the banded matrix in Figure 5.1. There are 6 equations and the semi-bandwidth is 2. In full matrix storage, 36 entries are stored including 20 entries "outside" the band that are known to be zero. If the matrix in Figure 5.1 is nonsymmetric (e.g., when solving the solute transport equation using equation 5.5 and 5.6) the entries within the band can be stored in a vector with length IJSIZE where

$$\text{IJSIZE} = (\text{NDOF})^2 - (\text{NDOF} - \text{SBW})(1 + \text{NDOF} - \text{SBW}) \tag{5.28}$$

where NDOF (for $\underline{N}$umber of $\underline{D}$egrees of $\underline{F}$reedom) is the number of equations and SBW is the semi-bandwidth. For the example in Figure 5.1, NDOF = 6, SBW = 2 and

$$\text{IJSIZE} = (6)^2 - (6-2)(1 + 6 - 2)$$
$$= 36 - 20 = \underline{16}$$

(a) Full Matrix Storage

$$[M] = \begin{bmatrix} m_{11} & m_{12} & 0 & 0 & 0 & 0 \\ m_{21} & m_{22} & m_{23} & 0 & 0 & 0 \\ 0 & m_{32} & m_{33} & m_{34} & 0 & 0 \\ 0 & 0 & m_{43} & m_{44} & m_{45} & 0 \\ 0 & 0 & 0 & m_{54} & m_{55} & m_{56} \\ 0 & 0 & 0 & 0 & m_{65} & m_{66} \end{bmatrix} \quad \begin{array}{l} \text{NDOF} = 6 \\ \text{SBW} = 2 \end{array}$$

36 entries
20 entries "outside" of band

Band

(b) Vector Storage
    Non-Symmetric Matrix

$$\{M\} = \begin{Bmatrix} m_{11} \\ m_{12} \\ m_{21} \\ m_{22} \\ m_{23} \\ m_{32} \\ m_{33} \\ m_{34} \\ m_{43} \\ m_{44} \\ m_{45} \\ m_{54} \\ m_{55} \\ m_{56} \\ m_{65} \\ m_{66} \end{Bmatrix}$$

(c) Vector Storage
    Symmetric Matrix

$$\{M\} = \begin{Bmatrix} m_{11} \\ m_{12} \\ m_{22} \\ m_{23} \\ m_{33} \\ m_{34} \\ m_{44} \\ m_{45} \\ m_{55} \\ m_{56} \\ m_{66} \end{Bmatrix}$$

Figure 5.1   Full matrix and vector storage schemes for non-symmetric and symmetric matrices.

Example

Compute the length of array required to store the banded, nonsymmetric matrix shown below in vector storage

$$[M] = \begin{bmatrix} 1 & 2 & 3 & 0 & 0 \\ 2 & 2 & 0 & 1 & 0 \\ 2 & 3 & 4 & 0 & 1 \\ 0 & 4 & 3 & 2 & 2 \\ 0 & 0 & 3 & 2 & 1 \end{bmatrix}$$

For this matrix NDOF = 5, SBW = 3 and from equation 5.28

$$\text{IJSIZE} = (5)^2 - (5-3)(1 + 5 - 3)$$
$$= 25 - 6 = \underline{19}$$

A typical entry m_{ij} in [M] (full matrix storage) can be assigned to an entry m_{IJ} in {M} (vector storage) using the algorithm in Figure 5.2. Note that the subscript IJ refers to a single index, the position of an entry in the vector {M}.

```
IJ = j
IF i>1 THEN
   IF SBW<NDOF THEN
      IF i>SBW THEN
         IJ = IJ + SBW - i
      ENDIF
      IJ = IJ + (i-1)(2SBW-1)
      L  = MIN(SBW,i)-1
      IJ = IJ - (L/2)[(SBW-1) + (SBW-L)]
      L  = i - NDOF + SBW - 2
      IF L>0 THEN
         IJ = IJ - L(L+1)/2
      ENDIF
   ELSE
      IJ = IJ + (i-1)NDOF
   ENDIF
ENDIF
```

Figure 5.2   Vector matrix storage for banded, nonsymmetric matrix. MIN(SBW, i) means take the minimum value of SBW and i.

The Choleski method can be used with nonsymmetric matrices in vector storage. The only difference is that the index IJ (in vector storage) must be computed using the algorithm in Figure 5.2 for each pair of indices (i,j) or (j,k) in equations 5.10, 5.14, and 5.16.

Example

Use the algorithm in Figure 5.2 to assign the entries of the nonsymmetric matrix shown below in full matrix storage to vector storage

$$\begin{bmatrix} 3 & 2 & 1 & 0 \\ 4 & 4 & 1 & 2 \\ 1 & 3 & 3 & 1 \\ 0 & 2 & 4 & 2 \end{bmatrix}$$

NDOF = 4
SBW = 3

$IJSIZE = (4)^2 - (4-3)(1+4-3) = 14$

```
i = 1    j = 1      IJ = j = 1
         j = 2      IJ = 2
         j = 3      IJ = 3
```

```
i = 2    j = 1      IJ = 1
                    2 > 1
                      3 < 4
                        2 < 3
                        IJ = 1 + (2-1)((2)(3)-1)    = 6
                        L  = MIN(3,2) - 1  = 2 - 1 = 1
                        IJ = 6 - 1/2[(3-1) + (3-1)]
                           = 6 - 2 = 4
                        L  = 2 - 4 + 3 - 2 = -1 < 0
                        ∴ IJ = 4
```

.
.
.

```
i = 4    j = 4      IJ = j = 4
                    i = 4 > 1
                      3 < 4
                      i = 4 > 3
                        IJ = 4 + 3 - 4 = 3
                        IJ = 3 + (4-1)(2(3) - 1)  = 18
                        L  = MIN(3,4) - 1 = 3 - 1 = 2
                        IJ = 18 - 2/2[(3-1) + (3-2)]
                           = 18 - 3 = 15
                        L  = 4 - 4 + 3 - 2 = 1 > 0
                        IJ = 15 - 1(1+1)/2 = 14
```

----

and

$$\{M\} = \begin{Bmatrix} 3 \\ 2 \\ 1 \\ 4 \\ 4 \\ 1 \\ 2 \\ 1 \\ 3 \\ 3 \\ 1 \\ 2 \\ 4 \\ 2 \end{Bmatrix}$$

IJ	i	j
1	1	1
2	1	2
3	1	3
4	2	1
5	2	2
6	2	3
7	2	4
8	3	1
9	3	2
10	3	3
11	3	4
12	4	2
13	4	3
14	4	4

5.1.4 Choleski Method for Symmetric Matrix in Vector Storage

If the matrix [M] is banded and symmetric (e.g., when solving the steady-state or transient groundwater flow equations) the entries within the band can be stored in a vector with length IJSIZE where

$$\text{IJSIZE} = \text{SBW}(\text{NDOF} - \text{SBW} + 1) + (\text{SBW} - 1)\left(\frac{\text{SBW}}{2}\right) \tag{5.29}$$

where NDOF is the number of equations and SBW is the semi-bandwidth (The quantity (SBW - 1)x(SBW) is always an even number). For the example in Figure 5.1, NDOF = 6, SBW = 2 and

$$\text{IJSIZE} = 2(6 - 2 + 1) + (2 - 1)\left(\frac{2}{2}\right)$$
$$= \underline{11}$$

A typical entry m_{ij} in [M] (full matrix storage) can be assigned to an entry m_{IJ} in {M} using the algorithm in Figure 5.3

```
IJ = j - i + 1
IF i>1 THEN
   IF SBW<NDOF THEN
      IJ = IJ + (i-1)SBW
      L  = i - NDOF + SBW - 2
      IF L>0 THEN
         IJ = IJ - L(L+1)/2
      ENDIF
   ELSE
      IJ = IJ + (i-1)(NDOF+(NDOF-i+2))/2
   ENDIF
ENDIF
```

Figure 5.3   Vector matrix storage for banded, symmetric matrix.

Example

Use the algorithm in Figure 5.3 to assign the entries of the symmetric matrix shown below in full matrix storage to vector storage

$$\begin{bmatrix} 3 & 2 & 0 & 0 \\ 2 & 4 & 2 & 0 \\ 0 & 2 & 4 & 2 \\ 0 & 0 & 2 & 3 \end{bmatrix}$$

NDOF = 4
SBW = 2

$$IJSIZE = 2(4-2+1) + (2-1)\left(\frac{2}{2}\right) = 7$$

```
i = 1     j = 1      IJ = 1 - 1 + 1 = 1
          j = 2      IJ = 2 - 1 + 1 = 2
```

```
i = 2     j = 2      IJ = 2 - 2 + 1 = 1
                     2 > 1
                       SBW < NDOF
                          IJ = 1 + (2-1)(2)   = 3
                          L  = 2 - 4 + 2 - 2 = -2 < 0
                          ∴ IJ = 3

          j = 3      IJ = 3 - 2 + 1 = 2
                     2 > 1
                       SBW < NDOF
                          IJ = 2 + (2-1)(2)   = 4
                          L  = 2 - 4 + 2 - 2 = -2 < 0
                          ∴ IJ = 4
```

```
                            ·
                            ·
                            ·
```

```
i = 4     j = 4      IJ = 4 - 4 + 1 = 1
                     4 > 1
                       SBW < NDOF
                          IJ = 1 + (4 - 1)(2)  = 7
                          L  = 4 - 4 + 2 - 2  = 0
                          ∴ IJ = 7
```

---

and

$$\{M\} = \begin{Bmatrix} 3 \\ 2 \\ 4 \\ 2 \\ 4 \\ 2 \\ 3 \end{Bmatrix}$$

IJ	i	j
1	1	1
2	1	2
3	2	2
4	2	3
5	3	3
6	3	4
7	4	4

The Choleski method can be used with symmetric matrices in vector storage. The only difference is that the index IJ (in vector storage) must be computed using the algorithm in Figure 5.2 for each pair of indices (i,j) or (j,k) in equations 5.18, 5.25, and 5.27.

5.2 APPLICATION OF CHOLESKI METHOD

5.2.1 Steady-State, Saturated Flow Equation

The matrix formulation for the steady-state, saturated flow equation (equation 5.1) can be solved by the Choleski method for symmetric matrices described in sections 5.1.2 and 5.1.4. We simply set $[M] = [K]$, the global conductance matrix; $\{X\} = \{h\}$, the unknown values of hydraulic head at each node, and $\{B\} = \{F\}$, the specified rates of groundwater flow at nodes representing sources and sinks (i.e., at Neumann nodes). The global conductance matrix is symmetric because the element matrices it contains are symmetric. Equation 5.1 is a system of linear equations because none of the entries of the global conductance matrix are functions of hydraulic head. Recall that the element conductance matrix $[K^{(e)}]$ is computed using the interpolation function derivatives for each node in the element and the components of saturated hydraulic conductivity for the element. The interpolation function derivatives depend only on the number of nodes in the element and the element's size and shape. The components of saturated hydraulic conductivity are constant within an element (but can vary from one element to the next) and do not depend on the value of hydraulic head at the element's nodes.

Because the global conductance matrix is symmetric and banded, equation 5.1 is conveniently assembled and solved in vector storage using the procedure described in sections 5.1.2 and 5.1.4. Decomposition of $[K]$ and $\{F\}$ is performed once and $\{h\}$ is obtained directly by backward substitution.

5.2.2 Transient, Saturated Flow Equation

The matrix formulation for the transient, saturated flow equation (equation 5.3) can be solved by the Choleski method for symmetric matrices described in sections 5.1.2 and 5.1.4. We set

$$[M] = [C] + \omega \Delta t \, [K] \tag{5.30a}$$

$$\{X\} = \{h\}_{t+\Delta t} \tag{5.30b}$$

$$\{B\} = ([C] - (1 - \omega)\Delta t \, [K])\{h_t\} + \Delta t \, ((1 - \omega)\{F\}_t + \omega\{F\}_{t+\Delta t}) \tag{5.30c}$$

where $[C]$ is the global capacitance matrix, $[K]$ is the global conductance matrix, ω is the relaxation factor, Δt is the time step, and $\{F\}$ are the specified rates of groundwater flow representing sources and sinks at Neumann nodes. $\{F\}$ is known at all time steps. $\{h\}_t$ is known from the initial conditions or from the solution obtained for the previous time step. $\{h\}_{t+\Delta t}$ are the unknown values of hydraulic head at time $t + \Delta t$.

$[M]$ is symmetric because $[C]$ and $[K]$ are symmetric. The choice of ω and Δt will have no effect on the symmetry of $[M]$. Equation 5.3 is a system of linear equations because none of the entries of $[C]$ or $[K]$ are functions of hydraulic head.

Because $[M]$ is symmetric and banded equation 5.3 is conveniently assembled and solved in vector storage using the procedures described in sections 5.1.2 and 5.1.4. Assembly and decomposition of $[M]$ is performed only once (unless ω or Δt changes from one time step to the next). However, assembly and decomposition of $\{B\}$ and backward substitution to obtain $\{h\}_{t+\Delta t}$ must be performed for each time step. Thus it is desirable to minimize the number of time steps used to solve a particular problem, suggesting that large time steps should be used. However, several numerical difficulties can arise during the solution process when Δt is large. The size of time step required to obtain useful results depends on the size and shape of the elements in the mesh, the values of specific storage and hydraulic conductivity for each element, whether a lumped or consistent formulation was used to compute the element capacitance matrices, the boundary conditions contained in {B}, and the value of the relaxation parameter ω.

The first type of difficulty that can occur when solving equation 5.2 occurs when calculated values of head violate "reality" (Segerlind, 1984). For example when computed values of head decrease near a point source or increase near a point sink. This situation can be avoided by choosing ω and Δt so that the coefficients in [M] (equation 5.30a) are positive for positions on the main diagonal and negative for off diagonal positions (see below) (Segerlind, 1984, Maadooliat, 1983).

The second type of difficulty that can occur when solving equation 5.2 is called *instability*. Instability occurs when the difference between the true solution and the numerical solution grows extremely large in a few time steps (Figure 5.4). Fortunately it is possible to avoid instability by setting $\omega \geq 1/2$. When this condition is met, the numerical solution for equation 5.3 will be unconditionally stable (Lapidus and Pinder, 1982, p.166).

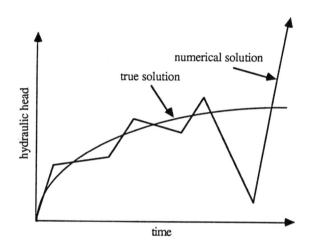

Figure 5.4   Numerical instability in computed value of hydraulic head.

The third type of difficulty that can occur when solving equation 5.3 for large values of Δt is called *numerical oscillation*. Numerical oscillation occurs when the computed values of hydraulic head fluctuate about the true solution. From one time step to the next the numerical solution is alternatively above and below the true solution (Figure 5.5).

Criteria to avoid numerical oscillations can be derived based on the properties of the matrix product $[M^{-1}][P]$ where

$$[P] = [C] - (1 - \omega)\,\Delta t\,[K] \tag{5.31}$$

To avoid numerical oscillations it is sufficient to require that all of the eigenvalues of $[M^{-1}]$ [P] be positive (Myers, 1971). An eigenvalue is defined as any number E that satisfies the matrix equation

$$\left| [M^{-1}][P] - E[I] \right| = 0 \tag{5.32}$$

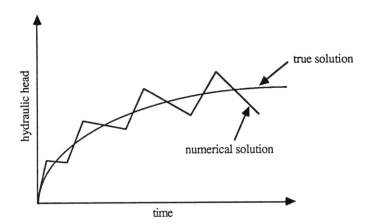

Figure 5.5   Numerical oscillation in computed value of hydraulic head.

where | | is the determinant and [I] is the identity matrix (see Appendix IV). Equation 5.32 is also satisfied (Segerlind, 1984) if

$$\left| [P] - E[M] \right| = 0 \tag{5.33}$$

In equation 5.33 E will be positive if both [P] and [M] are positive definite. Now [C] and [K] will always be positive definite. For transient, saturated flow

$$[M] = [C] + \omega \Delta t [K] \tag{5.34}$$

where ω and Δt are positive constants. [M] will always be positive definite because a matrix obtained by *adding* a positive definite matrix to a positive definite matrix multiplied by any positive constant is also positive definite (Myers, 1971).

However, [P] is not guaranteed to be positive definite because a matrix obtained by *subtracting* a portion of one positive definite matrix from another is not necessarily positive definite (Myers, 1971). The problem of avoiding numerical oscillations thus becomes one of selecting values of ω and Δt that insure that [P] is positive definite. In other words, we select values of ω and Δt such that (for any number β)

$$\left| [P] - \beta [I] \right| = 0 \tag{5.35}$$

[P] will be positive definite if no value of $\beta \leq 0$. Now the minimum value of any global matrix is greater than the minimum eigenvalue for all its component (or element) matrices (Fried, 1979). Thus we can develop criteria to select values of ω and t by using a form of equation 5.35 written for an arbitrary element e

$$\left| [P^{(e)}] - \beta^{(e)}[I] \right| = 0 \tag{5.36}$$

where

$$[P^{(e)}] = [C^{(e)}] - (1 - \omega)\Delta t [K^{(e)}] \tag{5.37}$$

and $[C^{(e)}]$ is the element capacitance matrix, $[K^{(e)}]$ is the element conductance matrix, and $\beta^{(e)}$ is an eigenvalue for element e. Now if the minimum eigenvalue for any element equals zero ($\beta^{(e)} = 0$), the minimum eigenvalue for the global matrix will be greater than 0 ($\beta \geq 0$) and numerical oscillations will not occur. Setting $\beta^{(e)} = 0$, equation 5.36 becomes

$$\left| [P^{(e)}] - 0\,[I] \right| = 0$$

$$\left| [P^{(e)}] \right| = 0$$

$$\left| [C^{(e)}] - (1 - \omega)\Delta t\,[K^{(e)}] \right| = 0 \tag{5.38}$$

Let $\alpha = (1 - \omega)\Delta t$. Then equation 5.38 becomes

$$\left| [C^{(e)}] - \alpha\,[K^{(e)}] \right| = 0 \tag{5.39}$$

and the criteria for stability becomes

$$\Delta t < \frac{\alpha}{1 - \omega}, \qquad \omega > 1 \tag{5.40}$$

where α is the *smallest* number (for any element) that satisfies equation 5.39 (the smallest number of α will occur in the smallest element in the mesh). For the case $\omega = 1$, it can be shown that no numerical oscillations will occur (Segerlind, 1984). In practice the following procedure can be used to avoid numerical oscillations when solving the transient, saturated flow equation:

1. Compute $[C^{(e)}]$ and $[K^{(e)}]$ for the smallest element in the mesh.

2. Solve equation 5.39 to obtain the minimum value of α for that element.

3. Use equation 5.40 to select a suitable value for Δt and ω.

Example

The element capacitance and conductance matrices for the smallest element in a mesh are given below. Find combinations of ω and Δt that do not violate reality and that prevent instability and numerical oscillations

$$[C^{(e)}] = \begin{bmatrix} 0.03 & 0 \\ 0 & 0.03 \end{bmatrix} \qquad\qquad [K^{(e)}] = \begin{bmatrix} 0.5 & -0.5 \\ -0.5 & 0.5 \end{bmatrix}$$

For a mesh consisting of this type of element the diagonal coefficients in [M] will be positive and the off-diagonal coefficients in [M] will be negative for all values of ω and Δt. From equation 5.39

$$\left| \begin{bmatrix} 0.03 & 0 \\ 0 & 0.03 \end{bmatrix} - \alpha \begin{bmatrix} 0.5 & -0.5 \\ -0.5 & 0.5 \end{bmatrix} \right| = 0$$

or

$$\left| \begin{matrix} (0.03 - \alpha 0.05) & (\alpha 0.05) \\ (\alpha 0.05) & (0.03 - \alpha 0.05) \end{matrix} \right| = 0$$

$$(0.03 - \alpha 0.05)(0.03 - \alpha 0.05) - (\alpha 0.05)(\alpha 0.05) = 0$$

or

$$0.0025\alpha^2 - 0.003\alpha + 0.0009 - 0.0025\alpha^2 = 0$$

or

$$\alpha = \frac{0.0009}{0.003} = \underline{0.3}$$

Instability will not occur if $\omega \geq 1/2$ and numerical oscillations will not occur if

$$\Delta t < \frac{0.3}{1 - \omega}$$

The values of Δt and ω that meet this criteria are plotted below. Any combination of Δt and ω in the shaded region will prevent instability and numerical oscillations.

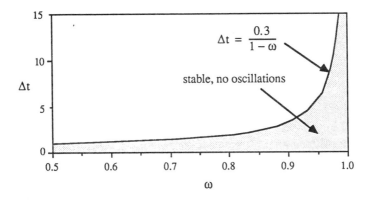

For linear elements with 4 nodes or less, equation 5.39 can be used to derive algebraic expressions for Δt that prevent numerical oscillations. For more complicated elements, equation 5.39 must be solved numerically. Consider first the linear bar element (Figure 4.5). The element conductance matrix is given by equation 4.15a

$$[K^{(e)}] = \frac{K_x^{(e)}}{L^{(e)}} \begin{bmatrix} 1 & -1 \\ -1 & 1 \end{bmatrix}$$

where $K_x^{(e)}$ is the saturated hydraulic conductivity and $L^{(e)}$ is the element length. If we use the consistent formulation the element capacitance matrix is given by equation 4.16a

$$[C^{(e)}] = \frac{S_s^{(e)} L^{(e)}}{6} \begin{bmatrix} 2 & 1 \\ 1 & 2 \end{bmatrix}$$

where $S_s^{(e)}$ is the specific storage.

For a mesh consisting of this type of element, the diagonal coefficients in [M] will be positive for all values of ω and Δt. The off-diagonal coefficients will be negative if

$$\frac{S_s^{(e)} L^{(e)}}{6} - \omega \Delta t \frac{K_x^{(e)}}{L^{(e)}} < 0$$

or

$$\Delta t < \frac{S_s^{(e)} L^{(e)^2}}{6 K_x^{(e)} \omega}$$

From equation 5.39

$$\left| \frac{S_s^{(e)} L^{(e)}}{6} \begin{bmatrix} 2 & 1 \\ 1 & 2 \end{bmatrix} - \frac{\alpha K_x^{(e)}}{L^{(e)}} \begin{bmatrix} 1 & -1 \\ -1 & 1 \end{bmatrix} \right| = 0$$

or

$$\left| \begin{matrix} \left(\dfrac{S_s^{(e)} L^{(e)}}{3} - \dfrac{\alpha K_x^{(e)}}{L^{(e)}} \right) & \left(\dfrac{S_s^{(e)} L^{(e)}}{6} + \dfrac{\alpha K_x^{(e)}}{L^{(e)}} \right) \\ \left(\dfrac{S_s^{(e)} L^{(e)}}{6} + \dfrac{\alpha K_x^{(e)}}{L^{(e)}} \right) & \left(\dfrac{S_s^{(e)} L^{(e)}}{3} - \dfrac{\alpha K_x^{(e)}}{L^{(e)}} \right) \end{matrix} \right| = 0$$

Evaluating the determinant we have

$$\left(\frac{S_s^{(e)} L^{(e)}}{3} - \frac{\alpha K_x^{(e)}}{L^{(e)}} \right)^2 - \left(\frac{S_s^{(e)} L^{(e)}}{6} + \frac{\alpha K_x^{(e)}}{L^{(e)}} \right)^2 = 0$$

which can be solved to give

$$\alpha = \frac{S_s^{(e)} L^{(e)^2}}{12 K_x^{(e)}}$$

so that the criteria for avoiding numerical oscillations for the linear bar element with a consistent formulation for $[C^{(e)}]$ is

$$\Delta t < \frac{S_s^{(e)} L^{(e)2}}{12 K_x^{(e)}(1-\omega)} \qquad (5.41)$$

If we use the lumped formulation the element capacitance matrix is given by equation 4.16b

$$[C^{(e)}] = \frac{S_s^{(e)} L^{(e)}}{2} \begin{bmatrix} 1 & 0 \\ 0 & 1 \end{bmatrix}$$

For a mesh consisting of this type of element the diagonal coefficients in [M] will be positive and the off-diagonal coefficients in [M] will be negative for all values of ω and Δt. Substituting $[C^{(e)}]$ and $[K^{(e)}]$ into equation 5.39 we have

$$\left| \frac{S_s^{(e)} L^{(e)}}{2} \begin{bmatrix} 1 & 0 \\ 0 & 1 \end{bmatrix} - \frac{\alpha K_x^{(e)}}{L^{(e)}} \begin{bmatrix} 1 & -1 \\ -1 & 1 \end{bmatrix} \right| = 0$$

or

$$\left| \begin{matrix} \left(\dfrac{S_s^{(e)} L^{(e)}}{2} - \dfrac{\alpha K_x^{(e)}}{L^{(e)}} \right) & \left(\dfrac{\alpha K_x^{(e)}}{L^{(e)}} \right) \\[3mm] \left(\dfrac{\alpha K_x^{(e)}}{L^{(e)}} \right) & \left(\dfrac{S_s^{(e)} L^{(e)}}{2} - \dfrac{\alpha K_x^{(e)}}{L^{(e)}} \right) \end{matrix} \right| = 0$$

Evaluating the determinant we have

$$\left(\frac{S_s^{(e)} L^{(e)}}{2} - \frac{\alpha K_x^{(e)}}{L^{(e)}} \right)^2 - \left(\frac{\alpha K_x^{(e)}}{L^{(e)}} \right)^2 = 0$$

which can be solved to give

$$\alpha = \frac{S_s^{(e)} (L^{(e)})^2}{4 K_x^{(e)}}$$

so that the criteria for avoiding numerical oscillations for the linear bar element with a lumped formulation for $[C^{(e)}]$ is

$$\Delta t < \frac{S_s^{(e)} (L^{(e)})^2}{4 K_x^{(e)}(1-\omega)} \qquad (5.42)$$

Equations similar to 5.41 and 5.42 can be derived for other element types (Segerlind, 1984)

1. Linear triangle element, lumped formulation

The element matrices are given by equations 4.20 and 4.22b. As an example consider the triangle element shown below. Node i is at (0,0), the element is a right triangle, and the side length is b

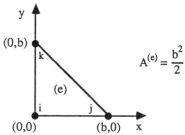

If we assume that $K_x^{(e)} = K_y^{(e)} = K^{(e)}$ we can use equation 5.39 to find

$$\alpha = \frac{2S_s^{(e)}A^{(e)}}{9K^{(e)}} \tag{5.43}$$

and

$$\Delta t < \frac{2S_s^{(e)}A^{(e)}}{9K^{(e)}(1-\omega)} \tag{5.44}$$

2. Linear rectangle element, lumped formulation

The element matrices are given by equations 4.26 and 4.27b. As an example consider a square element $(2a^{(e)} = 2b^{(e)})$. If we assume that $K_x^{(e)} = K_y^{(e)} = K^{(e)}$ we can use equation 5.39 to find

$$\alpha = \frac{S_s^{(e)}a^{(e)}b^{(e)}}{K^{(e)}} \tag{5.45}$$

and

$$\Delta t < \frac{S_s^{(e)}a^{(e)}b^{(e)}}{K^{(e)}(1-\omega)} \tag{5.46}$$

5.2.3 Solute Transport Equation

The matrix formulation for the solute transport equation (equation 5.5) can be solved by the Choleski method for nonsymmetric matrices described in sections 5.1.3 and 5.1.5. This is done by setting

$$[M] = [A] + \omega\Delta t\,[D] \tag{5.48a}$$

$$\{X\} = \{C\}_{t+\Delta t} \tag{5.48b}$$

$$\{B\} = (\{A\} - (1-\omega)\Delta t\,[D]\,)\{C\}_t + \Delta t\,((1-\omega)\{F\}_t + \omega\{F\}_{t+\Delta t}) \tag{5.48c}$$

where [A] is the global sorption matrix, [D] is the global advection-dispersion matrix, ω is the relaxation factor, Δt is the time step, and $\{F\}$ are the specified solute fluxes representing sources and sinks at Neumann nodes. $\{F\}$ is known at all time steps. $\{C\}_t$ is known from the initial conditions or from the solution obtained for the previous timestep. $\{C\}_{t+\Delta t}$ are the unknown values of solute concentration at time $t+\Delta t$.

[M] is nonsymmetric because [D] is nonsymmetric. The choice or ω and Δt will have no effect on the symmetry of [M]. Equation 5.5 is a system of linear equations because none of the entries of [A] or [D] are functions of solute concentration.

Because [M] is banded equation 5.5 is conveniently assembled and solved in vector storage (Section 5.1.5). Assembly and decomposition of [M] are usually performed only once. However, for solute transport in transient groundwater flow conditions, e.g. Section 3.7.3, [M] would be assembled and decomposed at each time step (because the components of apparent groundwater velocity, $v_x^{(e)}$, $v_y^{(e)}$, and $v_z^{(e)}$ used to compute $[D^{(e)}]$ are changing from one time step to the next in transient flow). [M] must also be assembled and decomposed if ω or Δt change during the solution process (e.g., it is common to use a small time step for the first few time steps, when solute concentrations are changing rapidly and then use a larger Δt when solute concentrations are changing more slowly).

The same criteria used to avoid problems of violating reality, instability, and numerical oscillations when solving the transient, saturated flow equation (Section 5.2.3) can be used when solving the solute transport equation. Thus to avoid problems of instability we set $\omega \geq 1/2$. The criteria for preventing numerical oscillations can be written (following the discussion in section 5.2.3)

$$\Delta t < \frac{\alpha}{1-\omega} \tag{5.49}$$

where α is the smallest number (for any element) that satisfies

$$\left| [A^{(e)}] - \alpha[D^{(e)}] \right| = 0 \tag{5.50}$$

where $[A^{(e)}]$ and $[D^{(e)}]$ are the element sorption matrix and advection–dispersion matrix, respectively.

For simple element types, equation 5.50 can be used to derive algebraic expressions for combinations of Δt and ω that prevent numerical oscillations. For more complicated elements, equation 5.50 must be evaluated numerically. As for the case of transient, saturated flow the smallest value of α will occur in the smallest element in the mesh.

Example

Determine the criteria for preventing numerical oscillations when solving the solute transport equation using linear bar elements (Figure 4.5).

We will use the lumped formulation for $[A^{(e)}]$ and $[D^{(e)}]$ (equations 4.19b and 4.18b) and set $\lambda = 0$

$$[A^{(e)}] = (\rho_b^{(e)} K_d^{(e)} + \theta^{(e)}) \frac{L^{(e)}}{2} \begin{bmatrix} 1 & 0 \\ 0 & 1 \end{bmatrix}$$

$$[D^{(e)}] = \frac{D_x^{(e)} \theta^{(e)}}{L^{(e)}} \begin{bmatrix} 1 & -1 \\ -1 & 1 \end{bmatrix} + \frac{v_x^{(e)}}{2} \begin{bmatrix} -1 & 1 \\ -1 & 1 \end{bmatrix}$$

where $\rho_b^{(e)}$ is bulk density, $K_d^{(e)}$ is the distribution coefficient, $\theta^{(e)}$ is volumetric water content, $L^{(e)}$ is the element length, $D_x^{(e)}$ is a dispersion coefficient, and $v_x^{(e)}$ is the apparent groundwater velocity. To simplify the algebra set

$$R^{(e)} = \rho_b^{(e)} K_d^{(e)} + \theta^{(e)}$$

Substituting into equation 5.50 we have

$$
\begin{vmatrix}
\dfrac{R^{(e)} L^{(e)}}{2} - \alpha\left(\dfrac{D_x^{(e)}\theta^{(e)}}{L^{(e)}} - \dfrac{v_x^{(e)}}{2}\right) & \alpha\left(\dfrac{D_x^{(e)}\theta^{(e)}}{2} - \dfrac{v_x^{(e)}}{2}\right) \\[3mm]
\alpha\left(\dfrac{D_x^{(e)}\theta^{(e)}}{L^{(e)}} + \dfrac{v_x^{(e)}}{2}\right) & \dfrac{R^{(e)} L^{(e)}}{2} - \alpha\left(\dfrac{D_x^{(e)}\theta^{(e)}}{L^{(e)}} + \dfrac{v_x^{(e)}}{2}\right)
\end{vmatrix} = 0
$$

Evaluating the determinant

$$
\left[\frac{R^{(e)} L^{(e)}}{2} - \alpha\left(\frac{D_x^{(e)}\theta^{(e)}}{L^{(e)}} - \frac{v_x^{(e)}}{2}\right)\right]\left[\frac{R^{(e)} L^{(e)}}{2} - \alpha\left(\frac{D_x^{(e)}\theta^{(e)}}{L^{(e)}} + \frac{v_x^{(e)}}{2}\right)\right]
$$

$$
-\left[\alpha\left(\frac{D_x^{(e)}\theta^{(e)}}{L^{(e)}} - \frac{v_x^{(e)}}{2}\right)\right]\left[\alpha\left(\frac{D_x^{(e)}\theta^{(e)}}{L^{(e)}} + \frac{v_x^{(e)}}{2}\right)\right] = 0
$$

or

$$(C_1 - \alpha C_2)(C_1 - \alpha C_3) - \alpha\,(C_2)\,\alpha\,(C_3) = 0$$

where $C_1 = \dfrac{R^{(e)} L^{(e)}}{2}$, etc.

$$C_1^2 - \alpha C_1 (C_3 + C_2) = 0$$

or

$$\alpha = \frac{C_1}{C_2 + C_3}$$

$$
= \frac{\dfrac{R^{(e)} L^{(e)}}{2}}{\dfrac{D_x^{(e)}\theta^{(e)}}{L^{(e)}} - \dfrac{v_x^{(e)}}{2} + \dfrac{D_x^{(e)}\theta^{(e)}}{L^{(e)}} + \dfrac{v_x^{(e)}}{2}}
$$

$$\alpha = \frac{R^{(e)}(L^{(e)})^2}{4 D_x^{(e)}\theta^{(e)}} \qquad\qquad (5.51a)$$

and

$$\Delta t < \frac{\alpha}{1 - \omega} \qquad\qquad (5.51b)$$

5.3   PROCEDURE FOR SOLVING SYSTEM OF NONLINEAR EQUATIONS

Equation 5.7 can be written in expanded form as

$$
\begin{bmatrix} m_{11}(x_1, \cdots, x_n) & \cdots & m_{1n}(x_1, \cdots, x_n) \\ \vdots & \cdots & \vdots \\ m_{n1}(x_1, \cdots, x_n) & \cdots & m_{nn}(x_1, \cdots, x_n) \end{bmatrix} \begin{Bmatrix} x_1 \\ \vdots \\ x_n \end{Bmatrix} = \begin{Bmatrix} b_1 \\ \vdots \\ b_n \end{Bmatrix}
\tag{5.52}
$$

where each m_{ij} is a function of one or more values of x_i. Although not shown explicitly in equation 5.52 the coefficients in $\{B\}$ are also sometimes functions of one or more x_i i.e., $b_i(x_1, ..., x_n)$. The nonlinear solution process begins by specifying an initial guess for $\{X\}$, $\{X^0\}$

$$
\{X^0\} = \begin{Bmatrix} x_1^0 \\ \vdots \\ x_n^0 \end{Bmatrix}
\tag{5.53}
$$

If $\{X^0\}$ was an exact solution, it would satisfy equation 5.52 exactly. In other words

$$
[M(X^0)]\{X^0\} = \{B\}
$$

or

$$
\{B\} - [M(X^0)]\{X^0\} = \{R^0\} = \{0\}
\tag{5.53}
$$

where $\{R^0\}$ are the residuals at each node (not to be confused with the residuals obtained in Chapter 3 using Galerkin's method). Equation 5.53 can also be written in expanded form

$$
\begin{Bmatrix} b_1 \\ \vdots \\ b_n \end{Bmatrix} - \begin{bmatrix} m_{11}(x_1^0, \cdots, x_n^0) & \cdots & m_{1n}(x_1^0, \cdots, x_n^0) \\ \vdots & \cdots & \vdots \\ m_{n1}(x_1^0, \cdots, x_n^0) & \cdots & m_{nn}(x_1^0, \cdots, x_n^0) \end{bmatrix} \begin{Bmatrix} x_1^0 \\ \vdots \\ x_n^0 \end{Bmatrix} = \begin{Bmatrix} r_1^0 \\ \vdots \\ r_n^0 \end{Bmatrix} = \begin{Bmatrix} 0 \\ \vdots \\ 0 \end{Bmatrix}
\tag{5.54}
$$

where r_i^0 are the entries of $\{R^0\}$. If the initial guess for $\{X\}$ is incorrect, the solution will not be exact and the residuals will not all equal zero, $\{R^0\} \neq \{0\}$. In practice, we usually only require the residuals to be "close" to zero

$$
|\max\{r_i^0\}| < \varepsilon
\tag{5.55}
$$

where max denotes the maximum value of the residual at any node and ε is a predetermined tolerance.

If the initial prediction $\{X^0\}$ does not satisfy equation 5.55 we search for an improved prediction $\{X^k\}$. The process continues in an iterative fashion until a sufficiently accurate solution is found. Several methods can be used to obtain the improved prediction $\{X^k\}$. We will consider only two methods: *Picard iteration* and the *Newton-Raphson* method.

5.3.1 Picard Iteration

In Picard iteration (also called the *substitution method*) we construct a sequence of
solutions $\{X^0\}$, $\{X^1\}$, $\{X^2\}$, etc. and each solution $\{X^k\}$ is calculated from the previous
solution $\{X^{k-1}\}$

$$\left[M(X^{k-1})\right]\{X^k\} = \{B\} \tag{5.56}$$

where $[M(X^{k-1})]$ represents the matrix of coefficients constructed from the previous
solution (Klute et al., 1965). Equation 5.56 can be written in expanded form as

$$\begin{bmatrix} m_{11}(x_1^{k-1}, \cdots, x_n^{k-1}) & \cdots & m_{1n}(x_1^{k-1}, \cdots, x_n^{k-1}) \\ \vdots & \cdots & \vdots \\ m_{n1}(x_1^{k-1}, \cdots, x_n^{k-1}) & \cdots & m_{nn}(x_1^{k-1}, \cdots, x_n^{k-1}) \end{bmatrix} \begin{Bmatrix} x_1^k \\ \vdots \\ x_n^k \end{Bmatrix} = \begin{Bmatrix} b_1 \\ \vdots \\ b_n \end{Bmatrix} \tag{5.57}$$

(If the coefficients in $\{B\}$ are functions of $\{X\}$ they would also be computed using
$\{X^{k-1}\}$). The procedure begins by specifying an initial guess $\{X^0\}$. Then we solve the
system of *linear* equations

$$\left[M(X^0)\right]\{X^1\} = \{B\} \tag{5.58}$$

to obtain the new solution $\{X^1\}$ (for example using Choleski's method). We then compute
the residuals

$$\{R^0\} = \{B\} - \left[M(X^0)\right]\{X^1\} \tag{5.59}$$

and determine if the r_i^0 are sufficiently close to zero. If they are, we can use the solution
$\{X^1\}$. If they are not we construct a new matrix of coefficients using $\{X^1\}$ and solve for
the next solution $\{X^2\}$

$$\left[M(X^1)\right]\{X^2\} = \{B\} \tag{5.60}$$

where $[M(X^1)]$ represents the matrix of coefficients constructed from solution $\{X^1\}$. This
process is repeated until the maximum value of a residual is smaller than a specified
tolerance ε

$$\left|\max\{r_i^k\}\right| < \varepsilon \tag{5.61}$$

Equation 5.61 is called a *test for convergence* or *convergence criterion* because it is a
measure of how close the approximate solution $\{X^k\}$ is to the unknown true solution $\{X\}$.
This test can also be performed on residuals computed from two consecutive solutions

$$\{R^k\} = \{X^k\} - \{X^{k-1}\} \tag{5.62}$$

which can be written in expanded form as

$$\left\{\begin{array}{c} r_1^k \\ \vdots \\ r_n^k \end{array}\right\} = \left\{\begin{array}{c} x_1^k \\ \vdots \\ x_n^k \end{array}\right\} - \left\{\begin{array}{c} x_1^{k-1} \\ \vdots \\ x_n^{k-1} \end{array}\right\} \tag{5.63}$$

The algorithm for Picard iteration is summarized in Figure 5.6.

1. Specify initial approximate solution $\{X^0\}$

2. For $k = 1, 2, 3, \ldots$ (each value of k is an iteration) do the following

   A. Construct the matrix of coefficients $[M(X^{k-1})]$

   B. Solve the system of linear equations

   $$\left[M(X^{k-1})\right]\{X^k\} = \{B\}$$

   for $\{X^k\}$

   C. Construct the vector of residuals $\{R^k\}$ using

   $$\{R^k\} = \{B\} - \left[M(X^{k-1})\right]\{X^k\}$$

   or

   $$\{R^k\} = \{X^k\} - \{X^{k-1}\}$$

   D. Test for convergence

   $$\left|\max\{r_i^k\}\right| < \varepsilon\ ?$$

   If convergence criterion is satisfied, use solution $\{X^k\}$, otherwise set $k = k + 1$ and repeat steps A, B, C and D

Figure 5.6   Algorithm for Picard iteration.

Example

Use Picard iteration to solve the following system of nonlinear equations (let $\varepsilon = 0.05$)

$$\begin{array}{rl} 5x_1^2 + 4x_1x_2 - 4x_2^2 & = 5 \\ -4x_1x_2 + 4x_2^2 + 3x_2x_3 - 3x_3^2 & = 4 \\ -3x_2x_3 + 3x_3^2 & = 3 \end{array}$$

The system of equations can be written in matrix form as

$$\begin{bmatrix} (5x_1 + 4x_2) & -4x_2 & 0 \\ -4x_2 & (4x_2 + 3x_3) & -3x_3 \\ 0 & -3x_3 & 3x_3 \end{bmatrix} \begin{Bmatrix} x_1 \\ x_2 \\ x_3 \end{Bmatrix} = \begin{Bmatrix} 5 \\ 4 \\ 3 \end{Bmatrix}$$

Following the algorithm in Figure 5.6, we specify an initial solution $\{X^0\}$ (usually selected arbitrarily)

$$\{X^0\} = \begin{Bmatrix} 1 \\ 1 \\ 1 \end{Bmatrix}$$

This can be used to construct the matrix of coefficients $[M(X^0)]$

$$\left[M(X^0) \right] = \begin{bmatrix} 5(1) + 4(1) & -4(1) & 0 \\ -4(1) & (4(1) + 3(1)) & -3(1) \\ 0 & -3(1) & 3(1) \end{bmatrix}$$

$$= \begin{bmatrix} 9 & -4 & 0 \\ -4 & 7 & -3 \\ 0 & -3 & 3 \end{bmatrix}$$

We can then construct the system of linear equations

$$\left[M(X^0) \right]\{X^1\} = \{B\}$$

$$\begin{bmatrix} 9 & -4 & 0 \\ -4 & 7 & -3 \\ 0 & -3 & 3 \end{bmatrix} \begin{Bmatrix} x_1^1 \\ x_2^1 \\ x_3^1 \end{Bmatrix} = \begin{Bmatrix} 5 \\ 4 \\ 3 \end{Bmatrix}$$

which can be solved to give

$$\{X^1\} = \begin{Bmatrix} x_1^1 \\ x_2^1 \\ x_3^1 \end{Bmatrix} = \begin{Bmatrix} 2.40 \\ 4.15 \\ 5.15 \end{Bmatrix}$$

The residual vector for the first iteration is

$$\{R^1\} = \{X^0\} - \{X^1\} = \begin{Bmatrix} 1 \\ 1 \\ 1 \end{Bmatrix} - \begin{Bmatrix} 2.40 \\ 4.15 \\ 5.15 \end{Bmatrix} = \begin{Bmatrix} -1.40 \\ -3.15 \\ -4.15 \end{Bmatrix}$$

and $\left| \max\{r_i^1\} \right| = 4.15 > 0.05$

For the next iteration ($k = 2$) the matrix of coefficients is

$$\left[M(X^1)\right] = \begin{bmatrix} (5(2.4) + 4(4.15)) & -4(4.15) & 0 \\ -4(4.15) & (4(4.15) + 3(5.15)) & -3(5.15) \\ 0 & -3(5.15) & 3(5.15) \end{bmatrix}$$

$$= \begin{bmatrix} 28.60 & -16.60 & 0 \\ -16.60 & 32.05 & -15.45 \\ 0 & -15.45 & 15.45 \end{bmatrix}$$

The resulting system of linear equations

$$\begin{bmatrix} 28.60 & -16.60 & 0 \\ -16.60 & 32.05 & -15.45 \\ 0 & -15.45 & 15.45 \end{bmatrix} \begin{Bmatrix} x_1^2 \\ x_2^2 \\ x_3^2 \end{Bmatrix} = \begin{Bmatrix} 5 \\ 4 \\ 3 \end{Bmatrix}$$

can be solved for $\{X^2\}$

$$\{X^2\} = \begin{Bmatrix} x_1^2 \\ x_2^2 \\ x_3^2 \end{Bmatrix} = \begin{Bmatrix} 1.00 \\ 1.42 \\ 1.62 \end{Bmatrix}$$

and

$$\{R^2\} = \{X^1\} - \{X^2\} = \begin{Bmatrix} 2.40 \\ 4.15 \\ 5.15 \end{Bmatrix} - \begin{Bmatrix} 1.00 \\ 1.42 \\ 1.62 \end{Bmatrix} = \begin{Bmatrix} 1.40 \\ 2.73 \\ 3.53 \end{Bmatrix}$$

with $|\max\{r_i^2\}| = 3.53 > 0.05$. The results for the remaining iterations are summarized below

k	x_1^k	x_2^k	x_3^k	$|\max\{r_i^k\}|$
0	1	1	1	-
1	2.40	4.15	5.15	4.15
2	1.40	1.42	1.62	3.53
3	1.54	2.77	3.39	1.77
4	1.56	2.19	2.49	0.90
5	1.54	2.34	2.74	0.25
6	1.56	2.31	2.67	0.07
7	1.54	2.30	2.67	0.02 < 0.05

and the solution is

$$\{X\} = \begin{Bmatrix} 1.54 \\ 2.30 \\ 2.67 \end{Bmatrix}$$

For some nonlinear problems (i.e., when the coefficients in $[M(X)]$ are sensitive to small changes in $\{X\}$) it is better to use a modification of Picard iteration based on an *incremental* solution procedure (sometimes called the *modified Newton-Raphson method*). As before, we specify an initial approximate solution $\{X^0\}$ and construct the matrix of coefficients $[M(X^0)]$. However, the residual vector $\{R^0\}$ is computed *before* the system of linear equations is solved i.e.,

$$\{R^0\} = \{B\} - \left[M(X^0)\right]\{X^0\} \tag{5.64}$$

We then test for convergence

$$\left| \max\{r_i^0\} \right| < \varepsilon ? \tag{5.65}$$

If a new solution is needed we construct a system of linear equations using $\{R^0\}$ as the right hand side

$$\left[M(X^0)\right]\{\Delta x^1\} = \{R^0\} \tag{5.66}$$

where $\{\Delta X^1\}$ is a vector of increments used to construct the next solution $\{X^1\}$

$$\{X^1\} = \{X^0\} + \omega^*\{\Delta x^1\} \tag{5.67}$$

and ω^* is a relaxation factor (usually determined by trial and error). We then compute a new residual vector $\{R^1\}$ and repeat the entire process until the convergence criterion is satisfied. The algorithm is summarized in Figure 5.7. This procedure can increase the rate of convergence compared to the algorithm in Figure 5.6 (i.e., a fewer number of iterations are required to reach the specified tolerance) for some problems but may also increase the total number of calculations because of the matrix multiplications required to obtain $\{R^k\}$.

Example

Use Picard iteration with an incremental solution procedure to solve the system of nonlinear equations in the previous example (let $\varepsilon = 0.05$ for max $\{\Delta x_i\}$ and use $\omega^* = 0.5$) Following the algorithm in Figure 5.7, we specify an initial solution

$$\{X^0\} = \begin{Bmatrix} 1 \\ 1 \\ 1 \end{Bmatrix}$$

The matrix of coefficients is the same as before

$$[M(X^0)] = \begin{bmatrix} 9 & -4 & 0 \\ -4 & 7 & -3 \\ 0 & -3 & 3 \end{bmatrix}$$

The residual vector is

1. Specify initial approximate solution $\{X^0\}$

2. For $k = 1, 2, 3, \ldots$ (each value of k is an iteration) do the following

   A. Construct the matrix of coefficients $[M(X^{k-1})]$

   B. Compute the entries of the residual vector $\{R^{k-1}\}$

   $$\{R^{k-1}\} = \{B\} - [M(X^{k-1})]\{X^{k-1}\}$$

   C. Test for convergence

   $$\left| \max\{r_i^{k-1}\} \right| < \varepsilon \ ?$$

   If convergence criterion is satisfied use solution $\{X^{k-1}\}$
   otherwise perform steps D, E, and F

   D. Solve the system of linear equations

   $$[M(X^{k-1})]\{\Delta x^k\} = \{R^{k-1}\}$$

   for $\{\Delta X^k\}$

   E. Construct the next solution $\{X^k\}$

   $$\{X^k\} = \{X^{k-1}\} + \omega^* \{\Delta X^k\}$$

   F. Set $k = k+1$ and go to step A.

Figure 5.7   Algorithm for Picard iteration with an incremental solution procedure.

$$\{R^0\} = \{B\} - \left[M(X^0) \right]\{X^0\}$$

$$= \begin{bmatrix} 5 \\ 4 \\ 3 \end{bmatrix} - \begin{bmatrix} 9 & -4 & 0 \\ -4 & 7 & -3 \\ 0 & -3 & 3 \end{bmatrix}\begin{Bmatrix} 1 \\ 1 \\ 1 \end{Bmatrix} = \begin{Bmatrix} 0 \\ 4 \\ 3 \end{Bmatrix}$$

and $\left| \max\{r_i^0\} \right| = 4$

We can then construct the system of linear equations

$$[M(X^0)]\{\Delta X^1\} = \{R^0\}$$

$$\begin{bmatrix} 9 & -4 & 0 \\ -4 & 7 & -3 \\ 0 & -3 & 3 \end{bmatrix}\begin{Bmatrix} \Delta x_1^1 \\ \Delta x_2^1 \\ \Delta x_3^1 \end{Bmatrix} = \begin{Bmatrix} 0 \\ 4 \\ 3 \end{Bmatrix}$$

and solve for $\{\Delta X^1\}$

$$\{\Delta X^1\} = \begin{Bmatrix} 1.40 \\ 3.15 \\ 4.15 \end{Bmatrix}$$

The next solution is given by

$$\{X^1\} = \{X^0\} + \omega^* \{\Delta X^1\} = \begin{Bmatrix} 1 \\ 1 \\ 1 \end{Bmatrix} + 0.5 \begin{Bmatrix} 1.40 \\ 3.15 \\ 4.15 \end{Bmatrix} = \begin{Bmatrix} 1.70 \\ 2.58 \\ 3.08 \end{Bmatrix}$$

The results for the remaining iterations are summarized below

k	x_1^k	x_2^k	x_3^k	$|max\{r_i^k\}|$	Δx_1^k	Δx_2^k	Δx_3^k	$|max\{\Delta x_i^k\}|$
0	1	1	1	4	1.40	3.15	4.15	4.15
1	1.70	2.58	3.08	1.62	-0.29	-0.49	-0.67	0.67
2	1.56	2.34	2.75	0.38	-0.02	-0.05	-0.10	0.10
3	1.55	2.32	2.70	0.13	-0.0026	-0.019	-0.029	0.029 < 0.5
4	1.549	2.311	2.686					

and the solution is

$$\{X\} = \begin{Bmatrix} 1.549 \\ 2.311 \\ 2.686 \end{Bmatrix}$$

In this example, the incremental solution procedure reduced the number of iterations required for Picard iteration from 7 to 4.

5.3.2 Newton-Raphson Method

For some problems, Picard iteration may converge slowly. In this case we may wish to use the Newton-Raphson method (Concus, 1967). This method is similar to the algorithm in Figure 5.7 in that we specify an initial solution $\{X^0\}$ and compute a series of new solutions $\{X^1\}$, $\{X^2\}$, $\{X^3\}$, ... where at each iteration k we compute the solution $\{X^k\}$ using equation 5.68

$$\{X^k\} = \{X^{k-1}\} + \omega^* \{\Delta X^k\} \tag{5.68}$$

As before, we continue to compute new solutions until the residuals are close to zero i.e., $\{R^k\} \approx \{0\}$. In the Newton–Raphson method we solve for $\{\Delta X^k\}$ by writing a Taylor series approximation for $\{R^k\}$ and setting it equal to zero

$$\{R^k\} = \{R^{k-1}\} + \frac{\partial \{R\}}{\partial \{X\}}\Bigg|_{\{X\}=\{X^{k-1}\}} \{\Delta X^k\}$$

$$+ \frac{\partial^2 \{R\}}{2! \partial \{X\}^2}\Bigg|_{\{X\}=\{X^{k-1}\}} (\{\Delta X^k\})^2 + \cdots = \{0\} \tag{5.69}$$

Neglecting the higher–order terms containing $(\{\Delta X^k\})^2$, $(\{\Delta X^k\})^3$, etc ... we have

$$\{R^k\} = \{R^{k-1}\} + \frac{\partial\{R\}}{\partial\{X\}}\Bigg|_{\{X\}=\{X^{k-1}\}}\{\Delta X^k\} = \{0\} \tag{5.70}$$

or

$$-\frac{\partial\{R\}}{\partial\{X\}}\Bigg|_{\{X\}=\{X^{k-1}\}}\{\Delta X^k\} = \{R^{k-1}\} \tag{5.71}$$

But

$$\{R^{k-1}\} = \{B\} - \left[M(X^{k-1})\right]\{X^{k-1}\} \tag{5.72}$$

and if we assume that the entries in $\{B\}$ are not functions of $\{X\}$ (for convenience only, this is not a requirement of the Newton-Raphson method) we can write (using the product rule)

$$-\frac{\partial\{R\}}{\partial\{X\}}\Bigg|_{\{X\}=\{X^{k-1}\}} = [M(X^{k-1})] + \frac{\partial[M(X)]}{\partial\{X\}}\Bigg|_{\{X\}=\{X^{k-1}\}}\{X^{k-1}\} \tag{5.73}$$

Substituting equation 5.73 into equation 5.71 gives a system of linear equations that can be solved for $\{\Delta X^k\}$

$$\left[[M(X^{k-1})] + \frac{\partial[M(X)]}{\partial\{X\}}\Bigg|_{\{X\}=\{X^{k-1}\}}\{X^{k-1}\}\right]\{\Delta X^k\} = \{R^{k-1}\} \tag{5.74a}$$

which can be written in expanded form as

$$\left(\begin{bmatrix} m_{11}(x_1^{k-1},...,x_n^{k-1}) & \cdots & m_{1n}(x_1^{k-1},...,x_n^{k-1}) \\ \vdots & \cdots & \vdots \\ m_{n1}(x_1^{k-1},...,x_n^{k-1}) & \cdots & m_{nn}(x_1^{k-1},...,x_n^{k-1}) \end{bmatrix} + x_1^{k-1}\begin{bmatrix} \frac{\partial m_{11}}{\partial x_1}(x_1^{k-1},...,x_n^{k-1}) & \cdots & \frac{\partial m_{1n}}{\partial x_1}(x_1^{k-1},...,x_n^{k-1}) \\ \vdots & \cdots & \vdots \\ \frac{\partial m_{n1}}{\partial x_1}(x_1^{k-1},...,x_n^{k-1}) & \cdots & \frac{\partial m_{nn}}{\partial x_1}(x_1^{k-1},...,x_n^{k-1}) \end{bmatrix}\right.$$

$$\left.+ \cdots + x_n^{k-1}\begin{bmatrix} \frac{\partial m_{11}}{\partial x_n}(x_1^{k-1},...,x_n^{k-1}) & \cdots & \frac{\partial m_{1n}}{\partial x_n}(x_1^{k-1},...,x_n^{k-1}) \\ \vdots & \cdots & \vdots \\ \frac{\partial m_{11}}{\partial x_n}(x_1^{k-1},...,x_n^{k-1}) & \cdots & \frac{\partial m_{nn}}{\partial x_n}(x_1^{k-1},...,x_n^{k-1}) \end{bmatrix}\right)\begin{Bmatrix} \Delta x_1^k \\ \vdots \\ \Delta x_n^k \end{Bmatrix} = \begin{Bmatrix} r_1^{k-1} \\ \vdots \\ r_n^{k-1} \end{Bmatrix} \tag{5.74b}$$

The algorithm is summarized in Figure 5.8. The major difficulty in implementing the Newton–Raphson method is computing the entries in $\frac{\partial[M(X)]}{\partial\{X\}}$. However, the rate of convergence is usually faster than Picard iteration and this can sometimes give an overall improvement in computational efficiency.

1. Specify initial approximate solution $\{X^0\}$

2. For $k = 1, 2, 3, \ldots$ (each value of k is an iteration) do the following

  A. Construct the matrices $[M(X^{k-1})]$ and $\left.\dfrac{\partial[M(X)]}{\partial\{X\}}\right|_{\{X\}=\{X^{k-1}\}}\{X^{k-1}\}$

  B. Compute the entries of the residual vector $\{R^{k-1}\}$

$$\{R^{k-1}\} = \{B\} - [M(X^{k-1})]\{X^{k-1}\}$$

  C. Test for convergence

$$\left|\max\{R_i^{k-1}\}\right| < \varepsilon\,?$$

    If convergence criterion is satisfied use solution $\{X^{k-1}\}$
    otherwise perform steps D, E, and F

  D. Solve the system of linear equations

$$\left[[M(X^{k-1})] + \left.\frac{\partial[M(X)]}{\partial\{X\}}\right|_{\{X\}=\{X^{k-1}\}}\{X^{k-1}\}\right]\{\Delta X^k\} = \{R^{k-1}\}$$

    for $\{\Delta X^k\}$

  E. Construct the next approximate solution $\{X^k\}$

$$\{X^k\} = \{X^{k-1}\} + \omega^*\{\Delta X^k\}$$

  F. Perform steps A through E until convergence criterion is satisfied

Figure 5.8   Algorithm for Newton-Raphson Method.

Example

Use the Newton-Raphson method to solve the system of equations in the previous example (let $\varepsilon = 0.05$ for max $\{\Delta x_i\}$ and use $\omega^* = 1.0$) Following the algorithm in Figure 5.8 we specify an initial solution $\{X^0\}$

$$\{X^0\} = \begin{Bmatrix} 1 \\ 1 \\ 1 \end{Bmatrix}$$

For $k = 1$, the matrix of coefficients $[M(X^{k-1})]$ and the residual vector $\{R^{k-1}\}$ are the same as in the previous example

$$[M(X^{k-1})] = [M(X^0)] = \begin{bmatrix} 9 & -4 & 0 \\ -4 & 7 & -3 \\ 0 & -3 & 3 \end{bmatrix}$$

$$\{R^{k-1}\} = \{R^0\} = \begin{Bmatrix} 0 \\ 4 \\ 3 \end{Bmatrix}$$

The derivatives of $[M(X)]$ are given by

$$\frac{\partial[M(X)]}{\partial x_1} = \begin{bmatrix} \frac{\partial}{\partial x_1}(5x_1 + 4x_2) & \frac{\partial}{\partial x_1}(-4x_2) & \frac{\partial}{\partial x_1}(0) \\ \frac{\partial}{\partial x_1}(-4x_2) & \frac{\partial}{\partial x_1}(4x_2 + 3x_3) & \frac{\partial}{\partial x_1}(-3x_3) \\ \frac{\partial}{\partial x_1}(0) & \frac{\partial}{\partial x_1}(-3x_3) & \frac{\partial}{\partial x_1}(3x_3) \end{bmatrix}$$

$$= \begin{bmatrix} 5 & 0 & 0 \\ 0 & 0 & 0 \\ 0 & 0 & 0 \end{bmatrix}$$

Similarly

$$\frac{\partial[M(X)]}{\partial x_2} = \begin{bmatrix} 4 & -4 & 0 \\ -4 & 4 & 0 \\ 0 & 0 & 0 \end{bmatrix} \quad \text{and} \quad \frac{\partial[M(X)]}{\partial x_3} = \begin{bmatrix} 0 & 0 & 0 \\ 0 & 3 & -3 \\ 0 & -3 & 3 \end{bmatrix}$$

$$\frac{\partial[M(X)]}{\partial\{X\}}\bigg|_{\{X\}=\{X^{k-1}\}}\{X^{k-1}\} = x_1^{k-1}\frac{\partial[M(X)]}{\partial x_1}\bigg|_{\{X\}=\{X^{k-1}\}}$$

$$+ x_2^{k-1}\frac{\partial[M(X)]}{\partial x_2}\bigg|_{\{X\}=\{X^{k-1}\}}$$

$$+ x_3^{k-1}\frac{\partial[M(X)]}{\partial x_3}\bigg|_{\{X\}=\{X^{k-1}\}}$$

$$= (1)\begin{bmatrix} 5 & 0 & 0 \\ 0 & 0 & 0 \\ 0 & 0 & 0 \end{bmatrix} + (1)\begin{bmatrix} 4 & -4 & 0 \\ -4 & 4 & 0 \\ 0 & 0 & 0 \end{bmatrix} + (1)\begin{bmatrix} 0 & 0 & 0 \\ 0 & 3 & -3 \\ 0 & -3 & 3 \end{bmatrix}$$

$$= \begin{bmatrix} 9 & -4 & 0 \\ -4 & 7 & -3 \\ 0 & -3 & 3 \end{bmatrix}$$

and referring to equation 5.74 we have

$$\left(\begin{bmatrix} 9 & -4 & 0 \\ -4 & 7 & -3 \\ 0 & -3 & 3 \end{bmatrix} + \begin{bmatrix} 9 & -4 & 0 \\ -4 & 7 & -3 \\ 0 & -3 & 3 \end{bmatrix}\right)\begin{Bmatrix} \Delta x_1^1 \\ \Delta x_2^1 \\ \Delta x_3^1 \end{Bmatrix} = \begin{Bmatrix} 0 \\ 4 \\ 3 \end{Bmatrix}$$

which can be solved for $\{\Delta X^1\}$

$$\{\Delta X^1\} = \begin{Bmatrix} \Delta x_1^1 \\ \Delta x_2^1 \\ \Delta x_3^1 \end{Bmatrix} = \begin{Bmatrix} 0.700 \\ 1.575 \\ 2.075 \end{Bmatrix}$$

and

$$\{X^1\} = \begin{Bmatrix} 1 \\ 1 \\ 1 \end{Bmatrix} + (1.0)\begin{Bmatrix} 0.700 \\ 1.575 \\ 2.075 \end{Bmatrix} = \begin{Bmatrix} 1.700 \\ 2.575 \\ 3.075 \end{Bmatrix}$$

For $k = 2$ we have

$$[M(X^1)] = \begin{bmatrix} 18.18 & -10.30 & 0 \\ -10.30 & 19.52 & -9.23 \\ 0 & -9.23 & 9.23 \end{bmatrix}$$

$$\{R^1\} = \begin{Bmatrix} 5 \\ 4 \\ 3 \end{Bmatrix} - \begin{Bmatrix} 18.80 & -10.30 & 0 \\ -10.30 & 19.52 & -9.23 \\ 0 & -9.23 & 9.23 \end{Bmatrix}\begin{Bmatrix} 1.700 \\ 2.575 \\ 3.075 \end{Bmatrix} = \begin{Bmatrix} -0.438 \\ -0.372 \\ -1.615 \end{Bmatrix}$$

and

$$\begin{bmatrix} 18.80 & -10.30 & 0 \\ -10.30 & 19.52 & -9.23 \\ 0 & -9.23 & 9.23 \end{bmatrix} + (1.700)\begin{bmatrix} 5 & 0 & 0 \\ 0 & 0 & 0 \\ 0 & 0 & 0 \end{bmatrix} + (2.575)\begin{bmatrix} 4 & -4 & 0 \\ -4 & 4 & 0 \\ 0 & 0 & 0 \end{bmatrix}$$

$$+ (3.075)\begin{bmatrix} 0 & 0 & 0 \\ 0 & 3 & -3 \\ 0 & -3 & 3 \end{bmatrix} = \begin{bmatrix} 37.60 & -20.60 & 0 \\ -20.60 & 39.05 & -18.46 \\ 0 & -18.46 & 18.46 \end{bmatrix}$$

$$\begin{bmatrix} 37.60 & -20.60 & 0 \\ -20.60 & 39.05 & -18.46 \\ 0 & -18.46 & 18.46 \end{bmatrix}\begin{Bmatrix} \Delta x_1^2 \\ \Delta x_2^2 \\ \Delta x_3^2 \end{Bmatrix} = \begin{Bmatrix} -0.143 \\ -0.239 \\ -0.327 \end{Bmatrix}$$

which can be solved to give

$$\{\Delta X^2\} = \begin{Bmatrix} -0.143 \\ -0.239 \\ -0.327 \end{Bmatrix}$$

and

$$\{X^2\} = \begin{Bmatrix} 1.700 \\ 2.575 \\ 3.075 \end{Bmatrix} + \begin{Bmatrix} -0.143 \\ -0.239 \\ -0.327 \end{Bmatrix} = \begin{Bmatrix} 1.557 \\ 2.336 \\ 2.748 \end{Bmatrix}$$

After one more iteration ($k = 3$) we find $\{X^3\} = \begin{Bmatrix} 1.54 \\ 2.32 \\ 2.71 \end{Bmatrix}$ with $|\max \{r_i^3\}| = 0.03$.

5.4   APPLICATION OF PICARD ITERATION

5.4.1 Steady-State, Unsaturated Flow Equation

The matrix formulation for the steady-state, unsaturated flow equation (equation 5.2) can be solved using Picard iteration. We simply set $[M(X)] = [K(\psi)]$, the global conductance matrix, $\{X\} = \{\psi\}$, the unknown values of pressure head at each node, and $\{B\} = \{F\}$, the specified rates of groundwater flow at nodes representing sources and sinks (i.e., at Neumann nodes). The global conductance matrix is symmetric because the element matrices it contains are symmetric. Equation 5.2 is a system of nonlinear equations because the entries of the element conductance matrices contain the hydraulic conductivity for the element and, in unsaturated flow, hydraulic conductivity is a function of pressure head. An example of the dependence of hydraulic conductivity on pressure head for three soils is in Figure 5.9. Hydraulic conductivity is maximum when the porous media is saturated i.e., $\psi = 0$ and decreases rapidly with increasing negative values of ψ. Unfortunately, the unsaturated hydraulic conductivity function is difficult to measure experimentally. $K(\psi)$ also displays *hysteresis* (i.e., the value of $K(\psi^0)$ for a fixed value of pressure head ψ^0 is usually different for conditions of wetting and drying).

To solve the steady-state, unsaturated flow equation using Picard iteration we can use the algorithm in Figure 5.6 or 5.7. To begin we specify an initial approximate solution $\{\psi^0\}$. We use these values to compute initial values of hydraulic conductivity for each element in the mesh and compute the element conductance matrices. The global conductance matrix is assembled and modified for specified values of ψ and the system of *linear* equations is solved for the new values of pressure head at each node $\{\psi^1\}$

$$[K(\psi^0)]\{\psi^1\} = \{F\}$$

The residuals are computed from

$$\{R^1\} = \{F\} - [K(\psi^0)]\{\psi^1\}$$

or

$$\{R^1\} = \{\psi^1\} - \{\psi^0\}$$

If the convergence criterion is satisfied we use the solution $\{\psi^1\}$. Otherwise we use $\{\psi^1\}$ to compute new values of hydraulic conductivity for each element, compute new element conductance matrices, assemble $[K(\psi^1)]$ and so on. Because of the need to determine values of hydraulic conductivity for each element for each iteration it is convenient to express measured values of $K(\psi)$ vs ψ in a simple analytical form. Several empirical equations have been proposed for this purpose e.g.,

$$K(\psi) = \frac{a}{b + \psi^m} \tag{5.76}$$

$$K(\psi) = K_s \exp(-a\,\psi) \tag{5.77}$$

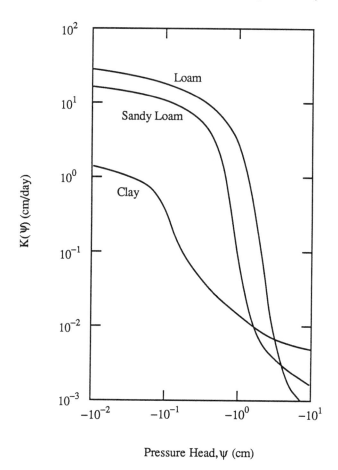

Figure 5.9   Dependence of hydraulic conductivity on pressure head for three soils.

where a, b, and m are empirical coefficients and K_S is the saturated value of hydraulic conductivity (i.e., at $\psi = 0$) (Gardner and Mayhugh, 1958). Other equations for $K(\psi)$ vs ψ are discussed in Mualem (1976), Raats and Gardner (1971), and Bear (1972). Although these equations are convenient for calculations they often provide a poor fit to experimental data. In many cases a simple "table lookup" interpolation scheme (i.e., using cubic splines) can be used to compute $K(\psi)$ for any value of ψ. ψ is computed at the nodes of the mesh so we must decide which value of ψ and $K(\psi)$ to use to compute the element conductance matrices. If analytical methods are used to compute $[K^{(e)}(\psi)]$ (e.g., for the linear bar, triangle, or rectangle elements) we can simply use the average value of ψ for the nodes of each element (see Chapter 6). If Gauss quadrature is used to compute $[K^{(e)}(\psi)]$ (e.g., for the isoparametric elements) we can also use the average value of ψ or we can compute the value of ψ and $K(\psi)$ at each Gauss point during the numerical integration (i.e., we can write $K^{(e)}(\psi(\varepsilon_i, \eta_i, \xi_i))$ etc. in equation 4.63. The vaule of ψ at the Gauss point can be computed using the element's interpolation functions and the values of ψ for each node in the element (see Chapter 6). The various steps in solving the steady-state, unsaturated flow equation are illustrated in the following example.

Example

Use Picard iteration to solve the steady-state, unsaturated flow equation for the mesh shown below. Assume that $K_x^{(e)}(\psi)$(cm/s) vs ψ(cm) is given by

$$K_x^{(e)}(\psi) = 0.01\exp(0.01\ \psi)$$

for all elements and use $\varepsilon = 0.05$

$$L^{(e)} = 5 \text{ cm for all elements}$$

The element conductance matrices are given by equation 4.15a

$$[K^{(e)}(\psi)] = \frac{K_x^{(e)}(\psi)}{L(e)}\begin{bmatrix} 1 & -1 \\ -1 & 1 \end{bmatrix} = K_x^{(e)}(\psi)\begin{bmatrix} 0.2 & -0.2 \\ -0.2 & 0.2 \end{bmatrix}$$

for all elements. The global conductance matrix can be assembled using the procedures in chapter 4 and the global system of equations can be written (after dividing by the common factor 0.2)

$$\begin{bmatrix} K_x^{(1)}(\psi) & -K_x^{(1)}(\psi) & 0 & 0 \\ -K_x^{(1)}(\psi) & (K_x^{(1)}(\psi) + K_x^{(2)}(\psi)) & -K_x^{(2)}(\psi) & 0 \\ 0 & -K_x^{(2)}(\psi) & (K_x^{(2)}(\psi) + K_x^{(3)}(\psi)) & -K_x^{(3)}(\psi) \\ 0 & 0 & -K_x^{(3)}(\psi) & K_x^{(3)}(\psi) \end{bmatrix} \begin{Bmatrix} \psi_1 \\ \psi_2 \\ \psi_3 \\ \psi_4 \end{Bmatrix} = \begin{Bmatrix} 0 \\ 0 \\ 0 \\ 0 \end{Bmatrix}$$

Modifying these equations for the boundary conditions $\psi_1 = -100$ and $\psi_4 = -80$ gives the system of nonlinear equations to be solved for ψ_2 and ψ_3

$$\begin{bmatrix} (K_x^{(1)}(\psi) + K_x^{(2)}(\psi)) & -K_x^{(2)}(\psi) \\ -K_x^{(2)}(\psi) & (K_x^{(2)}(\psi) + K_x^{(3)}(\psi)) \end{bmatrix} \begin{Bmatrix} \psi_2 \\ \psi_3 \end{Bmatrix} = \begin{Bmatrix} -100K_x^{(1)}(\psi) \\ -80K_x^{(3)}(\psi) \end{Bmatrix}$$

Following the algorithm in Figure 5.6 we specify an initial solution $\{\psi^0\}$

$$\{\psi^0\} = \begin{Bmatrix} -90 \\ -85 \end{Bmatrix}$$

The average value of ψ, $\bar{\psi}$ and $K(\bar{\psi})$ for each element can now be computed

$$\bar{\psi}^{(1)} = (\psi_1 + \psi_2)/2 = (-100 - 90)/2 = -95.0$$
$$\bar{\psi}^{(2)} = (\psi_2 + \psi_3)/2 = (-90 - 85)/2 = -87.5$$
$$\bar{\psi}^{(3)} = (\psi_3 + \psi_4)/2 = (-85 - 80)/2 = -82.5$$

$$K_x^{(1)}(\psi) = 0.01 \exp(0.01(-95)) = 0.00387 \text{ cm/s}$$
$$K_x^{(2)}(\psi) = 0.01 \exp(0.01(-87.5)) = 0.00417 \text{ cm/s}$$
$$K_x^{(3)}(\psi) = 0.01 \exp(0.01(-82.5)) = 0.00438 \text{ cm/s}$$

The system of linear equations for the first iteration ($k = 1$) is

$$\begin{bmatrix} (0.00387 + 0.00417) & -0.00417 \\ -0.00417 & (0.00417 + 0.00438) \end{bmatrix} \begin{Bmatrix} \psi_2^1 \\ \psi_3^1 \end{Bmatrix} = \begin{Bmatrix} -0.387 \\ -0.351 \end{Bmatrix}$$

and

$$\{\psi^1\} = \begin{Bmatrix} \psi_2^1 \\ \psi_3^1 \end{Bmatrix} = \begin{Bmatrix} -92.94 \\ -86.37 \end{Bmatrix}$$

The residual vector $\{R^1\}$ is

$$\{R^1\} = \{\psi^0\} - \{\psi^1\} = \begin{Bmatrix} -90 \\ -85 \end{Bmatrix} - \begin{Bmatrix} -92.94 \\ -86.36 \end{Bmatrix} = \begin{Bmatrix} 2.94 \\ 1.36 \end{Bmatrix}$$

and $|\max\{r_i^1\}| = 2.94$. The results for the remaining iterations are summarized below

k	ψ_2^k	ψ_3^k	$K_x^{(1)}(\bar{\psi}^{(1)})$	$K_x^{(2)}(\bar{\psi}^{(2)})$	$K_x^{(3)}(\bar{\psi}^{(3)})$	$\|\max\{r_i^k\}\|$
0	-90	-85	0.00387	0.00417	0.00438	–
1	-92.94	-86.36	0.00381	0.00408	0.00435	2.95
2	-92.88	-86.23	0.00381	0.00408	0.00436	0.13
3	-92.88	-86.22				0.01 < 0.05

and the solution is

$$\{\psi\} = \begin{Bmatrix} -100.00 \\ -92.88 \\ -86.22 \\ -80.00 \end{Bmatrix}$$

5.4.2 Transient, Unsaturated Flow Equation

The matrix formulation for the transient, unsaturated flow equation (equation 5.4) can be solved using Picard iteration. We simply set

$$[M(X)] = [C(\psi)] + \omega\Delta t\, [K(\psi)] \tag{5.78a}$$

$$\{X\} = \{\psi\}_{t+\Delta t} \tag{5.78b}$$

$$\{B\} = ([C(\psi)] - (1-\omega)\, \Delta t\, [K(\psi)]\,)\{\psi\}_t + \Delta t\, ((1-\omega)\{F\}_t + \omega\{F\}_{t+\Delta t}\,) \tag{5.78c}$$

in equation 5.7 where $[C(\psi)]$ is the global capacitance matrix, $[K(\psi)]$ is the global conductance matrix, ω is the relaxation factor, Δt is the time step, and $\{F\}$ are the specified rates of groundwater flow representing sources and sinks at Neumann nodes. $\{F\}$ is known at all time steps. $\{\psi\}_t$ is known from the initial conditions or from the solution for the previous time step. $\{\psi\}_{t+\Delta t}$ contains the unknown values of pressure head at time t+Δt. $[M(X)]$ is symmetric because $[C(\psi)]$ and $[K(\psi)]$ are symmetric.

Equation 5.7 is a system of nonlinear equations because the entries of the element conductance and capacitance matrices contain the hydraulic conductivity and specific moisture capacity for the element, and for unsaturated flow these are functions of pressure head. An example of the dependence of hydraulic conductivity on pressure head is in Figure 5.9. $C(\psi)$ is usually calculated by differentiating experimental curves of θ vs ψ, where θ is volumetric water content (Figure 5.10)

$$C(\psi) = \frac{\partial\theta}{\partial\psi} \tag{5.79}$$

Both $K(\psi)$ and $C(\psi)$ may display hysteresis.

To solve the transient, unsaturated flow equation using Picard iteration we use the algorithm in Figure 5.6 or 5.7 *at each time step*. To begin we specify an initial guess for $\{\psi^0\}_{t+\Delta t}$. We use these values to compute initial values of hydraulic conductivity and specific moisture capacity for each element in the mesh and compute the element conductance and capacitance matrices. The global matrices are assembled and modified for specified values of ψ and the system of linear equations is solved for the new values of pressure head at each node $\{\psi^1\}_{t+\Delta t}$

$$([C(\psi^0)] + \omega\Delta t\, [K(\psi^0)]\{\psi^1\}_{t+\Delta t} = ([C(\psi^0)] - (1-\omega)\, \Delta t\, [K(\psi^0)]\})\{\psi\}_t$$

$$+ \Delta t\, ((1-\omega)\{F\}_t + \omega\{F\}_{t+\Delta t}) \tag{5.80}$$

where all values of $\{\psi^0\}$ refer to time t+Δt. (The values of $\{\psi\}_t$ on the right-hand side of equation 5.80 are known from the solution for the previous solution and do <u>not</u> change during the iterations required to find $\{\psi^1\}_{t+\Delta t}$). The residuals are computed from

$$\{R^1\} = ([C(\psi^0)] - (1-\omega)\, \Delta t\, [K(\psi^0)])\{\psi\}_t + \Delta t\, ((1-\omega)\{F\}_t + \omega\{F\}_{t+\Delta t})$$

$$- ([C(\psi^0)] + \omega\Delta t\, [K(\psi^0)]\{\psi^1\}_{t+\Delta t}) \tag{5.81}$$

or

$$\{R^1\} = \{\psi^1\} - \{\psi^0\} \tag{5.82}$$

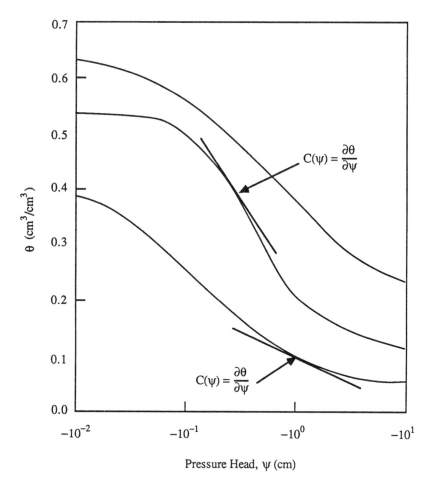

Figure 5.10   Dependence of volumetric water content on pressure head for three soils and definition fo specific moisture capacity.

If the convergence criterion is satisfied we use the solution $\{\psi^1\}_{t+\Delta t}$ for this time step, set $\{\psi\}_t = \{\psi\}_{t+\Delta t}$, and proceed to the next time step. Otherwise we compute new values of hydraulic conductivity and specific moisture capacity, assemble and modify $[K(\psi^1)]$ and $[C(\psi^1)]$ and solve the system of equations for $\{\psi^2\}_{t+\Delta t}$, and so on. We usually use the solution from the previous time step as the initial guess to begin the iterations for the next time step

$$\{\psi^0\}_{t+\Delta t} = \{\psi\}_t \tag{5.83}$$

The procedure for solving the transient, unsaturated flow equation using Picard iteration is in Figure 5.11. Similar algorithms can be written for Picard iteration with an incremental solution procedure and for the Newton-Raphson method.

Because of the need to determine values of hydraulic conductivity and specific moisture capacity for each element for each iteration it is convenient to express measured values of

$K(\psi)$ vs ψ and $C(\psi)$ vs ψ in simple analytical forms. Several empirical equations have been proposed for this purpose, although it is often preferable to use "table look up" interpolation schemes to compute $K(\psi)$ and $C(\psi)$ for any value of ψ. As with the case of steady-state, unsaturated flow we can use the average value of ψ within an element to compute the entries in $[K^{(e)}(\psi)]$ and $[C^{(e)}(\psi)]$ or we can compute the value of ψ, $K(\psi)$, and $C(\psi)$ at each Gauss point during the numerical integration.

For each time step do the following

1. Specify an initial guess solution for pressure head at the current time step $\{\psi^0\}_{t+\Delta t}$. Usually we use the values of pressure head computed from the previous time step

$$\{\psi^0\}_{t+\Delta t} = \{\psi\}_t$$

2. For $k = 1, 2, 3, \ldots$ (each value of k is an iteration) do the following

   A. Compute the values of $K^{(e)}(\psi^{k-1})$ and $C^{(e)}(\psi^{k-1})$ for each element. Construct the element conductance and capacitance matrices $[K^{(e)}(\psi^{k-1})]$ and $[C^{(e)}(\psi^{k-1})]$

   B. Solve the system of linear equations

   $$([C(\psi^{k-1})] + \omega\Delta t\,[K(\psi^{k-1})])\{\psi\}_{t+\Delta t}$$
   $$= ([C(\psi^{k-1})] - (1-\omega)\Delta t\,[K(\psi^{k-1})])\{\psi\}_t$$
   $$+ \Delta t\,((1-\omega)\{F\}_t + \omega\{F\}_{t+\Delta t})$$

   for $\{\psi^k\}_{t+\Delta t}$

   C. Construct the vector of residuals $\{R^k\}$ using

   $$\{R^k\} = ([C(\psi^{k-1})] - (1-\omega)\Delta t\,[K(\psi^{k-1})])\{\psi\}_t$$
   $$+ \Delta t\,((1-\omega)\{F\}_t + \omega\{F\}_{t+\Delta t})$$
   $$- ([C(\psi^{k-1})] + \omega\Delta t\,[K(\psi^{k-1})])\{\psi\}_{t+\Delta t}$$

   or

   $$\{R^k\} = \{\psi^{k-1}\} - \{\psi^k\}$$

   D. Test for convergence

   $$\left|\max\{r_i^k\}\right| < \varepsilon\,?$$

   If convergence criterion is satisfied use solution $\{\psi^k\}_{t+\Delta t}$. Set $\{\psi\}_t = \{\psi^k\}_{t+\Delta t}$ and proceed to the next time step. Otherwise, set $k = k + 1$ and repeat steps A, B, C, D

Figure 5.11   Procedure for solving the transient, unsaturated flow equation using Picard iteration.

5.4.3 Modification of Solution Procedure for Relatively Dry Porous Media

When the porous media is relatively dry, hydraulic conductivity can change rapidly with only a small change in pressure head. For example, in Figure 5.9 the hydraulic conductivity of the sandy loam soil decreases from 0.8 to 0.0017 cm/day as the pressure head decreases from -1 to -10 cm. This behavior can cause difficulties when solving the steady-state or transient, unsaturated flow equations using Picard iteration or the Newton-Raphson method. An alternative approach can be developed by rewriting equations 1.2 and 1.4 in terms of the volumetric water content θ. For steady-state, unsaturated flow we have

$$\frac{\partial}{\partial x}\left(D_x(\theta)\frac{\partial \theta}{\partial x}\right) + \frac{\partial}{\partial y}\left(D_y(\theta)\frac{\partial \theta}{\partial y}\right) + \frac{\partial}{\partial z}\left(D_z(\theta)\frac{\partial \theta}{\partial z}\right) + \frac{\partial K(\theta)}{\partial z} = 0 \tag{5.84}$$

where $D_x(\theta)$, $D_y(\theta,)$, and $D_z(\theta)$ are the components of the aquifers *diffusivity*

$$D(\theta) = K(\theta)\frac{\partial \psi}{\partial \theta} = \frac{K(\theta)}{C(\theta)}$$

and $K(\theta)$ and $C(\theta)$ are the unsaturated hydraulic conductivity and specific moisture capacity written as functions of θ. Empirical expressions can be developed for $K(\theta)$ and $C(\theta)$ or simiple "table-look up" or interpolation schemes (e.g. using the data in Figures 5.9 and 5.10) can be used. For example Gardner and Mayhugh (1958) proposed the equation

$$D(\theta) = a\,\exp(b\theta) \tag{5.85}$$

where a and b are empirical coefficients. In equations 5.84 the z coordinate axis is oriented vertically upward. For transient, unsaturated flow we have

$$\frac{\partial}{\partial x}\left(D_x(\theta)\frac{\partial \theta}{\partial x}\right) + \frac{\partial}{\partial y}\left(D_y(\theta)\frac{\partial \theta}{\partial y}\right) + \frac{\partial}{\partial z}\left(D_z(\theta)\frac{\partial \theta}{\partial z}\right) + \frac{\partial K(\theta)}{\partial z} = \frac{\partial \theta}{\partial t} \tag{5.86}$$

If the porous media is relatively dry it is often assumed that the effect of gravity on water flow is small. In this case, the last term on the right-hand side of equation 5.84 can be discarded. Following the procedures described in Chapter 3 and 4, matrix expressions similar to equations 5.2 and 5.4 can be developed for use in solving equations 5.84 and 5.86.

$$[C(\theta)]\{\overset{\circ}{\theta}\} + [D(\theta)]\{\theta\} = \{F\} \tag{5.87}$$

Equation 5.84 can be written

$$([C] + \omega\Delta t\,[D(\theta)])\,\{\theta\}_{t+\Delta t} = ([C] - (1-\omega)\,\Delta t\,[D(\theta)])\{\theta\}_t$$
$$+ \Delta t\,((1-\omega)\,\{F\}_t + \omega\{F\}_{t+\Delta t}) \tag{5.88}$$

where [C] is a *global capacitance matrix* that can be obtained by assembling the element *capacitance matrices* $[C^{(e)}]$ for all elements in the mesh (written here with a consistent element formulation)

$$[C^{(e)}] = \int\int\int_{V^{(e)}} \begin{bmatrix} N_1^{(e)} \\ \vdots \\ N_n^{(e)} \end{bmatrix} [N_1^{(e)} \cdots N_n^{(e)}] \, dx \, dy \, dz \qquad (5.89)$$

and $[D(\theta)]$ is a *global diffusivity matrix* that can be obtained by assembling the *element diffusivity matrices* $[D^{(e)}(\theta)]$ for all elements in the mesh

$$[D^{(e)}(\theta)] = \int\int\int \underbrace{\begin{bmatrix} \dfrac{\partial N_1^{(e)}}{\partial x} & \dfrac{\partial N_1^{(e)}}{\partial y} & \dfrac{\partial N_1^{(e)}}{\partial z} \\ \vdots & \vdots & \vdots \\ \dfrac{\partial N_n^{(e)}}{\partial x} & \dfrac{\partial N_n^{(e)}}{\partial y} & \dfrac{\partial N_n^{(e)}}{\partial z} \end{bmatrix}}_{n \times 3} \underbrace{\begin{bmatrix} D_x^{(e)}(\theta) & 0 & 0 \\ 0 & D_y^{(e)}(\theta) & 0 \\ 0 & 0 & D_z^{(e)}(\theta) \end{bmatrix}}_{3 \times 3}$$

$$\underbrace{\begin{bmatrix} \dfrac{\partial N_1^{(e)}}{\partial x} & \cdots & \dfrac{\partial N_n^{(e)}}{\partial x} \\ \dfrac{\partial N_1^{(e)}}{\partial y} & \cdots & \dfrac{\partial N_n^{(e)}}{\partial y} \\ \dfrac{\partial N_1^{(e)}}{\partial z} & \cdots & \dfrac{\partial N_n^{(e)}}{\partial z} \end{bmatrix}}_{3 \times n} dx \, dy \, dz$$

$$(5.90)$$

The procedures in Chapter 4 can be used to compute the element matrices, assemble the global system of equations, and modify the system of equations for known values of θ at Dirichlet nodes. Equation 5.88 is a system of nonlinear equations that can be solved using either Picard iteration or the Newton-Raphson method. Because the diffusivity varies less with changing water content than hydraulic conductivity varies with changing pressure head the numerical solutions should converge more rapidly when solving equation 5.86 than when solving equation 5.2. However, equation 5.86 can be extremely difficult to solve when the porous media is nonhomogeneous because large changes in water content can occur abruptly at boundaries between layers with different hydraulic properties (e.g. at the contact between sand and clay layers). In this case solutions based on equation 5.2 are preferred.

NOTES AND ADDITIONAL READING

1. A variety of numerical methods can be used to solve the system of linear equations represented by equations 5.1, 5.3, and 5.5. Some examples are Gauss elimination (Cook, 1981), Gauss-Seidel iteration (Wang and Anderson, 1979), and the wavefront method (Irons, 1970). The merits of these methods are compared in Meyer (1973) and Jensen and Parks (1970). There is no single "best" equation solver because the performance of an algorithm varies from one class of problem to the next. Unless the number of unknowns exceeds several hundred, choice of method has little impact on speed or accuracy of computation.

2. An alternative form of matrix storage that is widely used is called "skyline" matrix storage (Everstine, 1979). Skyline storage is useful for reducing storage requirements when the number of zero entries within the band is large.

3. A thorough discussion of instability and numerical oscillation for transient groundwater flow and solute transport is in Lapidus and Pinder (1982), Pinder and Gray (1977) and Remson et al. (1971, p.71 - 77).

4. A computer program that uses Picard iteration to solve the transient unsaturated groundwater flow equation using linear triangle elements is described in Davis and Neumann (1983). Comparisons of several computer programs for transient, unsaturated groundwater flow are in Matanga and Frind (1981), Bachmat et al. (1978), and Oster (1982).

Problems

1. Perform triangular decomposition on the following matrices

(a) $\begin{bmatrix} 1 & -1 & 0 \\ -1 & 2 & -1 \\ 0 & -1 & 1 \end{bmatrix}$

(b) $\begin{bmatrix} 2 & 1 & 2 \\ 2 & 4 & 1 \\ 2 & 1 & 2 \end{bmatrix}$

(c) $\begin{bmatrix} 4 & -1 & -2 & -1 \\ -1 & 4 & -1 & -2 \\ -2 & -1 & 4 & -1 \\ -1 & -2 & -1 & 4 \end{bmatrix}$

(d) $\begin{bmatrix} 4 & -2 & -2 & -1 \\ -1 & 4 & -2 & -2 \\ -2 & -1 & 4 & -1 \\ -1 & -2 & -1 & 4 \end{bmatrix}$

2. Solve the system of equations $[M]\{X\} = \{B_i\}$, $i = 1, 2$, using Choleski's method

(a) $[M] = \begin{bmatrix} 1 & -1 & 0 \\ -1 & 2 & -1 \\ 0 & -1 & 1 \end{bmatrix}$, $\{B_1\} = \begin{Bmatrix} 5 \\ 0 \\ 0 \end{Bmatrix}$, $\{B_2\} = \begin{Bmatrix} 10 \\ 0 \\ 0 \end{Bmatrix}$

(b) $[M] = \begin{bmatrix} 2 & 1 & 2 \\ 2 & 4 & 1 \\ 2 & 1 & 2 \end{bmatrix}$, $\{B_1\} = \begin{Bmatrix} 0 \\ 10 \\ 0 \end{Bmatrix}$, $\{B_2\} = \begin{Bmatrix} 10 \\ 10 \\ 10 \end{Bmatrix}$

(c) $[M] = \begin{bmatrix} 4 & -1 & -2 & -1 \\ -1 & 4 & -1 & -2 \\ -2 & -1 & 4 & -1 \\ -1 & -2 & -1 & 4 \end{bmatrix}$, $\{B_1\} = \begin{Bmatrix} 1 \\ 1 \\ 1 \\ 1 \end{Bmatrix}$, $\{B_2\} = \begin{Bmatrix} 1 \\ 0 \\ 0 \\ 0 \end{Bmatrix}$

(d) $[M] = \begin{bmatrix} 4 & -2 & -2 & -1 \\ -1 & 4 & -2 & -2 \\ -2 & -1 & 4 & -1 \\ -1 & -2 & -1 & 4 \end{bmatrix}$, $\{B_1\} = \begin{Bmatrix} 1 \\ 1 \\ 1 \\ 1 \end{Bmatrix}$, $\{B_2\} = \begin{Bmatrix} 0 \\ 0 \\ 0 \\ 1 \end{Bmatrix}$

3. Compute the length of vector $\{M\}$ required to store each matrix in problem 1 in vector storage.

4. Use the algorithms in Figures 5.2 and 5.3 to assign the entries of the following matrices to vector storage

(a)
$$\begin{bmatrix} 1 & 2 & 0 & 0 \\ 2 & 1 & 2 & 0 \\ 0 & 2 & 1 & 1 \\ 0 & 0 & 1 & 2 \end{bmatrix}$$

(b)
$$\begin{bmatrix} 1 & 0 & 2 & 0 \\ 0 & 1 & 0 & 0 \\ 2 & 0 & 1 & 2 \\ 0 & 0 & 2 & 1 \end{bmatrix}$$

(c)
$$\begin{bmatrix} 4 & 0 & 2 & 2 & 0 \\ 0 & 4 & 2 & 4 & 4 \\ 4 & 2 & 4 & 0 & 2 \\ 4 & 4 & 0 & 4 & 0 \\ 0 & 4 & 2 & 0 & 4 \end{bmatrix}$$

5. Find combinations of ψ and Δt that prevent instability and numerical oscillations when solving the transient, saturated flow equation using the elements shown below (use the consistent formulation for element capacitance matrices). $S_s^{(e)} = 0.1\ \mathrm{m}^{-1}$, $K_x^{(e)} = K_y^{(e)} = 0.1$ m/day.

(a)

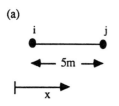

(b)

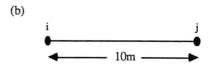

(c)   (d)

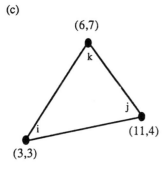
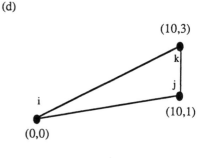

6. Repeat problem 5 if $K_x^{(e)} = 1$ m/day and $K_y^{(e)} = 0.05$ m/day.

7. Repeat problem 5 using a lumped formulation for the element capacitance matrices.

8. Repeat problem 6 using a lumped formulation for the element capacitance matrices.

9. Find combinations of ω and Δt that prevent instability and numerical oscillations when solving the solute transport equation using the elements in problem 5. Consider saturated groundwater flow and use the lumped formulation for element sorption and advection–dispersion matrices. $\rho_b^{(e)} = 1600$ kg/m^3 $K_d^{(e)} = \lambda = 0$, n = 0.35, $D_{xx}^{(e)} = D_{yy}^{(e)} = 1$ m^2/day, $D_{xy}^{(e)} = D_{yx}^{(e)} = 0.5$ m^2/day, $v_x^{(e)} = 1$ m/day, $v_y^{(e)} = 0.2$ m/day.

10. Solve the following systems of nonlinear equations using Picard iteration (Figure 5.6)

(a)
$$\begin{bmatrix} (2x_1 + 3x_2) & -3x_2 & 0 \\ -3x_2 & (3x_2 + 4x_3) & -4x_3 \\ 0 & -4x_3 & 4x_3 \end{bmatrix} \begin{Bmatrix} x_1 \\ x_2 \\ x_3 \end{Bmatrix} = \begin{Bmatrix} 10 \\ 5 \\ 1 \end{Bmatrix}$$

(b)
$$\begin{bmatrix} (x_1 + 2x_2) & -2x_2 & 0 \\ -2x_2 & (x_1 + 2x_2 + x_3) & (-2x_2 - x_3) \\ 0 & (-2x_2 - x_3) & (2x_2 + x_3) \end{bmatrix} \begin{Bmatrix} x_1 \\ x_2 \\ x_3 \end{Bmatrix} = \begin{Bmatrix} 6 \\ 4 \\ 2 \end{Bmatrix}$$

11. Repeat problem 10 using Picard iteration with an incremental solution procedure (Figure 5.7).

12. Repeat problem 10 using the Newton-Raphson Method.

Chapter 6

STEP 5: CALCULATE REQUIRED ELEMENT RESULTANTS

By solving the global system of equations we obtain values of the field variable (hydraulic head, pressure head, or solute concentration) at each node in the finite element mesh (and at each time step if we are solving a transient groundwater flow or solute transport problem). We also may wish to calculate certain additional quantities for each element that we will collectively refer to as *element resultants*. The three types of element resultants usually considered are

1) The value of the field variable at any specified *point* (not necessarily a nodal point) within an element,

2) The *average value* of the field variable within an element, and

3) The values of *derivatives* of the field variable at any specified point within an element

In most cases the computed values of the field variable at the nodal points are the only information required. However, it is possible to calculate the value of the field variable at any point in the mesh. For example, the value of hydraulic head may be required at a pumping well that is not located at a node.

6.1   LINEAR ELEMENTS

If the interpolation functions for an element are defined using a global coordinate system the procedure is very simple. We first specify the coordinates of the point of interest and determine which element in the mesh contains the point (If the point falls on the boundary between two elements either element can be considered to "contain" the point because the approximate solution is continuous from one element to the next). Let (x_0, y_0, z_0) refer to the global coordinates of the specified point and let e_0 refer to the element that contains the point. The value of head or concentration at the point can be obtained directly from the approximate solutions (see section 3.1)

$$\hat{h}^{(e_0)}(x_0,y_0,z_0) = \sum_{i=1}^{n} N_i^{(e_0)}(x_0,y_0,z_0) h_i \qquad (6.1a)$$

$$\hat{\psi}^{(e_0)}(x_0,y_0,z_0) = \sum_{i=1}^{n} N_i^{(e_0)}(x_0,y_0,z_0) \psi_i \qquad (6.1b)$$

$$\hat{C}^{(e_0)}(x_0,y_0,z_0) = \sum_{i=1}^{n} N_i^{(e_0)}(x_0,y_0,z_0) C_i \qquad (6.1c)$$

where $N_i^{(e_0)}$ are the interpolation functions for the element containing the point, n is the number of nodes in element e_0, and h_i, ψ_i. and C_i are the computed values of hydraulic head, pressure head, and solute concentration at each node in element e_0.

Example

Given the mesh shown below compute the value of pressure head ψ at points A and B.

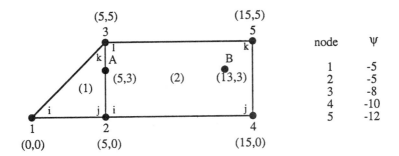

node	ψ
1	-5
2	-5
3	-8
4	-10
5	-12

We can compute the value of ψ at point A using the interpolation functions for either element.
For element 1 (see Figure 4.6):

$$N_i^{(1)}(5,3) = \frac{1}{2A^{(e)}} (a_i + b_i(5) + c_i(3))$$

$$N_j^{(1)}(5,3) = \frac{1}{2A^{(e)}} (a_j + b_j(5) + c_j(3))$$

$$N_k^{(1)}(5,3) = \frac{1}{2A^{(e)}} (a_k + b_k(5) + c_k(3))$$

$$2A^{(1)} = 5(5) = 25$$
$$a_j = x_k y_i - x_i y_k = (5)(0) - 0(5) = 0$$
$$b_j = y_k - y_i = 5 - 0 = 5$$
$$c_j = x_i - x_k = 0 - 5 = -5$$

$$N_i^{(1)}(5,3) = 0 \quad \text{(why ?)}$$

$$N_j^{(1)}(5,3) = \frac{1}{25} (0 + 5(5) - 5(3)) = 0.4$$

$$N_k^{(1)}(5,3) = 1 - 0.40 = 0.6 \quad \text{(why ?)}$$

$$\psi^{(1)}(5,3) = 0(-5) + 0.4(-5) + 0.6(-8) = \underline{-4.8}$$

For element 2 (see Figure 4.7):

$$N_i^{(2)} = \frac{(2b^{(e)}-s)(2a^{(e)}-t)}{4a^{(e)}b^{(e)}}$$

$$N_j^{(2)} = \frac{s(2a^{(e)}-t)}{4a^{(e)}b^{(e)}}$$

$$N_k^{(2)} = \frac{st}{4a^{(e)}b^{(e)}}$$

$$N_l^{(2)} = \frac{(2b^{(e)}-s)t}{4a^{(e)}b^{(e)}}$$

$$2b^{(e)} = 15 - 5 = 10 \qquad\qquad 2a^{(e)} = 5 - 0 = 5$$

At point A, $x = 5$, $y = 3$, $s = 0$, and $t = 3$

$$N_i^{(2)}(0,3) = \frac{(10-0)(5-3)}{(10)(5)} = 0.4$$

$$N_j^{(2)}(0,3) = 1 - 0.4 = 0.6 \quad \text{(why ?)}$$

$$N_k^{(2)}(0,3) = N_l^{(2)}(0,3) = 0$$

$$\psi^{(2)}(0,3) = 0.4(-5) + 0(-10) + 0(-12) + 0.6(-8) = \underline{-4.8} = \psi^{(1)}(5,3)$$

We compute the value of ψ at point B using the interpolation functions for element 2. At point B, $x = 13$, $y = 3$, $s = 8$, and $t = 3$

$$N_i^{(2)}(8,3) = \frac{(10-8)(5-3)}{(10)(5)} = 0.08$$

$$N_j^{(2)}(8,3) = \frac{8(10-3)}{(10)(5)} = 0.32$$

$$N_k^{(2)}(8,3) = \frac{8(3)}{4(5/2)(5)} = 0.48$$

$$N_l^{(2)}(8,3) = \frac{(10-8)(3)}{(10)(5)} = 0.12$$

$$\psi^{(2)}(8,3) = 0.08(-5) + 0.32(-10) + 0.48(-12) + 0.12(-8) = \underline{-10.3}$$

If the interpolation functions for an element are defined using a local coordinate system the same procedure is used but the approximate solution is given by

$$\hat{h}^{(e_0)}(\varepsilon_0,\eta_0,\zeta_0) = \sum_{i=1}^{n} N_i^{(e_0)}(\varepsilon_0,\eta_0,\zeta_0)h_i \tag{6.2a}$$

$$\hat{\psi}^{(e_0)}(\varepsilon_0,\eta_0,\zeta_0) = \sum_{i=1}^{n} N_i^{(e_0)}(\varepsilon_0,\eta_0,\zeta_0)\psi_i \tag{6.2b}$$

$$\hat{C}^{(e_0)}(\varepsilon_0,\eta_0,\zeta_0) = \sum_{i=1}^{n} N_i^{(e_0)}(\varepsilon_0,\eta_0,\zeta_0)C_i \tag{6.2c}$$

where $(\varepsilon_0,\eta_0,\zeta_0)$ are the local coordinates of the specified point.

Example
Compute the value of solute concentration at point A for the isoparametric, linear quadrilateral element shown below in the local (ε,η) coordinate system. The local coordinates of point A are $(2/3,1/3)$

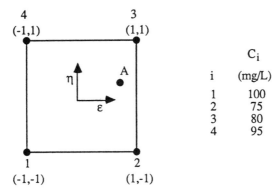

i	C_i (mg/L)
1	100
2	75
3	80
4	95

The interpolation functions for this type of element are in Figure 4.10. Using equation 6.2c we have

$$\hat{C}\left(\frac{2}{3},\frac{1}{3}\right) = \sum_{i=1}^{4} N_i\,(\varepsilon_0,\eta_0)\,C_i$$

$$= \frac{1}{4}\left[\left(1-\frac{2}{3}\right)\left(1-\frac{1}{3}\right)(100) + \left(1+\frac{2}{3}\right)\left(1-\frac{1}{3}\right)(75)\right.$$

$$\left. + \left(1+\frac{2}{3}\right)\left(1+\frac{1}{3}\right)(80) + \left(1-\frac{2}{3}\right)\left(1+\frac{1}{3}\right)(95)\right] = \underline{81\ mg/L}$$

(Note that the same results could be obtained by using the interpolation functions for the linear rectangle element in Figure 4.7).

If the coordinates of the specified point are given in a global coordinate system and the interpolation functions for the element containing the point are defined using a local coordinate system, a coordinate transformation is required to obtain the coordinates of the specified point in the local coordinate system

$$\varepsilon_0 = f_1(x_0) \tag{6.3a}$$

$$\eta_0 = f_2(y_0) \tag{6.3b}$$

$$\zeta_0 = f_3(z_0) \tag{6.3c}$$

where f_1, f_2, and f_3 are the coordinate transformation equations. Once the values of ε_0, η_0, and ζ_0 are obtained equation 6.2 can be used to obtain the value of $\hat{h}$, $\hat{\psi}$, or $\hat{C}$ at the specified point. Unfortunately, obtaining general coordinate transformation equations for many two-, and three-dimensional isoparametric elements is difficult because the interpolation functions are nonlinear. This approach is seldom used in practice. If isoparametric elements are used (and if the shape of the elements will be highly distorted) it is recommended that the analyst

1) insures that nodal points are placed at any special points of interest, and

2) uses a large enough number of nodal points so that contour lines, streamlines, etc. can be drawn using only the computed values of the field variable at the nodes.

The second type of element resultant we will consider is the average value of the field variable within an element. This information is needed, for example, when solving unsaturated groundwater flow problems. For unsaturated porous media, the components of hydraulic conductivity and specific moisture capacity are functions of the pressure head, ψ

$$K = K(\psi) \tag{6.4a}$$
$$C = C(\psi) \tag{6.4b}$$

We usually assume that the value of hydraulic conductivity and specific moisture capacity are constant within an element but can vary from one element to the next. A typical procedure is to compute an average value of pressure head for each element and then use this average value to obtain a value of $K(\psi)$ or $C(\psi)$. The average value of pressure head within element e, $\bar{\psi}^{(e)}$ is given by

$$\bar{\psi}^{(e)} = \frac{1}{V^{(e)}} \int_{V^{(e)}} \hat{\psi}^{(e)} \, dx \, dy \, dz$$

$$= \frac{1}{V^{(e)}} \int_{V^{(e)}} \left(\sum_{i=1}^{n} N_i^{(e)}(x,y,z)\psi_i \right) dx \, dy \, dz \tag{6.5}$$

where $V^{(e)}$ is the volume of element e. Similar equations can be written for one– and two–dimensional elements.

Evaluating the integral in equation 6.5 for isoparametric elements will usually require numerical integration and the procedures of Chapter 4 can be used for this purpose. However, as long as the element shape is not highly distorted and if pressure head is not changing rapidly within the element an acceptable approximation to equation 6.5 is given by

$$\bar{\psi}^{(e)} \approx \frac{1}{n} \sum_{i=1}^{n} \psi_i \tag{6.6}$$

where n is the number of nodes in element e. Equation 6.6 assigns equal weight to the value of ψ at each node. For two types of elements, the linear bar and linear triangle, the results obtained by equations 6.5 and 6.6 will be identical.

Example

Compute the average value of ψ, $\bar{\psi}$ for the element shown below using equations 6.5 and 6.6

$$\psi_i = -1 \qquad \overset{(e)}{\underset{}{}} \qquad \psi_j = -1.8$$

$$\underset{\substack{i \\ x_i = 1}}{\bullet} \rule{2cm}{0.4pt} \underset{\substack{j \\ x_j = 3}}{\bullet}$$

The one-dimensional form of equation 6.5 is

$$\bar{\psi}^{(e)} = \frac{1}{L^{(e)}} \int_{L^{(e)}} \left(\sum_{i=1}^{2} N_i^{(e)}(x)\psi_i \right) dx$$

$$= \frac{1}{L^{(e)}} \int_{x_i^{(e)}}^{x_j^{(e)}} (N_i^{(e)}(x)\psi_i + N_j^{(e)}(x)\psi_j) dx$$

$$= \frac{1}{L^{(e)}} \int_{1}^{3} \left\{ \left(\frac{x_j^{(e)} - x}{L^{(e)}} \right)\psi_i + \left(\frac{x - x_i^{(e)}}{L^{(e)}} \right)\psi_j \right\} dx$$

$$= \frac{1}{2} \int_{1}^{3} \left\{ \left(\frac{3-x}{2} \right)(-1) + \left(\frac{x-1}{2} \right)(-1.8) \right\} dx$$

$$= \frac{1}{2} \int_{1}^{3} \left(-\frac{3}{2} + \frac{x}{2} - 0.9x + 0.9 \right) dx$$

$$= \frac{1}{2} \int_{1}^{3} (-0.4x - 0.6) dx$$

$$= \frac{1}{2} \left(\frac{-0.4x^2}{2} - 0.6x \right)\Big|_{1}^{3} = \frac{1}{2} (-1.8 - 1.8 + 0.2 + 0.6) = \underline{-1.4}$$

Using equation 6.6 we have

$$\bar{\psi}^{(e)} = \frac{1}{2} (-1 + (-1.8)) = \underline{-1.4}$$

Example

Compute the average value of ψ, $\bar{\psi}$ for the element shown below using equations 6.5 and 6.6

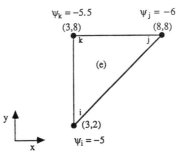

The two-dimensional form of equation 6.5 is

$$\bar{\psi}^{(e)} = \frac{1}{A^{(e)}} \int_{A^{(e)}} \left(\sum_{i=1}^{3} N_i^{(e)}(x,y)\psi_i \right) dx \, dy$$

$$= \frac{1}{A^{(e)}} \left(\int_{A^{(e)}} N_i^{(e)}(x,y)\psi_i \, dx \, dy + \int_{A^{(e)}} N_j^{(e)}(x,y)\psi_j \, dx \, dy + \int_{A^{(e)}} N_k^{(e)}(x,y)\psi_k \, dx \, dy \right)$$

Performing the integrations one at a time using the integration formulas for the linear triangle element described in Chapter 4 we have

$$\int_{A^{(e)}} N_i^{(e)}(x,y)\psi_i \, dx \, dy = \frac{A^{(e)}}{3}\psi_i = \frac{15}{3}(-5) = -25$$

$$\int_{A^{(e)}} N_j^{(e)}(x,y)\psi_j \, dx \, dy = \frac{A^{(e)}}{3}\psi_j = \frac{15}{3}(-6) = -30$$

$$\int_{A^{(e)}} N_k^{(e)}(x,y)\psi_k \, dx \, dy = \frac{A^{(e)}}{3}\psi_k = \frac{15}{3}(-5.5) = -27.5$$

and

$$\bar{\psi}^{(e)} = \frac{1}{15}(-25 - 30 - 27.5) = \underline{-5.5}$$

Using equation 6.6 we have

$$\bar{\psi}^{(e)} = \frac{1}{3}(-5 - 6 - 5.5) = \underline{-5.5}$$

The third type of element resultant that we will need to compute is the derivative of the field variable at a specified point within an element. In groundwater flow problems the derivatives of hydraulic head or pressure head are needed to compute the components of apparent groundwater velocity using Darcy's Law. The velocity components can then be used to draw streamlines, showing the direction and magnitude of apparent groundwater velocity throughout the mesh. In a solute transport problem, the components of apparent groundwater velocity are needed to compute the element advection-dispersion matrices. Derivatives of solute concentration can also be used to compute the components of solute flux although this is rarely done in practice.

The components of apparent groundwater velocity are computed using Darcy's Law which can be written

$$v_x = -K_x \frac{\partial h}{\partial x} \qquad (6.7a)$$

$$v_y = -K_y \frac{\partial h}{\partial y} \qquad (6.7b)$$

$$v_z = -K_z \frac{\partial h}{\partial z} \qquad (6.7c)$$

for saturated flow and

$$v_x = -K_x(\psi)\frac{\partial\psi}{\partial x} \tag{6.8a}$$

$$v_y = -K_y(\psi)\frac{\partial\psi}{\partial y} \tag{6.8b}$$

$$v_z = -K_z(\psi)\left(\frac{\partial\psi}{\partial z} + 1\right) \tag{6.8c}$$

for unsaturated flow. In equation 6.8 it assumed that the z axis is directed vertically downward. Remember that hydraulic head = pressure head + elevation head

$$h = \psi + z^* \tag{6.9}$$

where the elevation head z^* is defined with respect to an arbitrarily placed datum (Appendix I). In unsaturated flow in a horizontal direction (in our case in the plane of the x and y coordinate axes) equation 6.8a is obtained from equation 6.7a

$$
\begin{aligned}
v_x &= -K_x\frac{\partial h}{\partial x} = -K_x(\psi)\frac{\partial h}{\partial x} \\
&= -K_x(\psi)\frac{\partial}{\partial x}(\psi + z^*)^{\,0} \\
&= -K_x(\psi)\left(\frac{\partial\psi}{\partial x} + \frac{\partial z}{\partial x}\right) \\
&= -K_x(\psi)\frac{\partial\psi}{\partial x}
\end{aligned}
$$

Similarly for v_y. Since the z axis is directed vertically downward $\frac{\partial z^*}{\partial z} = 1$ and

$$
\begin{aligned}
v_z &= -K_z(\psi)\frac{\partial}{\partial z}(\psi + z^*) \\
&= -K_z(\psi)\left(\frac{\partial\psi}{\partial z} + \frac{\partial z}{\partial z}\right)^{1} \\
&= -K_z(\psi)\left(\frac{\partial\psi}{\partial z} + 1\right)
\end{aligned}
$$

Once we have computed the values of hydraulic head or pressure head for each node in the mesh we can compute the derivatives of head at any specified point in the mesh. Let (x_0,y_0,z_0) refer to the coordinates (in a global coordinate system) of the specified point and let e_0 refer to the element that contains the point. The values of the derivatives of head at the specified point can be obtained by evaluating the derivatives of the approximate solution (equation 6.1) at the point. For hydraulic head the approximate solution is

$$
\begin{aligned}
\hat{h}^{(e_0)}(x_0,y_0,z_0) &= \sum_{i=1}^{n} N_i^{(e_0)}(x_0,y_0,z_0)h_i \\
&= [\, N_i^{(e_0)}(x_0,y_0,z_0) \cdots N_n^{(e_0)}(x_0,y_0,z_0)]\begin{Bmatrix} h_1 \\ \vdots \\ h_n \end{Bmatrix}
\end{aligned} \tag{6.9}
$$

The derivatives are given by

$$\frac{\partial \hat{h}^{(e_0)}}{\partial x}(x_0,y_0,z_0) = \left[\frac{\partial N_1^{(e_0)}}{\partial x}(x_0,y_0,z_0) \cdots \frac{\partial N_n^{(e_0)}}{\partial x}(x_0,y_0,z_0)\right]\begin{Bmatrix} h_1 \\ \vdots \\ h_n \end{Bmatrix} \qquad (6.10a)$$

$$\frac{\partial \hat{h}^{(e_0)}}{\partial y}(x_0,y_0,z_0) = \left[\frac{\partial N_1^{(e_0)}}{\partial y}(x_0,y_0,z_0) \cdots \frac{\partial N_n^{(e_0)}}{\partial y}(x_0,y_0,z_0)\right]\begin{Bmatrix} h_1 \\ \vdots \\ h_n \end{Bmatrix} \qquad (6.10b)$$

$$\frac{\partial \hat{h}^{(e_0)}}{\partial z}(x_0,y_0,z_0) = \left[\frac{\partial N_1^{(e_0)}}{\partial z}(x_0,y_0,z_0) \cdots \frac{\partial N_n^{(e_0)}}{\partial z}(x_0,y_0,z_0)\right]\begin{Bmatrix} h_1 \\ \vdots \\ h_n \end{Bmatrix} \qquad (6.10c)$$

Similar expressions can be written for the derivatives of pressure head

$$\frac{\partial \hat{\psi}^{(e_0)}}{\partial x}(x_0,y_0,z_0) = \left[\frac{\partial N_1^{(e_0)}}{\partial x}(x_0,y_0,z_0) \cdots \frac{\partial N_n^{(e_0)}}{\partial x}(x_0,y_0,z_0)\right]\begin{Bmatrix} \psi_1 \\ \vdots \\ \psi_n \end{Bmatrix} \qquad (6.11a)$$

$$\frac{\partial \hat{\psi}^{(e_0)}}{\partial y}(x_0,y_0,z_0) = \left[\frac{\partial N_1^{(e_0)}}{\partial y}(x_0,y_0,z_0) \cdots \frac{\partial N_n^{(e_0)}}{\partial y}(x_0,y_0,z_0)\right]\begin{Bmatrix} \psi_1 \\ \vdots \\ \psi_n \end{Bmatrix} \qquad (6.11b)$$

$$\frac{\partial \hat{\psi}^{(e_0)}}{\partial z}(x_0,y_0,z_0) = \left[\frac{\partial N_1^{(e_0)}}{\partial z}(x_0,y_0,z_0) \cdots \frac{\partial N_n^{(e_0)}}{\partial z}(x_0,y_0,z_0)\right]\begin{Bmatrix} \psi_1 \\ \vdots \\ \psi_n \end{Bmatrix} \qquad (6.11c)$$

The components of apparent groundwater velocity at the specified point are given by

$$v_x^{(e_0)}(x_0,y_0,z_0) = -K_x^{(e_0)}\frac{\partial \hat{h}^{(e_0)}}{\partial x}(x_0,y_0,z_0) \qquad (6.12a)$$

$$v_y^{(e_0)}(x_0,y_0,z_0) = -K_y^{(e_0)}\frac{\partial \hat{h}^{(e_0)}}{\partial y}(x_0,y_0,z_0) \qquad (6.12b)$$

$$v_z^{(e_0)}(x_0,y_0,z_0) = -K_z^{(e_0)}\frac{\partial \hat{h}^{(e_0)}}{\partial z}(x_0,y_0,z_0) \qquad (6.12c)$$

for saturated flow and

$$v_x^{(e_0)}(x_0,y_0,z_0) = -K_x^{(e_0)}(\psi)\frac{\partial \hat{\psi}^{(e_0)}}{\partial x}(x_0,y_0,z_0) \qquad (6.13a)$$

$$v_y^{(e_0)}(x_0,y_0,z_0) = -K_y^{(e_0)}(\psi)\frac{\partial \hat{\psi}^{(e_0)}}{\partial y}(x_0,y_0,z_0) \qquad (6.13b)$$

$$v_z^{(e_0)}(x_0,y_0,z_0) = -K_z^{(e_0)}(\psi)\left(\frac{\partial \hat{\psi}^{(e_0)}}{\partial z}(x_0,y_0,z_0) + 1\right) \qquad (6.13a)$$

for unsaturated flow, where, for example, $v_x^{(e_0)}(x_0, y_0, z_0)$ is the component of apparent groundwater velocity in the x coordinate direction at the point (x_0,y_0,z_0) within element e_0. The saturated and unsaturated hydraulic conductivities have also been superscripted with e_0 because they can vary from one element to the next. The use of equations 6.10 to 6.13 is illustrated by the following examples.

Example

Compute v_x at $x_0 = 2.5$ cm for the element shown below. Let $K_x = 0.02$ cm/d, $h_i = 1$ cm, and $h_j = 1.6$ cm

The one-dimensional form of equation 6.10a is (for a two-node element n = 2)

$$\frac{\partial \hat{h}^{(e_0)}}{\partial x}(x_0) = \left[\frac{\partial N_i^{(e_0)}}{\partial x}(x_0) \quad \frac{\partial N_j^{(e_0)}}{\partial x}(x_0) \right] \begin{Bmatrix} h_i \\ h_j \end{Bmatrix}$$

For this type of element (see Figure 4.5)

$$\frac{\partial N_i^{(e_0)}}{\partial x}(x_0) = -\frac{1}{L^{(e_0)}} = -\frac{1}{2}$$

$$\frac{\partial N_j^{(e_0)}}{\partial x}(x_0) = \frac{1}{L^{(e_0)}} = \frac{1}{2}$$

$$\frac{\partial \hat{h}^{(e_0)}}{\partial x}(2.5) = \left[-\frac{1}{2} \quad \frac{1}{2} \right] \begin{Bmatrix} 1 \\ 1.6 \end{Bmatrix} = 0.3$$

and

$$v_x^{(e_0)}(x_0) = v_x^{(e_0)} = -(0.02)(0.3) = \underline{-0.006 cm/d}$$

The negative value means that flow is in the negative x direction (to the left in the figure).

Example

Compute v_x and v_y at $x_0 = 5$ and $y_0 = 7$ for the element show below. Let $K_x = 0.03$ cm/d and $K_y = 0.015$ cm/d.

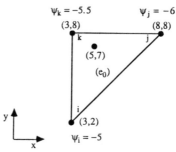

The two-dimensional forms of equations 6.10a and 6.10b are

$$\frac{\partial \hat{h}^{(e_0)}}{\partial x}(x_0,y_0) = \left[\frac{\partial N_i^{(e_0)}}{\partial x}(x_0,y_0) \quad \frac{\partial N_j^{(e_0)}}{\partial x}(x_0,y_0) \quad \frac{\partial N_k^{(e_0)}}{\partial x}(x_0,y_0)\right] \begin{Bmatrix} \psi_i \\ \psi_j \\ \psi_k \end{Bmatrix}$$

$$\frac{\partial \hat{h}^{(e_0)}}{\partial y}(x_0,y_0) = \left[\frac{\partial N_i^{(e_0)}}{\partial y}(x_0,y_0) \quad \frac{\partial N_j^{(e_0)}}{\partial y}(x_0,y_0) \quad \frac{\partial N_k^{(e_0)}}{\partial y}(x_0,y_0)\right] \begin{Bmatrix} \psi_i \\ \psi_j \\ \psi_k \end{Bmatrix}$$

For this type of element (see Figure 4.6 and the example following equation 4.25)

$$\frac{\partial N_i^{(e_0)}}{\partial x} = \frac{b_i}{2A^{(e_0)}} = 0 \qquad \frac{\partial N_j^{(e_0)}}{\partial x} = \frac{b_j}{2A^{(e_0)}} = \frac{6}{30} \qquad \frac{\partial N_k^{(e_0)}}{\partial x} = \frac{b_k}{2A^{(e_0)}} = -\frac{6}{30}$$

$$\frac{\partial N_i^{(e_0)}}{\partial y} = \frac{c_i}{2A^{(e_0)}} = -\frac{5}{30} \qquad \frac{\partial N_j^{(e_0)}}{\partial y} = \frac{c_j}{2A^{(e_0)}} = 0 \qquad \frac{\partial N_k^{(e_0)}}{\partial y} = \frac{c_k}{2A^{(e_0)}} = \frac{5}{30}$$

$$\frac{\partial h^{(e_0)}}{\partial x}(5,7) = \left(\frac{1}{30}\right)[\,0 \quad 6 \quad -6]\begin{Bmatrix} -5 \\ -6 \\ -5.5 \end{Bmatrix} = -\frac{3}{30} = -0.010$$

$$\frac{\partial h^{(e_0)}}{\partial y}(5,7) = \left(\frac{1}{30}\right)[-5 \quad 0 \quad 5]\begin{Bmatrix} -5 \\ -6 \\ -5.5 \end{Bmatrix} = -\frac{2.5}{30} = -0.083$$

and

$$v_x^{(e_0)}(5,7) = v_x^{(e_0)} = -(0.03)(-0.100) \quad = \underline{0.0030 \text{ cm/d}}$$

$$v_y^{(e_0)}(5,7) = v_y^{(e_0)} = -(0.015)(-0.083) \quad = \underline{0.0013 \text{ cm/d}}$$

The positive values mean that flow is in the positive x and y directions (up and to the right in the figure).

6.2   ISOPARAMETRIC ELEMENTS

Note that for the two types of elements in the previous examples the components of apparent groundwater velocity within the elements were constant (did not vary from point-to-point within the element). This is so because the derivatives of the interpolation functions for these types of elements are not functions of either x or y.

For isoparametric elements, the values of the derivatives of head at a specified point in the local coordinate system $(\varepsilon_0, \eta_0, \zeta_0)$ can be obtained by evaluating the derivatives of the approximate solution at the point. The derivatives (with respect to the local coordinate directions) are

$$\frac{\partial \hat{h}^{(e_0)}}{\partial \varepsilon}(\varepsilon_0,\eta_0,\zeta_0) = \left[\frac{\partial N_1^{(e_0)}}{\partial \varepsilon}(\varepsilon_0,\eta_0,\zeta_0) \cdots \frac{\partial N_n^{(e_0)}}{\partial \varepsilon}(\varepsilon_0,\eta_0,\zeta_0)\right]\left\{\begin{array}{c}h_1\\ \vdots \\ h_n\end{array}\right\} \qquad (6.14a)$$

$$\frac{\partial \hat{h}^{(e_0)}}{\partial \eta}(\varepsilon_0,\eta_0,\zeta_0) = \left[\frac{\partial N_1^{(e_0)}}{\partial \eta}(\varepsilon_0,\eta_0,\zeta_0) \cdots \frac{\partial N_n^{(e_0)}}{\partial \eta}(\varepsilon_0,\eta_0,\zeta_0)\right]\left\{\begin{array}{c}h_1\\ \vdots \\ h_n\end{array}\right\} \qquad (6.14b)$$

$$\frac{\partial \hat{h}^{(e_0)}}{\partial \zeta}(\varepsilon_0,\eta_0,\zeta_0) = \left[\frac{\partial N_1^{(e_0)}}{\partial \zeta}(\varepsilon_0,\eta_0,\zeta_0) \cdots \frac{\partial N_n^{(e_0)}}{\partial \zeta}(\varepsilon_0,\eta_0,\zeta_0)\right]\left\{\begin{array}{c}h_1\\ \vdots \\ h_n\end{array}\right\} \qquad (6.14c)$$

for hydraulic head and

$$\frac{\partial \hat{\psi}^{(e_0)}}{\partial \varepsilon}(\varepsilon_0,\eta_0,\zeta_0) = \left[\frac{\partial N_1^{(e_0)}}{\partial \varepsilon}(\varepsilon_0,\eta_0,\zeta_0) \cdots \frac{\partial N_n^{(e_0)}}{\partial \varepsilon}(\varepsilon_0,\eta_0,\zeta_0)\right]\left\{\begin{array}{c}\psi_1\\ \vdots \\ \psi_n\end{array}\right\} \qquad (6.15a)$$

$$\frac{\partial \hat{\psi}^{(e_0)}}{\partial \eta}(\varepsilon_0,\eta_0,\zeta_0) = \left[\frac{\partial N_1^{(e_0)}}{\partial \eta}(\varepsilon_0,\eta_0,\zeta_0) \cdots \frac{\partial N_n^{(e_0)}}{\partial \eta}(\varepsilon_0,\eta_0,\zeta_0)\right]\left\{\begin{array}{c}\psi_1\\ \vdots \\ \psi_n\end{array}\right\} \qquad (6.15b)$$

$$\frac{\partial \hat{\psi}^{(e_0)}}{\partial \zeta}(\varepsilon_0,\eta_0,\zeta_0) = \left[\frac{\partial N_1^{(e_0)}}{\partial \zeta}(\varepsilon_0,\eta_0,\zeta_0) \cdots \frac{\partial N_n^{(e_0)}}{\partial \zeta}(\varepsilon_0,\eta_0,\zeta_0)\right]\left\{\begin{array}{c}\psi_1\\ \vdots \\ \psi_n\end{array}\right\} \qquad (6.15c)$$

for pressure head. To obtain the values of the derivatives for the point in the global coordinate system, a coordinate transformation can be used (see Chapter 4). The coordinate transformation equations are

$$\left\{\begin{array}{c}\dfrac{\partial \hat{h}^{(e_0)}}{\partial x}(\varepsilon_0,\eta_0,\zeta_0)\\[2mm] \dfrac{\partial \hat{h}^{(e_0)}}{\partial y}(\varepsilon_0,\eta_0,\zeta_0)\\[2mm] \dfrac{\partial \hat{h}^{(e_0)}}{\partial z}(\varepsilon_0,\eta_0,\zeta_0)\end{array}\right\} = [J^{-1}(\varepsilon_0,\eta_0,\zeta_0)]\left[\begin{array}{ccc}\dfrac{\partial N_1^{(e_0)}}{\partial \varepsilon}(\varepsilon_0,\eta_0,\zeta_0) & \cdots & \dfrac{\partial N_n^{(e_0)}}{\partial \varepsilon}(\varepsilon_0,\eta_0,\zeta_0)\\[2mm] \dfrac{\partial N_1^{(e_0)}}{\partial \eta}(\varepsilon_0,\eta_0,\zeta_0) & \cdots & \dfrac{\partial N_n^{(e_0)}}{\partial \eta}(\varepsilon_0,\eta_0,\zeta_0)\\[2mm] \dfrac{\partial N_1^{(e_0)}}{\partial \zeta}(\varepsilon_0,\eta_0,\zeta_0) & \cdots & \dfrac{\partial N_n^{(e_0)}}{\partial \zeta}(\varepsilon_0,\eta_0,\zeta_0)\end{array}\right]\left\{\begin{array}{c}h_1\\ \vdots \\ h_n\end{array}\right\}$$

$$\quad\;\; 3\times1 \qquad\qquad\qquad 3\times3 \qquad\qquad\qquad\qquad\qquad 3\times n \qquad\qquad\qquad\qquad n\times1$$

$$(6.16)$$

for three-dimensional elements,

$$
\left\{
\begin{array}{c}
\dfrac{\partial \hat{h}^{(e_0)}}{\partial x}(\varepsilon_0,\eta_0) \\[2ex]
\dfrac{\partial \hat{h}^{(e_0)}}{\partial y}(\varepsilon_0,\eta_0)
\end{array}
\right\}
=
[J^{-1}(\varepsilon_0,\eta_0)]
\left[
\begin{array}{ccc}
\dfrac{\partial N_1^{(e_0)}}{\partial \varepsilon}(\varepsilon_0,\eta_0) & \cdots & \dfrac{\partial N_n^{(e_0)}}{\partial \varepsilon}(\varepsilon_0,\eta_0) \\[2ex]
\dfrac{\partial N_1^{(e_0)}}{\partial \eta}(\varepsilon_0,\eta_0) & \cdots & \dfrac{\partial N_n^{(e_0)}}{\partial \eta}(\varepsilon_0,\eta_0)
\end{array}
\right]
\left\{
\begin{array}{c}
h_1 \\ \vdots \\ h_n
\end{array}
\right\}
\qquad (6.17)
$$

$$
\underset{2\times 1}{}\qquad\qquad \underset{2\times 2}{}\qquad\qquad\qquad\qquad \underset{2\times n}{}\qquad\qquad\qquad \underset{n\times 1}{}
$$

for two-dimensional elements, and

$$
\left\{
\dfrac{\partial \hat{h}^{(e_0)}}{\partial x}(\varepsilon_0)
\right\}
=
[J^{-1}(\varepsilon_0)]
\left\{
\begin{array}{ccc}
\dfrac{\partial N_1^{(e_0)}}{\partial \varepsilon}(\varepsilon_0) & \cdots & \dfrac{\partial N_n^{(e_0)}}{\partial \varepsilon}(\varepsilon_0)
\end{array}
\right]
\left\{
\begin{array}{c}
h_1 \\ \vdots \\ h_n
\end{array}
\right\}
\qquad (6.18)
$$

$$
\underset{1\times 1}{}\qquad\qquad \underset{1\times 1}{}\qquad\qquad\qquad \underset{1\times n}{}\qquad\qquad \underset{n\times 1}{}
$$

for one-dimensional elements.

For unsaturated flow we have

$$
\left\{
\begin{array}{c}
\dfrac{\partial \hat{\psi}^{(e_0)}}{\partial x}(\varepsilon_0,\eta_0,\zeta_0) \\[2ex]
\dfrac{\partial \hat{\psi}^{(e_0)}}{\partial y}(\varepsilon_0,\eta_0,\zeta_0) \\[2ex]
\dfrac{\partial \hat{\psi}^{(e_0)}}{\partial z}(\varepsilon_0,\eta_0,\zeta_0)
\end{array}
\right\}
=
[J^{-1}(\varepsilon_0,\eta_0,\zeta_0)]
\left[
\begin{array}{ccc}
\dfrac{\partial N_1^{(e_0)}}{\partial \varepsilon}(\varepsilon_0,\eta_0,\zeta_0) & \cdots & \dfrac{\partial N_n^{(e_0)}}{\partial \varepsilon}(\varepsilon_0,\eta_0,\zeta_0) \\[2ex]
\dfrac{\partial N_1^{(e_0)}}{\partial \eta}(\varepsilon_0,\eta_0,\zeta_0) & \cdots & \dfrac{\partial N_n^{(e_0)}}{\partial \eta}(\varepsilon_0,\eta_0,\zeta_0) \\[2ex]
\dfrac{\partial N_1^{(e_0)}}{\partial \zeta}(\varepsilon_0,\eta_0,\zeta_0) & \cdots & \dfrac{\partial N_n^{(e_0)}}{\partial \zeta}(\varepsilon_0,\eta_0,\zeta_0)
\end{array}
\right]
\left\{
\begin{array}{c}
\psi_1 \\ \vdots \\ \psi_n
\end{array}
\right\}
$$

$$
\underset{3\times 1}{}\qquad\qquad \underset{3\times 3}{}\qquad\qquad\qquad\qquad \underset{3\times n}{}\qquad\qquad\qquad\quad \underset{}{(6.19)}
$$

for three-dimensional elements,

$$
\left\{
\begin{array}{c}
\dfrac{\partial \hat{\psi}^{(e_0)}}{\partial x}(\varepsilon_0,\eta_0) \\[2ex]
\dfrac{\partial \hat{\psi}^{(e_0)}}{\partial y}(\varepsilon_0,\eta_0)
\end{array}
\right\}
=
[J^{-1}(\varepsilon_0,\eta_0)]
\left[
\begin{array}{ccc}
\dfrac{\partial N_1^{(e_0)}}{\partial \varepsilon}(\varepsilon_0,\eta_0) & \cdots & \dfrac{\partial N_n^{(e_0)}}{\partial \varepsilon}(\varepsilon_0,\eta_0) \\[2ex]
\dfrac{\partial N_1^{(e_0)}}{\partial \eta}(\varepsilon_0,\eta_0) & \cdots & \dfrac{\partial N_n^{(e_0)}}{\partial \eta}(\varepsilon_0,\eta_0)
\end{array}
\right]
\left\{
\begin{array}{c}
\psi_1 \\ \vdots \\ \psi_n
\end{array}
\right\}
\qquad (6.20)
$$

$$
\underset{2\times 1}{}\qquad\qquad \underset{2\times 2}{}\qquad\qquad\qquad\qquad \underset{2\times n}{}\qquad\qquad\qquad \underset{n\times 1}{}
$$

for two-dimensional elements, and

$$
\left\{
\dfrac{\partial \hat{\psi}^{(e_0)}}{\partial x}(\varepsilon_0)
\right\}
=
[J^{-1}(\varepsilon_0)]
\left\{
\begin{array}{ccc}
\dfrac{\partial N_1^{(e_0)}}{\partial \varepsilon}(\varepsilon_0) & \cdots & \dfrac{\partial N_n^{(e_0)}}{\partial \varepsilon}(\varepsilon_0)
\end{array}
\right]
\left\{
\begin{array}{c}
\psi_1 \\ \vdots \\ \psi_n
\end{array}
\right\}
\qquad (6.21)
$$

$$
\underset{1\times 1}{}\qquad\qquad \underset{1\times 1}{}\qquad\qquad\qquad \underset{1\times n}{}\qquad\qquad \underset{n\times 1}{}
$$

for one-dimensional elements.

The entries of the Jacobian matrix [J] are given in equations 4.32, 4.35, and 4.37 for one-, two-, and three-dimensional elements, respectively. The entries in $[J^{-1}]$ can be computed using the equations in Appendix IV.

The components of apparent groundwater velocity at the specified point are given by

$$v_x^{(e_0)}(\varepsilon_0, \eta_0, \zeta_0) = -K_x^{(e_0)} \frac{\partial \hat{h}^{(e_0)}}{\partial x}(\varepsilon_0, \eta_0, \zeta_0) \tag{6.22a}$$

$$v_y^{(e_0)}(\varepsilon_0, \eta_0, \zeta_0) = -K_y^{(e_0)} \frac{\partial \hat{h}^{(e_0)}}{\partial y}(\varepsilon_0, \eta_0, \zeta_0) \tag{6.22b}$$

$$v_z^{(e_0)}(\varepsilon_0, \eta_0, \zeta_0) = -K_z^{(e_0)} \frac{\partial \hat{h}^{(e_0)}}{\partial z}(\varepsilon_0, \eta_0, \zeta_0) \tag{6.22c}$$

for saturated flow and

$$v_x^{(e_0)}(\varepsilon_0, \eta_0, \zeta_0) = -K_x^{(e_0)}(\psi) \frac{\partial \hat{\psi}^{(e_0)}}{\partial x}(\varepsilon_0, \eta_0, \zeta_0) \tag{6.23a}$$

$$v_y^{(e_0)}(\varepsilon_0, \eta_0, \zeta_0) = -K_y^{(e_0)}(\psi) \frac{\partial \hat{\psi}^{(e_0)}}{\partial y}(\varepsilon_0, \eta_0, \zeta_0) \tag{6.23b}$$

$$v_z^{(e_0)}(\varepsilon_0, \eta_0, \zeta_0) = -K_z^{(e_0)}(\psi) \left(\frac{\partial \hat{\psi}^{(e_0)}}{\partial z}(\varepsilon_0, \eta_0, \zeta_0) + 1 \right) \tag{6.23c}$$

for unsaturated flow where, for example, $v_x^{(e_0)}(\varepsilon_0, \eta_0, \zeta_0)$ is the component of apparent groundwater velocity in the x coordinate direction at the point $(\varepsilon_0, \eta_0, \zeta_0)$ within element e_0. For two-dimensional elements we have

$$v_x^{(e_0)}(\varepsilon_0, \eta_0) = -K_x^{(e_0)} \frac{\partial \hat{h}^{(e_0)}}{\partial x}(\varepsilon_0, \eta_0) \tag{6.24a}$$

$$v_y^{(e_0)}(\varepsilon_0, \eta_0) = -K_y^{(e_0)} \frac{\partial \hat{h}^{(e_0)}}{\partial y}(\varepsilon_0, \eta_0) \tag{6.24b}$$

for saturated flow and

$$v_x^{(e_0)}(\varepsilon_0, \eta_0) = -K_x^{(e_0)}(\psi) \frac{\partial \hat{\psi}^{(e_0)}}{\partial x}(\varepsilon_0, \eta_0) \tag{6.25a}$$

$$v_y^{(e_0)}(\varepsilon_0, \eta_0) = -K_y^{(e_0)}(\psi) \frac{\partial \hat{\psi}^{(e_0)}}{\partial y}(\varepsilon_0, \eta_0) \tag{6.25b}$$

for unsaturated flow. For one-dimensional elements we have

$$v_x^{(e_0)}(\varepsilon_0) = -K_x^{(e_0)} \frac{\partial \hat{h}^{(e_0)}}{\partial x}(\varepsilon_0) \tag{6.26a}$$

for saturated flow and

$$v_x^{(e_0)}(\varepsilon_0) = -K_x^{(e_0)}(\psi) \frac{\partial \hat{\psi}^{(e_0)}}{\partial x}(\varepsilon_0)$$ (6.26b)

for unsaturated flow.
The use of equations 6.14 to 6.26 is illustrated by the following examples.

Example

Compute v_x and v_y at $\left(\varepsilon = \frac{1}{2}, \eta = \frac{1}{3}\right)$ and at $(\varepsilon = 0, \eta = 0)$ for the element show below. Let $K_x = 0.03$ cm/d and $K_y = 0.015$ cm/d.

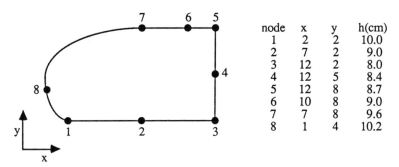

node	x	y	h(cm)
1	2	2	10.0
2	7	2	9.0
3	12	2	8.0
4	12	5	8.4
5	12	8	8.7
6	10	8	9.0
7	7	8	9.6
8	1	4	10.2

The interpolation function derivatives for this type of element are in Figure 4.11. At the point $(\varepsilon = 1/2, \eta = 1/3)$ the derivatives are

$$\frac{\partial N_1}{\partial \varepsilon} = \frac{1}{4}\left[1 + (-1)\left(\frac{1}{3}\right)\right]\left\{2(-1)^2\left(\frac{1}{2}\right) + (-1)(-1)\left(\frac{1}{3}\right)\right\} = \frac{2}{9}$$

$$\frac{\partial N_2}{\partial \varepsilon} = \left(-\frac{1}{2}\right)\left[1 + (-1)\left(\frac{1}{3}\right)\right] = -\frac{1}{3}$$

$$\frac{\partial N_3}{\partial \varepsilon} = \frac{1}{4}\left[1 + (-1)\left(\frac{1}{3}\right)\right]\left[2(1)^2\left(\frac{1}{2}\right) + (-1)(1)\left(\frac{1}{3}\right)\right] = \frac{1}{9}$$

$$\frac{\partial N_4}{\partial \varepsilon} = \frac{(1)}{2}\left[1 - \left(\frac{1}{3}\right)^2\right] = \frac{4}{9}$$

$$\frac{\partial N_5}{\partial \varepsilon} = \frac{1}{4}\left[1 + (1)\left(\frac{1}{3}\right)\right]\left[2(1)^2\left(\frac{1}{2}\right) + (1)(1)\left(\frac{1}{3}\right)\right] = \frac{4}{9}$$

$$\frac{\partial N_6}{\partial \varepsilon} = \left(-\frac{1}{2}\right)\left[1 + (1)\left(\frac{1}{3}\right)\right] = -\frac{2}{3}$$

$$\frac{\partial N_7}{\partial \varepsilon} = \frac{1}{4}\left[1 + (1)\left(\frac{1}{3}\right)\right]\left[2(-1)^2\left(\frac{1}{2}\right) + (1)(-1)\left(\frac{1}{3}\right)\right] = \frac{2}{9}$$

$$\frac{\partial N_8}{\partial \varepsilon} = \frac{(-1)}{2}\left[1 - \left(\frac{1}{3}\right)^2\right] = -\frac{4}{9}$$

$$\frac{\partial N_1}{\partial \eta} = \frac{1}{4}\left[1 + (-1)\left(\frac{1}{2}\right)\right]\left[2(-1)^2\left(\frac{1}{3}\right) + (-1)(-1)\left(\frac{1}{2}\right)\right] = \frac{7}{48}$$

$$\frac{\partial N_2}{\partial \eta} = \frac{(-1)}{2}\left[1 - \left(\frac{1}{2}\right)^2\right] = -\frac{3}{8}$$

$$\frac{\partial N_3}{\partial \eta} = \frac{1}{4}\left[1 + (1)\left(\frac{1}{2}\right)\right]\left[2(-1)^2\left(\frac{1}{3}\right) + (-1)(1)\left(\frac{1}{2}\right)\right] = \frac{1}{16}$$

$$\frac{\partial N_4}{\partial \eta} = -\left(\frac{1}{3}\right)\left[1 + (1)\left(\frac{1}{2}\right)\right] = -\frac{1}{2}$$

$$\frac{\partial N_5}{\partial \eta} = \frac{1}{4}\left[1 + (1)\left(\frac{1}{2}\right)\right]\left[2(1)^2\left(\frac{1}{3}\right) + (1)(1)\left(\frac{1}{2}\right)\right] = \frac{7}{16}$$

$$\frac{\partial N_6}{\partial \eta} = \frac{(1)}{2}\left[1 - \left(\frac{1}{2}\right)^2\right] = \frac{3}{8}$$

$$\frac{\partial N_7}{\partial \eta} = \frac{1}{4}\left[1 + (-1)\left(\frac{1}{2}\right)\right]\left[2(1)^2\left(\frac{1}{3}\right) + (1)(-1)\left(\frac{1}{2}\right)\right] = \frac{1}{48}$$

$$\frac{\partial N_8}{\partial \eta} = -\left(\frac{1}{3}\right)\left[1 + (-1)\left(\frac{1}{2}\right)\right] = -\frac{1}{6}$$

The Jacobian matrix at the point is given by equation 4.36

$$\left[J\left(\frac{1}{2},\frac{1}{3}\right)\right] = \begin{bmatrix} \frac{2}{9} & -\frac{1}{3} & \frac{1}{9} & \frac{4}{9} & \frac{4}{9} & -\frac{2}{3} & \frac{2}{9} & -\frac{4}{9} \\ \frac{7}{48} & -\frac{3}{8} & \frac{1}{16} & -\frac{1}{2} & \frac{7}{16} & \frac{3}{8} & \frac{1}{48} & -\frac{1}{6} \end{bmatrix} \begin{Bmatrix} 2 & 2 \\ 7 & 2 \\ 12 & 2 \\ 12 & 5 \\ 12 & 8 \\ 10 & 8 \\ 7 & 8 \\ 1 & 4 \end{Bmatrix}$$

$$= \begin{bmatrix} 4.556 & 0.444 \\ 1.396 & 3.167 \end{bmatrix}$$

and

$$\left[J^{-1}\left(\frac{1}{2},\frac{1}{3}\right)\right] = \begin{bmatrix} 0.229 & -0.032 \\ -0.101 & 0.330 \end{bmatrix}$$

Using equation 6.17 we have

$$\begin{bmatrix} \dfrac{\partial \hat{h}^{(e_o)}}{\partial x}\left(\dfrac{1}{2},\dfrac{1}{3}\right) \\[2ex] \dfrac{\partial \hat{h}^{(e_o)}}{\partial y}\left(\dfrac{1}{2},\dfrac{1}{3}\right) \end{bmatrix} = \begin{bmatrix} 0.229 & -0.032 \\ -0.101 & 0.330 \end{bmatrix} \begin{bmatrix} \dfrac{2}{9} & -\dfrac{1}{3} & \dfrac{1}{9} & \dfrac{4}{9} & \dfrac{4}{9} & -\dfrac{2}{3} & \dfrac{2}{9} & -\dfrac{4}{9} \\[2ex] \dfrac{7}{48} & -\dfrac{3}{8} & \dfrac{1}{16} & -\dfrac{1}{2} & \dfrac{7}{16} & \dfrac{3}{8} & \dfrac{1}{48} & -\dfrac{1}{6} \end{bmatrix} \begin{Bmatrix} 10.0 \\ 9.0 \\ 8.0 \\ 8.4 \\ 8.7 \\ 9.0 \\ 9.6 \\ 10.2 \end{Bmatrix}$$

$$= \begin{bmatrix} 0.229 & -0.032 \\ -0.101 & 0.330 \end{bmatrix} \begin{Bmatrix} -0.689 \\ 0.065 \end{Bmatrix} = \begin{Bmatrix} -0.160 \\ 0.091 \end{Bmatrix}$$

and

$$v_x^{(e_o)}\left(\frac{1}{2},\frac{1}{3}\right) = -(0.03)(-0.160) = 4.8 \times 10^{-3} \text{ cm/d}$$

$$v_y^{(e_o)}\left(\frac{1}{2},\frac{1}{3}\right) = -(0.015)(0.091) = -1.4 \times 10^{-3} \text{ cm/d}$$

At the point ($\varepsilon = 0, \eta = 0$) the derivatives are

$$\frac{\partial N_1}{\partial \varepsilon} = 0 \qquad \frac{\partial N_2}{\partial \varepsilon} = 0 \qquad \frac{\partial N_3}{\partial \varepsilon} = 0 \qquad \frac{\partial N_4}{\partial \varepsilon} = \frac{1}{2}$$

$$\frac{\partial N_5}{\partial \varepsilon} = 0 \qquad \frac{\partial N_6}{\partial \varepsilon} = 0 \qquad \frac{\partial N_7}{\partial \varepsilon} = 0 \qquad \frac{\partial N_8}{\partial \varepsilon} = -\frac{1}{2}$$

$$\frac{\partial N_1}{\partial \eta} = 0 \qquad \frac{\partial N_2}{\partial \varepsilon} = -\frac{1}{2} \qquad \frac{\partial N_3}{\partial \varepsilon} = 0 \qquad \frac{\partial N_4}{\partial \varepsilon} = 0$$

$$\frac{\partial N_5}{\partial \varepsilon} = 0 \qquad \frac{\partial N_6}{\partial \varepsilon} = \frac{1}{2} \qquad \frac{\partial N_7}{\partial \varepsilon} = 0 \qquad \frac{\partial N_8}{\partial \varepsilon} = 0$$

The Jacobian matrix at the point is

$$[J(0,0)] = \begin{bmatrix} 0 & 0 & 0 & 1/2 & 0 & 0 & 0 & -1/2 \\ 0 & -1/2 & 0 & 0 & 0 & 1/2 & 0 & 0 \end{bmatrix} \begin{Bmatrix} 2 & 2 \\ 7 & 2 \\ 12 & 2 \\ 12 & 5 \\ 12 & 8 \\ 10 & 8 \\ 7 & 8 \\ 1 & 4 \end{Bmatrix}$$

$$= \begin{bmatrix} 5.50 & 0.50 \\ 1.50 & 3.00 \end{bmatrix}$$

and

$$[J^{-1}(0,0)] = \begin{bmatrix} 0.190 & -0.032 \\ -0.095 & 0.349 \end{bmatrix}$$

Using equation 6.17

$$\left[\begin{array}{c} \dfrac{\partial \hat{h}^{(e_o)}}{\partial x}(0,0) \\[2mm] \dfrac{\partial \hat{h}^{(e_o)}}{\partial y}(0,0) \end{array}\right] = \left[\begin{array}{cc} 0.190 & -0.032 \\ -0.095 & 0.349 \end{array}\right]\left[\begin{array}{cccccccc} 0 & 0 & 0 & 1/2 & 0 & 0 & 0 & -1/2 \\ 0 & -1/2 & 0 & 0 & 0 & 1/2 & 0 & 0 \end{array}\right]\left\{\begin{array}{c} 10.0 \\ 9.0 \\ 8.0 \\ 8.4 \\ 8.7 \\ 9.0 \\ 9.6 \\ 10.2 \end{array}\right\}$$

$$= \left[\begin{array}{cc} 0.190 & -0.032 \\ -0.095 & 0.349 \end{array}\right]\left\{\begin{array}{c} -0.900 \\ 0.000 \end{array}\right\} = \left\{\begin{array}{c} -0.171 \\ 0.086 \end{array}\right\}$$

and

$$v_x^{(e_o)}(0,0) = -(0.03)(-0.171) = 5.13 \times 10^{-3} \text{ cm/d}$$

$$v_y^{(e_o)}(0,0) = -(0.015)(0.086) = 1.29 \times 10^{-3} \text{ cm/d}$$

Example

Compute v_x, v_y, and v_z at ($\varepsilon = -1$, $\eta = 1$, $\zeta = 1$) and at ($\varepsilon = 0$, $\eta = 0$, $\zeta = 0$) for the element shown below. Let $K_x = 0.01$ cm/d, $K_y = 0.02$ cm/d, and $K_z = 0.03$ cm/d.

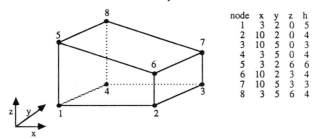

node	x	y	z	h
1	3	2	0	5
2	10	2	0	4
3	10	5	0	3
4	3	5	0	4
5	3	2	6	6
6	10	2	3	4
7	10	5	3	3
8	3	5	6	4

The interpolation function derivatives for this type of element are in Figure 4.13. At the point ($\varepsilon = -1$, $\eta = 1$, $\zeta = 1$) the derivatives are

$$\frac{\partial N_1}{\partial \varepsilon} = -\frac{1}{8}[1 + (-1)(1)][1 + (-1)(1)] = 0$$

$$\frac{\partial N_2}{\partial \varepsilon} = \frac{1}{8}[1 + (-1)(1)][1 + (-1)(1)] = 0$$

$$\frac{\partial N_3}{\partial \varepsilon} = \frac{1}{8}[1 + (1)(1)][1 + (-1)(1)] = 0$$

$$\frac{\partial N_4}{\partial \varepsilon} = -\frac{1}{8}[1 + (1)(1)][1 + (-1)(1)] = 0$$

$$\frac{\partial N_5}{\partial \varepsilon} = -\frac{1}{8}[1 + (-1)(1)][1 + (1)(1)] = 0$$

$$\frac{\partial N_6}{\partial \varepsilon} = \frac{1}{8}[1 + (-1)(1)][1 + (1)(1)] = 0$$

$$\frac{\partial N_7}{\partial \varepsilon} = \frac{1}{8}[1+(1)(1)][1+(1)(1)] = \frac{1}{2}$$

$$\frac{\partial N_8}{\partial \varepsilon} = -\frac{1}{8}[1+(1)(1)][1+(1)(1)] = -\frac{1}{2}$$

$$\frac{\partial N_1}{\partial \eta} = -\frac{1}{8}[1+(-1)(-1)][1+(-1)(1)] = 0$$

$$\frac{\partial N_2}{\partial \eta} = -\frac{1}{8}[1+(1)(-1)][1+(-1)(1)] = 0$$

$$\frac{\partial N_3}{\partial \eta} = \frac{1}{8}[1+(1)(-1)][1+(-1)(1)] = 0$$

$$\frac{\partial N_4}{\partial \eta} = \frac{1}{8}[1+(-1)(-1)][1+(-1)(1)] = 0$$

$$\frac{\partial N_5}{\partial \eta} = -\frac{1}{8}[1+(-1)(-1)][1+(1)(1)] = -\frac{1}{2}$$

$$\frac{\partial N_6}{\partial \eta} = -\frac{1}{8}[1+(1)(-1)][1+(1)(1)] = 0$$

$$\frac{\partial N_7}{\partial \eta} = \frac{1}{8}[1+(1)(-1)][1+(1)(1)] = 0$$

$$\frac{\partial N_8}{\partial \eta} = \frac{1}{8}[1+(-1)(-1)][1+(1)(1)] = \frac{1}{2}$$

$$\frac{\partial N_1}{\partial \zeta} = -\frac{1}{8}[1+(-1)(-1)][1+(-1)(1)] = 0$$

$$\frac{\partial N_2}{\partial \zeta} = -\frac{1}{8}[1+(1)(-1)][1+(-1)(1)] = 0$$

$$\frac{\partial N_3}{\partial \zeta} = -\frac{1}{8}[1+(1)(-1)][1+(1)(1)] = 0$$

$$\frac{\partial N_4}{\partial \zeta} = -\frac{1}{8}[1+(-1)(-1)][1+(1)(1)] = -\frac{1}{2}$$

$$\frac{\partial N_5}{\partial \zeta} = \frac{1}{8}[1+(-1)(-1)][1+(-1)(1)] = 0$$

$$\frac{\partial N_6}{\partial \zeta} = \frac{1}{8}[1+(1)(-1)][1+(-1)(1)] = 0$$

$$\frac{\partial N_7}{\partial \zeta} = \frac{1}{8}[1+(1)(-1)][1+(1)(1)] = 0$$

$$\frac{\partial N_8}{\partial \zeta} = \frac{1}{8}[1+(-1)(-1)][1+(1)(1)] = \frac{1}{2}$$

The Jacobian matrix at the point is given by equation 4.38

$$
[J(-1,1,1)] = \begin{bmatrix} 0 & 0 & 0 & 0 & 0 & 0 & 1/2 & -1/2 \\ 0 & 0 & 0 & 0 & -1/2 & 0 & 0 & 1/2 \\ 0 & 0 & 0 & -1/2 & 0 & 0 & 0 & 1/2 \end{bmatrix} \begin{Bmatrix} 3 & 2 & 0 \\ 10 & 2 & 0 \\ 10 & 5 & 0 \\ 3 & 5 & 0 \\ 3 & 2 & 6 \\ 10 & 2 & 3 \\ 10 & 5 & 3 \\ 3 & 5 & 6 \end{Bmatrix}
$$

$$
= \begin{bmatrix} 3.50 & 0.00 & -1.50 \\ 0.00 & 1.50 & 0.00 \\ 0.00 & 0.00 & 3.00 \end{bmatrix}
$$

and

$$
[J^{-1}(-1,1,1)] = \begin{bmatrix} 0.29 & 0.00 & 0.14 \\ 0.00 & 0.67 & 0.00 \\ 0.00 & 0.00 & 0.33 \end{bmatrix}
$$

Using equation 6.16 we have

$$
\begin{Bmatrix} \dfrac{\partial \hat{h}^{(e_0)}}{\partial x}(-1,1,1) \\[2mm] \dfrac{\partial \hat{h}^{(e_0)}}{\partial y}(-1,1,1) \\[2mm] \dfrac{\partial \hat{h}^{(e_0)}}{\partial z}(-1,1,1) \end{Bmatrix} = \begin{bmatrix} 0.29 & 0.00 & 0.14 \\ 0.00 & 0.67 & 0.00 \\ 0.00 & 0.00 & 0.33 \end{bmatrix} \begin{bmatrix} 0 & 0 & 0 & 0 & 0 & 0 & 1/2 & -1/2 \\ 0 & 0 & 0 & 0 & -1/2 & 0 & 0 & 1/2 \\ 0 & 0 & 0 & -1/2 & 0 & 0 & 0 & 1/2 \end{bmatrix} \begin{Bmatrix} 5 \\ 4 \\ 3 \\ 4 \\ 6 \\ 4 \\ 3 \\ 4 \end{Bmatrix}
$$

$$
= \begin{bmatrix} 0.29 & 0.00 & 0.14 \\ 0.00 & 0.67 & 0.00 \\ 0.00 & 0.00 & 0.33 \end{bmatrix} \begin{Bmatrix} -0.50 \\ -1.00 \\ 0.00 \end{Bmatrix} = \begin{Bmatrix} -0.15 \\ -0.67 \\ 0.00 \end{Bmatrix}
$$

and

$$
v_x^{(e_0)}(-1,1,1) = -(0.01)(-0.15) = 1.50 \times 10^{-3} \text{ cm/d}
$$

$$
v_y^{(e_0)}(-1,1,1) = -(0.02)(-0.67) = 1.34 \times 10^{-2} \text{ cm/d}
$$

$$
v_z^{(e_0)}(-1,1,1) = -(0.03)(0.00) = 0.00 \text{ cm/d}
$$

At the point $(\varepsilon = 0, \eta = 0, \zeta = 0)$ the derivatives are

$$\frac{\partial N_1}{\partial \varepsilon} = -\frac{1}{8} \qquad \frac{\partial N_2}{\partial \varepsilon} = \frac{1}{8} \qquad \frac{\partial N_3}{\partial \varepsilon} = \frac{1}{8} \qquad \frac{\partial N_4}{\partial \varepsilon} = -\frac{1}{8}$$

$$\frac{\partial N_5}{\partial \varepsilon} = -\frac{1}{8} \qquad \frac{\partial N_6}{\partial \varepsilon} = \frac{1}{8} \qquad \frac{\partial N_7}{\partial \varepsilon} = \frac{1}{8} \qquad \frac{\partial N_8}{\partial \varepsilon} = -\frac{1}{8}$$

$$\frac{\partial N_1}{\partial \eta} = -\frac{1}{8} \qquad \frac{\partial N_2}{\partial \eta} = -\frac{1}{8} \qquad \frac{\partial N_3}{\partial \eta} = \frac{1}{8} \qquad \frac{\partial N_4}{\partial \eta} = \frac{1}{8}$$

$$\frac{\partial N_5}{\partial \eta} = -\frac{1}{8} \qquad \frac{\partial N_6}{\partial \eta} = -\frac{1}{8} \qquad \frac{\partial N_7}{\partial \eta} = \frac{1}{8} \qquad \frac{\partial N_8}{\partial \eta} = \frac{1}{8}$$

$$\frac{\partial N_1}{\partial \zeta} = -\frac{1}{8} \qquad \frac{\partial N_2}{\partial \zeta} = -\frac{1}{8} \qquad \frac{\partial N_3}{\partial \zeta} = -\frac{1}{8} \qquad \frac{\partial N_4}{\partial \zeta} = -\frac{1}{8}$$

$$\frac{\partial N_5}{\partial \zeta} = \frac{1}{8} \qquad \frac{\partial N_6}{\partial \zeta} = \frac{1}{8} \qquad \frac{\partial N_7}{\partial \zeta} = \frac{1}{8} \qquad \frac{\partial N_8}{\partial \zeta} = \frac{1}{8}$$

The Jacobian matrix at the point is

$$[J(0,0,0)] = \begin{bmatrix} -1/8 & 1/8 & 1/8 & -1/8 & -1/8 & 1/8 & 1/8 & -1/8 \\ -1/8 & -1/8 & 1/8 & 1/8 & -1/8 & -1/8 & 1/8 & 1/8 \\ -1/8 & -1/8 & -1/8 & -1/8 & 1/8 & 1/8 & 1/8 & 1/8 \end{bmatrix} \begin{Bmatrix} 3 & 2 & 0 \\ 10 & 2 & 0 \\ 10 & 5 & 0 \\ 3 & 5 & 0 \\ 3 & 2 & 6 \\ 10 & 2 & 3 \\ 10 & 5 & 3 \\ 3 & 5 & 6 \end{Bmatrix}$$

$$= \begin{bmatrix} 3.50 & 0.00 & -0.75 \\ 0.00 & 1.50 & 0.00 \\ 0.00 & 0.00 & 2.25 \end{bmatrix}$$

and

$$[J^{-1}(0,0,0)] = \begin{bmatrix} 0.29 & 0.00 & 0.10 \\ 0.00 & 0.67 & 0.00 \\ 0.00 & 0.00 & 0.44 \end{bmatrix}$$

Using equation 6.16

$$\begin{Bmatrix} \frac{\partial \hat{h}^{(\varphi)}}{\partial x}(0,0,0) \\ \frac{\partial \hat{h}^{(\varphi)}}{\partial y}(0,0,0) \\ \frac{\partial \hat{h}^{(\varphi)}}{\partial z}(0,0,0) \end{Bmatrix} = \begin{bmatrix} 0.29 & 0.00 & 0.10 \\ 0.00 & 0.67 & 0.00 \\ 0.00 & 0.00 & 0.44 \end{bmatrix} \begin{bmatrix} -1/8 & 1/8 & 1/8 & -1/8 & -1/8 & 1/8 & 1/8 & -1/8 \\ -1/8 & -1/8 & 1/8 & 1/8 & -1/8 & -1/8 & 1/8 & 1/8 \\ -1/8 & -1/8 & -1/8 & -1/8 & 1/8 & 1/8 & 1/8 & 1/8 \end{bmatrix} \begin{Bmatrix} 5 \\ 4 \\ 3 \\ 4 \\ 6 \\ 4 \\ 3 \\ 4 \end{Bmatrix}$$

$$= \begin{bmatrix} 0.29 & 0.00 & 0.10 \\ 0.00 & 0.67 & 0.00 \\ 0.00 & 0.00 & 0.44 \end{bmatrix} \begin{Bmatrix} -0.625 \\ -0.625 \\ 0.125 \end{Bmatrix} = \begin{Bmatrix} -0.169 \\ -0.419 \\ 0.055 \end{Bmatrix}$$

and

$$v_x^{(e_0)}(0,0,0) = -(0.01)(-0.169) = 1.69 \times 10^{-3} \text{ cm/d}$$

$$v_y^{(e_0)}(0,0,0) = -(0.02)(-0.419) = 8.38 \times 10^{-3} \text{ cm/d}$$

$$v_z^{(e_0)}(0,0,0) = -(0.03)(0.055) = -1.65 \times 10^{-3} \text{ cm/d}$$

The previous examples have shown that, for some types of elements, the computed components of apparent groundwater velocity vary from point-to-point within the element. This is true for the linear rectangle element (because the derivatives of the interpolation functions are functions of s and t, see Figure 4.7) and for all the isoparametric elements except the one-dimensional, linear bar element (because the derivatives of the interpolation functions are functions of ϵ, η, and ζ, see Figure 4.9b to 4.15).

The previous examples have also shown why it is common practice, when using isoparametric elements, to compute the components of apparent groundwater velocities at the *center* of the element. When the derivatives are evaluated at the center of the element $((\epsilon = 0)$, $(\epsilon = 0, \eta = 0)$, or $(\epsilon = 0, \eta = 0, \zeta = 0)$ for one- two-, and three- dimensional elements respectively) the calculations are greatly simplified. Expressions giving the values of the derivatives at the center of each type of isoparametric element are in Figures 4.9 to 4.15.

Once the components of apparent groundwater velocity have been computed, the magnitude and direction of apparent groundwater velocity can be computed and plotted for each element in the mesh.

Example

Compute and plot the magnitude and direction of apparent groundwater flow for the mesh shown below.

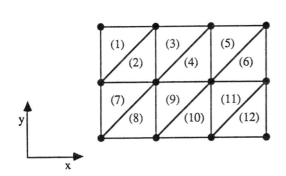

element	v_x	v_y
1	2	-2
2	3	-1
3	3	0
4	3	1
5	2	2
6	2	3
7	3	-1
8	3	0
9	3	0
10	2	1
11	2	1
12	2	2

For element 1

$$\bar{v} = \sqrt{v_x^2 + v_y^2} = \sqrt{(2)^2 + (-2)^2} = 2.83$$

$$\theta = \tan^{-1}\frac{v_y}{v_x} = \tan^{-1}\left(\frac{-2}{2}\right) = -45.0°$$

where $\bar{v}$ is the magnitude of apparent groundwater velocity and θ is an angle between $\bar{v}$ and x-axis. Plotting these at the center of the element we have

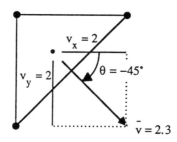

The results for the remaining elements are

element	$\bar{v}$	θ
1	2.83	-45.0
2	3.16	-18.4
3	3.00	0.0
4	3.16	18.4
5	2.83	45.0
6	3.61	56.3
7	3.16	-18.4
8	3.00	0.0
9	3.00	0.0
10	2.24	26.6
11	2.24	26.6
12	2.83	45.0

which are plotted below

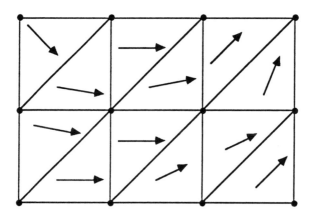

Problems

Note: In the following problems, the units of nodal point coordinates and hydraulic head are centimeters.

1. Compute v_x for the elements shown below ($K_x^{(1)} = 2$ cm/s, $K_x^{(2)} = 1.5$ cm/s).

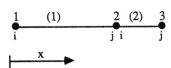

node	x	h
1	-3	200
2	1	100
3	2	50

2. Compute v_x and v_y for the elements shown below ($K_x^{(1)} = K_x^{(2)} = 1$cm/s, $K_y^{(1)} = 1$ cm/s, $K_y^{(2)} = 3$ cm/s).

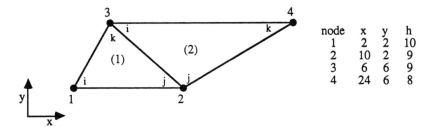

node	x	y	h
1	2	2	10
2	10	2	9
3	6	6	9
4	24	6	8

3. Compute h, v_x and v_y at point A, B, and C for the element shown below ($K_x^{(1)} = 0.5$ cm/s, $K_y^{(1)} = 2$ cm/s).

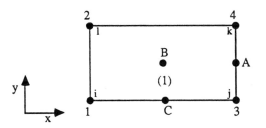

node or point	x	y	h
1	0	0	8
2	0	4	7
3	6	0	6
4	6	4	5
A	6	2	-
B	3	2	-
C	3	0	-

4. Compute h, v_x and v_y at point A for the elements shown below ($K_x^{(1)} = 1$ cm/s, $K_y^{(1)} = 0.05$ cm/s, $K_x^{(2)} = 0.5$ cm/s, $K_y^{(2)} = 0.01$ cm/s).

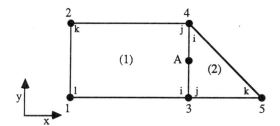

node or point	x	y	h
1	1	1	9
2	7	1	10
3	1	5	8
4	7	5	7
5	12	1	6
A	7	3	-

5. Compute h, v_x and v_y at point A, B and C for the element shown below ($K_x = 0.05$ cm/s $K_y = 0.01$ cm/s).

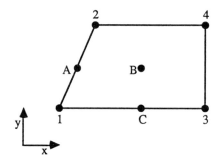

node	x	y	h
1	0	0	10
2	2	6	9
3	8	0	9
4	8	6	8

point	ε	η
A	-1	0
B	0	0
C	0	-1

6. Compute h, v_x and v_y at node 5 for the elements shown below ($K_x = 0.05$ cm/s, and $K_y = 0.01$ cm/s for all elements).

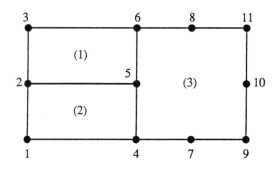

node	x	y	h
1	0	0	1
2	0	2	2
3	0	4	6
4	4	0	6
5	4	2	7
6	4	4	8
7	6	0	7
8	6	4	9
9	8	0	8
10	8	2	9
11	8	4	10

7. Compute h, v_x and v_y at point A and B for the element shown below ($K_x = 1$ cm/s, K_y = 2 cm/s).

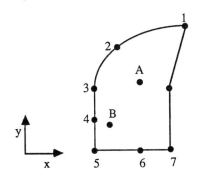

node	x	y	h
1	5	6	5
2	1.5	5.2	4
3	0	3.5	1
4	0	2.0	0.5
5	0	0	0
6	2	0	0.5
7	4	0	1
8	4	3	4

point	ε	η
A	0	0
B	-1/2	-1/2

8. Compute h, v_x and v_y at point A ($\varepsilon = 0$, $\eta = 0$) for the element shown below ($K_x = 0.5$ cm/s, $K_y = 1$ cm/s).

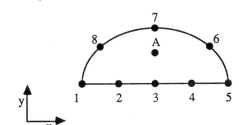

node	x	y	h
1	2	2	2
2	4	2	2
3	6	2	2
4	8	2	2
5	10	2	2
6	8.5	3	3
7	6	5	4
8	3	3	2

9. Compute h, v_x, v_y and v_z at point A ($\varepsilon = 0$, $\eta = 0$, $\zeta = 0$) and B ($\varepsilon = 1$, $\eta = 0$, $\zeta = 1$) for the element shown below ($K_x = K_y = K_z = 1$ cm/s).

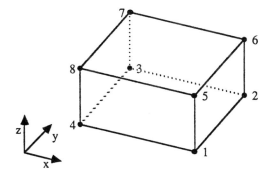

node	x	y	z	h
1	4	0	0	2
2	4	2	0	2
3	0	2	0	6
4	0	0	0	6
5	4	0	2	2
6	4	2	2	2
7	0	4	2	6
8	0	0	2	6

10. Repeat problem 9 if the values of hydraulic head are

node	h	node	h
1	10	5	9
2	9	6	8
3	8	7	7
4	9	8	8

11. Compute v_x and v_y for each element shown below and sketch the magnitude and direction of groundwater flow in each element ($K_x = K_y = 1$ cm/s for all elements). Hint: the interpolation function derivatives need only be computed for one element.

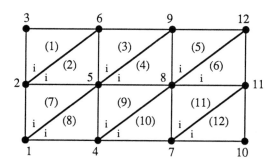

node	x	y	h
1	0	0	8.6
2	0	3	9.2
3	0	6	10
4	4	0	8.2
5	4	3	8.5
6	4	6	9
7	8	0	7.8
8	8	3	7.5
9	8	6	7.8
10	12	0	6.8
11	12	3	7.8
12	12	6	7.3

PART TWO

COMPUTER IMPLEMENTATION

Chapter 7

FINITE ELEMENT COMPUTER PROGRAMS

7.1 INTRODUCTION

In previous chapters, the steps involved in solving a groundwater flow or solute transport problem by the finite element method were illustrated with simple hand calculations for meshes with a very few nodes and elements. In practice we will need to solve problems with many hundreds or even thousands of nodes and this requires the use of a computer program. Implementing the various steps of the finite element method in a computer program is not difficult; a simple program may consist of less than a hundred lines of code. However, as we increase the versatility of a particular computer program, for example, by giving the user a choice of several element types, we will inevitably increase it's size. It is not uncommon for programs to consist of several hundred thousands of lines of code. The programs presented in this chapter were developed primarily for educational purposes but nevertheless contain many advanced features that make them suitable for use in solving many problems encountered in practice. Because the programs are written in a "modular" form (i.e., the computations are performed in a set of sub-programs) the reader will find it easy to modify the programs or to use portions of the code to develop other programs. The programs are written in FORTRAN-77 and were initially intended for use on microcomputers. However, they have successfully been compiled on a variety of mini and mainframe computers as well. Many arrays and variables are defined in the INCLUDE file "COMALL" (Figure 7.1). By editing this file the user can adjust program data requirements to match the memory capacity of the particular computer used.

An attempt has been made to choose FORTRAN variable names that are suggestive of the variables and symbols used in the text. References are also given in the code to equation or figure numbers in the text as needed to explain a computation or procedure.

Five computer programs are presented: GW1, GW2, GW3, GW4, and ST1. Program GW1 solves the steady-state, saturated groundwater flow equation. Program GW2 solves the steady-state, unsaturated flow equation (neglecting gravitational effects). Program GW3 solves the transient, saturated flow equation. Program GW4 solves the transient, unsaturated flow equation (neglecting gravitational effects). Program ST1 solves the solute transport equation for steady-state, saturated groundwater flow. All programs are capable of solving one-, two-, and three-dimensional problems as well as problems with axisymmetry. The programs will accommodate up to thirteen different element types. The element matrices $[K^{(e)}]$, $[C^{(e)}]$, $[A^{(e)}]$, and $[D^{(e)}]$ are computed in a set of subroutines. Because of space limitations subroutines are not provided for all element matrices and element types. However, examples of each are included and the reader should have no difficulty in coding a particular subroutine using the examples as a guide.

```
C*********************************************************************
C                       FILE "COMALL"
C
C     THIS FILE DIMENSIONS ARRAYS FOR  THOSE VARIABLES SHARED
C     BY THE SUBROUTINES IN THE TEXT.  WHEN THE SUBROUTINES ARE
C     COMPILED, THE STATEMENT
C                   $INCLUDE: 'COMALL'
C     THAT APPEARS IN EACH SUBROUTINE WILL DIRECT THE COMPILER TO
C     PROCEED AS THOUGH THE SPECIFIED FILE (COMALL) WERE INSERTED
C     AT THE POINT OF THE $INCLUDE (SOME COMPILERS MAY USE A
C     A SLIGHTLY DIFFERENT FORM OF THE INCLUDE STATEMENT).  BY
C     CHANGING THE PARAMETER STATEMENTS IN THIS FILE THE USER CAN
C     EASILY MODIFY THE SUBROUTINES FOR USE ON ANY COMPUTER SYSTEM.
C
C     DEFINITION OF PARAMETERS:
C
C     INF  = UNIT SPECIFICATION FOR INPUT DATA FILE
C     OUTF = UNIT SPECIFICATION FOR OUTPUT FILE
C     MAX1 = MAXIMUM NUMBER OF NODES
C     MAX2 = MAXIMUM NUMBER OF ELEMENTS
C     MAX3 = MAXIMUM NUMBER OF NODES PER ELEMENT
C     MAX4 = MAXIMUM NUMBER OF MATERIAL SETS
C     MAX5 = MAXIMUM NUMBER OF MATERIAL PROPERTIES PER MATERIAL SET
C     MAX6 = MAXIMUM VALUE OF SEMI-BANDWIDTH
C     MAX7 = MAXIMUM NUMBER OF DIFFERENT TIME STEP INCREMENTS
C     MAX8 = MAXIMUM SIZE OF MODIFIED GLOBAL CONDUCTANCE MATRIX IN
C            VECTOR STORAGE
C
C*********************************************************************
          REAL M,B1
          INTEGER OUTF,DIM,ELEMTYP,SBW,E,DTSTEP
          LOGICAL SYMM
          CHARACTER*20 LABEL1,LABEL2
          PARAMETER (INF=5,OUTF=6)
          PARAMETER (MAX1=200,MAX2=200,MAX3=32,MAX4=200,
     1               MAX5=32,MAX6=200,MAX7=20,MAX8=40000)
          COMMON /COM1/ DIM,NUMNOD,NUMELM,NUMMAT,NUMPROP,
     1                  NDN,NNN,NDOF,SBW,ICH(MAX1),LCH(MAX1),
     2                  X(MAX1),FLUX(MAX1),B(MAX1),X1(MAX1),
     3                  X2(MAX1),X3(MAX1),SYMM,LABEL1,LABEL2
          COMMON /COM2/ IN(MAX2,MAX3),ELEMTYP(MAX2),V1(MAX2),
     1                  V2(MAX2),V3(MAX2)
          COMMON /COM3/ MATSET(MAX2),PROP(MAX4,MAX5)
          COMMON /TFUNC/ FC(MAX1),DTSTEP(MAX7),DELTAT(MAX7),
     1                   TIME(MAX7),GT(MAX7),OMEGA,OMOMEGA,
     2                   MXSTEP,T,IDT,IGT,IGTDT
          COMMON /GLOBAL/ M(MAX8)
          COMMON /GLOB/ B1(MAX8)
```

Figure 7.1 Source code listing for COMALL.

7.2   STEADY-STATE, SATURATED GROUNDWATER FLOW, PROGRAM GW1

The steps used to solve the steady-state, saturated groundwater flow equation by the finite element method were described in Part 1. The computer program GW1 implements these steps in a set of eight FORTRAN subroutines: NODES, ELEMENT, MATERL, BOUND, ASMBK, DECOMP, SOLVE, VELOCITY, and DUMP. The operations performed by each subroutine are described briefly in the source code listing for GW1 (Figure 7.2); detailed information about these operations is contained in subsequent chapters (e.g., the operations performed by subroutine NODES are described in Chapter 8). Thirteen different element types are provided (see Chapter 12). Material properties (i.e, the components of saturated hydraulic conductivity) can be different for each element in the mesh (see Chapter 10). Both Dirichlet (specified hydraulic head) and Neumann (specified flow) boundary conditions can be prescribed. Hydraulic head is computed at each node and the components of apparent groundwater velocity are computed at the center of each element in the mesh. Program GW1 reads the problem description (i.e., node numbers and coordinates, element numbers, etc.) from a single input file. This information is written to an output file followed by computed values of hydraulic head and apparent groundwater velocity (Figure 7.3). The components of apparent groundwater velocity for each element are also written to an additional output file for use with the solute transport program ST1 (see Section 7.6). Arrays and variables can also be written to additional user-defined output files using subroutine DUMP (see Chapter 15).

```
          PROGRAM GW1
C*****************************************************************
C       THIS PROGRAM SOLVES STEADY-STATE, SATURATED
C       GROUNDWATER FLOW PROBLEMS.
C*****************************************************************
$INCLUDE: 'COMALL'
          DIMENSION XX(MAX1)
          INTEGER HDF,VLF
          LOGICAL LOOP
          CHARACTER*20 INFILE,OUTFILE
          CHARACTER*80 TITLE
C
          SYMM = .TRUE.
          LOOP = .FALSE.
          LABEL1 = '   HYDRAULIC HEAD'
          LABEL2 = '  GROUNDWATER FLOW'
          WRITE(*,10) ' ENTER THE NAME OF THE INPUT DATA FILE: '
   10     FORMAT(A\)
          READ(*,20) INFILE
   20     FORMAT(A)
          WRITE(*,10) ' ENTER THE NAME OF THE OUTPUT FILE: '
          READ(*,20) OUTFILE
          OPEN(INF,FILE=INFILE)
          OPEN(OUTF,FILE=OUTFILE,STATUS='NEW')
          READ(INF,20) TITLE
          WRITE(OUTF,20) TITLE
C
C   1.   INPUT NODE NUMBERS AND COORDINATES
C
```

```
          READ(INF,*) DIM
          CALL NODES
C
C   2.  INPUT ELEMENT NUMBERS, TYPES, AND NODE NUMBERS
C
          CALL ELEMENT
C
C   3.  INPUT MATERIAL PROPERTIES FOR EACH ELEMENT
C
          CALL MATERL
C
C   4.  INPUT BOUNDARY CONDITIONS
C
          CALL BOUND
C
C   5.  ASSEMBLE AND MODIFY THE GLOBAL SYSTEM OF EQUATIONS
C
          CALL ASMBK
C
C   6.  SOLVE THE SYSTEM OF EQUATIONS
C
          CALL DECOMP(NDOF,SBW,SYMM,M)
          CALL SOLVE(NDOF,SBW,SYMM,M,B,XX)
C
C   7.  WRITE OUT COMPUTED HYDRAULIC HEAD VALUES
C
          WRITE(OUTF,30) LABEL1,LABEL1
   30     FORMAT(//70('*')//16X,'COMPUTED VALUES OF ',A/
      1          16X,39('-')//19X,'NODE NO.',10X,A/)
          J = 0
          DO 50 I = 1, NUMNOD
            IF (ICH(I) .EQ. 0) THEN
              J = J + 1
              X(I) = XX(J)
            ENDIF
            IF (ICH(I) .EQ. 0) THEN
              WRITE(OUTF,40) I,X(I),' '
            ELSE
              WRITE(OUTF,40) I,X(I),'*'
            ENDIF
   40       FORMAT(19X,I5,12X,F15.4,A)
   50     CONTINUE
          WRITE(OUTF,60)
   60     FORMAT(/40X,'* = SPECIFIED VALUE')
C
C   8.  COMPUTE GROUNDWATER VELOCITIES FOR EACH ELEMENT
C
          CALL VELOCITY
C
C   9.  WRITE OUT CONTENTS OF ARRAYS (IF REQUESTED)
C
          CALL DUMP(LOOP,HDF,VLF)
          END
```

Figure 7.2 Source code listing for program GW1.

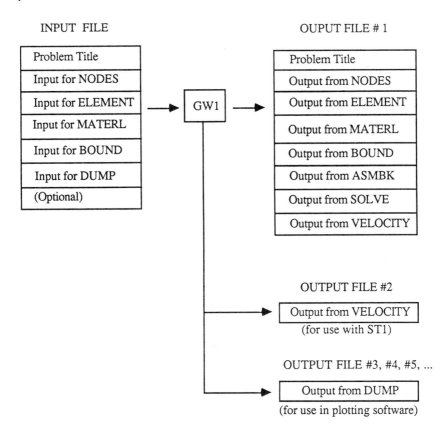

Figure 7.3   Input and output file structure for program GW1.

The use of program GW1 is best illustrated with an example. The mesh in Figure 7.4 is being used to solve a two-dimensional (plan view) groundwater flow problem in a confined aquifer. The aquifer consists of two types of material: silty sand and sandy gravel. Aquifer recharge is occurring along the constant head boundary and groundwater is being pumped from the aquifer at a single well. All other boundaries are considered impermeable. The input and output files are in Figures 7.5 and 7.6.

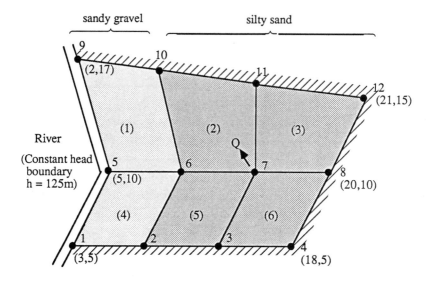

Figure 7.4   Example problem for program GW1.

EXAMPLE PROBLEM FOR PROGRAM GW1 (SEE FIGURE 7.4)   ] Problem Title
```
2                              (Problem Dimension)
1    1    3000    5000         (Node Coordinates)
4    1   18000    5000
5    1    5000   10000
8    1   20000   10000
9    1    2000   17000
12   1   21000   15000
-1  -1   -1      -1
1    6    1    5    6   10    9      (Element Node Numbers)
3    6    1    7    8   12   11
4    6    1    1    2    6    5
6    6    1    3    4    8    7
-1  -1   -1   -1   -1   -1   -1
```

- Problem Title
- Input for NODES
- Input for ELEMENT

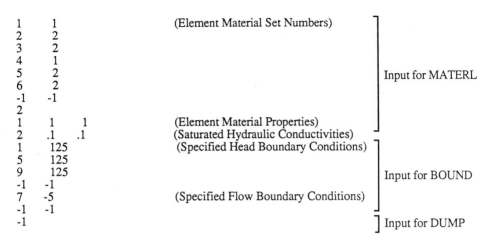

```
1    1                    (Element Material Set Numbers)      ⌉
2    2                                                        │
3    2                                                        │
4    1                                                        │
5    2                                                        │  Input for MATERL
6    2                                                        │
-1   -1                                                       │
2                                                             │
1    1    1              (Element Material Properties)        │
2    .1   .1             (Saturated Hydraulic Conductivities) ⌋
1    125                (Specified Head Boundary Conditions)  ⌉
5    125                                                       │
9    125                                                       │
-1   -1                                                        │  Input for BOUND
7    -5                 (Specified Flow Boundary Conditions)   │
-1   -1                                                        │
-1                                                             ⌋ Input for DUMP
```

Figure 7.5   Example input file for program GW1.

EXAMPLE PROBLEM FOR PROGRAM GW1 (SEE FIGURE 7.4)

NODE NUMBER	NODAL COORDINATES	
	X	Y
1	3000.0000	5000.0000
2	8000.0000	5000.0000
3	13000.0000	5000.0000
4	18000.0000	5000.0000
5	5000.0000	10000.0000
6	10000.0000	10000.0000
7	15000.0000	10000.0000
8	20000.0000	10000.0000
9	2000.0000	17000.0000
10	8333.3333	16333.3300
11	14666.6700	15666.6700
12	21000.0000	15000.0000

ELEMENT NO.	ELEMENT TYPE	NODE NUMBERS			
1	6	5	6	10	9
2	6	6	7	11	10
3	6	7	8	12	11
4	6	1	2	6	5
5	6	2	3	7	6
6	6	3	4	8	7

ELEMENT NO.	MATERIAL SET NUMBER
1	1
2	2
3	2
4	1
5	2
6	2

ELEMENT SET NO.	MATERIAL PROPERTIES	
1	1.000000E+00	1.000000E+00
2	1.000000E -01	1.000000E -01

ELEMENT NO.	SPECIFIED HYDRAULIC HEAD
1	125.0000
5	125.0000
9	125.0000

NUMBER OF NODES WITH SPECIFIED   HYDRAULIC HEAD = 3

NODE NO.	SPECIFIED GROUNDWATER FLOW
7	-5.0000

NUMBER OF NODES WITH SPECIFIED   GROUNDWATER FLOW = 1

NUMBER OF DEGREES OF FREEDOM IN MODIFIED K MATRIX = 9

SEMI-BANDWIDTH OF MODIFIED K MATRIX = 5

****

COMPUTED VALUES OF	HYDRAULIC HEAD

NODE NO.	HYDRAULIC HEAD
1	125.0000*
2	123.5652
3	108.8910
4	94.8066
5	125.0000*
6	122.0316
7	88.5567
8	97.6736
9	125.0000*
10	123.5305
11	106.4210
12	94.3300

* = SPECIFIED VALUE

****

COMPUTED VALUES OF APPARENT GROUNDWATER VELOCITY

ELEMENT	VX	VY
1	3.930606E-04	2.514829E-05
2	4.411657E-04	-8.783318E-05
3	1.816742E-05	-1.372678E-04
4	4.403232E-04	-2.276382E-05
5	4.814902E-04	2.608370E-05
6	4.967537E-05	1.548034E-04

Figure 7.6   Example output file from program GW1.

7.3   STEADY-STATE, UNSATURATED GROUNDWATER FLOW, PROGRAM GW2

The computer program GW2 solves the steady-state, unsaturated groundwater flow equation (equation 1.2) (Figure 7.7). The effect of gravity is not included (although it can easily be added if necessary). The program is almost identical to GW1 except that subroutine INITIAL (see Chapter 16) is used to read in initial values of pressure head for each node in the mesh (Figure 7.8) and Picard iteration is used to solve the system of nonlinear equations.

```
          PROGRAM GW2
C*************************************************************************
C      THIS PROGRAM SOLVES STEADY-STATE, UNSATURATED GROUNDWATER FLOW
C      FLOW PROBLEMS.  SUBROUTINE ASMBK, KBAR2, KBAR3, ETC. MUST BE
C      MODIFIED FOR USE IN UNSATURATED FLOW PROBLEMS
C*************************************************************************
$INCLUDE: 'COMALL'
          DIMENSION XX(MAX1)
          INTEGER HDF,VLF
          LOGICAL LOOP,CONVRGE
          CHARACTER*20 INFILE,OUTFILE
          CHARACTER*80 TITLE
          DATA MAXIT/20/,TOLRNCE/0.01/
C
          SYMM = .TRUE.
          LOOP = .FALSE.
          LABEL1 = '   PRESSURE HEAD'
          LABEL2 = ' GROUNDWATER FLOW'
          WRITE(*,10) ' ENTER THE NAME OF THE INPUT DATA FILE: '
   10     FORMAT(A\)
          READ(*,20) INFILE
   20     FORMAT(A)
          WRITE(*,10) ' ENTER THE NAME OF THE OUTPUT FILE: '
          READ(*,20) OUTFILE
          OPEN(INF,FILE=INFILE)
          OPEN(OUTF,FILE=OUTFILE,STATUS='NEW')
          READ(INF,20) TITLE
          WRITE(OUTF,20) TITLE
C
C   1. INPUT NODE NUMBERS AND COORDINATES
C
          READ(INF,*) DIM
          CALL NODES
C
C   2. INPUT ELEMENT NUMBERS, TYPES, AND NODE NUMBERS
C
          CALL ELEMENT
C
C   3. INPUT MATERIAL PROPERTIES FOR EACH ELEMENT
C
          CALL MATERL
C
C   4. INPUT BOUNDARY CONDITIONS
C
          CALL BOUND
C
C   5. INPUT INITIAL CONDITIONS
C
          CALL INITIAL
C
C   6. BEGIN PICARD ITERATION
C
          DO 40 ITER = 1, MAXIT
C
```

```
C   7. ASSEMBLE AND MODIFY THE GLOBAL SYSTEM OF EQUATIONS
C
        CALL ASMBK
C
C   8. SOLVE THE SYSTEM OF EQUATIONS
C
        CALL DECOMP(NDOF,SBW,SYMM,M)
        CALL SOLVE(NDOF,SBW,SYMM,M,B,XX)
        CONVRGE = .TRUE.
        II = NDOF
        DO 30 I = NUMNOD, 1, -1
          IF (ICH(I) .EQ. 0) THEN
C
C   9. CHECK FOR CONVERGENCE
C
            IF (ABS((X(I) - XX(II)) / X(I)) .GE. TOLRNCE)
     1        CONVRGE = .FALSE.
            X(I) = XX(II)
            II = II - 1
          ENDIF
   30   CONTINUE
        IF (CONVRGE) GOTO 50
   40   CONTINUE
        WRITE(OUTF,20) ' *** MAXIMUM NUMBER OF ITERATIONS EXCEEDED ***'
C
C  10. WRITE OUT COMPUTED PRESSURE HEAD VALUES
C
   50   WRITE(OUTF,60) LABEL1,LABEL1
   60   FORMAT(//70('*')//16X,'COMPUTED VALUES OF ',A/
     1       16X,39('-')//19X,'NODE NO.',10X,A/)
        DO 80 I = 1, NUMNOD
          IF (ICH(I) .EQ. 0) THEN
            WRITE(OUTF,70) I,X(I),' '
          ELSE
            WRITE(OUTF,70) I,X(I),'*'
          ENDIF
   70     FORMAT(19X,I5,12X,F15.4,A)
   80   CONTINUE
        WRITE(OUTF,90)
   90   FORMAT(/40X,'* = SPECIFIED VALUE')
C
C  11. COMPUTE GROUNDWATER VELOCITIES FOR EACH ELEMENT
C
        CALL VELOCITY
C
C  12. WRITE OUT CONTENTS OF ARRAYS (IF REQUESTED)
C
        CALL DUMP(LOOP,HDF,VLF)
        END
```

Figure 7.7   Source code listing for program GW2.

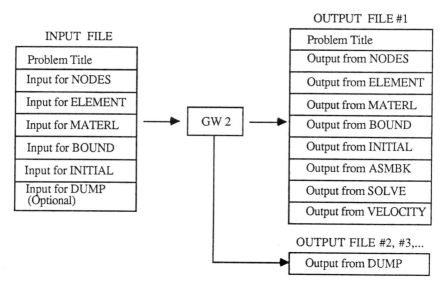

Figure 7.8   Input and output file structure for program GW2.

The user must modify the calculation of the element conductance matrix $[K^{(e)}]$ performed in subroutines KBAR2, KTRI3, etc. called by ASMBK (see Chapter 12). Currently these subroutines use a single fixed-value of hydraulic conductivity read by subroutine MATERL (Chapter 10) i.e.,

$$KXE = PROP (MATSET (E), 1)$$
$$KYE = PROP (MATSET (E), 2)$$
$$KZE = PROP (MATSET (E), 3)$$

These statements will have to be modified to compute the function $K_x(\psi)$, $K_y(\psi)$, and $K_z(\psi)$ after each iteration. The value of pressure head for each node in the mesh is recomputed at the end of each iteration and stored in the array X. The user could put the $K_x(\psi)$, $K_y(\psi)$, and $K_z(\psi)$ functions in a subroutine PSIK

$$CALL\ PSIK\ (E, KXE, KYE, KZE)$$

where E is the element number and KXE, KYE, and KZE are the <u>computed</u> values of unsaturated hydraulic conductivity for that element.

7.4   TRANSIENT, SATURATED GROUNDWATER FLOW, PROGRAM GW3

The computer program GW3 solves the transient, saturated groundwater flow equation using a set of eleven subroutines : NODES, ELEMENT, MATERL, BOUND, INITIAL, ASMBKC, RHS, DECOMP, SOLVE, VELOCITY, and DUMP (Figure 7.9). Detailed information about the operation of these subroutines is contained in subsequent chapters. Program GW3 reads the program description from a single input file (Figure 7.10). This information is written to an output file followed by computed values of hydraulic head and apparent groundwater velocity at each time step. Arrays and variables can also be written to additional user-defined files using subroutine DUMP (see Chapter 15). Example input and output files are in Figures 7.12 and 7.13 for the mesh in Figure 7.11. In the example, a single well is pumping in a confined aquifer (homogeneous and isotropic). The results are compared with the Jacob approximation to the Theiss solution for one point in Figure 7.15. The coding of input data for the mesh in Figure 7.14 is left as an exercise.

```
      PROGRAM GW3
C**********************************************************************
C     THIS PROGRAM SOLVES TRANSIENT, SATURATED
C     GROUNDWATER FLOW PROBLEMS.
C**********************************************************************
$INCLUDE: 'COMALL'
      DIMENSION XX(MAX1),V(MAX2,3)
      INTEGER HDF,VLF
      LOGICAL LOOP
      CHARACTER*20 INFILE,OUTFILE
      CHARACTER*80 TITLE
      EQUIVALENCE (V1,V(1,1)),(V2,V(1,2)),(V3,V(1,3))
C
      SYMM = .TRUE.
      LOOP = .TRUE.
      LABEL1 = '   HYDRAULIC HEAD'
      LABEL2 = '  GROUNDWATER FLOW'
      WRITE(*,10) ' ENTER THE NAME OF THE INPUT DATA FILE: '
  10  FORMAT(A\)
      READ(*,20) INFILE
  20  FORMAT(A)
      WRITE(*,10) ' ENTER THE NAME OF THE OUTPUT FILE: '
      READ(*,20) OUTFILE
      OPEN(INF,FILE=INFILE)
      OPEN(OUTF,FILE=OUTFILE,STATUS='NEW')
      READ(INF,20) TITLE
      WRITE(OUTF,20) TITLE
C
C  1. INPUT NODE NUMBERS AND COORDINATES
C
      READ(INF,*) DIM
      CALL NODES
C
C  2. INPUT ELEMENT NUMBERS, TYPES, AND NODE NUMBERS
C
      CALL ELEMENT
C
C  3. INPUT MATERIAL PROPERTIES FOR EACH ELEMENT
C
      CALL MATERL
C
```

```
C  4. INPUT BOUNDARY CONDITIONS
C
      CALL BOUND
C
C  5. INPUT INITIAL CONDITIONS
C
      CALL INITIAL
C
C  6. WRITE OUT CONTENTS OF ARRAYS (IF REQUESTED)
C
      CALL DUMP(LOOP,HDF,VLF)
C
C  7. INITIALIZE COUNTERS
C
      IF (DIM .LE. 3) THEN
        IDIM = DIM
      ELSE
        IDIM = 2
      ENDIF
      IDT = 0
      IGT = 1
      IGTDT = 1
      T = 0.
C
C  8. FOR EACH TIME STEP. . .
C
      DO 90 ISTEP = 1, MXSTEP
C     IF SIZE OF TIME STEP CHANGES REASSEMBLE GLOBAL MATRICES
         IF (ISTEP .EQ. 1 .OR. ISTEP .GT. DTSTEP(IDT)) THEN
            IDT = IDT + 1
C
C  9. ASSEMBLE AND MODIFY THE GLOBAL SYSTEM OF EQUATIONS
C
            CALL ASMBKC
C
C 10. DECOMPOSE THE MODIFIED GLOBAL SYSTEM OF EQUATIONS
C
            CALL DECOMP(NDOF,SBW,SYMM,M)
         ENDIF
C
C 11. CALCULATE THE RIGHT HAND SIDE VECTOR FOR THIS TIME STEP
C
      CALL RHS
C
C 12. SOLVE THE SYSTEM OF EQUATIONS AND OUTPUT NODAL VALUES
C
      CALL SOLVE(NDOF,SBW,SYMM,M,B,XX)
      WRITE(OUTF,30) LABEL1,LABEL1
  30  FORMAT(//70('*')//16X,'COMPUTED VALUES OF ',A/
     1       16X,39('-')//19X,'NODE NO.',10X,A/)
      J = 0
      DO 50 I = 1, NUMNOD
        IF (ICH(I) .EQ. 0) THEN
          J = J + 1
          X(I) = XX(J)
        ENDIF
        IF (ICH(I) .EQ. 0) THEN
          WRITE(OUTF,40) I,X(I),' '
          ELSE
            WRITE(OUTF,40) I,X(I),'*'
        ENDIF
```

```
40          FORMAT(19X,I5,12X,F15.4,A)
            IF (HDF .NE. 0) WRITE(HDF,*) I,X(I)
50          CONTINUE
            WRITE(OUTF,60)
60          FORMAT(/40X,'* = SPECIFIED')
            WRITE(OUTF,70) T
70          FORMAT(/19X,'*** RESULTS FOR TIME =',F7.2,' ***')
C
C   13.  COMPUTE VELOCITIES
C
            CALL VELOCITY
            WRITE(OUTF,70) T
            IF (VLF .NE. 0) THEN
              DO 80 I = 1, NUMELM
                WRITE(VLF,*) I,(V(I,J),J=1,IDIM)
80            CONTINUE
            ENDIF
90          CONTINUE
            END
```

Figure 7.9   Source code listing for program GW3.

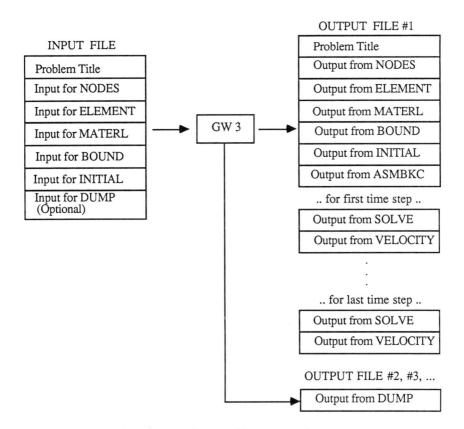

Figure 7.10   Input and output file structure for program GW3.

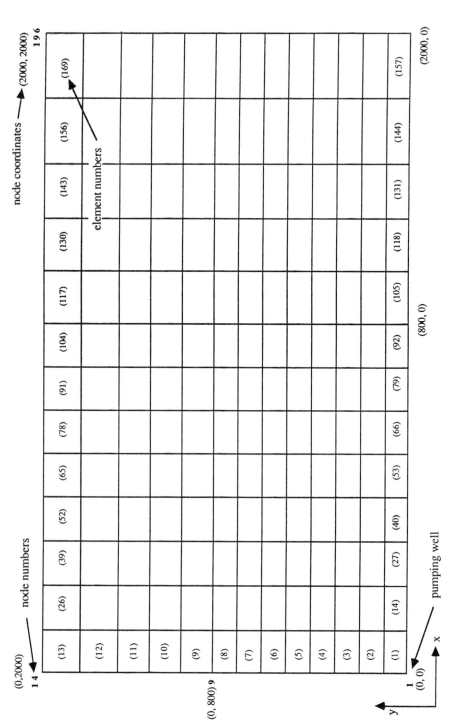

Figure 7.11   Example problem for program GW3 using linear rectangle elements.

EXAMPLE PROBLEM FOR PROGRAM GW3 (SEE FIGURE 7.11)

2				(Problem Dimension)
1	1	0	0	(Node Numbers and
9	1	0	800	Coordinates)
10	1	0	1000	
12	1	0	1400	
13	1	0	1700	
14	1	0	2000	
15	1	100	0	
23	1	100	800	
24	1	100	1000	
26	1	100	1400	
27	1	100	1700	
28	1	100	2000	
29	1	200	0	
37	1	200	800	
38	1	200	1000	
40	1	200	1400	
41	1	200	1700	
42	1	200	2000	
43	1	300	0	
51	1	300	800	
52	1	300	1000	
54	1	300	1400	
55	1	300	1700	
56	1	300	2000	
57	1	400	0	
65	1	400	800	
.	.	.	.	
.	.	.	.	
.	.	.	.	
183	1	2000	0	
191	1	2000	800	
192	1	2000	1000	
194	1	2000	1400	
195	1	2000	1700	
196	1	2000	2000	

-1	-1	-1	-1						(Element
1	5 1	1	15	16	2				Number,
13	5 1					13	27	28	14	Types, and
14	5 1	15	29	30	16				Node
26	5 1					27	41	42	28	Numbers)
27	5 1	29	43	44	30				
39	5 1					41	55	56	42	
40	5 1	43	57	58	44				
52	5 1					55	69	70	56	

Figure continued →

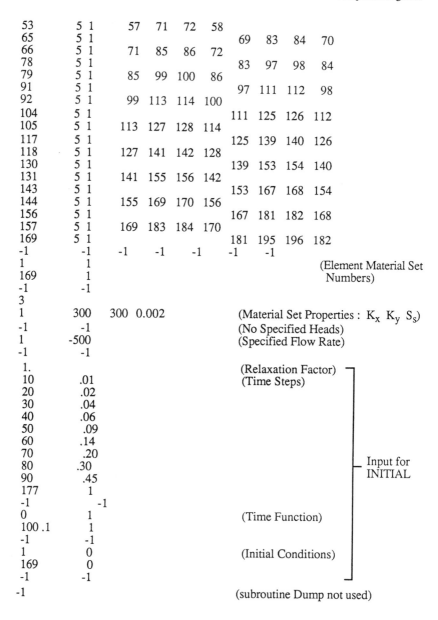

53	5 1	57	71	72	58				
65	5 1					69	83	84	70
66	5 1	71	85	86	72				
78	5 1					83	97	98	84
79	5 1	85	99	100	86				
91	5 1					97	111	112	98
92	5 1	99	113	114	100				
104	5 1					111	125	126	112
105	5 1	113	127	128	114				
117	5 1					125	139	140	126
118	5 1	127	141	142	128				
130	5 1					139	153	154	140
131	5 1	141	155	156	142				
143	5 1					153	167	168	154
144	5 1	155	169	170	156				
156	5 1					167	181	182	168
157	5 1	169	183	184	170				
169	5 1					181	195	196	182
-1	-1	-1	-1	-1	-1	-1			

```
1        1
169      1                                    (Element Material Set
-1      -1                                       Numbers)
3
1       300     300  0.002                    (Material Set Properties : Kx  Ky  Ss)
-1       -1                                   (No Specified Heads)
1      -500                                   (Specified Flow Rate)
-1       -1

1.                                            (Relaxation Factor)
10      .01                                   (Time Steps)
20      .02
30      .04
40      .06
50      .09
60      .14
70      .20
80      .30
90      .45
177      1
-1           -1
0        1                                    (Time Function)
100 .1   1
-1      -1
1        0                                    (Initial Conditions)
169      0
-1      -1
-1                                            (subroutine Dump not used)
```

Figure 7.12   Example input file for program GW3.

EXAMPLE PROBLEM FOR PROGRAM GW3 (SEE FIGURE 7.11)

NODE NUMBER	NODAL COORDINATES X	NODAL COORDINATES Y
1	.0000	.0000
2	.0000	100.0000
3	.0000	200.0000
4	.0000	300.0000
5	.0000	400.0000
6	.0000	500.0000
7	.0000	600.0000
8	.0000	700.0000
9	.0000	800.0000
10	.0000	1000.0000
11	.0000	1200.0000
12	.0000	1400.0000
13	.0000	1700.0000
14	.0000	2000.0000
.	.	.
.	.	.
.	.	.
183	2000.0000	.0000
184	2000.0000	100.0000
185	2000.0000	200.0000
186	2000.0000	300.0000
187	2000.0000	400.0000
188	2000.0000	500.0000
189	2000.0000	600.0000
190	2000.0000	700.0000
191	2000.0000	800.0000
192	2000.0000	1000.0000
193	2000.0000	1200.0000
194	2000.0000	1400.0000
195	2000.0000	1700.0000
196	2000.0000	2000.0000

ELEMENT NO.	ELEMENT TYPE	NODE NUMBERS			
1	5	1	15	16	2
2	5	2	16	17	3
3	5	3	17	18	4
4	5	4	18	19	5
5	5	5	19	20	6

ELEMENT NO.	ELEMENT TYPE	NODE NUMBERS			
6	5	6	20	21	7
7	5	7	21	22	8
8	5	8	22	23	9
9	5	9	23	24	10
10	5	10	24	25	11
.	.	.	.	.	.
.	.	.	.	.	.
.	.	.	.	.	.
160	5	172	186	187	173
161	5	173	187	188	174
162	5	174	188	189	175
163	5	175	189	190	176
164	5	176	190	191	177
165	5	177	191	192	178
166	5	178	192	193	179
167	5	179	193	194	180
168	5	180	194	195	181
169	5	181	195	196	182

ELEMENT NO.	MATERIAL SET NUMBER
1	1
2	1
3	1
4	1
5	1
6	1
7	1
8	1
9	1
10	1
.	.
.	.
.	.
162	1
163	1
164	1
165	1
166	1
167	1
168	1
169	1

ELEMENT SET NO.	MATERIAL PROPERTIES		
1	3.000000E+02	3.000000E+02	2.000000E-03

NUMBER OF NODES WITH SPECIFIED   HYDRAULIC HEAD = 0

NODE NO.	SPECIFIED GROUNDWATER FLOW
1	-500.000

NUMBER OF NODES WITH SPECIFIED   GROUNDWATER FLOW = 1

OMEGA = 1.0000

START	END	DELTA T
1	10	.0100
11	20	.0200
21	30	.0400
31	40	.0600
41	50	.0900
51	60	.1400
61	70	.2000
71	80	.3000
81	90	.4500
91	177	1.0000

TOTAL TIME = 100.1000

TIME T	G(T)
.0000	1.0000
100.1000	1.0000

COMPUTED VALUES OF   HYDRAULIC HEAD

NODE NO.	HYDRAULIC HEAD
1	.0000
2	.0000
3	.0000
4	.0000
5	.0000
6	.0000
7	.0000

COMPUTED VALUES OF   HYDRAULIC HEAD

NODE NO.   HYDRAULIC HEAD

8	.0000
9	.0000
10	.0000
.	.
.	.
.	.
190	.0000
191	.0000
192	.0000
193	.0000
194	.0000
195	.0000
196	.0000

* = SPECIFIED VALUE

NUMBER OF DEGREES OF FREEDOM IN MODIFIED,
GLOBAL COMBINED CONDUCTANCE AND CAPACITANCE MATRIX = 196

SEMI-BANDWIDTH OF MODIFIED,
GLOBAL COMBINED CONDUCTANCE AND CAPACITANCE MATRIX = 16

*** RESULTS FOR TIME = .01 ***
.
.
.

*** RESULTS FOR TIME = .02 ***
.
.
.

*** RESULTS FOR TIME = .03 ***
.
.

*** RESULTS FOR TIME = .04 ***

****
COMPUTED VALUES OF   HYDRAULIC HEAD

NODE NO.   HYDRAULIC HEAD

1	-10.6991
2	-8.2799
3	-7.6519

COMPUTED VALUES OF   HYDRAULIC HEAD

NODE NO.	HYDRAULIC HEAD
4	-7.2344
5	-6.9411
6	-6.7157
7	-6.5350
8	-6.3858
9	-6.2606
10	-6.0636
11	-5.9199
12	-5.8166
13	-5.7229
14	-5.6928
15	-8.2799
16	-8.1215
17	-7.5490
18	-7.1863
19	-6.9122
20	-6.6969
.	.
.	.
.	.
180	-5.5604
181	-5.5323
182	-5.5224
183	-5.6928
184	-5.6916
185	-5.6881
186	-5.6823
187	-5.6744
188	-5.6647
189	-5.6535
190	-5.6410
191	-5.6276
192	-5.5996
193	-5.5722
194	-5.5481
195	-5.5224
196	-5.5133

* = SPECIFIED VALUE

.
.
.

ETC.

Figure 7.13   Example output file for program GW3.

Figure 7.14   Example problem for program GW3, linear triangle and linear rectangle elements.

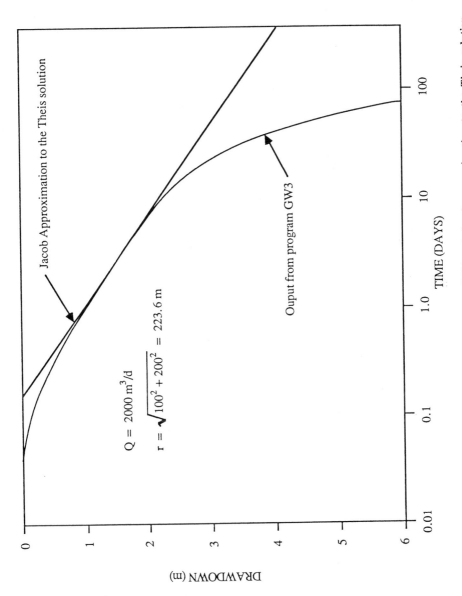

Jacob Approximation to the Theis solution

Ouput from program GW3

$Q = 2000 \text{ m}^3/\text{d}$

$r = \sqrt{100^2 + 200^2} = 223.6 \text{ m}$

TIME (DAYS)

DRAWDOWN (m)

Figure 7.15   Comparison between output from program GW4 and Jacob approximation to the Theis solution.

7.5   TRANSIENT, UNSATURATED GROUNDWATER FLOW, PROGRAM GW4

The computer program GW4 solves the transient, unsaturated groundwater flow equation (equation 1.4) (Figure 7.16). The program is similar to GW3 except that Picard iteration is used to solve the system of nonlinear equations at each time step. The input and output file structure are in Figure 7.17.

```
       PROGRAM GW4
C************************************************************************
C       THIS PROGRAM SOLVES TRANSIENT, UNSATURATED
C       GROUNDWATER FLOW PROBLEMS.  SUBROUTINES ASMBKC,
C       KBAR2, CBAR2, ETC. WILL HAVE TO BE MODIFIED FOR
C       UNSATURATED CONDITIONS.
C************************************************************************
$INCLUDE: 'COMALL'
       DIMENSION XX(MAX1),V(MAX2,3)
       INTEGER HDF,VLF
       LOGICAL LOOP,CONVRGE
       CHARACTER*20 INFILE,OUTFILE
       CHARACTER*80 TITLE
       EQUIVALENCE (V1,V(1,1)),(V2,V(1,2)),(V3,V(1,3))
       DATA MAXIT/20/,TOLRNCE/0.01/
C
       SYMM = .TRUE.
       LOOP = .TRUE.
       LABEL1 = '   PRESSURE HEAD'
       LABEL2 = '  GROUNDWATER FLOW'
       WRITE(*,10) ' ENTER THE NAME OF THE INPUT DATA FILE:
 10    FORMAT(A\)
       READ(*,20) INFILE
 20    FORMAT(A)
       WRITE(*,10) ' ENTER THE NAME OF THE OUTPUT FILE:
       READ(*,20) OUTFILE
       OPEN(INF,FILE=INFILE)
       OPEN(OUTF,FILE=OUTFILE,STATUS='NEW')
       READ(INF,20) TITLE
       WRITE(OUTF,20) TITLE
C
C  1. INPUT NODE NUMBERS AND COORDINATES
C
       READ(INF,*) DIM
       CALL NODES
C
C  2. INPUT ELEMENT NUMBERS, TYPES, AND NODE NUMBERS
C
       CALL ELEMENT
C
```

```
C  3. INPUT MATERIAL PROPERTIES FOR EACH ELEMENT
C
      CALL MATERL
C
C  4. INPUT BOUNDARY CONDITIONS
C
      CALL BOUND
C
C  5. INPUT INITIAL CONDITIONS
C
      CALL INITIAL
C
C  6. WRITE OUT CONTENTS OF ARRAYS (IF REQUESTED)
C
      CALL DUMP(LOOP,HDF,VLF)
C
C  7. INITIALIZE COUNTERS
C
      IF (DIM .LE. 3) THEN
        IDIM = DIM
      ELSE
        IDIM = 2
      ENDIF
      IDT = 0
      IGT = 1
      IGTDT = 1
      T = 0.
C
C  8. FOR EACH TIME STEP. . .
C
      DO 120 ISTEP = 1, MXSTEP
        DO 40 ITER = 1, MAXIT
C       IF SIZE OF TIME STEP CHANGES REASSEMBLE GLOBAL MATRICES
          IF ((ITER .GT. 1) .OR. (ISTEP .EQ. 1) .OR.
     1        (ISTEP .GT. DTSTEP(IDT))) THEN
            IF (ITER .EQ. 1) IDT = IDT + 1
C
C  9. ASSEMBLE AND MODIFY THE GLOBAL SYSTEM OF EQUATIONS
C
            CALL ASMBKC
C
C 10. DECOMPOSE THE MODIFIED GLOBAL SYSTEM OF EQUATIONS
C
            CALL DECOMP(NDOF,SBW,SYMM,M)
          ENDIF
C
C 11. CALCULATE THE RIGHT HAND SIDE VECTOR FOR THIS TIME STEP
C
          CALL RHS
C
C 12. BEGIN PICARD ITERATION
C
```

Figure continued →

```
          CALL SOLVE(NDOF,SBW,SYMM,M,B,XX)
          CONVRGE = .TRUE.
          II = NDOF
          DO 30 I = NUMNOD, 1, -1
            IF (ICH(I) .EQ. 0) THEN
              IF (ABS((X(I) - XX(II)) / X(I)) .GT. TOLRNCE)
     1           CONVRGE = .FALSE.
              X(I) = X(II)
              II = II - 1
            ENDIF
  30      CONTINUE

          IF (CONVRGE) GOTO 50
  40      CONTINUE
          WRITE(*,20) ' *** EXCEEDS MAXIMUM NUMBER OF ITERATIONS ***'
  50      WRITE(OUTF,60) LABEL1,LABEL1
  60      FORMAT(//70('*')//16X,'COMPUTED VALUES OF ',A/
     1           16X,39('-')//19X,'NODE NO.',10X,A/)
          DO 80 I = 1, NUMNOD
            IF (ICH(I) .EQ. 0) THEN
              WRITE(OUTF,70) I,X(I),' '
            ELSE
              WRITE(OUTF,70) I,X(I),'*'
            ENDIF
  70        FORMAT(19X,I5,12X,F15.4,A)
            IF (HDF .NE. 0) WRITE(HDF,*) I,X(I)
  80      CONTINUE
          WRITE(OUTF,90)
  90      FORMAT(/40X,'* = SPECIFIED')
          WRITE(OUTF,100) T
 100      FORMAT(/19X,'*** RESULTS FOR TIME =',F7.2,' ***')
C
C  13. COMPUTE VELOCITIES
C
          CALL VELOCITY
          WRITE(OUTF,100) T
          IF (VLF .NE. 0) THEN
          DO 110 I = 1, NUMELM
            WRITE(VLF,*) I,(V(I,J),J=1,IDIM)
 110      CONTINUE
          ENDIF
 120      CONTINUE
          END
```

Figure 7.16   Source code listing for program GW4.

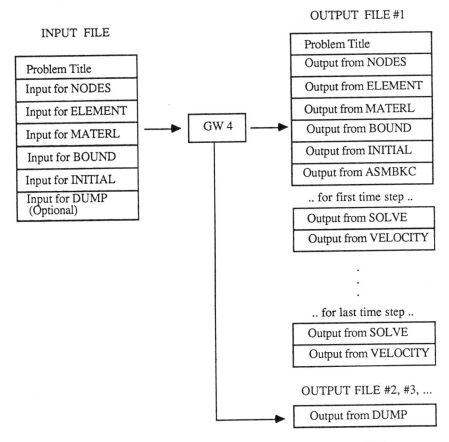

Figure 7.17   Input and output file structure for program GW4.

7.6 SOLUTE TRANSPORT, PROGRAM ST1

The steps used to solve the solute transport equation for steady-state, saturated groundwater flow were described in Part 1. The computer program ST1 implements these steps in a set of ten FORTRAN subroutines (NODES, ELEMENT, MATERL, BOUND, INITIAL, ASMAD, DECOMP, RHS, SOLVE, and DUMP) (Figure 7.18).   The operations performed by each subroutine are described briefly in the source code listing for ST1; additional details are in subsequent chapters. The input and output file structure is similar to program GW3 (Figure 7.19). The components of apparent groundwater velocity for each element are read from a file created by program GW 1 (using subroutine DUMP, see Chapter 15). Example input and output files for the mesh in Figure 7.20 are in Figures 7.21 and 7.22. A comparison between solute concentrations computed with ST1 and with an analytical solution from Bear (1979) is in Figure 7.23. Example input and output files for the mesh in Figure 7.24 are in Figures 7.25 and 7.26. The coding of the mesh in Figure 7.27 is left as an exercise.

```
      PROGRAM ST1
C**********************************************************************
C     THIS PROGRAM SOLVES SOLUTE TRANSPORT PROBLEMS FOR
C     STEADY-STATE, SATURATED GROUNDWATER FLOW.
C**********************************************************************
$INCLUDE: 'COMALL'
      DIMENSION XX(MAX1),V(MAX2,3)
      INTEGER HDF,VLF
      LOGICAL LOOP
      CHARACTER*20 INFILE,OUTFILE,VELFILE
      CHARACTER*80 TITLE
      EQUIVALENCE (V1,V(1,1)),(V2,V(1,2)),(V3,V(1,3))
C
      SYMM = .FALSE.
      LOOP = .TRUE.
      LABEL1 = 'SOLUTE CONCENTRATION'
      LABEL2 = '    SOLUTE FLUX'
      WRITE(*,10) ' ENTER THE NAME OF THE INPUT DATA FILE: '
   10 FORMAT(A\)
      READ(*,20) INFILE
   20 FORMAT(A)
      WRITE(*,10) ' ENTER THE NAME OF THE VELOCITY FILE: '
      READ(*,20) VELFILE
      WRITE(*,10) ' ENTER THE NAME OF THE OUTPUT FILE: '
      READ(*,20) OUTFILE
      OPEN(INF,FILE=INFILE)
      VLF = 2
      OPEN(VLF,FILE=VELFILE)
      OPEN(OUTF,FILE=OUTFILE,STATUS='NEW')
      READ(INF,20) TITLE
      WRITE(OUTF,20) TITLE
C
C  1. INPUT NODE NUMBERS AND COORDINATES
C
      READ(INF,*) DIM
      CALL NODES
C
C  2. INPUT ELEMENT NUMBERS, TYPES, AND NODE NUMBERS
C
      CALL ELEMENT
C
C  3. INPUT MATERIAL PROPERTIES FOR EACH ELEMENT
C
      CALL MATERL
C
C  4. INPUT BOUNDARY CONDITIONS
C
      CALL BOUND
C
C  5. INPUT INITIAL CONDITIONS
C
      CALL INITIAL
C
C  6. WRITE OUT CONTENTS OF ARRAYS (IF REQUESTED)
C
      CALL DUMP(LOOP,HDF,IDMY)
C
C  7. INITIALIZE COUNTERS
C
      IF (DIM .LE. 3) THEN
         IDIM = DIM
```

```
      ELSE
        IDIM = 2
      ENDIF
      DO 30 I = 1, NUMELM
        READ(VLF,*) J,(V(J,K),K=1,IDIM)
 30   CONTINUE
      IDT = 0
      IGT = 1
      IGTDT = 1
      T = 0.
C
C   8. FOR EACH TIME STEP. . .
C
      DO 90 ISTEP = 1, MXSTEP
C     IF SIZE OF TIME STEP CHANGES REASSEMBLE GLOBAL MATRICES
        IF (ISTEP .EQ. 1 .OR. ISTEP .GT. DTSTEP(IDT)) THEN
          IDT = IDT + 1
C
C   9. ASSEMBLE AND MODIFY THE GLOBAL SYSTEM OF EQUATIONS
C
          CALL ASMBAD
C
C 10. DECOMPOSE THE MODIFIED GLOBAL SYSTEM OF EQUATIONS
C
          CALL DECOMP(NDOF,SBW,SYMM,M)
        ENDIF
C
C 11. CALCULATE THE RIGHT HAND SIDE VECTOR FOR THIS TIME STEP
C
        CALL RHS
C
C 12. SOLVE THE SYSTEM OF EQUATIONS AND OUTPUT NODAL VALUES
C
        CALL SOLVE(NDOF,SBW,SYMM,M,B,XX)
        WRITE(OUTF,40) LABEL1,LABEL1
 40     FORMAT(//70('*')//16X,'COMPUTED VALUES OF ',A/
     1         16X,39('-')//19X,'NODE NO.',10X,A/)
        J = 0
        DO 60 I = 1, NUMNOD
          IF (ICH(I) .EQ. 0) THEN
            J = J + 1
            X(I) = XX(J)
          ENDIF
          IF (ICH(I) .EQ. 0) THEN
            WRITE(OUTF,50) I,X(I),' '
          ELSE
            WRITE(OUTF,50) I,X(I),'*'
          ENDIF
 50       FORMAT(19X,I5,12X,F15.4,A)
          IF (HDF .GT. 0) WRITE(HDF,*) I,X(I)
 60     CONTINUE
        WRITE(OUTF,70)
 70     FORMAT(/40X,'* = SPECIFIED')
        WRITE(OUTF,80) T
 80     FORMAT(/19X,'*** RESULTS FOR TIME =',F7.2,' ***')
 90   CONTINUE
      END
```

Figure 7.18   Source code listing for program ST1.

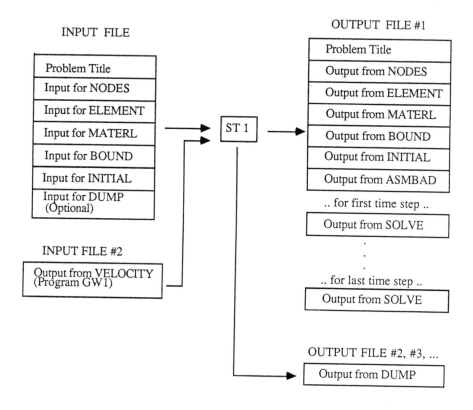

Figure 7.19   Input and output file structure for program ST1.

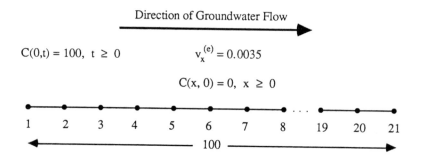

Figure 7.20   Example problem for program ST1 using linear bar elements.

EXAMPLE PROBLEM FOR PROGRAM ST1 (SEE FIGURE 7.20)

1								(Problem Dimension)
1	1	0						(Nodal Point Data)
21	1	100						
-1	-1	-1						
1	1	1	1	2				(Element Numbers, Types,
20	1	1	20	21				and Node Numbers)
-1	-1	-1	-1	-1				
1	1							(Element Material Set Data)
20	1							
-1	-1							
7								(Material Properties)
1	100	0	1	0	.35	1	1	
1	100							(Specified Concentration)
-1	-1							
-1	-1							(No Specified Solute Fluxes)
1.								(Relaxation Factor)
11	50							(Take 11 Time Steps of 50 Days Each)
-1	-1							
0	1							(Time Function)
1	1							
11	1							
-1	-1							
-1	-1							
-1								

Figure 7.21   Example input file for program ST1.

EXAMPLE PROBLEM FOR PROGRAM ST1 (SEE FIGURE 7.20)

NODE NUMBER	NODAL COORDINATES X
1	.0000
2	5.0000
3	10.0000
4	15.0000
5	20.0000
6	25.0000
7	30.0000
8	35.0000
9	40.0000
10	45.0000
11	50.0000
12	55.0000
13	60.0000
14	65.0000
15	70.0000
16	75.0000
17	80.0000
18	85.0000
19	90.0000
20	95.0000
21	100.0000

ELEMENT NO.	ELEMENT TYPE	NODE NUMBERS	
1	1	1	2
2	1	2	3
3	1	3	4
4	1	4	5
5	1	5	6
6	1	6	7
.	.	.	.
.	.	.	.
18	1	18	19
19	1	19	20
20	1	20	21

ELEMENT NO.	MATERIAL SET NUMBER
1	1
2	1
3	1
.	.
.	.
18	1
19	1
20	1

MATERIAL SET NO.	MATERIAL PROPERTIES			
1	1.000000E+02	0.000000E+00	1.000000E+00	0.000000E+00
	3.500000E-01	1.000000E+00	1.000000E+00	

NODE NO.	SPECIFIED SOLUTE CONCENTRATION
1	100.0000

NUMBER OF NODES WITH SPECIFIED SOLUTE CONCENTRATION = 1

NUMBER OF NODES WITH SPECIFIED   SOLUTE FLUX = 0

OMEGA = 1.0000

START	END	DELTA T
1	11	50.0000

TOTAL TIME = 550.0000

TIME T	G(T)
----------	-------
.0000	1.0000
1.0000	1.0000
550.0000	1.0000

INITIAL VALUES OF SOLUTE CONCENTRATION

NODE NO.	SOLUTE CONCENTRATION
--------------	------------------------------------
1	100.0000*
2	.0000
3	.0000
.	.
.	.
18	.0000
19	.0000
20	.0000
21	.0000

* = SPECIFIED

NUMBER OF DEGREES OF FREEDOM IN MODIFIED,
GLOBAL COMBINED SORPTION AND ADVECTION-DISPERSION MATRIX = 20

SEMI-BANDWIDTH OF MODIFIED,
GLOBAL COMBINED SORPTION AND ADVECTION-DISPERSION MATRIX = 2

****

COMPUTED VALUES OF SOLUTE CONCENTRATION

NODE NO.	SOLUTE CONCENTRATION
1	100.0000*
2	54.2676
3	27.0555
4	13.4887
5	6.7249
6	3.3527
7	1.6715
8	.8333
9	.4155
10	.2071
11	.1033
12	.0515
13	.0257
14	.0128
15	.0064
16	.0032
17	.0016
18	.0008
19	.0004
20	.0002
21	.0002

* = SPECIFIED

*** RESULTS FOR TIME = 50.00 ***

****

COMPUTED VALUES OF SOLUTE CONCENTRATION

NODE NO.   SOLUTE CONCENTRATION

1	100.0000*
2	69.8320
3	44.9692
4	27.4820
5	16.2252
6	9.3475
7	5.2876
8	2.9489
9	1.6261
10	.8885
11	.4817
12	.2595
13	.1390
14	.0741
15	.0393
16	.0208
17	.0110
18	.0059
19	.0032
20	.0020
21	.0016

* = SPECIFIED

*** RESULTS FOR TIME = 100.00 ***

****

COMPUTED VALUES OF SOLUTE CONCENTRATION

NODE NO.   SOLUTE CONCENTRATION

1	100.0000*
2	83.1162
3	66.6319
4	51.5875
5	38.6600
6	28.1214
7	19.9123
8	13.7633
9	9.3098
10	6.1769
11	4.0280
12	2.5863
13	1.6376
14	1.0241
15	.6335
16	.3884

COMPUTED VALUES OF SOLUTE CONCENTRATION

NODE NO.	SOLUTE CONCENTRATION
17	.2369
18	.1453
19	.0921
20	.0645
21	.0559

* = SPECIFIED

*** RESULTS FOR TIME = 250.00 ***

****

COMPUTED VALUES OF SOLUTE CONCENTRATION

NODE NO.	SOLUTE CONCENTRATION
1	100.0000*
2	84.9458
3	70.0360
4	56.0643
5	43.6218
6	33.0407
7	24.4070
8	17.6170
9	12.4487
10	8.6274
11	5.8739
12	3.9350
13	2.5975
14	1.6919
15	1.0891
16	.6945
17	.4407
18	.2814
19	.1857
20	.1349
21	.1189

* = SPECIFIED

*** RESULTS FOR TIME = 500.00 ***

****

COMPUTED VALUES OF SOLUTE CONCENTRATION

NODE NO.	SOLUTE CONCENTRATION
1	100.0000*
2	89.7406
3	79.2830
4	68.9309

Figure continued →

COMPUTED VALUES OF SOLUTE CONCENTRATION

NODE NO.	SOLUTE CONCENTRATION
5	58.9663
6	49.6288
7	41.1007
8	33.5004
9	26.8832
10	21.2485
11	16.5506
12	12.7118
13	9.6349
14	7.2147
15	5.3466
16	3.9339
17	2.8918
18	2.1504
19	1.6557
20	1.3713
21	1.2781

* = SPECIFIED

Figure 7.22   Example output file from program ST1.

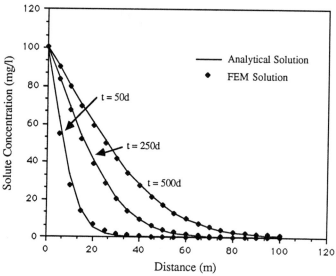

Figure 7.23   Comparison of analytical and FEM solutions for program ST1.

EXAMPLE PROBLEM FOR PROGRAM ST1 (SEE FIGURE 7.24)

2				(PROBLEM DIMENSION)
1	1	0	0	(NODAL POINT DATA)
2	1	0	40	
3	1	0	60	
7	1	0	100	
9	1	0	140	
10	1	0	180	

11	1	10	0						
.	.	.	.						
.	.	.	.						
111	1	220	0						
112	1	220	40						
113	1	220	60						
117	1	220	100						
119	1	220	140						
120	1	220	180						
-1	-1	-1	-1					(ELEMENT DATA)	
1	4	10	1	11	2				
11	4	10	101	111	102				
12	4	10	2	11	12				
22	4	10	102	111	112				
23	4	10	2	12	3				
33	4	10	102	112	103				
.	.	.	.	.	.				
177	4	10	9	19	20				
187	4	10	109	119	120				
188	4	10	10	9	20				
198	4	10	110	109	120				
-1	-1	-1	-1	-1	-1	-1			
1	1								
198	1								
-1	-1							(MATERIAL PROPERTIES)	
7									
1	2	1	0	2	0	.25	1	(SPECIFIED CONCENTRATIONS)	
1	0								
2	0								
3	0								
4	0								
5	300								
6	0								
7	0								
8	0								
.	.								
.	.								
116	0								
117	0								
118	0								
119	0								
120	0								
-1	-1							(SPECIFIED SOLUTE FLUX)	
5	300								
-1	-1							(RELAXATION FACTOR)	
1.								(TAKE 20 TIME STEPS OF 10 EACH)	
20	10								
-1	-1							(TIME FACTOR)	
0	1								
200	1								
-1	-1							(INITIAL CONCENTRATIONS)	
1	0								
120	0								
-1	-1								
-1									

Figure 7.24   Example input file for program ST1.

EXAMPLE PROBLEM FOR PROGRAM ST1 (SEE FIGURE 7.24)

NODE NUMBER	NODAL COORDINATES X	Y
1	.0000	.0000
2	.0000	40.0000
3	.0000	60.0000
4	.0000	70.0000
5	.0000	80.0000
6	.0000	90.0000
7	.0000	100.0000
8	.0000	120.0000
9	.0000	140.0000
10	.0000	180.0000
.	.	.
.		
110	160.0000	180.0000
111	220.0000	.0000
112	220.0000	40.0000
113	220.0000	60.0000
114	220.0000	70.0000
115	220.0000	80.0000
116	220.0000	90.0000
117	220.0000	100.0000
118	220.0000	120.0000
119	220.0000	140.0000
120	220.0000	180.0000

ELEMENT NO.	ELEMENT TYPE	NODE NUMBERS		
1	4	1	11	2
2	4	11	21	12
3	4	21	31	22
4	4	31	41	32
5	4	41	51	42
6	4	51	61	52
7	4	61	71	62
8	4	71	81	72
9	4	81	91	82
10	4	91	101	92
.	.	.	.	.
.				
157	4	28	38	39
158	4	38	48	49
159	4	48	58	59
160	4	58	68	69
161	4	68	78	79
162	4	78	88	89
163	4	88	98	99
164	4	98	108	109
165	4	108	118	119
166	4	9	8	19
167	4	19	18	29

ELEMENT NO.	ELEMENT TYPE	NODE NUMBERS		
168	4	29	28	39
169	4	39	38	49
170	4	49	48	59
171	4	59	58	69
172	4	69	68	79
173	4	79	78	89
174	4	89	88	99
175	4	99	98	109
176	4	109	108	119
177	4	9	19	20
178	4	19	29	30
179	4	29	39	40
180	4	39	49	50
181	4	49	59	60
182	4	59	69	70
183	4	69	79	80
184	4	79	89	90
185	4	89	99	100
186	4	99	109	110
187	4	109	119	120
188	4	10	9	20
189	4	20	19	30
190	4	30	29	40
191	4	40	39	50
192	4	50	49	60
193	4	60	59	70
194	4	70	69	80
195	4	80	79	90
196	4	90	89	100
197	4	100	99	110
198	4	110	109	120

ELEMENT NO.	MATERIAL SET NUMBER
1	1
2	1
3	1
4	1
5	1
6	1
7	1
8	1
9	1
10	1
.	.
.	.
194	1
195	1
196	1
197	1
198	1

MATERIAL SET NO.	MATERIAL PROPERTIES			
1	2.000000E+00	1.000000E+00	0.000000E+00	2.000000E+00
	0.000000E+00	2.500000E-01	1.000000E+00	

NODE NO.	SPECIFIED SOLUTE CONCENTRATION
1	.0000
2	.0000
3	.0000
4	.0000
5	300.0000
6	.0000
7	.0000
8	.0000
9	.0000
10	.0000
20	.0000
30	.0000
40	.0000
50	.0000
60	.0000
70	.0000
80	.0000
90	.0000
100	.0000
110	.0000
111	.0000
112	.0000
113	.0000
114	.0000
115	.0000
116	.0000
117	.0000
118	.0000
119	.0000
120	.0000

NUMBER OF NODES WITH SPECIFIED SOLUTE CONCENTRATION = 30

NODE NO.	SPECIFIED SOLUTE FLUX
5	300.0000

NUMBER OF NODES WITH SPECIFIED   SOLUTE FLUX = 1

OMEGA = 1.0000

START	END	DELTA T
-----------	------	-------------
1	20	10.0000

TOTAL TIME = 200.0000

TIME T	G(T)
-----------	------
.0000	1.0000
200.0000	1.0000

INITIAL VALUES OF SOLUTE CONCENTRATION

NODE NO.	SOLUTE CONCENTRATION
---------------	------------------------------------
1	.0000*
2	.0000*
3	.0000*
4	.0000*
5	300.0000*
6	.0000*
.	.
.	
117	.0000*
118	.0000*
119	.0000*
120	.0000*

* = SPECIFIED

NUMBER OF DEGREES OF FREEDOM IN MODIFIED,
GLOBAL COMBINED SORPTION AND ADVECTION-DISPERSION MATRIX = 90

SEMI-BANDWIDTH OF MODIFIED,
GLOBAL COMBINED SORPTION AND ADVECTION-DISPERSION MATRIX = 11

*** RESULTS FOR TIME = 10.00 ***

*** RESULTS FOR TIME = 20.00 ***

*** RESULTS FOR TIME = 30.00 ***

*** RESULTS FOR TIME = 40.00 ***

*** RESULTS FOR TIME = 50.00 ***

*** RESULTS FOR TIME = 60.00 ***

*** RESULTS FOR TIME = 70.00 ***

*** RESULTS FOR TIME = 80.00 ***

*** RESULTS FOR TIME = 90.00 ***

*** RESULTS FOR TIME = 100.00 ***

*** RESULTS FOR TIME = 110.00 ***

*** RESULTS FOR TIME = 120.00 ***

*** RESULTS FOR TIME = 130.00 ***

****

COMPUTED VALUES OF SOLUTE CONCENTRATION

NODE NO.	SOLUTE CONCENTRATION
1	.0000*
2	.0000*
3	.0000*
4	.0000*
5	300.0000*
6	.0000*
7	.0000*
8	.0000*
9	.0000*
10	.0000*
11	-.4024
12	.5881
13	-3.3477
14	42.5015
15	228.2954
16	42.4914
17	-3.2888
18	.9138
19	-.1811
20	.0000*
21	-.1427
22	.0470
23	-.6714
24	62.3467
25	182.7819
26	62.3475
27	-.5372
28	.2034
29	-.0590
30	.0000*
31	.2244
32	-.6711
33	3.7866
34	68.3779
35	149.0262
36	68.3933

COMPUTED VALUES OF SOLUTE CONCENTRATION

NODE NO.	SOLUTE CONCENTRATION
37	3.9406
38	-.7868
39	.0968
40	.0000*
41	.4454
42	-1.1199
43	7.3474
44	64.3831
45	117.9321
46	64.4055
47	7.4753
48	-1.4140
49	.1748
50	.0000*
51	.5042
52	-1.2155
53	8.6262
54	53.3718
55	89.8220
56	53.3947
57	8.7133
58	-1.5534
59	.1859
60	.0000*
61	.4447
62	-1.0155
63	7.5597
64	36.9853
65	56.5931
66	37.0008
67	7.6051
68	-1.3035
69	.1395
70	.0000*
71	.3103
72	-.5886
73	3.0124
74	13.5022
75	19.7430
76	13.5078
77	3.0336
78	-.7412
79	.0969
80	.0000*
81	.1952
82	-.2791
83	.4304
84	3.0719
85	5.0566
86	3.0735
87	.4379

Figure continued ⟶

COMPUTED VALUES OF SOLUTE CONCENTRATION

NODE NO.	SOLUTE CONCENTRATION
88	-.3506
89	.0688
90	.0000*
91	.0746
92	-.0640
93	-.0802
94	.2728
95	.6212
96	.2730
97	-.0784
98	-.0779
99	.0251
100	.0000*
101	.0102
102	.0051
103	-.0254
104	-.0480
105	-.0048
106	-.0481
107	-.0255
108	.0088
109	.0042
110	.0000*
111	.0000*
112	.0000*
113	.0000*
114	.0000*
115	.0000*
116	.0000*
117	.0000*
118	.0000*
119	.0000*
120	.0000*

* = SPECIFIED

Figure 7.25   Example output file from program ST1.

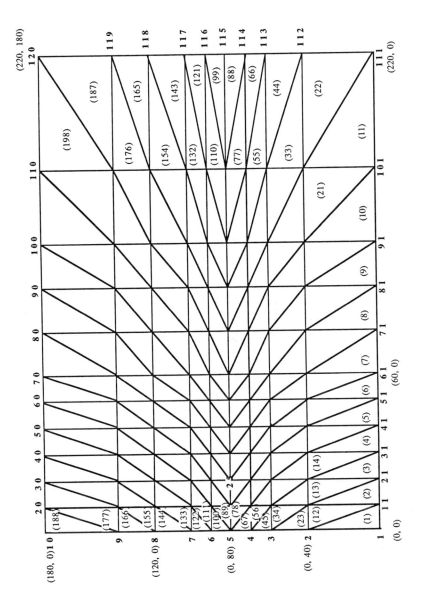

Figure 7.26   Example problem for program ST1, linear triangle elements.

302

Finite Element Computer Programs

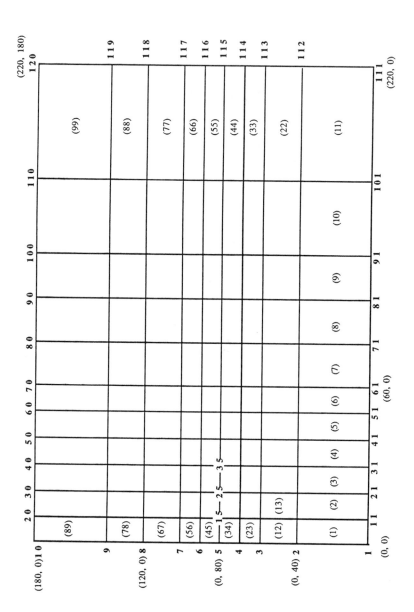

Figure 7.27   Example problem for program ST1, linear rectangle or linear quadrilateral elements.

Chapter 8

SUBROUTINE NODES

8.1 PURPOSE

Subroutine NODES inputs the node numbers and coordinates for each node in the finite element mesh. The subroutine can be used for one-, two-, or three-dimensional problems or for problems with axisymmetry.

8.2 INPUT

Node numbers and coordinates are read "free-format" from the user-supplied file assigned to unit "INF". The operation of the subroutine is controlled by the variable "DIM" (see Section 8.5 for a description of program usage). Both INF and DIM are passed to the subroutine through labeled common blocks contained in the file "COMALL" (see Chapter 7 for a listing of COMALL).

8.3 OUTPUT

Node numbers and coordinates are written to the user-defined file assigned to unit "OUTF". OUTF is passed to the subroutine through a labeled common block in COMALL. Column headings are added to the list of node numbers and coordinates written to OUTF. The number of nodes in the mesh, variable "NUMNOD", and the coordinates for each node in the mesh, arrays "X1", "X2", and "X3", are stored in labeled common blocks (contained in the file COMALL) for use by other subroutines.

8.4 DEFINITIONS OF VARIABLES

DIM   = Type of coordinate system used in this problem (Figure 8.1).
          = 1, problem is one-dimensional.
          = 2, problem is two-dimensional.
          = 3, problem is three-dimensional.
          = 4, problem is two-dimensional (axisymmetric).

INC   = Node number increment used to generate "missing" node numbers.

NUMNOD = Number of nodes in the mesh.

X1(I)   = x coordinate for node I if DIM = 1, 2, or 3.
          = r coordinate for node I if DIM = 4.

X2(I)   = not used if DIM = 1.
          = y coordinate for node I if DIM = 2 or 3.
          = z coordinate for node I if DIM = 4.

X3(I)   = not used if DIM = 1, 2, or 4.
          = z coordinate for node I if DIM = 3.

DIM = 1

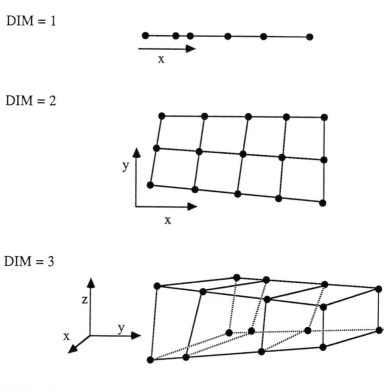

DIM = 2

DIM = 3

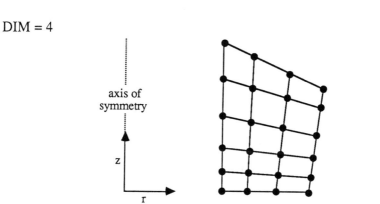

DIM = 4

Figure 8.1   Correct values of DIM to use with different types of finite element meshes.

8.5   USAGE

Example input data and output for subroutine NODES are in Figures 8.2 to 8.5. Each line of input contains the node number, node number increment, and the coordinates of a node. The mesh in Figure 8.2 is two-dimensional (DIM = 2). Node number 1 has a node number increment of 1 (INC = 1), an x coordinate of 0 (X1(1) = 0) and a y coordinate of 0 (X2(1) = 0). Node number 4 has a node number increment of 1 (INC = 1), an x coordinate of 2 (X1(4) = 2) and a y coordinate of 9 (X2(4) = 9). The subroutine has the capability to "generate" the node numbers and coordinates for nodes "missing" from the

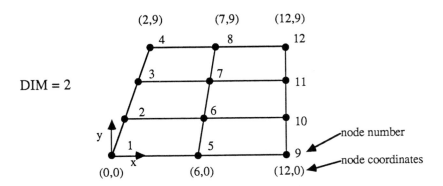

Input Data :

1	1	0.	0.
4	1	2.	9.
5	1	6.	0.
8	1	7.	9.
9	1	12.	0.
12	1	12.	9.
−1	−1	−1	−1

Output :

NODE NUMBER	NODAL COORDINATES	
	X	Y
1	.0000	.0000
2	.6667	3.0000
3	1.3333	6.0000
4	2.0000	9.0000
5	6.0000	.0000
6	6.3333	3.0000
7	6.6667	6.0000
8	7.0000	9.0000
9	12.0000	.0000
10	12.0000	3.0000
11	12.0000	6.0000
12	12.0000	9.0000

Figure 8.2   Example input data and output for subroutine NODES. ·

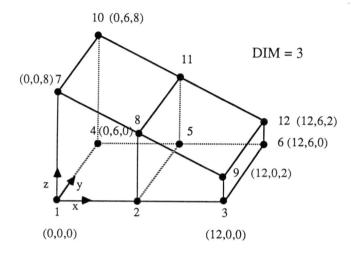

Input Data :

1	1	0.	0.	0.
3	1	12.	0.	0.
4	1	0.	6.	0.
6	1	12.	6.	0.
7	1	0.	0.	8.
9	1	12.	0.	2.
10	1	0.	6.	8.
12	1	12.	6.	2.
−1	−1	−1	−1	−1

Output :

NODE NUMBER	NODAL COORDINATES X	Y	Z
1	.0000	.0000	.0000
2	6.0000	.0000	.0000
3	12.0000	.0000	.0000
4	.0000	6.0000	.0000
5	6.0000	6.0000	.0000
6	12.0000	6.0000	.0000
7	.0000	.0000	8.0000
8	6.0000	.0000	5.0000
9	12.0000	.0000	2.0000
10	.0000	6.0000	8.0000
11	6.0000	6.0000	5.0000
12	12.0000	6.0000	2.0000

Figure 8.3   Example input data and output for subroutine NODES.

input file, and this feature can be used to greatly simplify data input for portions of the mesh where nodes are regularly spaced. In Figure 8.2, nodes 2 and 3 are equally spaced between nodes 1 and 4. Therefore it is necessary only to enter the node numbers, node number increments, and coordinates for nodes 1 and 4. The subroutine computes the node numbers and coordinates for the two "missing" nodes (nodes 2 and 3) using the node number coordinates for nodes 1 and 4 and the node number increment for node 4. Linear interpolation (in this case between nodes 1 and 4) is used to compute the coordinates of the "missing" nodes. The computed coordinates of the missing nodes are also stored in the arrays X1 and X2 (X1(2) = 0.6667, X2(2) = 3.0000, X1(3) = 1.3333, X2(3) = 6.0000). This process is repeated for nodes 5 and 8 (nodes 6 and 7 are "missing") and for nodes 9 and 12 (nodes 10 and 11 are "missing"). Input is terminated by placing a -1 in all fields.

The mesh in Figure 8.3 is three-dimensional (DIM = 3), so three coordinates (x, y, and z) are read for each node. For example, node 9 has a node number increment of 1, an x coordinate of 12 (X1(9) = 12), a y coordinate of 0 (X2(9) = 0, and a z coordinate of 2 (X3(9) = 2). Node numbers and coordinates for the "missing" nodes 2, 5, 8, and 11 are computed by the subroutine.

DIM = 4

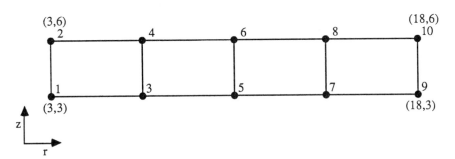

Input Data :
1	2	3.	3.
9	2	18.	3.
2	2	3.	6.
10	2	18.	6.
-1	-1	-1	-1

Output :

NODE NUMBER	NODAL COORDINATES R	Z
1	3.0000	3.0000
2	3.0000	6.0000
3	6.7500	3.0000
4	6.7500	6.0000
5	10.5000	3.0000
6	10.5000	6.0000
7	14.2500	3.0000
8	14.2500	6.0000
9	18.0000	3.0000
10	18.0000	6.0000

Figure 8.4   Example input data and output for subroutine NODES.

The nodes do not have to be entered sequentially and the node number increments can be assigned any positive integer value. Two examples are in Figures 8.4 and 8.5. The mesh in Figure 8.4 is for an axisymmetric problem (DIM = 4), so two coordinates (r and z) are read for each node. For example, node 2 has a node number increment of 2, an r coordinate of 3 (X1(2) = 3) and a z coordinate of 6 (X2(2) = 6). Node numbers and coordinates for the "missing" nodes 3, 4, 5, 6, 7, and 8 are computed by the subroutine. The mesh in Figure 8.5 is two-dimensional (DIM = 2) and a node number increment of 3 is used to generate the "missing" nodes 4, 7, 10, 5, 8, 11, 6, 9, and 12.

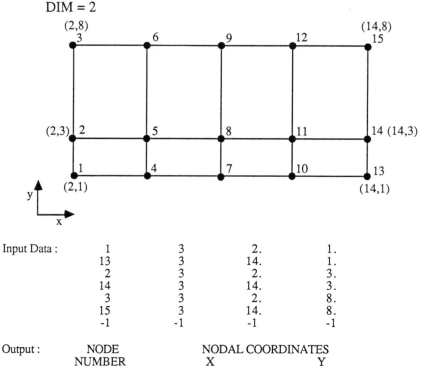

Input Data :

1	3	2.	1.
13	3	14.	1.
2	3	2.	3.
14	3	14.	3.
3	3	2.	8.
15	3	14.	8.
-1	-1	-1	-1

Output :

NODE NUMBER	NODAL COORDINATES X	Y
1	2.0000	1.0000
2	2.0000	3.0000
3	2.0000	8.0000
4	5.0000	1.0000
5	5.0000	3.0000
6	5.0000	8.0000
7	8.0000	1.0000
8	8.0000	3.0000
9	8.0000	8.0000
10	11.0000	1.0000
11	11.0000	3.0000
12	11.0000	8.0000
13	14.0000	1.0000
14	14.0000	3.0000
15	14.0000	8.0000

Figure 8.5   Example input data and output for subroutine NODES.

8.6   SOURCE CODE LISTING

```
          SUBROUTINE NODES
C**********************************************************************
C
C  8.1  PURPOSE:
C          TO INPUT NODE NUMBERS AND COORDINATES
C
C  8.2  INPUT:
C          NODE NUMBERS AND COORDINATES ARE READ FROM THE USER-
C          SUPPLIED FILE ASSIGNED TO UNIT "INF"
C
C  8.3  OUTPUT:
C          NODE NUMBERS AND COORDINATES ARE WRITTEN TO THE USER-
C          DEFINED FILE ASSIGNED TO UNIT "OUTF"
C
C  8.4  DEFINITIONS OF VARIABLES:
C             DIM = COORDINATE SYSTEM TYPE
C             INC = NODE NUMBER INCREMENT
C          NUMNOD = NUMBER OF NODES READ
C           X1(I) = X COORDINATE FOR NODE I IF DIM = 1, 2, OR 3
C                 = R COORDINATE FOR NODE I IF DIM = 4
C           X2(I) = IS NOT USED IF DIM = 1
C                 = Y COORDINATE FOR NODE I IF DIM = 2 OR 3
C                 = Z COORDINATE FOR NODE I IF DIM = 4
C           X3(I) = IS NOT USED IF DIM = 1, 2, OR 4
C                 = Z COORDINATE FOR NODE I IF DIM = 3
C
C  8.5  USAGE:
C          NODE NUMBERS AND COORDINATES ARE READ, ONE NODE
C          PER LINE, BEGINNING WITH NODE 1.  NODE NUMBERS
C          FOR "MISSING" NODES ARE GENERATED BY THE
C          SUBROUTINE BY ADDING THE NODE NUMBER INCREMENT
C          TO THE NODE NUMBER FOR THE PRECEEDING NODE.
C          COORDINATES FOR "MISSING" NODES ARE COMPUTED BY
C          THE SUBROUTINE USING LINEAR INTERPOLATION.  TO
C          TERMINATE INPUT, PLACE A -1 IN ALL FIELDS OF
C          THE INPUT FILE.
C
C          SUBROUTINES CALLED:
C            NONE
C
C          REFERENCES:
C            ISTOK,J.D. GROUNDWATER FLOW AND SOLUTE TRANSPORT
C            MODELING BY THE FINITE ELEMENT METHOD, CHAPTER 8.
C
C**********************************************************************
      $INCLUDE:'COMALL'
          DIMENSION XYZ(MAX1,3)
          EQUIVALENCE (X1,XYZ(1,1)),(X2,XYZ(1,2)),(X3,XYZ(1,3))
          INTEGER CNODE,OLDNOD
C
          NUMNOD = 0
          IDIM = DIM
          IF (DIM .EQ. 4) IDIM = 2
          OLDNOD = MAX1
C         READ FROM INPUT FILE: NODE NUMBER, NODE NUMBER INCREMENT,
C         AND NODAL COORDINATES
   10     READ(INF,*) CNODE,INC,(XYZ(CNODE,I),I=1,IDIM)
          IF (CNODE .EQ. -1) GOTO 40
          IF (CNODE .GT. NUMNOD) NUMNOD = CNODE
```

```
C          GENERATE NODE NUMBERS AND COORDINATES FOR "MISSING" NODES
           NGENP1 = (CNODE - OLDNOD) / INC
           IF (NGENP1 .GT. 0) THEN
              DO 30 I = 1, IDIM
                 XYZINC = (XYZ(CNODE,I) - XYZ(OLDNOD,I)) / FLOAT(NGENP1)
                 DO 20 J = OLDNOD + INC, CNODE - INC, INC
                    XYZ(J,I) = XYZ(J-INC,I) + XYZINC
 20              CONTINUE
 30           CONTINUE
           ENDIF
           OLDNOD = CNODE
           GOTO 10
C          WRITE NODE NUMBERS AND NODAL COORDINATES TO OUTPUT FILE
 40        IF (NUMNOD .GT. 0) THEN
              IF (DIM .EQ. 1) THEN
                 WRITE(OUTF,50)
 50              FORMAT(3X,'NODE',10X,'NODAL COORDINATES'/
     1                  2X,'NUMBER',18X,'X'/
     2                  2X,6('-'),8X,20('-'))
              ELSEIF (DIM .EQ. 2) THEN
                 WRITE(OUTF,60)
 60              FORMAT(3X,'NODE',21X,'NODAL COORDINATES'/
     1                  2X,'NUMBER',18X,'X',20X,'Y'/
     2                  2X,6('-'),8X,20('-'),1X,20('-'))
              ELSEIF (DIM .EQ. 3) THEN
                 WRITE(OUTF,70)
 70              FORMAT(3X,'NODE',31X,'NODAL COORDINATES'/
     1                  2X,'NUMBER',18X,'X',20X,'Y',20X,'Z'/
     2                  2X,6('-'),8X,20('-'),1X,20('-'),1X,20('-'))
              ELSEIF (DIM .EQ. 4) THEN
                 WRITE(OUTF,80)
 80              FORMAT(3X,'NODE',21X,'NODAL COORDINATES'/
     1                  2X,'NUMBER',18X,'R',20X,'Z'/
     2                  2X,6('-'),8X,20('-'),1X,20('-'))
              ENDIF
              DO 100 I = 1, NUMNOD
                 WRITE(OUTF,90) I,(XYZ(I,J),J=1,IDIM)
 90              FORMAT(I6,10X,3(F15.4,6X))
 100          CONTINUE
           ELSE
              WRITE(OUTF,110)
 110          FORMAT(' NO NODAL POINT DATA READ.')
           ENDIF
           RETURN
           END
```

Chapter 9

SUBROUTINE ELEMENT

9.1 PURPOSE

Subroutine ELEMENT inputs the element number, element type, and element node numbers for each element in the finite element mesh. The subroutine can be used for one-, two-, or three-dimensional problems or for problems with axisymmetry.

9.2 INPUT

Element numbers, element types, and element node numbers are read "free-format" from the user-supplied file assigned to unit "INF". INF is passed to the subroutine through a labeled common block contained in the file "COMALL" (see Chapter 7 for a listing of COMALL).

9.3 OUTPUT

Element numbers, element types, and element node numbers are written to the user-defined file assigned to unit "OUTF". Column headings are added to the list of element numbers, element types, and element numbers written to OUTF. The number of elements in the mesh (variable "NUMELM"), element types (array "ELEMTYP"), and element node numbers (array "IN"), are stored in labeled common blocks in COMALL for use by other subroutines.

9.4 DEFINITIONS OF VARIABLES

ELEMTYP(I)   =   Element type for element I (Table 9.1).

IN(I,J)   =   Node number J for element I.

INC   =   Element node number increment.

NODETBL(I)   =   Number of nodes in element type I (Table 9.1).

NUMELM   =   Number of elements in the mesh.

9.5 USAGE

Example input and output for subroutine ELEMENT are in Figures 9.1 to 9.3. Each line of input contains an element number, and the element type, element node number increment, and node numbers for the element. The mesh in Figure 9.1 is one-dimensional (DIM = 1) and contains five linear bar elements. Element 1 has an element type of 1 (ELEMTYP(1) = 1), an element node number increment of 1 (INC = 1), and node numbers 1 and 2 (IN(1,1) = 1 and IN(1,2) = 2). Element 5 has an element type of 1

Table 9.1 Element types used in subroutine ELEMENT.

Element Type	Number of Nodes	Coordinate System Type	Element Description	Reference
1	2	One-dimensional (DIM = 1)	Linear bar	Figure 4.9a
2	3	"	Quadratic bar	Figure 4.9b
3	4	"	Cubic bar	Figure 4.9c
4	3	Two-dimensional (DIM = 2)	Linear triangle	Figure 4.6
5	4	"	Linear rectangle	Figure 4.7
6	4	"	Linear quadrilateral	Figure 4.10
7	8	"	Quadratic quadrilateral	Figure 4.11
8	12	"	Cubic quadrilateral	Figure 4.12
9	8	Three-dimensional (DIM = 3)	Linear parallelepiped	Figure 4.13
10	20	"	Quadratic parallelepiped	Figure 4.14
11	32	"	Cubic parallelepiped	Figure 4.15
12	3	Axisymmetric (DIM = 4)	Linear triangle (axisymmetric)	—[1]
13	4	"	Linear rectangle (axisymmetric)	—[1]

[1] See notes following Chapter 4.

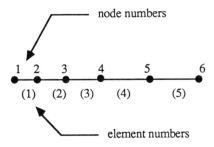

Input Data :

1	1	1	1	2
5	1	1	5	6
−1	−1	−1	−1	−1

Output :

ELEMENT NO.	ELEMENT TYPE	NODE NUMBERS	
1	1	1	2
2	1	2	3
3	1	3	4
4	1	4	5
5	1	5	6

Figure 9.1   Example input data and output for subroutine ELEMENT.

(ELEMTYP(5) = 1), an element node number increment of 1 (INC = 1), and node numbers 5 and 6 (IN(5,1) = 5 and IN(5,2) = 6). The subroutine has the capability to "generate" element numbers, element types, and element node numbers for elements "missing" from the input file, and this feature can be used to greatly simplify data input. In Figure 9.1, for example, the element node numbers for element 2 are generated within ELEMENT by adding the node number increment for node 5 to the node numbers for element 1 :

$$IN(2,1) = IN(1,1) + INC = 1 + 1 = 2$$
$$IN(2,2) = IN(1,2) + INC = 2 + 1 = 3$$

Similarly the element node numbers for elements 3 and 4 are generated within ELEMENT by adding the node number increment for node 5 to the node numbers for element 2 and 3. For element 3 :

$$IN(3,1) = IN(2,1) + INC = 2 + 1 = 3$$
$$IN(3,2) = IN(2,2) + INC = 3 + 1 = 4$$

For element 4 :

$$IN(4,1) = IN(3,1) + INC = 3 + 1 = 4$$
$$IN(4,2) = IN(4,2) + INC = 4 + 1 = 5$$

Input is terminated by placing a -1 in all fields.

The mesh in Figure 9.2 is two-dimensional (DIM = 2) and contains ten linear quadrilateral elements. Element 1 has an element type of 6 (ELEMTYP(1) = 6), an element node number increment of 3 (INC = 3), and element node numbers 1, 4, 5, and 2 (IN(1,1) = 1, IN(1,2) = 4, IN(1,3) = 5, IN(1,4) = 2). Element 5 has an element type of 6 (ELEMTYP(5) = 6), an element node number increment of 3 (INC = 3), and element node numbers 13, 16, 17, and 14 (IN(5,1) = 13, IN(5,7) = 16, IN(5,3) = 17, and IN(5,4) = 14). The node number increment for element 5 is used to "generate" the element node numbers for the "missing" elements 2, 3, and 4.

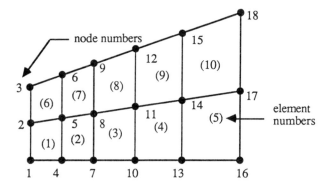

Input Data :

1	6	3	1	4	5	2
5	6	3	13	16	17	14
6	6	3	2	5	6	3
10	6	3	14	17	18	15
-1	-1	-1	-1	-1	-1	-1

Output :

ELEMENT NO.	ELEMENT TYPE	NODE NUMBERS		
1	6	1	4	5	2
2	6	4	7	8	5
3	6	7	10	11	8
4	6	10	13	14	11
5	6	13	16	17	14
6	6	2	5	6	3
7	6	5	8	9	6
8	6	8	11	12	9
9	6	11	14	15	12
10	6	14	17	18	15

Figure 9.2   Example input data and output for subroutine ELEMENT.

For element 2 :
$$\begin{aligned}
IN(2,1) &= IN(1,1) + INC = 1 + 3 = 4 \\
IN(2,2) &= IN(1,2) + INC = 4 + 3 = 7 \\
IN(2,3) &= IN(1,3) + INC = 5 + 3 = 8 \\
IN(2,4) &= IN(1,4) + INC = 2 + 3 = 5
\end{aligned}$$

For element 3 :
$$\begin{aligned}
IN(3,1) &= IN(2,1) + INC = 4 + 3 = 7 \\
IN(3,2) &= IN(2,2) + INC = 7 + 3 = 10 \\
IN(3,3) &= IN(2,3) + INC = 8 + 3 = 11 \\
IN(3,4) &= IN(2,4) + INC = 5 + 3 = 8
\end{aligned}$$

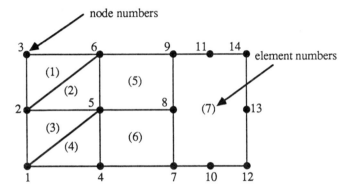

Input Data :

1	4	1	2	6	3					
2	4	1	2	5	6					
3	4	1	1	5	2					
4	4	1	4	5	1					
5	5	1	5	8	9	6				
6	5	1	4	7	8	5				
7	7	1	7	10	12	13	14	11	9	8
-1	-1	-1	-1	-1	-1	-1	-1	-1	-1	-1

Output :

ELEMENT NO.	ELEMENT TYPE	NODE NUMBERS							
1	4	2	6	3					
2	4	2	5	6					
3	4	1	5	2					
4	4	4	5	1					
5	5	5	8	9	6				
6	5	4	7	8	5				
7	7	7	10	12	13	14	11	9	8

Figure 9.3   Example input data and output for subroutine ELEMENT.

for element 4 :

$$IN(4,1) = IN(3,1) + INC = 7 + 3 = 10$$
$$IN(4,2) = IN(3,2) + INC = 10 + 3 = 13$$
$$IN(4,3) = IN(3,3) + INC = 11 + 3 = 14$$
$$IN(4,4) = IN(3,4) + INC = 8 + 3 = 11$$

The same procedure is used to "generate" the "missing" elements between elements 6 and 10.

Subroutine ELEMENT allows the user to use any combination of compatible element types in a mesh. Referring to Table 9.1 a one-dimensional mesh can be designed using any combination of one-dimensional elements (Types 1, 2, or 3). Similarly, any combination of element types 3, 4, 5, 6, 7, or 8 can be used in a two-dimensional mesh, any combination of element types 9, 10, or 11 can be used in a three-dimensional mesh, and any combination of element types 12 or 13 can be used in an axisymmetric mesh. An example of a two-dimensional mesh containing a three element types is in Figure 9.3. Elements 1, 2, 3, and 4 are linear triangle elements (Type 4), elements 5 and 6 are linear rectangle elements (Type 5), and element 7 is a quadratic quadrilateral element (Type 7).

Note that the element node numbers must be entered in a specific order. The proper order is shown in the definition sketch for the interpolation functions for each element type. For example, by referring to Figure 4.10 we see that the node numbers for linear quadrilateral elements must be entered in a "counter clockwise" fashion.

9.6   SOURCE CODE LISTING

```
      SUBROUTINE ELEMENT
C*********************************************************************
C
C  9.1  PURPOSE:
C           TO INPUT ELEMENT NUMBERS, TYPES,  NODE NUMBERS
C
C  9.2  INPUT:
C           ELEMENT NUMBERS, TYPES, AND NODE NUMBERS ARE READ
C           FROM THE USER-SUPPLIED FILE ASSIGNED TO UNIT "INF"
C
C  9.3  OUTPUT:
C           ELEMENT NUMBERS, TYPES, AND NODE NUMBERS ARE WRITTEN
C           TO THE USER-DEFINED FILE ASSIGNED TO UNIT "OUTF"
C
C  9.4  DEFINITIONS OF VARIABLES:
C           ELEMTYP(I) = ELEMENT TYPE FOR ELEMENT I
C              IN(I,J) = NODE NUMBER J FOR ELEMENT I
C                  INC = NODE NUMBER INCREMENT
C           NODETBL(I) = NUMBER OF NODES IN ELEMENT TYPE I
C               NUMELM = NUMBER OF ELEMENTS IN MESH
C
C  9.5  USAGE:
C           ELEMENT DATA (ELEMENT NUMBER, TYPE, AND NODE NUMBERS)
C           ARE READ SEQUENTIALLY, SET OF ELEMENT DATA PER LINE.
C           ELEMENT NUMBERS, TYPES, AND NODE NUMBERS FOR "MISSING"
C           ELEMENTS ARE GENERATED BY THE SUBROUTINE.  TO TERMINATE
C           INPUT, PLACE A -1 IN ALL FIELDS OF THE INPUT FILE.
C
C           SUBROUTINES CALLED:
C              NONE
C
C           REFERENCES:
C              ISTOK,J.D. GROUNDWATER FLOW AND SOLUTE TRANSPORT
C              MODELING BY THE FINITE ELEMENT METHOD, CHAPTER 9.
C
C*********************************************************************
```

```
$INCLUDE:'COMALL'
         INTEGER OLDELM,ELM,TYPE
         DIMENSION NODETBL(13)
         DATA NODETBL/2,3,4,3,4,4,8,12,8,20,32,3,4/
C
         MAXNOD=0
         OLDELM=MAX2
         NUMELM=0
C        READ FROM INPUT FILE: ELEMENT NUMBER, ELEMENT TYPE,
C        AND ELEMENT NODE NUMBERS
  10     READ(INF,*) ELM,TYPE,INC,(IN(ELM,I),I=1,NODETBL(ABS(TYPE)))
         IF (ELM .EQ. -1) GOTO 40
         ELEMTYP(ELM) = TYPE
         IF (ELM .GT. NUMELM) NUMELM = ELM
         IF (NODETBL(TYPE) .GT. MAXNOD) MAXNOD = NODETBL(TYPE)
C        GENERATE THE MISSING ELEMENTS
         IF (ELM .GT. OLDELM+1) THEN
            DO 30 I = OLDELM + 1, ELM -1
              IM1 = I - 1
              DO 20 J = 1, NODETBL(TYPE)
                  IN(I,J) = IN(IM1,J) + INC
  20          CONTINUE
              ELEMTYP(I) = TYPE
  30        CONTINUE
         ENDIF
         OLDELM = ELM
         GOTO 10
C        WRITE ELEMENT NUMBERS AND ELEMENT NODE NUMBERS TO OUTPUT FILE
  40     IF (NUMELM .GT. 0) THEN
            IF (MAXNOD .EQ. 2) THEN
               WRITE(OUTF,50) (' ',I=1,2)
            ELSEIF (MAXNOD .EQ. 3) THEN
               WRITE(OUTF,50) ('      ',I=1,2)
            ELSEIF (MAXNOD .EQ. 4) THEN
               WRITE(OUTF,50) ('         ',I=1,2)
            ELSEIF (MAXNOD .GT. 4) THEN
               WRITE(OUTF,50) ('                         ',I=1,2)
            ENDIF
  50        FORMAT(/,2(2X,'ELEMENT',4X)/4X,'NO.',10X,'TYPE',6X,A,
        1          'NODE NUMBERS'/2(2X,'-------',4X),1X,A,'------------')
            DO 70 I = 1, NUMELM
               WRITE(OUTF,60) I,ELEMTYP(I),
        1                      (IN(I,J),J=1,NODETBL(ELEMTYP(I)))
  60           FORMAT(I7,I13,6X,8I6:4(/26X,8I6))
  70        CONTINUE
         ELSE
            WRITE(OUTF,80)
  80        FORMAT(' NO ELEMENT DATA READ.')
         ENDIF
         RETURN
         END
```

Chapter 10

SUBROUTINE MATERL

10.1 PURPOSE

Subroutine MATERL inputs the element material set numbers for each element in the finite element mesh and a set of material properties for each material set. The term "material property" refers to any physical or chemical property for an element. For example, in solving the steady-state, saturated groundwater flow equation we use the components of saturated hydraulic conductivity to compute a conductance matrix for each element . It is convenient to assign all elements with the same saturated hydraulic conductivity to a single material set, the material properties for that set would be the components of saturated hydraulic conductivity for the elements in the set. Other examples of material properties are specific storage, porosity, bulk density, dispersivities, distribution coefficients, or a table of values of unsaturated hydraulic conductivity as a function of pressure head.

10.2 INPUT

Element material set numbers are read "free-format" from the user-supplied file assigned to unit "INF". INF is passed to the subroutine through a labeled common block contained in the file "COMALL" (see Chapter 7 for a listing of COMALL). Then the number of material properties in a material set and a list of material properties for each material set are read from INF.

10.3 OUTPUT

Element material set numbers and the list of material properties for elements in each material set are written to the user-defined file assigned to unit "OUTF". Column headings are added to the list of material set numbers and material properties written to OUTF. The number of element material sets (variable "NUMMAT"), the material set numbers for each element (array "MATSET"), the number of material properties in a material set (variable "NUMPROP"), and the material properties for each material set, array "PROP", are stored in labeled common blocks in COMALL for use by other subroutines.

10.4 DEFINITIONS OF VARIABLES

MATSET(I)   =   Material set number for element I.

NUMMAT   =   Number of material sets in the mesh.

NUMPROP   =   Number of material properties in a material set.

PROP(I,J)   =   Material property J for material set I.

Table 10.1   Material property order for MATERL.

Computer Program	Problem Dimension	Material Properties, PROP (I,J)					
		J = 1	2	3	4	5	6
GW 1	1	K_x	-----	-----	not used	-----	-----
	2	K_x	K_y				
	3	K_x	K_y	K_z			
	4	K_r	K_z				
GW 3	1	K_x	S_s				
	2	K_x	K_y	S_s			
	3	K_x	K_y	K_z	S_s		
ST 1	1	a_L	λ	ρb	K_d	n	
	2	a_L	a_T	λ	ρ	K_d	n
	3	a_L	a_T	λ	ρb	K_d	n

10.5 USAGE

Material set numbers are read in sequentially from element one to element NUMELM, one element and material set number per line. The subroutine has the capability to "generate" the material set numbers for elements "missing" from the input file. In the mesh in Figure 10.1, the material set numbers for all elements are the same. In this case we need only list the material set numbers for elements 1 and 10 (MATSET(1) = 1 and

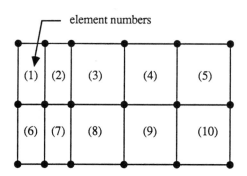

element numbers

All elements have identical material properties
$$K_x = K_y = 10, \ S_s = 0.0002$$

Input Data :

```
 1     1
10     1
-1    -1
 3
 1    10.        10.    0.0002
```

Output :

ELEMENT NO.	MATERIAL SET NUMBER
1	1
2	1
3	1
4	1
5	1
6	1
7	1
8	1
9	1
10	1

MATERIAL SET NO.	MATERIAL PROPERTIES		
1	.100000E+2	.100000E+2	.200000E-3

Figure 10.1   **Example input data and output for subroutine MATERL.**

MATSET(10) = 1). The subroutine "generates" the material set numbers for the eight "missing" elements (elements 2, 3, 4, 5, 6, 7, 8, and 9), MATSET(2) = MATSET(3) = ⋯ MATSET(9) = 1. Input of material set number is terminated by placing a -1 in both fields.

The number of material properties in a material set is read next. Note that all material sets must have the same number of material properties. For the mesh in Figure 10.1 NUMMAT = 3 (the two components of hydraulic conductivity and specific storage). The material properties for each material set are read next. The correct order to use for entering material properties is in Table 10.1. Any consistent system of units can be used (If in doubt see the source code listing for the subroutines that compute the element matrices, e.g. KBARZ, KQUA4, DQUA4, etc.). For the mesh in Figure 10.1, PROP(1,1) = 10,

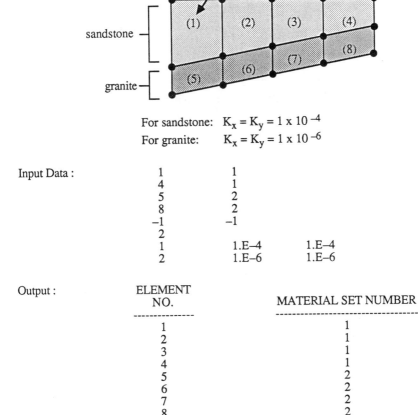

element numbers

sandstone

granite

For sandstone: $K_x = K_y = 1 \times 10^{-4}$

For granite: $K_x = K_y = 1 \times 10^{-6}$

Input Data :

1	1
4	1
5	2
8	2
-1	-1
2	
1	1.E–4 1.E–4
2	1.E–6 1.E–6

Output :

ELEMENT NO.	MATERIAL SET NUMBER
1	1
2	1
3	1
4	1
5	2
6	2
7	2
8	2

MATERIAL SET NO.	MATERIAL PROPERTIES	
1	.100000E–03	.100000E–03
2	.100000E–05	.100000E–05

Figure 10.2   Example input data and output for subroutine MATERL.

PROP(1,2) = 10, and PROP(1,3) = 0.0002. Input of material properties is terminated automatically by the subroutine.

There are two material sets in the mesh in Figure 10.2. Elements 1, 2, 3, and 4 are in material set 1 and elements 5, 6, 7, and 8 are in material set 2. Each material set has two properties, in this case the two components saturated hydraulic conductivity.

Material properties can also be used to represent a table, for example a table of values of unsaturated hydraulic conductivity as a function of pressure head. If an unsaturated flow problem was being solved, the value of unsaturated hydraulic conductivity at any value of pressure head could be obtained from such a list by interpolation (alternative approaches to describing unsaturated hydraulic conductivity were described in Chapter 5). An example of the use of material property data for this purpose is in Figure 10.3.

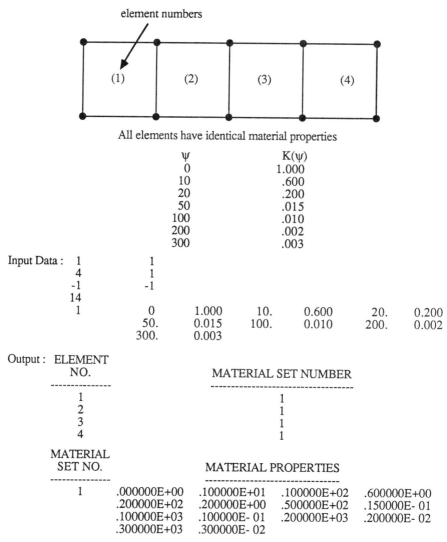

element numbers

All elements have identical material properties

ψ	$K(\psi)$
0	1.000
10	.600
20	.200
50	.015
100	.010
200	.002
300	.003

Input Data :

1	1
4	1
-1	-1
14	

1							
	0	1.000	10.	0.600	20.	0.200	
	50.	0.015	100.	0.010	200.	0.002	
	300.	0.003					

Output :

ELEMENT NO.	MATERIAL SET NUMBER
---------------	------------------------------------
1	1
2	1
3	1
4	1

MATERIAL SET NO.	MATERIAL PROPERTIES		
---------------	----------------------------------		
1	.000000E+00	.100000E+01	.100000E+02	.600000E+00
	.200000E+02	.200000E+00	.500000E+02	.150000E- 01
	.100000E+03	.100000E- 01	.200000E+03	.200000E- 02
	.300000E+03	.300000E- 02		

Figure 10.3   Example input data and output for subroutine **MATERL**.

10.6 SOURCE CODE LISTING

```
      SUBROUTINE MATERL
C*********************************************************************
C
C 10.1  PURPOSE:
C          TO INPUT ELEMENT MATERIAL SET NUMBERS AND MATERIAL
C          PROPERTIES FOR EACH MATERIAL SET
C
C 10.2  INPUT:
C          ELEMENT MATERIAL SET NUMBERS AND MATERIAL PROPERTIES
C          FOR EACH MATERIAL SET ARE READ FROM THE USER-SUPPLIED
C          FILE ASSIGNED TO UNIT "INF"
C
C 10.3  OUTPUT:
C          ELEMENT MATERIAL SET NUMBERS AND MATERIAL PROPERTIES
C          FOR EACH MATERIAL SET ARE WRITTEN TO THE USER-DEFINED
C          FILE ASSIGNED TO UNIT "OUTF"
C
C 10.4  DEFINITIONS OF VARIABLES:
C          MATSET(I) = MATERIAL SET NUMBER FOR ELEMENT I
C             NUMMAT = NUMBER OF MATERIAL SETS
C            NUMPROP = NUMBER OF MATERIAL PROPERTIES IN EACH
C                      MATERIAL SET
C          PROP(I,J) = MATERIAL PROPERTY J FOR MATERIAL SET I
C
C 10.5  USAGE:
C          ELEMENT MATERIAL SET NUMBERS ARE READ IN SEQUENTIALLY,
C          ONE ELEMENT NUMBER AND MATERIAL SET NUMBER PER LINE.
C          MATERIAL SET NUMBERS FOR "MISSING" ELEMENTS ARE
C          GENERATED BY THE SUBROUTINE BY ASSIGNING THE MATERIAL
C          SET NUMBER OF THE PRECEEDING ELEMENT TO EACH "MISSING"
C          ELEMENT.  TO TERMINATE INPUT OF ELEMENT MATERIAL SET
C          NUMBERS, PLACE A -1 IN ALL FIELDS OF THE INPUT FILE.
C          THE PROGRAM THEN READS THE NUMBER OF MATERIAL SET
C          PROPERTIES IN EACH MATERIAL SET (THE NUMBER OF MATERIAL
C          PROPERTIES IN EACH MATERIAL SET IS THE SAME).  THEN
C          THE MATERIAL PROPERTIES FOR EACH MATERIAL SET ARE READ IN
C          SEQUENTIALLY, ONE MATERIAL SET NUMBER AND ONE SET OF
C          MATERIAL PROPERTIES PER LINE.  INPUT IS TERMINATED
C          AUTOMATICALLY WHEN THE LAST MATERIAL SET PROPERTIES ARE
C          READ.
C
C       SUBROUTINES CALLED:
C          NONE
C
C       REFERENCES:
C          ISTOK,J.D. GROUNDWATER FLOW AND SOLUTE TRANSPORT
C          MODELING BY THE FINITE ELEMENT METHOD, CHAPTER 10.
C
C*********************************************************************
$INCLUDE:'COMALL'
      INTEGER OLDELM,ELM,SETNUM
C
      OLDELM = MAX4
      NUMMAT = 0
C     READ FROM INPUT FILE: ELEMENT NUMBER, AND MATERIAL SET NUMBER
   10 READ(INF,*) ELM,MATSET(ELM)
      IF (ELM .EQ. -1) GOTO 30
C     DETERMINE THE NUMBER OF MATERIAL SETS
      IF (MATSET(ELM) .GT. NUMMAT) NUMMAT = MATSET(ELM)
```

```
C          GENERATE THE MATERIAL SET NUMBER FOR EACH "MISSING" ELEMENT
           IF (ELM .GT. OLDELM + 1) THEN
              DO 20 I = OLDELM + 1, ELM - 1
                 MATSET(I) = MATSET(I-1)
  20          CONTINUE
           END IF
           OLDELM = ELM
           GOTO 10
C          WRITE THE MATERIAL SET NUMBER FOR EACH ELEMENT TO OUTPUT FILE
  30       IF (NUMELM .GT. 0) THEN
              WRITE(OUTF,40)
  40          FORMAT(//2X,'ELEMENT'/4X,'NO.',9X,'MATERIAL SET NUMBER'/
     1              2X,'-------',7X,'--------------------')
              DO 60 I = 1, NUMELM
                 WRITE(OUTF,50) I,MATSET(I)
  50             FORMAT(I6,I20)
  60          CONTINUE
C          READ FROM INPUT FILE: THE NUMBER OF PROPERTIES IN EACH
MATERIAL SET
              READ(INF,*) NUMPROP
              IF (NUMPROP .EQ. 1) THEN
                 WRITE(OUTF,70) (' ',I=1,2)
              ELSEIF (NUMPROP .EQ. 2) THEN
                 WRITE(OUTF,70) ('          ',I=1,2)
              ELSEIF (NUMPROP .EQ. 3) THEN
                 WRITE(OUTF,70) ('                    ',I=1,2)
              ELSEIF (NUMPROP .GE. 4) THEN
                 WRITE(OUTF,70) ('                              ',I=1,2)
              ENDIF
  70          FORMAT(//2X,'MATERIAL'/3X,'SET NO.',3X,A,
     1               'MATERIAL  PROPERTIES'/2X,'--------',3X,A,
     2               '--------------------')
C          WRITE MATERIAL PROPERTIES INFORMATION TO OUTPUT FILE
              DO 90 I = 1, NUMMAT
                 READ(INF,*)  SETNUM,(PROP(SETNUM,J),J=1,NUMPROP)
                 WRITE(OUTF,80) SETNUM,(PROP(SETNUM,J),J=1,NUMPROP)
  80             FORMAT(I7,7X,8(1P4E15.6/14X))
  90          CONTINUE
           ELSE
              WRITE(OUTF,100)
 100          FORMAT(' NO ELEMENT MATERIAL PROPERTY DATA READ.')
           ENDIF
           RETURN
           END
```

Chapter 11

SUBROUTINE BOUND

11.1 PURPOSE

Subroutine BOUND inputs specified values of the field variable (hydraulic head, pressure head, or solute concentration) and specified values of either groundwater flow or solute flux, for selected nodes in mesh. These values are used to represent Dirichlet and Neumann boundary conditions for the groundwater flow and solute transport equations.

11.2 INPUT

Specified values of the field variable are read "free-format" from the user-supplied file assigned to unit "INF". INF is passed to the subroutine through a labeled common block contained in the file "COMALL" (See Chapter 7 for a listing of COMALL). Then specified values of either groundwater or solute flux are read from INF.

11.3 OUTPUT

Specified values of the field variable (hydraulic head, pressure head, or solute concentration) and either groundwater flow or solute flux are written to the user-defined file assigned to unit "OUTF". Column headings are added to the list of specified values written to OUTF. The character variables "LABEL1" and "LABEL2" are used to label the column headings on OUTF as "HYDRAULIC HEAD", "PRESSURE HEAD", or "SOLUTE CONCENTRATION" and "GROUNDWATER FLOW" or "SOLUTE FLUX".

11.4 DEFINITIONS OF VARIABLES

FLUX(I)   = Specified value of groundwater flow or solute flux at node I.

X(I)   = Specified value of the field variable (hydraulic head, pressure head or solute concentration) at node I.

ICH(I)   = 1 if the value of the field variable is specified for node I.
            = 0, otherwise.

LCH(I)   = $\sum_{k=1}^{I} ICH(k)$. The arrays ICH and LCH are used in subroutines ASMBK, ASMBKC, and ASMBAD to modify global system of equations for specified values of the field variable.

LABEL1   = Character variable used to label column headings for specified values of the field variable on the file assigned to unit OUTF.

LABEL1 = "HYDRAULIC HEAD", "PRESSURE HEAD", or "SOLUTE CONCENTRATION".

LABEL2   = Character variable used to label column headings for specified values of groundwater flow or solute flux on the file assigned to unit OUTF. LABEL2 = "GROUNDWATER FLOW" or "SOLUTE FLUX".

NDN   = Number of nodes with specified values of the field variable (named for "Number of Dirichlet Nodes").

NDOF   = Number of nodes where the value of the field variable are unknown (named for "Number of Degrees of Freedom).
          = NUMNOD - NDN.

NNN   = Number of nodes with specified values of groundwater flow or solute flux (named for "Number of Neumann Nodes).

11.5 USAGE

Specified values of the field variable are read first, one node number and the specified value of the field variable at that node per line. The node numbers and specified values can be listed in any order on the input file. The value of the field variable must be specified for at least one node in the mesh. Input is terminated by placing a -1 in both fields. Specified values of groundwater flow or solute flux are read next, one node number and the specified value of groundwater flow or solute flux at that node per line. The node numbers and specified values can be listed in any order on the input file. Input is terminated by placing a -1 in both fields.

The mesh in Figure 11.1 is for an unsaturated groundwater flow problem. In this case, the calling program would assign the character strings "PRESSURE HEAD" and "GROUNDWATER FLOW", to the character variables "LABEL1" and "LABEL2", respectively. Four values of pressure head are specified (at nodes 1, 2, 5, and 6) and there are no specified values of groundwater flow. Specified values of pressure head are assigned to the array "X", the remaining entries of X are arbitrarily assigned a value of zero (the values of pressure head at these nodes will be computed by subroutine SOLVE, see Chapter 13). The entries of the arrays "ICH" and "LCH" are assigned and, since there are no specified values of groundwater flow, the entries of the array are assigned a value of zero. After BOUND is executed these arrays would contain the following :

I	X(I)	ICH(I)	LCH(I)	FLUX(I)
1	-10.	1	1	0.
2	-10.	1	2	0.
3	0.	0	2	0.
4	0.	0	2	0.
5	-5.	1	3	0.
6	-5.	1	4	0.

The mesh in Figure 11.2 is for a saturated groundwater flow problem. In this case the calling program would assign the character strings "HYDRAULIC HEAD" and "GROUNDWATER FLOW" to the character variables "LABEL1" and "LABEL2", respectively. Five values of hydraulic head are specified (at nodes 3, 5, 8, 10, and 13) and a pumping well is located at node 7. After BOUND is executed the entries of arrays X, ICH, LCH, and FLUX would contain the following :

I	X(I)	ICH(I)	LCH(I)	FLUX(I)
1	0.	0	0	0.
2	0.	0	0	0.
3	10.	1	1	0.
4	0.	0	1	0.
5	10.	1	2	0.
6	0.	0	2	0.
7	0.	0	2	-10.
8	10.	1	3	0.
9	0.	0	3	0.
10	10.	1	4	0.
11	0.	0	4	0.
12	0.	0	4	0.
13	10.	1	5	0.

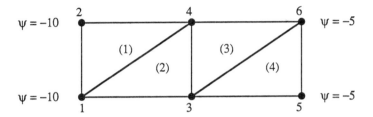

Input Data :

1	-10.
2	-10.
5	-5.
6	-5.
-1	-1
-1	-1

Output :

NODE NO.	SPECIFIED PRESSURE HEAD
1	-10.00
2	-10.00
5	-5.00
6	-5.00

NUMBER OF NODES WITH SPECIFIED   PRESSURE HEAD   =   4

NUMBER OF NODES WITH SPECIFIED   GROUNDWATER FLOW   =   0

Figure 11.1   Example input data and output for subroutine BOUND.

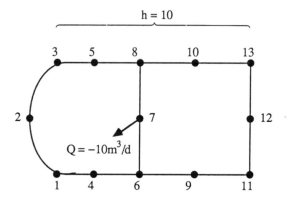

Input Data :

3	10.
5	10.
8	10.
10	10.
13	10.
-1	-1
7	-10.
-1	-1

Output :

NODE NO.	SPECIFIED HYDRAULIC HEAD
3	10.00
5	10.00
8	10.00
10	10.00
13	10.00

NUMBER OF NODES WITH SPECIFIED   HYDRAULIC HEAD   =   5

NODE NO.	SPECIFIED GROUNDWATER FLOW
7	-10.00

NUMBER OF NODES WITH SPECIFIED   GROUNDWATER FLOW   =   1

Figure 11.2   Example input data and output for subroutine BOUND.

11.6 SOURCE CODE LISTING

```
      SUBROUTINE BOUND
C*********************************************************************
C
C   11.1  PURPOSE:
C           TO INPUT SPECIFIED VALUES OF THE FIELD VARIABLE
C           (HYDRAULIC HEAD, PRESSURE HEAD, OR SOLUTE CONCENTRATION)
C           AND SPECIFIED VALUES OF GROUNDWATER FLOW OR SOLUTE FLUX,
C           FOR SELECTED NODES.
C
C   11.2  INPUT:
C           SPECIFIED VALUES OF THE FIELD VARIABLE AND SPECIFIED
C           VALUES OF GROUNDWATER FLOW OR SOLUTE FLUX ARE READ
C           FROM THE USER-SUPPLIED FILE ASSIGNED TO UNIT "INF".
C
C   11.3  OUTPUT:
C           SPECIFIED VALUES OF THE FIELD VARIABLE AND SPECIFIED
C           VALUES OF GROUNDWATER FLOW OR SOLUTE FLUX ARE WRITTEN
C           TO THE USER-DEFINED FILE ASSIGNED TO UNIT "OUTF".
C
C   11.4  DEFINITIONS OF VARIABLES:
C           FLUX(I) = SPECIFIED VALUE OF GROUNDWATER FLOW OR
C                     SOLUTE FLUX AT NODE I
C            ICH(I) = 1 IF THE VALUE OF THE FIELD VARIABLE IS
C                     SPECIFIED FOR NODE I,
C                   = 0 OTHERWISE
C            LCH(I) = ICH(I) + ICH(I-1) + ICH(I-2) + ...
C                     THE ARRAYS ICH AND LCH ARE USED TO MODIFY
C                     GLOBAL SYSTEM OF EQUATIONS IN SUBROUTINES
C                     ASMBK, ASMBKC, AND ASMBAD
C            LABEL1 = CHARACTER VARIABLE USED TO LABEL COLUMN
C                     HEADINGS FOR SPECIFIED VALUES OF THE FIELD
C                     VARIABLE ON FILE ASSIGNED TO UNIT OUTF.
C                     LABEL1 = "HYDRAULIC HEAD", "PRESSURE HEAD"
C                     OR "SOLUTE CONCENTRATION"
C            LABEL2 = CHARACTER VARIABLE USED TO LABEL COLUMN HEADINGS
C                     FOR SPECIFIED VALUES OF GROUNDWATER FLOW OR
C                     SOLUTE FLUX ON FILE ASSIGNED TO UNIT OUTF.
C                     LABEL2 = "GROUNDWATER FLOW" OR "SOLUTE FLUX"
C               NDN = NUMBER OF NODES WITH SPECIFIED VALUES OF THE
C                     FIELD VARIABLE (NAMED FOR NUMBER OF DIRICHLET
C                     NODES)
C              NDOF = NUMBER OF NODES WHERE THE VALUE OF THE FIELD
C                     VARIABLE IS UNKNOWN (NAMED FOR NUMBER OF DEGREES
C                     OF FREEDOM)
C               NNN = NUMBER OF NODES WITH SPECIFIED VALUES OF
C                     GROUNDWATER FLOW OR SOLUTE FLUX (NAMED FOR NUMBER
C                     OF NEUMANN NODES)
C              X(I) = SPECIFIED VALUE OF THE FIELD VARIABLE
C                     (HYDRAULIC HEAD, PRESSURE HEAD, OR
C                     SOLUTE CONCENTRATION) AT NODE I
C
C   11.5  USAGE:
C           SPECIFIED VALUES OF THE FIELD VARIABLE ARE READ FIRST, ONE
C           NODE NUMBER AND THE SPECIFIED VALUE OF THE FIELD VARIABLE
C           AT THAT NODE PER LINE.  THE NODE NUMBERS CAN BE LISTED IN
C           ANY ORDER ON THE INPUT FILE.  THE VALUE OF THE FIELD
C           VARIABLE MUST BE SPECIFIED FOR AT LEAST ONE NODE IN THE
C           MESH.  INPUT IS TERMINATED BY PLACING A -1 IN BOTH FIELDS.
C           SPECIFIED VALUES OF GROUNDWATER FLOW OR SOLUTE FLUX ARE
```

```
C             READ NEXT, ONE NODE NUMBER AND THE SPECIFIED VALUE OF
C             GROUNDWATER FLOW OR SOLUTE FLUX AT THAT NODE PER LINE.
C             THE NODE NUMBERS AND SPECIFIED VALUES CAN BE LISTED IN ANY
C             ORDER ON THE INPUT FILE.  INPUT IS TERMINATED BY PLACING
C             A -1 IN BOTH FIELDS.
C
C          SUBROUTINES CALLED:
C             NONE
C
C          REFERENCES:
C             ISTOK,J.D. GROUNDWATER FLOW AND SOLUTE TRANSPORT
C             MODELING BY THE FINITE ELEMENT METHOD, CHAPTER 11.
C
C*********************************************************************
$INCLUDE: 'COMALL'
C
C          INITIALIZATION
           DO 10 I = 1, NUMNOD
             ICH(I) = 0
             FLUX(I) = 0.
  10       CONTINUE
           NDN = 0
C          READ FROM INPUT FILE: NODE NUMBER AND SPECIFIED VALUE OF
C          FIELD VARIABLE
  20       READ(INF,*) I,X(I)
           IF (I .NE. -1) THEN
             IF(NDN .EQ. 0) WRITE(OUTF,30) LABEL1
  30         FORMAT(//3X,'NODE',15X,'SPECIFIED'/4X,'NO.',10X,A/
      1             2X,'------',9X,'--------------------')
             NDN = NDN + 1
             ICH(I) = 1
C            WRITE INFORMATION JUST READ TO OUTPUT FILE
             WRITE(OUTF,40) I,X(I)
  40         FORMAT(I6,10X,F15.4)
             GOTO 20
           ENDIF
           WRITE(OUTF,50) LABEL1,NDN
  50       FORMAT(//' NUMBER OF NODES WITH SPECIFIED ',A,' =',I7)
           NNN = 0
C          READ FROM INPUT FILE: NODE NUMBER AND SPECIFIED VALUE OF
C          GROUNDWATER FLOW OR SOLUTE FLUX
  60       READ(INF,*) I,FLUX(I)
           IF (I .NE. -1) THEN
             IF (NNN .EQ. 0) WRITE(OUTF,30) LABEL2
             NNN = NNN + 1
C            WRITE THE INFORMATION JUST READ TO OUTPUT FILE
             WRITE(OUTF,40) I,FLUX(I)
             GOTO 60
           ENDIF
           WRITE(OUTF,50) LABEL2,NNN
           LCH(1) = ICH(1)
           DO 70 I = 2, NUMNOD
             LCH(I) = LCH(I-1) + ICH(I)
  70       CONTINUE
           NDOF = NUMNOD - NDN
           RETURN
           END
```

Chapter 12

SUBROUTINE ASMBK

12.1 PURPOSE

Subroutine ASMBK assembles the global conductance matrix [K] and the global specified flow matrix {F} (equation 5.1). The global matrices are modified during the assembly process to account for specified values of the field variable (hydraulic head or pressure head) and groundwater flow. ASMBK also computes the semi-bandwidth and the number of degrees of freedom for the modified system of equations.

12.2 INPUT

None

12.3 OUTPUT

The semi-bandwidth and number of degrees of freedom for the modified system of equations are written to the user-defined file assigned to unit "OUTF".

12.4 DEFINITIONS OF VARIABLES

B(I)	=	Modified specified flow matrix.
E	=	Element number.
ELEMTYP(I)	=	Element type for element I (see Table 9.1 for a list of element types).
FLUX(I)	=	Specified value of groundwater flow at node I.
ICH(I)	=	1 if the value of the field variable is specified for node I,
	=	0 otherwise.
IJSIZE	=	Length of array M.
KE(I,J)	=	Conductance matrix for element e in full matrix storage.
LCH(I)	=	$\sum_{k=1}^{I} ICH(k)$. The arrays ICH and LCH are used to modify the global system of equations for specified values of the field variable.
M(IJ)	=	Modified global conductance matrix in vector storage.

NDOF   = Number of nodes where the value of the field variable is
                  unknown.

NODETBL(I)   = Number of nodes in element type I.

NUMELM   = Number of elements in mesh.

SBW   = Semi-bandwidth of modified global conductance matrix.

X(I)   = Value of the field variable (hydraulic head or pressure head) at
                  node I.

12.5 USAGE

Subroutine ASMBK assembles the global conductance matrix [K] and the global
specified groundwater flow matrix {F}. [K] and {F} are modified to account for specified
values of the field variable (hydraulic head or pressure head) during the assembly process,
using the procedures in Chapter 4. The global conductance matrix is assembled and
modified in vector storage in the array M. The modified, global specified flow matrix is
stored in the array B. Arrays M and B can be passed to subroutines DECOMP and SOLVE
(see Chapter 13) to obtain the remaining unknown values of head.

The number of degrees of freedom (number of unknown values of the field variable),
NDOF is computed in subroutine BOUND as

$$NDOF = NUMNOD - NDN \tag{12.2}$$

where NUMNOD is the number of nodes in the mesh (Chapter 8) and NDN is the number
of nodes with specified values of the field variable (Number of Dirichlet Nodes) (Chapter
11). The semi-bandwidth, SBW for the modified system of equations is computed in
ASMBK using

$$SBW = R + 1 \tag{12.3}$$

where R is the maximum difference in node numbers for any two nodes within any element
in the mesh. However, if the value of the field variable is specified for a node, that node is
not used in the calculation of R (because the row in [K] for that node will be eliminated
when [K] is modified for the specified value of head).

The element conductance matrices are computed in a set of subroutines, one subroutine
for each element type (Table 12.1). Each subroutine name in this set begins with the letter
"K" (for the element conductance matrix $[K^{(e)}]$) followed by three or four letters that
identify the element type and the number of nodes in elements of that type. For example,
subroutine KBAR2 computes the element conductance matrix for one-dimensional, linear
bar elements and subroutine KPAR20 computes the element conductance matrix for three-
dimensional, quadratic parallelepiped elements. Subroutines KTRI3A and KREC4A
compute the element conductance matrix for two-dimensional (axisymmetric) linear triangle
and linear rectangle elements, respectively.

The source code listing for each element conductance matrix subroutine gives the
figure containing the interpolation functions and the equation used to compute $[K^{(e)}]$ for that
element type. Subroutines KBAR2, KTRI3, KREC4, and KTRI3A use analytical methods
to compute $[K^{(e)}]$ (Section 4.3.1). The rest of the subroutines use numerical methods

Table 12.1   Subroutines used to compute element conductance matrices in ASMBK.

Element Type	Description	Subroutine Name	DIM
1	Linear bar	KBAR2	1
2	Quadratic bar	KBAR3	1
3	Cubic bar	KBAR4	1
4	Linear triangle	KTRI3	2
5	Linear rectangle	KREC4	2
6	Linear quadrilateral	KQUA4	2
7	Quadratic quadrilateral	KQUA8	2
8	Cubic quadrilateral	KQUA12	2
9	Linear parallelepiped	KPAR8	3
10	Quadratic parallelepiped	KPAR20	3
11	Cubic parallelepiped	KPAR32	3
12	Linear triangle (axisymmetric)	KTRI3A*	4
13	Linear rectangle (axisymmetric)	KREC4A*	4

*Source code listing not provided for these subroutines.

(Section 4.3.2). An attempt has been made to choose FORTRAN variable names that are suggestive of the symbols used in the equations in Chapter 4. A list of the most important FORTRAN variable names and their symbols are in Table 12.2.

The operation of ASMBK is most easily explained by considering specific examples. The mesh in Figure 12.1 contains four nodes (NUMNOD = 4) and three elements (NUMELM = 3).

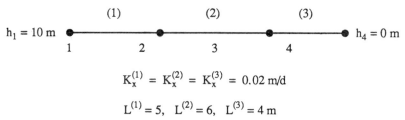

$$K_x^{(1)} = K_x^{(2)} = K_x^{(3)} = 0.02 \text{ m/d}$$

$$L^{(1)} = 5, \quad L^{(2)} = 6, \quad L^{(3)} = 4 \text{ m}$$

Figure 12.1   Example mesh for ASMBK.

The value of hydraulic head is specified at nodes one and four (ICH(1) = ICH(4) = 1, NDN = 2) and we are to compute the head at nodes two and three. NDOF = 4 - 2 = 2 and SBW = 1+1 = 2. All elements are the same type, ELEMTYP(1) = ELEMTYP(2) = ELEMTYP(3) = 1, corresponding to a linear bar element type (Table 12.1). This element type has two nodes (NODETBL(1) = 2) and the element conductance matrix for this element type are computed using subroutine KBAR2. The results are:

for element 1

$$E = 1 \qquad KE = \begin{bmatrix} 0.0040 & -0.0040 \\ -0.0040 & 0.0040 \end{bmatrix}$$

for element 2

$$E = 2 \qquad KE = \begin{bmatrix} 0.0033 & -0.0033 \\ -0.0033 & 0.0033 \end{bmatrix}$$

for element 3

$$E = 3 \qquad KE = \begin{bmatrix} 0.0050 & -0.0050 \\ -0.0050 & 0.0050 \end{bmatrix}$$

The global system of equations for this problem is

$$\begin{bmatrix} 0.0040 & -0.0040 & 0 & 0 \\ -0.0040 & 0.0073 & -0.0033 & 0 \\ 0 & -0.0033 & 0.0083 & -0.0050 \\ 0 & 0 & -0.0050 & 0.0050 \end{bmatrix} \begin{Bmatrix} h_1 = 10 \\ h_2 \\ h_3 \\ h_4 = 0 \end{Bmatrix} = \begin{Bmatrix} 0 \\ 0 \\ 0 \\ 0 \end{Bmatrix}$$

which can be modified to give

$$\begin{bmatrix} 0.0073 & -0.0033 \\ -0.0033 & 0.0083 \end{bmatrix} \begin{Bmatrix} h_2 \\ h_3 \end{Bmatrix} = \begin{Bmatrix} 0.04 \\ 0 \end{Bmatrix} \qquad (12.4)$$

Equation 12.4 is stored in three arrays in ASMBK: M, X, and B

$$M = \begin{Bmatrix} 0.0073 \\ -0.0033 \\ 0.0083 \end{Bmatrix} \qquad X = \begin{Bmatrix} h_2 \\ h_3 \end{Bmatrix} \qquad B = \begin{Bmatrix} 0.04 \\ 0 \end{Bmatrix}$$

These arrays can be passed to subroutines DECOMP and SOLVE (see Chapters 13) to obtain h_2 and h_3.

For another example consider the mesh in Figure 12.2.

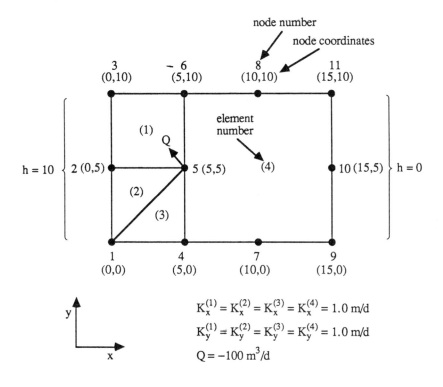

$$K_x^{(1)} = K_x^{(2)} = K_x^{(3)} = K_x^{(4)} = 1.0 \text{ m/d}$$

$$K_y^{(1)} = K_y^{(2)} = K_y^{(3)} = K_y^{(4)} = 1.0 \text{ m/d}$$

$$Q = -100 \text{ m}^3/\text{d}$$

Element Number	Node Numbers							
	1	2	3	4	5	6	7	8
1	2	5	6	3	-	-	-	-
2	1	5	2	-	-	-	-	-
3	1	4	5	-	-	-	-	-
4	4	7	9	10	11	8	6	5

Figure 12.2   Example mesh for ASMBK.

Table 12.2 Definitions for selected variables in element conductance matrix subroutines.

FORTRAN Variable or Array Name	Definition	Symbol(s) in Text
AE4	$4 \times$ (Area of element)	$4A^{(e)}$
DETJAC	Determinant of Jacobian matrix at Gauss point	$\|J(\epsilon_i)\|$ or $\|J(\epsilon_i, \eta_j)\|$ or $\|J(\epsilon_i, \eta_j, \zeta_k)\|$
DNDXI(I)	Partial derivative of interpolation function for node I with respect to ϵ at Gausspoint	$\dfrac{\partial N_i}{\partial \epsilon}(\epsilon_i)$ or $\dfrac{\partial N_i}{\partial \epsilon}(\epsilon_i, \eta_j)$ or $\dfrac{\partial N_i}{\partial \epsilon}(\epsilon_i, \eta_j, \zeta_k)$
DNETA(I)	Partial derivative of interpolation function for node I with respect to η at Gausspoint	$\dfrac{\partial N_i}{\partial \eta}(\epsilon_i, \eta_j)$ or $\dfrac{\partial N_i}{\partial \eta}(\epsilon_i, \eta_j, \zeta_k)$
DNDZETA(I)	Partial derivative of interpolation function for node I with respect to ζ at Gausspoint	$\dfrac{\partial N_i}{\partial \zeta}(\epsilon_i, \eta_j, \zeta_k)$
DNDX(I)	Partial derivative of interpolation function for node I with respect to x at Gausspoint	$\dfrac{\partial N_i}{\partial x}(\epsilon_i)$ or $\dfrac{\partial N_i}{\partial x}(\epsilon_i, \eta_j)$ or $\dfrac{\partial N_i}{\partial x}(\epsilon_i, \eta_j, \zeta_k)$

Symbol	Description	Notation
DNDY(I)	Partial derivative of interpolation function for node I with respect to y at Gausspoint	$\dfrac{\partial N_i}{\partial y}(\epsilon_i,\eta_j)$ or $\dfrac{\partial N_i}{\partial y}(\epsilon_i,\eta_j,\zeta_k)$
DNDZ(I)	Partial derivative of interpolation function for node I with respect to z at Gausspoint	$\dfrac{\partial N_i}{\partial z}(\epsilon_i,\eta_j,\zeta_k)$
JAC(I,J)	Jacobian matrix at Gauss point	$[J(\epsilon_i)]$ or $[J(\epsilon_i,\eta_j)]$ or $[J(\epsilon_i,\eta_j,\zeta_k)]$
JACINV(I,J)	Inverse of Jacobian matrix at Gauss point	$[J(\epsilon_i)^{-1}]$ or $[J(\epsilon_i,\eta_j)^{-1}]$ or $[J(\epsilon_i,\eta_j,\zeta_k)^{-1}]$
KE(I,J)	Conductance matrix for element e	$[K^{(e)}]$
KXE	Hydraulic conductivity in x direction for element e	$K_x^{(e)}$
KYE	Hydraulic conductivity in y direction for element e	$K_y^{(e)}$
KZE	Hydraulic conductivity in z direction for element e	$K_z^{(e)}$
LE	Length of element e	$L^{(e)}$
W(I)	Weight for Gauss point I	
XI(I), ETA(I), ZETA(I)	Location of Gauss points in ϵ, η, and ζ coordinate directions	$\epsilon,\ \eta,\ \zeta$
SIGN1(I),SIGN2(I), SIGN3(I)	Algebraic signs of terms in interpolation functions and derivatives for node I.	$\epsilon_i,\ \eta_j,\ \zeta_k$

The mesh contains eleven nodes (NUMNOD = 11) and four elements (NUMNOD = 4). The value of hydraulic head is specified at nodes 1, 2, 3, 9, 10, and 11 (NDN = 6) and we are to compute the head at nodes 4, 5, 6, 7, and 8. NDOF = 11 - 6 = 5 and SBW = 4 + 1 = 5, where the maximum difference in node numbers occurs in element 4 (nodes 1, 2, 3, 9, 10, and 11 are not used to calculate SBW). Element 1 is a linear rectangle (ELEMTYP(1) = 5, with NODETBL(5) = 4), elements 2 and 3 are linear triangles (ELEMTYP(2) = ELEMTYP(3) = 4, with NODETB(4) = 3), and element 4 is a quadratic quadrilateral (ELEMTYP(4) = 7, with NODETBL(7) = 8). The element conductance matrix for element 1 is computed by subroutine KREC4:

$$
[K^{(1)}] =
\begin{matrix}
& 2 & 5 & 6 & 3 \\
\begin{bmatrix}
0.666 & -0.166 & -0.333 & -0.166 \\
-0.166 & 0.666 & -0.166 & -0.333 \\
-0.333 & -0.166 & 0.666 & -0.166 \\
-0.166 & -0.333 & -0.166 & 0.666
\end{bmatrix}
& \begin{matrix} 2 \\ 5 \\ 6 \\ 3 \end{matrix}
\end{matrix}
$$

the element conductance matrices for elements 2 and 3 are computed by subroutine KTRI3:

$$
[K^{(2)}] =
\begin{matrix}
& 1 & 5 & 2 \\
\begin{bmatrix}
0.500 & 0.000 & -0.500 \\
0.000 & 0.500 & -0.500 \\
-0.500 & -0.500 & 1.000
\end{bmatrix}
& \begin{matrix} 1 \\ 5 \\ 2 \end{matrix}
\end{matrix}
$$

$$
[K^{(3)}] =
\begin{matrix}
& 1 & 4 & 5 \\
\begin{bmatrix}
0.500 & -0.500 & 0.000 \\
-0.500 & 1.000 & -0.500 \\
0.000 & -0.500 & 0.500
\end{bmatrix}
& \begin{matrix} 1 \\ 4 \\ 5 \end{matrix}
\end{matrix}
$$

and the element conductance matrix for element 4 is computed by subroutine KQUA8:

$$
[K^{(4)}] =
\begin{matrix}
& 4 & 7 & 9 & 10 & 11 & 8 & 6 & 5 \\
\begin{bmatrix}
1.555 & -0.822 & 0.500 & -0.511 & 0.511 & -0.511 & 0.500 & -0.822 \\
-0.822 & 2.311 & -0.822 & 0.000 & -0.511 & 0.355 & -0.511 & 0.000 \\
0.500 & -0.822 & 1.155 & -0.822 & 0.500 & -0.511 & 0.511 & -0.511 \\
-0.511 & 0.000 & -0.822 & 2.311 & -0.822 & 0.000 & -0.511 & 0.355 \\
0.511 & -0.511 & 0.500 & -0.822 & 1.152 & -0.822 & 0.500 & -0.511 \\
-0.511 & 0.355 & -0.511 & 0.000 & -0.822 & 2.311 & -0.822 & 0.000 \\
0.500 & -0.511 & 0.511 & -0.511 & 0.500 & -0.822 & 1.155 & -0.822 \\
-0.822 & 0.000 & -0.511 & 0.355 & -0.511 & 0.000 & -0.822 & 2.311
\end{bmatrix}
& \begin{matrix} 4 \\ 7 \\ 9 \\ 10 \\ 11 \\ 8 \\ 6 \\ 5 \end{matrix}
\end{matrix}
$$

The global system of equations for the problem is

$$
\begin{array}{c}
1\\2\\3\\4\\5\\6\\7\\8\\9\\10\\11
\end{array}
\begin{bmatrix}
1.000 & -0.500 & 0.000 & -0.500 & 0.000 & 0.000 & 0.000 & 0.000 & 0.000 & 0.000 & 0.000 \\
-0.500 & 1.666 & -0.166 & 0.000 & -0.666 & -0.333 & 0.000 & 0.000 & 0.000 & 0.000 & 0.000 \\
0.000 & -0.166 & 0.666 & 0.000 & -0.333 & -0.166 & 0.000 & 0.000 & 0.000 & 0.000 & 0.000 \\
-0.500 & 0.000 & 0.000 & 2.555 & -1.322 & 0.500 & -0.822 & -0.511 & 0.500 & -0.511 & 0.511 \\
0.000 & -0.666 & -0.333 & -1.322 & 3.977 & -0.988 & 0.000 & 0.000 & -0.511 & 0.355 & -0.511 \\
0.000 & -0.333 & -0.166 & 0.500 & -0.988 & 1.821 & -0.511 & -0.822 & 0.511 & -0.511 & 0.500 \\
0.000 & 0.000 & 0.000 & -0.822 & 0.000 & -0.511 & 2.311 & 0.355 & -0.822 & 0.000 & -0.511 \\
0.000 & 0.000 & 0.000 & -0.511 & 0.000 & -0.822 & 0.355 & 2.311 & -0.511 & 0.000 & -0.822 \\
0.000 & 0.000 & 0.000 & 0.500 & -0.511 & 0.511 & -0.822 & -0.511 & 1.155 & -0.822 & 0.500 \\
0.000 & 0.000 & 0.000 & -0.511 & 0.355 & -0.511 & 0.000 & 0.000 & -0.822 & 2.311 & -0.822 \\
0.000 & 0.000 & 0.000 & 0.511 & -0.511 & 0.500 & -0.511 & -0.822 & 0.500 & -0.822 & 1.155
\end{bmatrix}
\begin{Bmatrix}
h_1=10\\h_2=10\\h_3=10\\h_4\\h_5\\h_6\\h_7\\h_8\\h_9=0\\h_{10}=0\\h_{11}=0
\end{Bmatrix}
=
\begin{Bmatrix}
0\\0\\0\\0\\-100\\0\\0\\0\\0\\0\\0
\end{Bmatrix}
$$

Figure 12.3   Global system of equations for the mesh in Figure 12.3.

After modifying the global system for the six specified values of hydraulic head we have

$$
\begin{bmatrix}
2.555 & -1.322 & 0.500 & -0.822 & -0.511 \\
-1.322 & 3.977 & -0.988 & 0.000 & 0.000 \\
0.500 & -0.988 & 1.821 & -0.511 & -0.822 \\
-0.822 & 0.000 & -0.511 & 2.311 & 0.355 \\
-0.511 & 0.000 & -0.822 & 0.355 & 2.311
\end{bmatrix}
\begin{Bmatrix}
h_4\\h_5\\h_6\\h_7\\h_8
\end{Bmatrix}
=
\begin{Bmatrix}
-5\\-90\\0\\0\\0
\end{Bmatrix}
\qquad (12.5)
$$

Equation 12.5 is stored in arrays M, X, and B

$$
M = \begin{Bmatrix}
2.555\\-1.322\\0.500\\-0.822\\-0.511\\3.977\\-0.988\\0.000\\0.000\\1.821\\-0.511\\-0.822\\2.311\\0.355\\2.311
\end{Bmatrix}
\qquad
X = \begin{Bmatrix}
h_4\\h_5\\h_6\\h_7\\h_8
\end{Bmatrix}
\qquad
B = \begin{Bmatrix}
-5\\-90\\0\\0\\0
\end{Bmatrix}
$$

These arrays can be passed to subroutines DECOMP and SOLVE to obtain h_4, h_5, h_6, h_7 and h_8.

12.6 SOURCE CODE LISTING

```
      SUBROUTINE ASMBK
C**********************************************************************
C
C 12.1  PURPOSE:
C          SUBROUTINE ASMBK ASSEMBLES THE GLOBAL CONDUCTANCE MATRIX
C          AND THE GLOBAL SPECIFIED FLOW MATRIX.  THE GLOBAL MATRICES
C          ARE MODIFIED DURING THE ASSEMBLY PROCESS TO ACCOUNT FOR
C          SPECIFIED VALUES OF THE FIELD VARIABLE AND GROUNDWATER
C          FLOW DURING THE ASSEMBLY PROCESS. THE GLOBAL CONDUCTANCE
C          MATRIX IS ASSEMBLED AND MODIFIED IN VECTOR STORAGE.  ASMBK
C          ALSO COMPUTES THE SEMI-BANDWIDTH AND THE NUMBER OF DEGREES
C          OF FREEDOM FOR THE MODIFIED GLOBAL CONDUCTANCE MATRIX.
C
C 12.2  INPUT:
C          NONE
C
C 12.3  OUTPUT:
C          THE SEMI-BANDWIDTH AND NUMBER OF DEGREES OF FREEDOM FOR
C          THE MODIFIED GLOBAL CONDUCTANCE MATRIX ARE WRITTEN TO THE
C          USER-DEFINED FILE ASSIGNED TO UNIT "OUTF"
C
C 12.4  DEFINITIONS OF VARIABLES:
C               B(I) = MODIFIED SPECIFIED FLOW MATRIX
C                  E = ELEMENT NUMBER
C          ELEMTYP(I) = ELEMENT TYPE FOR ELEMENT I (SEE TABLE 9.1
C                       FOR A LIST OF ELEMENT TYPES)
C            FLUX(I) = SPECIFIED VALUE OF GROUNDWATER FLOW
C                       AT NODE I
C             ICH(I) = 1 IF THE VALUE OF THE FIELD VARIABLE IS
C                       SPECIFIED FOR NODE I,
C                    = 0 OTHERWISE
C             IJSIZE = LENGTH OF ARRAY M
C            KE(I,J) = CONDUCTANCE MATRIX FOR ELEMENT E IN
C                       FULL MATRIX STORAGE
C             LCH(I) = ICH(I) + ICH(I-1) + ICH(I-2) + ...
C                       THE ARRAYS ICH AND LCH ARE USED TO MODIFY
C                       GLOBAL SYSTEM OF EQUATIONS FOR SPECIFIED
C                       VALUES OF THE FIELD VARIABLES
C              M(IJ) = MODIFIED GLOBAL CONDUCTANCE MATRIX
C                       IN VECTOR STORAGE
C               NDOF = NUMBER OF NODES WHERE THE VALUE OF THE
C                       FIELD VARIABLE IS UNKNOWN
C          NODETBL(I) = NUMBER OF NODES IN ELEMENT TYPE I
C             NUMELM = NUMBER OF ELEMENTS IN MESH
C                SBW = SEMI-BANDWIDTH OF MODIFIED GLOBAL
C                       CONDUCTANCE MATRIX
C               X(I) = VALUE OF THE FIELD VARIABLE (HYDRAULIC
C                       HEAD OR PRESSURE HEAD) AT NODE I
C
C 12.5  USAGE:
C          THE SEMI-BANDWIDTH OF THE GLOBAL CONDUCTANCE MATRIX
C          IS COMPUTED FIRST.  THEN THE ENTRIES OF THE ELEMENT
C          CONDUCTANCE MATRIX ARE COMPUTED IN A SET OF SUBROUTINES,
C          ONE SUBROUTINE FOR EACH ELEMENT TYPE.  THE GLOBAL
C          CONDUCTANCE MATRIX FOR THE MESH IS ASSEMBLED BY ADDING
C          THE CORRESPONDING ENTRIES OF THE ELEMENT CONDUCTANCE
C          MATRICES TO THE GLOBAL CONDUCTANCE MATRIX.  DURING THE
C          ASSEMBLY PROCESS THE GLOBAL CONDUCTANCE MATRIX IS MODIFIED
```

```
C            FOR SPECIFIED VALUES OF HEAD.  SPECIFIED VALUES OF
C            GROUNDWATER FLOW ARE ADDED TO THE GLOBAL FLOW MATRIX.
C
C         SUBROUTINES CALLED:
C            KBAR2,KBAR3,KBAR4,KTRI3,KREC4,KQUA4,KQUA8,KQUA12,KPAR8,
C            KPAR20,KBAR32,KTRI3A,KREC4A
C            LOC
C
C         REFERENCES:
C            ISTOK,J.D. GROUNDWATER FLOW AND SOLUTE TRANSPORT
C            MODELING BY THE FINITE ELEMENT METHOD, CHAPTER 12.
C
C****************************************************************
      $INCLUDE:'COMALL'
            REAL KE(MAX3,MAX3)
            INTEGER NODETBL(13)
            DATA NODETBL/2,3,4,3,4,4,8,12,8,20,32,3,4/
C
C         COMPUTE THE SEMI-BANDWIDTH
C
            SBW = 1
            DO 30 E = 1, NUMELM
              DO 20 I = 1, NODETBL(ELEMTYP(E))
                KI = IN(E,I)
                IF (ICH(KI) .EQ. 0 .AND. I .LT. NODETBL(ELEMTYP(E))) THEN
                  II = KI - LCH(KI)
                  DO 10 J = I + 1, NODETBL(ELEMTYP(E))
                    KJ = IN(E,J)
                    IF (ICH(KJ) .EQ. 0) THEN
                      JJ = ABS(KJ - LCH(KJ) - II) + 1
                      IF (JJ .GT. SBW) SBW = JJ
                    ENDIF
 10               CONTINUE
                ENDIF
 20           CONTINUE
 30         CONTINUE
            WRITE(OUTF,40) NDOF,SBW
 40         FORMAT(//' NUMBER OF DEGREES OF FREEDOM IN MODIFIED K MATRIX =',
          1          I5///' SEMI-BANDWIDTH OF MODIFIED K MATRIX =',I5)
            IF (SBW .GT. MAX6) STOP'** EXCEEDS MAXIMUM SEMI-BAND WIDTH **'
C           INITIALIZE ENTRIES OF GLOBAL CONDUCTANCE MATRIX TO ZERO
            IJSIZE = SBW * (NDOF - SBW + 1) + (SBW - 1) * SBW / 2
            DO 50 IJ = 1, IJSIZE
              M(IJ) = 0.0
 50         CONTINUE
            DO 60 I = 1, NUMNOD
              IF (ICH(I) .EQ. 0) B(I-LCH(I)) = FLUX(I)
 60         CONTINUE
C           LOOP ON THE NUMBER OF ELEMENTS
            DO 90 E = 1, NUMELM
C             COMPUTE THE ELEMENT CONDUCTANCE MATRIX FOR THIS ELEMENT TYPE
              IF (ELEMTYP(E) .EQ. 1) THEN
C               ELEMENT IS A ONE-DIMENSIONAL, LINEAR BAR
                CALL KBAR2(E,KE)
              ELSEIF (ELEMTYP(E) .EQ .2) THEN
C               ELEMENT IS A ONE-DIMENSIONAL, QUADRATIC BAR
                CALL KBAR3(E,KE)
              ELSEIF (ELEMTYP(E) .EQ. 3) THEN
C               ELEMENT IS A ONE-DIMENSIONAL, CUBIC BAR
                CALL KBAR4(E,KE)
              ELSEIF (ELEMTYP(E) .EQ. 4) THEN
```

```
C             ELEMENT IS A TWO-DIMENSIONAL, LINEAR TRIANGLE
              CALL KTRI3(E,KE)
           ELSEIF (ELEMTYP(E) .EQ. 5) THEN
C             ELEMENT IS A TWO-DIMENSIONAL, LINEAR RECTANGLE
              CALL KREC4(E,KE)
           ELSEIF (ELEMTYP(E) .EQ. 6) THEN
C             ELEMENT IS A TWO-DIMENSIONAL, LINEAR QUADRILATERAL
              CALL KQUA4(E,KE)
           ELSEIF (ELEMTYP(E) .EQ. 7) THEN
C             ELEMENT IS A TWO-DIMENSIONAL, QUADRATIC QUADRILATERAL
              CALL KQUA8(E,KE)
           ELSEIF (ELEMTYP(E) .EQ. 8) THEN
C             ELEMENT IS A TWO-DIMENSIONAL, CUBIC QUADRILATERAL
              CALL KQUA12(E,KE)
           ELSEIF (ELEMTYP(E) .EQ. 9) THEN
C             ELEMENT IS A THREE-DIMENSIONAL, LINEAR PARALLELEPIPED
              CALL KPAR8(E,KE)
           ELSEIF (ELEMTYP(E) .EQ. 10) THEN
C             ELEMENT IS A THREE-DIMENSIONAL, QUADRATIC PARALLELEPIPED
              CALL KPAR20(E,KE)
           ELSEIF (ELEMTYP(E) .EQ. 11) THEN
C             ELEMENT IS A THREE-DIMENSIONAL, CUBIC PARALLELEPIPED
              CALL KPAR32(E,KE)
           ELSEIF (ELEMTYP(E) .EQ. 12) THEN
C             ELEMENT IS A TWO-DIMENSIONAL, LINEAR TRIANGLE (AXISYMMETRIC)
              CALL KTRI3A(E,KE)
           ELSEIF (ELEMTYP(E) .EQ. 13) THEN
C             ELEMENT IS A TWO-DIMENSIONAL, LINEAR RECTANGLE
(AXISYMMETRIC)
              CALL KREC4A(E,KE)
           ENDIF

C
C         ADD THE ELEMENT CONDUCTANCE MATRIX FOR THIS ELEMENT
C         TO THE GLOBAL CONDUCTANCE MATRIX
C         KE(I,J) ----------->     M(IJ)        <=>          M(KI,KJ)
C (FULL MATRIX STORAGE)   (VECTOR MATRIX STORAGE)      (FULL MATRIX
STORAGE)
C
           DO 80 I = 1, NODETBL(ELEMTYP(E))
             KI = IN(E,I)
             IF (ICH(KI) .EQ. 0) THEN
               II = KI - LCH(KI)
               DO 70 J = 1, NODETBL(ELEMTYP(E))
                 KJ = IN(E,J)
                 IF (ICH(KJ) .NE. 0) THEN
                   B(II) = B(II) - KE(I,J) * X(KJ)
                 ELSEIF (J .GE. I) THEN
                   JJ = KJ - LCH(KJ)
                   CALL LOC(II,JJ,IJ,NDOF,SBW,SYMM)
                   M(IJ) = M(IJ) + KE(I,J)
                 ENDIF
70             CONTINUE
             ENDIF
80         CONTINUE
90       CONTINUE

         RETURN
         END
```

```
      SUBROUTINE LOC(I,J,IJ,NDOF,SBW,SYMM)
C***********************************************************************
C      PURPOSE:
C        SUBROUTINE LOC COMPUTES THE LOCATION IN VECTOR STORAGE
C        OF A SPECIFIED ROW AND COLUMN OF A MATRIX (SYMMETRIC OR
C        NONSYMMETRIC) IN FULL MATRIX STORAGE
C
C      DEFINITIONS OF VARIABLES:
C          I = SPECIFIED ROW OF MATRIX IN FULL MATRIX STORAGE
C          J = SPECIFIED COLUMN OF MATRIX IN FULL MATRIX
C              STORAGE
C         IJ = LOCATION IN VECTOR STORAGE CORRESPONDING TO
C              SPECIFIED ROW AND COLUMN IN FULL MATRIX STORAGE
C       NDOF = NUMBER OF DEGREES OF FREEDOM OF MATRIX
C        SBW = SEMI-BANDWIDTH OF MATRIX
C
C      REFERENCES:
C       ISTOK,J.D. GROUNDWATER FLOW AND SOLUTE TRANSPORT
C       MODELING BY THE FINITE ELEMENT METHOD, CHAPTER 5,
C       SECTIONS 5.1.3 AND 5.1.4
C***********************************************************************
      INTEGER SBW
      LOGICAL SYMM
C
      IF (SYMM) THEN
C        M IS A SYMMETRIC MATRIX
         II = I
         JJ = J
         IF (I .GT. J) THEN
           K = I
           I = J
           J = K
         ENDIF
         IJ = J - I + 1
         IF (I .GT. 1) THEN
           IF (SBW .LT. NDOF) THEN
             IJ = IJ + (I - 1) * SBW
             L = I - NDOF + SBW - 2
             IF (L .GT. 0) IJ = IJ - L * (L + 1) / 2
           ELSE
             IJ = IJ + (I - 1) * (NDOF + (NDOF - I + 2)) / 2
           ENDIF
         ENDIF
         I = II
         J = JJ
      ELSE
C        M IS A NONSYMMETRIC MATRIX
         IJ = J
         IF (I .GT. 1) THEN
           IF (SBW .LT. NDOF) THEN
             IF (I .GT. SBW) IJ = IJ + SBW - I
             IJ = IJ + (I - 1) * (2 * SBW - 1)
             L = MIN(SBW,I) - 1
             IJ = IJ - L * ((SBW - 1) + (SBW - L)) / 2
             L = I - NDOF + SBW -2
             IF (L .GT. 0) IJ = IJ - L * (L + 1) / 2
           ELSE
             IJ = IJ + (I - 1) * NDOF
           ENDIF
         ENDIF
      ENDIF
      RETURN
      END
```

```
       SUBROUTINE KBAR2(E,KE)
C**********************************************************************
C
C      PURPOSE:
C         TO COMPUTE THE ELEMENT CONDUCTANCE MATRIX FOR A
C         ONE-DIMENSIONAL, LINEAR BAR ELEMENT
C
C      DEFINITIONS OF VARIABLES:
C           E = ELEMENT NUMBER
C       KE(I,J) = ELEMENT CONDUCTANCE MATRIX
C          KXE = HYDRAULIC CONDUCTIVITY IN X COORDINATE DIRECTION
C           LE = ELEMENT LENGTH
C
C      REFERENCES:
C         ISTOK,J.D. GROUNDWATER FLOW AND SOLUTE TRANSPORT
C         MODELING BY THE FINITE ELEMENT METHOD, CHAPTER 4,
C         FIGURE 4.5, EQUATION 4.15.
C
C**********************************************************************
$INCLUDE:'COMALL'
       REAL KE(MAX3,MAX3),KXE,LE
C
       KXE = PROP(MATSET(E),1)
       LE = ABS(X1(IN(E,2)) - X1(IN(E,1)))
       KE(1,1) =  KXE / LE
       KE(1,2) = -KE(1,1)
       KE(2,1) = -KE(1,1)
       KE(2,2) =  KE(1,1)
       RETURN
       END




       SUBROUTINE KBAR3(E,KE)
C**********************************************************************
C
C      PURPOSE:
C         TO COMPUTE THE ELEMENT CONDUCTANCE MATRIX FOR ONE-
C         DIMENSIONAL, QUADRATIC BAR ELEMENT
C
C      DEFINITIONS OF VARIABLES:
C         DETJAC = DETERMINANT OF JACOBIAN MATRIX
C        DNDXI(I) = PARTIAL DERIVATIVE OF INTERPOLATION
C                   FUNCTION WITH RESPECT TO XI AT NODE I
C         DNDX(I) = PARTIAL DERIVATIVE OF INTERPOLATION
C                   FUNCTION WITH RESPECT TO X AT NODE I
C           E = ELEMENT NUMBER
C          JAC = JACOBIAN MATRIX
C        JACINV = INVERSE OF JACOBIAN MATRIX
C       KE(I,J) = ELEMENT CONDUCTANCE MATRIX
C          KXE = HYDRAULIC CONDUCTIVITY
C                   IN X COORDINATE DIRECTION
C          W(I) = WEIGHT FOR GAUSS POINT I
C         XI(I) = LOCATION OF GAUSS POINT I
C                   IN XI COORDINATE DIRECTION
C
C      REFERENCES:
C         ISTOK,J.D. GROUNDWATER FLOW AND SOLUTE TRANSPORT
C         MODELING BY THE FINITE ELEMENT METHOD, CHAPTER 4,
C         FIGURE 4.9, EQUATION 4.61
C
C**********************************************************************
```

```
$INCLUDE: 'COMALL'
      REAL JAC,JACINV,KE(MAX3,MAX3),DNDXI(3),DNDX(3),W(2),
     1      XI(2),KXE
C
      XI(1) =  1 / SQRT(3.)
      XI(2) = -XI(1)
      W(1)= 1.
      W(2)= 1.
      KXE = PROP(MATSET(E),1)
      DO 30 I = 1, 3
        DO 20 J = 1, 3
          KE(I,J) = 0.
  20    CONTINUE
  30 CONTINUE
      DO 80 I = 1, 2
        DNDXI(1)= -0.5 + XI(I)
        DNDXI(2)= -2.0 * XI(I)
        DNDXI(3)=  0.5 + XI(I)
        JAC = 0
        DO 40 J = 1, 3
          JAC = JAC + DNDXI(J) * X1(IN(E,J))
  40    CONTINUE
        JACINV = 1 / JAC
        DETJAC = JAC
        DO 50 J = 1, 3
          DNDX(J) = JACINV * DNDXI(J)
  50    CONTINUE
        DO 70 J = 1, 3
          DO 60 K = 1, 3
            KE(J,K) = KE(J,K) + W(I) * KXE * DNDX(J)
     1                 * DNDX(K) * DETJAC
  60      CONTINUE
  70    CONTINUE
  80 CONTINUE
      RETURN
      END




      SUBROUTINE KBAR4(E,KE)
C*********************************************************************
C
C     PURPOSE:
C        TO COMPUTE THE ELEMENT CONDUCTANCE MATRIX FOR ONE-
C        DIMENSIONAL, CUBIC BAR ELEMENT
C
C     DEFINITIONS OF VARIABLES:
C        DETJAC = DETERMINANT OF JACOBIAN MATRIX
C        DNDXI(I) = PARTIAL DERIVATIVE OF INTERPOLATION
C                   FUNCTION WITH RESPECT TO XI AT NODE I
C        DNDX(I) = PARTIAL DERIVATIVE OF INTERPOLATION
C                   FUNCTION WITH RESPECT TO X AT NODE I
C             E = ELEMENT NUMBER
C           JAC = JACOBIAN MATRIX
C        JACINV = INVERSE OF JACOBIAN MATRIX
C        KE(I,J) = ELEMENT CONDUCTANCE MATRIX
C           KXE = HYDRAULIC CONDUCTIVITY IN X
C                   COORDINATE DIRECTION
C          W(I) = WEIGHT FOR GAUSS POINT I
C         XI(I) = LOCATION OF GAUSS POINT I IN
C                   XI COORDINATE DIRECTION
```

```
C
C         REFERENCES:
C            ISTOK,J.D. GROUNDWATER FLOW AND SOLUTE TRANSPORT
C            MODELING BY THE FINITE ELEMENT METHOD, CHAPTER 4,
C            FIGURE 4.9B, EQUATION 4.61
C
C************************************************************************
$INCLUDE: 'COMALL'
      REAL JAC,JACINV,KE(MAX3,MAX3),DNDXI(4),DNDX(4),W(3),
     1      XI(3),KXE
C
      XI(1) =  0.
      XI(2) =  SQRT(3. / 5.)
      XI(3) = -XI(2)
      W(1) = 8. / 9.
      W(2) = 5. / 9.
      W(3) = W(2)
      KXE = PROP(MATSET(E),1)
      DO 30 I = 1, 4
        DO 20 J = 1, 4
          KE(I,J) = 0.
   20   CONTINUE
   30 CONTINUE
      DO 80 I = 1, 3
        DNDXI(1) = -(9. / 16.) * (3. * (XI(I)**2) -
     1                2. * XI(I) - 1. / 9.)
        DNDXI(2) =  (27. / 16.) * (3. * (XI(I)**2) -
     1                (2. / 3.) * XI(I) - 1.)
        DNDXI(3) = -(27. / 16.) * (3. * (XI(I)**2) +
     1                (2. / 3.) * XI(I) - 1.)
        DNDXI(4) =  (9. / 16.) * (3. * (XI(I)**2) +
     1                2. * XI(I) - 1. / 9.)
        JAC = 0
        DO 40 J = 1, 4
          JAC = JAC + DNDXI(J) * X1(IN(E,J))
   40   CONTINUE
        JACINV = 1 / JAC
        DETJAC = JAC
        DO 50 J = 1, 4
          DNDX(J) = JACINV * DNDXI(J)
   50   CONTINUE
        DO 70 J = 1, 4
          DO 60 K = 1, 4
            KE(J,K) = KE(J,K) + W(I) * KXE * DNDX(J)
     1                  * DNDX(K) * DETJAC
   60     CONTINUE
   70   CONTINUE
   80 CONTINUE
      RETURN
      END


      SUBROUTINE KTRI3(E,KE)
C************************************************************************
C
C         PURPOSE:
C            TO COMPUTE THE ELEMENT CONDUCTANCE MATRIX FOR TWO-
C            DIMENSIONAL, LINEAR TRIANGLE ELEMENT
C
C         DEFINITIONS OF VARIABLES:
C              AE4 = FOUR TIMES ELEMENT AREA
C                E = ELEMENT NUMBER
```

```
C                KE(I,J) = ELEMENT CONDUCTANCE MATRIX
C                    KXE = HYDRAULIC CONDUCTIVITY IN X
C                          COORDINATE DIRECTION
C                    KYE = HYDRAULIC CONDUCTIVITY IN Y
C                          COORDINATE DIRECTION
C
C          REFERENCES:
C            ISTOK,J.D. GROUNDWATER FLOW AND SOLUTE TRANSPORT
C            MODELING BY THE FINITE ELEMENT METHOD, CHAPTER 4,
C            FIGURE 4.6, EQUATION 4.20
C
C****************************************************************************

$INCLUDE: 'COMALL'
      REAL KE(MAX3,MAX3),KXE,KYE,BE(3),CE(3)
C
      KXE = PROP(MATSET(E),1)
      KYE = PROP(MATSET(E),2)
      BE(1) = X2(IN(E,2)) - X2(IN(E,3))
      BE(2) = X2(IN(E,3)) - X2(IN(E,1))
      BE(3) = X2(IN(E,1)) - X2(IN(E,2))
      CE(1) = X1(IN(E,3)) - X1(IN(E,2))
      CE(2) = X1(IN(E,1)) - X1(IN(E,3))
      CE(3) = X1(IN(E,2)) - X1(IN(E,1))
      AE4 = 2 * (X1(IN(E,2)) * X2(IN(E,3)) + X1(IN(E,1)) *
     1      X2(IN(E,2)) + X2(IN(E,1)) * X1(IN(E,3)) -
     2      X2(IN(E,3)) * X1(IN(E,1)) - X1(IN(E,3)) *
     3      X2(IN(E,2)) - X1(IN(E,2)) * X2(IN(E,1)))
      DO 20 I = 1, 3
        DO 10 J = 1, 3
          KE(I,J) = (KXE * BE(I) * BE(J) + KYE * CE(I) * CE(J)) / AE4
   10   CONTINUE
   20 CONTINUE
      RETURN
      END




      SUBROUTINE KREC4(E,KE)
C****************************************************************************
C
C          PURPOSE:
C            TO COMPUTE THE ELEMENT CONDUCTANCE MATRIX FOR TWO-
C            DIMENSIONAL, LINEAR RECTANGLE ELEMENT
C
C          DEFINITIONS OF VARIABLES:
C                      E = ELEMENT NUMBER
C                KE(I,J) = ELEMENT CONDUCTANCE MATRIX
C                    KXE = HYDRAULIC CONDUCTIVITY IN X
C                          COORDINATE DIRECTION
C                    KYE = HYDRAULIC CONDUCTIVITY IN Y
C                          COORDINATE DIRECTION
C
C          REFERENCES:
C            ISTOK,J.D. GROUNDWATER FLOW AND SOLUTE TRANSPORT
C            MODELING BY THE FINITE ELEMENT METHOD, CHAPTER 4,
C            FIGURE 4.7, EQUATION 4.26
C
C****************************************************************************
```

```
$INCLUDE: 'COMALL'
      REAL KE(MAX3,MAX3),KXE,KYE
C
      KXE = PROP(MATSET(E),1)
      KYE = PROP(MATSET(E),2)
      AE = ABS(X2(IN(E,1))-X2(IN(E,3))) / 2.
      BE = ABS(X1(IN(E,1)) - X1(IN(E,3))) / 2.
      CX = KXE * AE / (6. * BE)
      CY = KYE * BE / (6. * AE)
      KE(1,1) =  2. * CX + 2. * CY
      KE(1,2) = -2. * CX + CY
      KE(1,3) = -CX - CY
      KE(1,4) =  CX - 2. * CY
      KE(2,1) = KE(1,2)
      KE(2,2) =  2. * CX + 2. * CY
      KE(2,3) =  CX - 2. * CY
      KE(2,4) = -CX - CY
      KE(3,1) = KE(1,3)
      KE(3,2) = KE(2,3)
      KE(3,3) =  2. * CX + 2. * CY
      KE(3,4) = -2. * CX + CY
      KE(4,1) = KE(1,4)
      KE(4,2) = KE(2,4)
      KE(4,3) = KE(3,4)
      KE(4,4) = 2. * CX + 2. * CY
      RETURN
      END


      SUBROUTINE KQUA4(E,KE)
C*******************************************************************
C
C        PURPOSE:
C          TO COMPUTE THE ELEMENT CONDUCTANCE MATRIX FOR TWO-
C          DIMENSIONAL, LINEAR QUADRILATERAL ELEMENT
C
C        DEFINITIONS OF VARIABLES:
C              DETJAC = DETERMINANT OF JACOBIAN MATRIX
C            DNDXI(I) = PARTIAL DERIVATIVE OF INTERPOLATION
C                       FUNCTION WITH RESPECT TO XI AT NODE I
C             DNDX(I) = PARTIAL DERIVATIVE OF INTERPOLATION
C                       FUNCTION WITH RESPECT TO X AT NODE I
C           DNDETA(I) = PARTIAL DERIVATIVE OF INTERPOLATION
C                       FUNCTION WITH RESPECT TO ETA AT NODE I
C             DNDY(I) = PARTIAL DERIVATIVE OF INTERPOLATION
C                       FUNCTION WITH RESPECT TO Y AT NODE I
C                   E = ELEMENT NUMBER
C              ETA(I) = LOCATION OF GAUSS POINT IN ETA COORDINATE
C                       DIRECTION
C            JAC(I,J) = JACOBIAN MATRIX
C         JACINV(I,J) = INVERSE OF JACOBIAN MATRIX
C             KE(I,J) = ELEMENT CONDUCTANCE MATRIX
C                 KXE = HYDRAULIC CONDUCTIVITY IN X
C                       COORDINATE DIRECTION
C                 KYE = HYDRAULIC CONDUCTIVITY IN Y
C                       COORDINATE DIRECTION
C                W(I) = WEIGHT FOR GAUSS POINT I
C          X1(IN(E,I)) = X COORDINATE FOR NODE I, ELEMENT E
C          X2(IN(E,I)) = Y COORDINATE FOR NODE I, ELEMENT E
C               XI(I) = LOCATION OF GAUSS POINT IN XI COORDINATE
C                       ·DIRECTION
C
```

```
C          REFERENCES:
C               ISTOK,J.D. GROUNDWATER FLOW AND SOLUTE TRANSPORT
C               MODELING BY THE FINITE ELEMENT METHOD, FIGURE 4.10,
C               EQUATION 4.62
C
C********************************************************************
$INCLUDE: 'COMALL'
      REAL JAC(2,2),JACINV(2,2),KE(MAX3,MAX3),DNDXI(4),DNDX(4),
     1     DNDETA(4),DNDY(4),W(2),XI(2),ETA(2),SIGN1(4),SIGN2(4),
     2     KXE,KYE
      DATA SIGN1/-1.,1.,1.,-1./
      DATA SIGN2/-1.,-1.,1.,1./
C
      XI(1) =  1. / SQRT(3.)
      XI(2) = -XI(1)
      ETA(1) = XI(1)
      ETA(2) = XI(2)
      W(1) = 1.
      W(2) = 1.
      KXE = PROP(MATSET(E),1)
      KYE = PROP(MATSET(E),2)

      DO 30 K = 1, 4
        DO 20 N = 1, 4
          KE(K,N) = 0.
20      CONTINUE
30    CONTINUE

      DO 120 I = 1, 2
        DO 110 J = 1, 2

          DO 50 K = 1, 2
            DO 40 N = 1, 2
              JAC(K,N) = 0.
40        CONTINUE
50      CONTINUE

          DO 60 N = 1, 4
            DNDXI(N)  = 0.25 * SIGN1(N) * (1. + SIGN2(N) * ETA(J))
            DNDETA(N) = 0.25 * SIGN2(N) * (1. + SIGN1(N) * XI(I))
60        CONTINUE
          DO 70 N = 1, 4
            JAC(1,1) = JAC(1,1) + DNDXI(N) * X1(IN(E,N))
            JAC(1,2) = JAC(1,2) + DNDXI(N) * X2(IN(E,N))
            JAC(2,1) = JAC(2,1) + DNDETA(N) * X1(IN(E,N))
            JAC(2,2) = JAC(2,2) + DNDETA(N) * X2(IN(E,N))
70        CONTINUE
          DETJAC = JAC(1,1) * JAC(2,2) - JAC(1,2) * JAC(2,1)
          JACINV(1,1) =  JAC(2,2) / DETJAC
          JACINV(1,2) = -JAC(1,2) / DETJAC
          JACINV(2,1) = -JAC(2,1) / DETJAC
          JACINV(2,2) =  JAC(1,1) / DETJAC
          DO 80 N = 1, 4
            DNDX(N) = JACINV(1,1) * DNDXI(N) + JACINV(1,2) * DNDETA(N)
            DNDY(N) = JACINV(2,1) * DNDXI(N) + JACINV(2,2) * DNDETA(N)
80        CONTINUE
          DO 100 K = 1, 4
            DO 90 N = 1, 4
              KE(K,N) = KE(K,N) + W(I) * W(J) * (KXE * DNDX(K) *
     1                  DNDX(N) + KYE * DNDY(K) * DNDY(N)) * DETJAC
90        CONTINUE
```

```
100     CONTINUE
110     CONTINUE
120   CONTINUE
      RETURN
      END



      SUBROUTINE KQUA8(E,KE)
C*********************************************************************
C
C       PURPOSE:
C         TO COMPUTE THE ELEMENT CONDUCTANCE MATRIX FOR TWO-
C         DIMENSIONAL, QUADRATIC QUADRILATERAL ELEMENT
C
C       DEFINITIONS OF VARIABLES:
C             DETJAC = DETERMINANT OF JACOBIAN MATRIX
C           DNDXI(I) = PARTIAL DERIVATIVE OF INTERPOLATION
C                      FUNCTION WITH RESPECT TO XI AT NODE I
C            DNDX(I) = PARTIAL DERIVATIVE OF INTERPOLATION
C                      FUNCTION WITH RESPECT TO X AT NODE I
C          DNDETA(I) = PARTIAL DERIVATIVE OF INTERPOLATION
C                      FUNCTION WITH RESPECT TO ETA AT NODE I
C            DNDY(I) = PARTIAL DERIVATIVE OF INTERPOLATION
C                      FUNCTION WITH RESPECT TO Y AT NODE I
C                  E = ELEMENT NUMBER
C             ETA(I) = LOCATION OF GAUSS POINT I IN ETA
C                      COORDINATE DIRECTION
C            JAC(I,J) = JACOBIAN MATRIX
C         JACINV(I,J) = INVERSE OF JACOBIAN MATRIX
C             KE(I,J) = ELEMENT CONDUCTANCE MATRIX
C                 KXE = HYDRAULIC CONDUCTIVITY IN X
C                      COORDINATE DIRECTION
C                 KYE = HYDRAULIC CONDUCTIVITY IN Y
C                      COORDINATE DIRECTION
C                W(I) = WEIGHT FOR GAUSS POINT I
C               XI(I) = LOCATION OF GAUSS POINT I IN XI
C                      COORDINATE DIRECTION
C
C       REFERENCES:
C         ISTOK,J.D. GROUNDWATER FLOW AND SOLUTE TRANSPORT
C         MODELING BY THE FINITE ELEMENT METHOD, FIGURE 4.11,
C         EQUATION 4.62
C
C*********************************************************************
$INCLUDE: 'COMALL'
      REAL JAC(2,2),JACINV(2,2),KE(MAX3,MAX3),DNDXI(8),
     1     DNDX(8),DNDETA(8),DNDY(8),W(3),XI(8),ETA(8),
     2     SIGN1(8),SIGN2(8),KXE,KYE
      DATA SIGN1/-1.,0.,1.,1.,1.,0.,-1.,-1./
      DATA SIGN2/-1.,-1.,-1.,0.,1.,1.,1.,0./
C
      XI(1) =  0.
      XI(2) =  SQRT(3. / 5.)
      XI(3) = -XI(2)
      ETA(1) = XI(1)
      ETA(2) = XI(2)
      ETA(3) = XI(3)
```

```
      W(1) = 8. / 9.
      W(2) = 5. / 9.
      W(3) = W(2)
      KXE = PROP(MATSET(E),1)
      KYE = PROP(MATSET(E),2)
      DO 30 K = 1, 8
        DO 20 N = 1, 8
          KE(K,N) = 0.
20      CONTINUE
30    CONTINUE
      DO 120 I = 1, 3
        DO 110 J = 1, 3
          DO 50 K = 1, 2
            DO 40 N = 1, 2
              JAC(K,N) = 0.
40          CONTINUE
50        CONTINUE
          DO 60 N = 1, 8
            IF ((N .EQ. 1) .OR. (N .EQ. 3) .OR.
     1         (N .EQ. 5) .OR. (N .EQ. 7)) THEN
               DNDXI(N) = 0.25 * (1. + SIGN2(N) * ETA(J)) *
     1                    (2. * SIGN1(N)**2 * XI(I) + SIGN2(N) *
     2                    SIGN1(N) * ETA(J))
               DNDETA(N) = 0.25 * (1. + SIGN1(N) * XI(I)) *
     1                     (2. * SIGN2(N)**2 * ETA(J) + SIGN2(N) *
     2                     SIGN1(N) * XI(I))
            ELSEIF ((N .EQ. 2) .OR. (N .EQ. 6)) THEN
               DNDXI(N) = -XI(I) * (1. + SIGN2(N) * ETA(J))
               DNDETA(N) =  0.5 * SIGN2(N) * (1. - XI(I)**2)
            ELSEIF ((N .EQ. 4) .OR. (N .EQ. 8)) THEN
               DNDXI(N) =  0.5 * SIGN1(N) * (1. - ETA(J)**2)
               DNDETA(N) = -ETA(J) * (1. + SIGN1(N) * XI(I))
            ENDIF
60        CONTINUE
          DO 70 N = 1, 8
            JAC(1,1) = JAC(1,1) + DNDXI(N) * X1(IN(E,N))
            JAC(1,2) = JAC(1,2) + DNDXI(N) * X2(IN(E,N))
            JAC(2,1) = JAC(2,1) + DNDETA(N) * X1(IN(E,N))
            JAC(2,2) = JAC(2,2) + DNDETA(N) * X2(IN(E,N))
70        CONTINUE
          DETJAC = JAC(1,1) * JAC(2,2) - JAC(1,2) * JAC(2,1)
          JACINV(1,1) =  JAC(2,2) / DETJAC
          JACINV(1,2) = -JAC(1,2) / DETJAC
          JACINV(2,1) = -JAC(2,1) / DETJAC
          JACINV(2,2) =  JAC(1,1) / DETJAC
          DO 80 N = 1, 8
            DNDX(N) = JACINV(1,1) * DNDXI(N) + JACINV(1,2) * DNDETA(N)
            DNDY(N) = JACINV(2,1) * DNDXI(N) + JACINV(2,2) * DNDETA(N)
80        CONTINUE
          DO 100 K = 1, 8
            DO 90 N = 1, 8
              KE(K,N) = KE(K,N) + W(I) * W(J) * (KXE * DNDX(K) *
     1                  DNDX(N) + KYE * DNDY(K) * DNDY(N)) * DETJAC
90          CONTINUE
100       CONTINUE
110     CONTINUE
120   CONTINUE
      RETURN
      END
```

```
      SUBROUTINE KQUA12(E,KE)
C*********************************************************************
C
C     PURPOSE:
C       TO COMPUTE THE ELEMENT CONDUCTANCE MATRIX FOR TWO-
C       DIMENSIONAL, CUBIC QUADRILATERAL ELEMENT
C
C     DEFINITIONS OF VARIABLES:
C           DETJAC = DETERMINANT OF JACOBIAN MATRIX
C         DNDXI(I) = PARTIAL DERIVATIVE OF INTERPOLATION
C                    FUNCTION WITH RESPECT TO XI AT NODE I
C          DNDX(I) = PARTIAL DERIVATIVE OF INTERPOLATION
C                    FUNCTION WITH RESPECT TO X AT NODE I
C        DNDETA(I) = PARTIAL DERIVATIVE OF INTERPOLATION
C                    FUNCTION WITH RESPECT TO ETA AT NODE I
C          DNDY(I) = PARTIAL DERIVATIVE OF INTERPOLATION
C                    FUNCTION WITH RESPECT TO Y AT NODE I
C                E = ELEMENT NUMBER
C           ETA(I) = LOCATION OF GAUSS POINT I IN ETA
C                    COORDINATE DIRECTION
C          JAC(I,J) = JACOBIAN MATRIX
C        JACINV(I,J) = INVERSE OF JACOBIAN MATRIX
C           KE(I,J) = ELEMENT CONDUCTANCE MATRIX
C              KXE = HYDRAULIC CONDUCTIVITY IN X
C                    COORDINATE DIRECTION
C              KYE = HYDRAULIC CONDUCTIVITY IN Y
C                    COORDINATE DIRECTION
C             W(I) = WEIGHT FOR GAUSS POINT I
C            XI(I) = LOCATION OF GAUSS POINT I IN XI
C                    COORDINATE DIRECTION
C
C     REFERENCES:
C       ISTOK,J.D. GROUNDWATER FLOW AND SOLUTE TRANSPORT
C       MODELING BY THE FINITE ELEMENT METHOD, CHAPTER 4,
C       FIGURE 4.12, EQUATION 4.62
C
C*********************************************************************
$INCLUDE: 'COMALL'
      REAL JAC(2,2),JACINV(2,2),KE(MAX3,MAX3),DNDXI(12),
     1     DNDX(12),DNDETA(12),DNDY(12),W(4),XI(12),ETA(12),
     2     SIGN1(12),SIGN2(12),KXE,KYE,KZE
      DATA SIGN1/-1.,-1.,1.,1.,1.,1.,1.,1.,-1.,-1.,-1.,-1./
      DATA SIGN2/-1.,-1.,-1.,-1.,-1.,1.,1.,1.,1.,1.,1.,-1./
C
      XI(1) = SQRT((3. - 2. * SQRT(6. / 5.)) / 7.)
      XI(2) = -XI(1)
      XI(3) = SQRT((3. + 2. * SQRT(6. / 5.)) / 7.)
      XI(4) = -XI(3)
      ETA(1) = XI(1)
      ETA(2) = XI(2)
      ETA(3) = XI(3)
      ETA(4) = XI(4)
      W(1) = 0.5 + 1. / (6. * SQRT(6. / 5.))
      W(2) = W(1)
      W(3) = 0.5 - 1. / (6. * SQRT(6. / 5.))
      W(4) = W(3)
      KXE = PROP(MATSET(E),1)
      KYE = PROP(MATSET(E),2)
      DO 20 K = 1, 12
        DO 10 N = 1, 12
          KE(K,N) = 0.
```

```
10       CONTINUE
20     CONTINUE
       DO 110 I = 1, 4
         DO 100 J = 1, 4
           DO 40 K = 1, 2
             DO 30 N = 1, 2
               JAC(K,N) = 0.
30           CONTINUE
40         CONTINUE
           DO 50 N = 1, 12
             IF ((N .EQ. 1) .OR. (N .EQ. 4) .OR.
     1          (N .EQ. 7) .OR. (N .EQ. 10)) THEN
               DNDXI(N) = (1. / 32.) * (1. + SIGN2(N) * ETA(J)) *
     1                    (18. * XI(I) + 27. * SIGN1(N) * XI(I)**2 +
     2                    9. * SIGN1(N) * ETA(J)**2 - 10. * SIGN1(N))
               DNDETA(N) = (1. / 32.) * (1.+ SIGN1(N) * XI(I)) *
     1                    (18. * ETA(J) + 27. * SIGN2(N) * ETA(J)**2 +
     2                    9. * SIGN2(N) * XI(I)**2 - 10. * SIGN2(N))
             ELSEIF ((N .EQ. 2) .OR. (N .EQ. 3) .OR.
     1          (N .EQ. 8) .OR. (N .EQ. 9)) THEN
               DNDXI(N) = (9. / 32.) * (1. + SIGN2(N) * ETA(J)) *
     1                    (9. * SIGN1(N)/3. - 2. * XI(I) - 27. *
     2                    SIGN1(N)/3. * XI(I)**2)
               DNDETA(N) = (9. / 32.) * (1. - XI(I)**2) * (SIGN2(N) +
     1                    9. * SIGN1(N)/3. * SIGN2(N) * XI(I))
             ELSEIF ((N .EQ. 5) .OR. (N .EQ. 6) .OR.
     1          (N .EQ. 11) .OR. (N .EQ. 12)) THEN
               DNDXI(N) = (9. / 32.) * (1. - ETA(J)**2) * (SIGN1(N) +
     1                    9. * SIGN2(N)/3. * SIGN1(N) * ETA(J))
               DNDETA(N) = (9. / 32.) * (1. + SIGN1(N) * XI(I)) *
     1                    (9. * SIGN2(N)/3. - 2. * ETA(J) - 27. *
     2                    SIGN2(N)/3. * ETA(J)**2)
             ENDIF
50         CONTINUE
           DO 60 N = 1, 12
             JAC(1,1) = JAC(1,1) + DNDXI(N) * X1(IN(E,N))
             JAC(1,2) = JAC(1,2) + DNDXI(N) * X2(IN(E,N))
             JAC(2,1) = JAC(2,1) + DNDETA(N) * X1(IN(E,N))
             JAC(2,2) = JAC(2,2) + DNDETA(N) * X2(IN(E,N))
60         CONTINUE
           DETJAC = JAC(1,1) * JAC(2,2) - JAC(1,2) * JAC(2,1)
           JACINV(1,1) =  JAC(2,2) / DETJAC
           JACINV(1,2) = -JAC(1,2) / DETJAC
           JACINV(2,1) = -JAC(2,1) / DETJAC
           JACINV(2,2) =  JAC(1,1) / DETJAC
           DO 70 N = 1, 12
             DNDX(N) = JACINV(1,1) * DNDXI(N) + JACINV(1,2) * DNDETA(N)
             DNDY(N) = JACINV(2,1) * DNDXI(N) + JACINV(2,2) * DNDETA(N)
70         CONTINUE
           DO 90 K = 1, 12
             DO 80 N = 1, 12
               KE(K,N) = KE(K,N) + W(I) * W(J) * (KXE * DNDX(K) *
     1                   DNDX(N) + KYE * DNDY(K) * DNDY(N)) * DETJAC
80           CONTINUE
90         CONTINUE
100      CONTINUE
110    CONTINUE
       RETURN
       END
```

```
      SUBROUTINE KPAR8(E,KE)
C**********************************************************************
C        PURPOSE:
C           TO COMPUTE THE ELEMENT CONDUCTANCE MATRIX FOR THREE-
C           DIMENSIONAL, LINEAR PARALLELEPIPED ELEMENT
C
C        DEFINITIONS OF VARIABLES:
C                 DETJAC = DETERMINANT OF JACOBIAN MATRIX
C               DNDXI(I) = PARTIAL DERIVATIVE OF INTERPOLATION
C                          FUNCTION WITH RESPECT TO XI AT NODE I
C                DNDX(I) = PARTIAL DERIVATIVE OF INTERPOLATION
C                          FUNCTION WITH RESPECT TO X AT NODE I
C              DNDETA(I) = PARTIAL DERIVATIVE OF INTERPOLATION
C                          FUNCTION WITH RESPECT TO ETA AT NODE I
C                DNDY(I) = PARTIAL DERIVATIVE OF INTERPOLATION
C                          FUNCTION WITH RESPECT TO Y AT NODE I
C             DNDZETA(I) = PARTIAL DERIVATIVE OF INTERPOLATION
C                          FUNCTION WITH RESPECT TO ZETA AT NODE I
C                DNDZ(I) = PARTIAL DERIVATIVE OF INTERPOLATION
C                          FUNCTION WITH RESPECT TO Z AT NODE I
C                      E = ELEMENT NUMBER
C                 ETA(I) = LOCATION OF GAUSS POINT IN ETA
C                          COORDINATE DIRECTION
C               JAC(I,J) = JACOBIAN MATRIX
C            JACINV(I,J) = INVERSE OF JACOBIAN MATRIX
C                KE(I,J) = ELEMENT CONDUCTANCE MATRIX
C                    KXE = HYDRAULIC CONDUCTIVITY IN X
C                          COORDINATE DIRECTION
C                    KYE = HYDRAULIC CONDUCTIVITY IN Y
C                          COORDINATE DIRECTION
C                    KZE = HYDRAULIC CONDUCTIVITY IN Z
C                          COORDINATE DIRECTION
C                   W(I) = WEIGHT FOR GAUSS POINT I
C                  XI(I) = LOCATION OF GAUSS POINT IN XI
C                          COORDINATE DIRECTION
C                ZETA(I) = LOCATION OF GAUSS POINT IN ZETA
C                          COORDINATE DIRECTION
C        REFERENCES:
C         ISTOK,J.D. GROUNDWATER FLOW AND SOLUTE TRANSPORT
C         MODELING BY THE FINITE ELEMENT METHOD, CHAPTER 4,
C         FIGURE 4.13, EQUATION 4.63
C**********************************************************************
$INCLUDE: 'COMALL'
      REAL JAC(3,3),JACINV(3,3),KE(MAX3,MAX3),DNDXI(8),DNDX(8),
     1     DNDETA(8),DNDY(8),DNDZETA(8),DNDZ(8),W(2),XI(8),
     2     ETA(8),ZETA(8),SIGN1(8),SIGN2(8),SIGN3(8),KXE,KYE,KZE
      DATA SIGN1/-1.,1.,1.,-1.,-1.,1.,1.,-1./
      DATA SIGN2/-1.,-1.,1.,1.,-1.,-1.,1.,1./
      DATA SIGN3/-1.,-1.,-1.,-1.,1.,1.,1.,1./
C
      XI(1)  = 1. / SQRT(3.)
      XI(2)  = -XI(1)
      ETA(1) = XI(1)
      ETA(2) = XI(2)
      ZETA(1) = XI(1)
      ZETA(2) = XI(2)
      W(1) = 1.
      W(2) = 1.
      KXE = PROP(MATSET(E),1)
      KYE = PROP(MATSET(E),2)
      KZE = PROP(MATSET(E),3)
```

```
      DO 20 K = 1, 8
        DO 10 N = 1, 8
          KE(K,N) = 0.
10      CONTINUE
20    CONTINUE
      DO 120 I = 1, 2
        DO 110 J = 1, 2
          DO 100 K = 1, 2
            DO 40 L = 1, 3
              DO 30 N = 1, 3
                JAC(L,N) = 0.
30          CONTINUE
40          CONTINUE
            DO 50 N = 1, 8
              DNDXI(N)   = 0.125 * SIGN1(N) * (1. + SIGN2(N) *
     1                     ETA(J)) * (1. + SIGN3(N) * ZETA(K))
              DNDETA(N)  = 0.125 * SIGN2(N) * (1. + SIGN1(N) *
     1                     XI(I)) * (1. + SIGN3(N) * ZETA(K))
              DNDZETA(N) = 0.125 * SIGN3(N) * (1. + SIGN1(N) *
     1                     XI(I)) * (1. + SIGN2(N) * ETA(J))
50          CONTINUE
            DO 60 N = 1, 8
              JAC(1,1) = JAC(1,1) + DNDXI(N)   * X1(IN(E,N))
              JAC(1,2) = JAC(1,2) + DNDXI(N)   * X2(IN(E,N))
              JAC(1,3) = JAC(1,3) + DNDXI(N)   * X3(IN(E,N))
              JAC(2,1) = JAC(2,1) + DNDETA(N)  * X1(IN(E,N))
              JAC(2,2) = JAC(2,2) + DNDETA(N)  * X2(IN(E,N))
              JAC(2,3) = JAC(2,3) + DNDETA(N)  * X3(IN(E,N))
              JAC(3,1) = JAC(3,1) + DNDZETA(N) * X1(IN(E,N))
              JAC(3,2) = JAC(3,2) + DNDZETA(N) * X2(IN(E,N))
              JAC(3,3) = JAC(3,3) + DNDZETA(N) * X3(IN(E,N))
60          CONTINUE
            DETJAC = JAC(1,1) * (JAC(2,2) * JAC(3,3) - JAC(3,2) *
     1               JAC(2,3)) - JAC(1,2) * (JAC(2,1) * JAC(3,3) -
     2               JAC(3,1) * JAC(2,3)) - JAC(1,3) * (JAC(2,1) *
     3               JAC(3,2) - JAC(3,1) * JAC(2,2))
            JACINV(1,1) = ( JAC(2,2) * JAC(3,3) - JAC(2,3) *
     1                     JAC(3,2)) / DETJAC
            JACINV(1,2) = (-JAC(2,1) * JAC(3,3) + JAC(2,3) *
     1                     JAC(3,1)) / DETJAC
            JACINV(1,3) = ( JAC(2,1) * JAC(3,2) - JAC(3,1) *
     1                     JAC(2,2)) / DETJAC
            JACINV(2,1) = (-JAC(1,2) * JAC(3,3) + JAC(1,3) *
     1                     JAC(3,2)) / DETJAC
            JACINV(2,2) = ( JAC(1,1) * JAC(3,3) - JAC(1,3) *
     1                     JAC(3,1)) / DETJAC
            JACINV(2,3) = (-JAC(1,1) * JAC(3,2) + JAC(1,2) *
     1                     JAC(3,1)) / DETJAC
            JACINV(3,1) = ( JAC(1,2) * JAC(2,3) - JAC(1,3) *
     1                     JAC(2,2)) / DETJAC
            JACINV(3,2) = (-JAC(1,1) * JAC(2,3) + JAC(1,3) *
     1                     JAC(2,1)) / DETJAC
            JACINV(3,3) = ( JAC(1,1) * JAC(2,2) - JAC(1,2) *
     1                     JAC(2,1)) / DETJAC
            DO 70 N = 1, 8

              DNDX(N) = JACINV(1,1) * DNDXI(N) + JACINV(1,2) *
     1                  DNDETA(N) + JACINV(1,3) * DNDZETA(N)
              DNDY(N) = JACINV(2,1) * DNDXI(N) + JACINV(2,2) *
     1                  DNDETA(N) + JACINV(2,3) * DNDZETA(N)
              DNDZ(N) = JACINV(3,1) * DNDXI(N) + JACINV(3,2) *
     1                  DNDETA(N) + JACINV(3,3) * DNDZETA(N)
```

```
70          CONTINUE
            DO 90 L = 1, 8
              DO 80 N = 1,8
                KE(L,N) = KE(L,N) + W(I) * W(J) * W(K) * (KXE *
     1                    DNDX(L) * DNDX(N) + KYE * DNDY(L) *
     2                    DNDY(N) + KZE * DNDZ(L) * DNDZ(N)) * DETJAC
80          CONTINUE
90          CONTINUE
100       CONTINUE
110     CONTINUE
120   CONTINUE
      RETURN
      END



      SUBROUTINE KPAR20(E,KE)
C***********************************************************************
C
C       PURPOSE:
C         TO COMPUTE THE ELEMENT CONDUCTANCE MATRIX FOR THREE-
C         DIMENSIONAL, QUADRATIC PARALLELEPIPED ELEMENT
C
C       DEFINITIONS OF VARIABLES:
C             DETJAC = DETERMINANT OF JACOBIAN MATRIX
C           DNDXI(I) = PARTIAL DERIVATIVE OF INTERPOLATION
C                      FUNCTION WITH RESPECT TO XI AT NODE I
C            DNDX(I) = PARTIAL DERIVATIVE OF INTERPOLATION
C                      FUNCTION WITH RESPECT TO X AT NODE I
C          DNDETA(I) = PARTIAL DERIVATIVE OF INTERPOLATION
C                      FUNCTION WITH RESPECT TO ETA AT NODE I
C            DNDY(I) = PARTIAL DERIVATIVE OF INTERPOLATION
C                      FUNCTION WITH RESPECT TO Y AT NODE I
C         DNDZETA(I) = PARTIAL DERIVATIVE OF INTERPOLATION
C                      FUNCTION WITH RESPECT TO ZETA AT NODE I
C            DNDZ(I) = PARTIAL DERIVATIVE OF INTERPOLATION
C                      FUNCTION WITH RESPECT TO Z AT NODE I
C                  E = ELEMENT NUMBER
C             ETA(I) = LOCATION OF GAUSS POINT IN ETA
C                      COORDINATE DIRECTION
C            JAC(I,J) = JACOBIAN MATRIX
C          JACINV(I,J) = INVERSE OF JACOBIAN MATRIX
C            KE(I,J) = ELEMENT CONDUCTANCE MATRIX
C                KXE = HYDRAULIC CONDUCTIVITY IN X
C                      COORDINATE DIRECTION
C                KYE = HYDRAULIC CONDUCTIVITY IN Y
C                      COORDINATE DIRECTION
C                KZE = HYDRAULIC CONDUCTIVITY IN Z
C                      COORDINATE DIRECTION
C               W(I) = WEIGHT FOR GAUSS POINT I
C              XI(I) = LOCATION OF GAUSS POINT IN XI
C                      COORDINATE DIRECTION
C            ZETA(I) = LOCATION OF GAUSS POINT IN ZETA
C                      COORDINATE DIRECTION
C
C       REFERENCES:
C         ISTOK,J.D. GROUNDWATER FLOW AND SOLUTE TRANSPORT
C         MODELING BY THE FINITE ELEMENT METHOD, CHAPTER 4,
C         FIGURE 4.14, EQUATION 4.63
C
C***********************************************************************
```

```
$INCLUDE: 'COMALL'
      REAL JAC(3,3),JACINV(3,3),KE(MAX3,MAX3),DNDXI(20),DNDX(20),
     1     DNDETA(20),DNDY(20),DNDZETA(20),DNDZ(20),W(3),XI(20),
     2     ETA(20),ZETA(20),SIGN1(20),SIGN2(20),SIGN3(20),KXE,KYE,KZE
      DATA SIGN1/-1.,0.,1.,1.,1.,0.,-1.,-1.,-1.,1.,1.,-1.,-1.,
     1           0.,1.,1.,1.,0.,-1.,-1./
      DATA SIGN2/-1.,-1.,-1.,0.,1.,1.,1.,0.,-1.,-1.,1.,1.,-1.,
     1           -1.,-1.,0.,1.,1.,1.,0./
      DATA SIGN3/-1.,-1.,-1.,-1.,-1.,-1.,-1.,-1.,0.,0.,0.,0.,1.,
     1           1.,1.,1.,1.,1.,1.,1./
C
      XI(1) = 0.
      XI(2) = 1. / SQRT(3. / 5.)
      XI(3) = -XI(2)
      ETA(1) = XI(1)
      ETA(2) = XI(2)
      ETA(3) = XI(3)
      ZETA(1) = XI(1)
      ZETA(2) = XI(2)
      ZETA(3) = XI(3)
      W(1) = 8. / 9.
      W(2) = 5. / 9.
      W(3) = W(2)
      KXE = PROP(MATSET(E),1)
      KYE = PROP(MATSET(E),2)
      KZE = PROP(MATSET(E),3)
      DO 20 K = 1, 20
         DO 10 L = 1, 20
            KE(K,L) = 0.
10       CONTINUE
20    CONTINUE
      DO 120 I = 1, 3
         DO 110 J = 1, 3
            DO 100 K = 1, 3
               DO 40 L = 1, 3
                  DO 30 N = 1, 3
                     JAC(L,N) = 0.
30                CONTINUE
40             CONTINUE
               DO 50 N = 1, 20
                  IF ((N .EQ. 1) .OR. (N .EQ. 3) .OR. (N .EQ. 5) .OR.
     1               (N .EQ. 7) .OR. (N .EQ. 13) .OR. (N .EQ. 15) .OR.
     2               (N .EQ. 17) .OR. (N .EQ. 19)) THEN
                     DNDXI(N) = 0.125 * SIGN1(N) * (1. + SIGN2(N) *
     1                  ETA(J)) * (1. + SIGN3(N) * ZETA(K)) *
     2                  (2. * SIGN1(N) * XI(I) + SIGN2(N) *
     3                  ETA(J) + SIGN3(N) * ZETA(K) -1.)
                     DNDETA(N) = 0.125 * SIGN2(N) * (1. + SIGN1(N) *
     1                  XI(I)) * (1. + SIGN3(N) * ZETA(K)) *
     2                  (2. * SIGN2(N) * ETA(J) + SIGN1(N) *
     3                  XI(I) + SIGN3(N) * ZETA(K) -1.)
                     DNDZETA(N) = 0.125 * SIGN3(N) * (1. + SIGN1(N) *
     1                  XI(I)) * (1. + SIGN2(N) * ETA(J)) *
     2                  (2. * SIGN3(N) * ZETA(K) + SIGN1(N) *
     3                  XI(I) + SIGN2(N) * ETA(J) -1.)
                  ELSEIF ((N .EQ. 2) .OR. (N .EQ. 6) .OR.
     1                  (N .EQ. 14) .OR. (N .EQ. 18)) THEN
                     DNDXI(N) = -0.5 * XI(I) * (1. + SIGN2(N) *
     1                  ETA(J)) * (1. + SIGN3(N) * ZETA(K))
```

```
            DNDETA(N) =  0.25 * SIGN2(N) * (1. - XI(I)**2) *
     1                   (1. + SIGN3(N) * ZETA(K))
            DNDZETA(N) = 0.25 * SIGN3(N) * (1. - XI(I)**2) *
     1                   (1. + SIGN2(N) * ETA(J))
          ELSEIF ((N .EQ. 4) .OR. (N .EQ. 8) .OR.
     1            (N .EQ. 16) .OR. (N .EQ. 20)) THEN
            DNDXI(N) =   0.25 * SIGN1(N) * (1. - ETA(J)**2) *
     1                   (1. + SIGN3(N) * ZETA(K))
            DNDETA(N) = -0.5 * ETA(J) * (1. + SIGN1(N) * XI(I)) *
     1                   (1. + SIGN3(N) * ZETA(K))
            DNDZETA(N) = 0.25 * SIGN3(N) * (1. + SIGN1(N) *
     1                   XI(I)) * (1. - ETA(J)**2)
          ELSEIF ((N .GE. 9) .AND. (N .LE. 12)) THEN
            DNDXI(N) =   0.25 * SIGN1(N) * (1 - ZETA(K)**2) *
     1                   (1. + SIGN2(N) * ETA(J))
            DNDETA(N) =  0.25 * SIGN2(N) * (1. - ZETA(K)**2) *
     1                   (1. + SIGN1(N) * XI(I))
            DNDZETA(N) = -0.5 * ZETA(K) * (1. + SIGN1(N) *
     1                   XI(I)) * (1. + SIGN2(N) * ETA(J))
          ENDIF
   50     CONTINUE
          DO 60 N = 1, 20
            JAC(1,1) = JAC(1,1) + DNDXI(N) * X1(IN(E,N))
            JAC(1,2) = JAC(1,2) + DNDXI(N) * X2(IN(E,N))
            JAC(1,3) = JAC(1,3) + DNDXI(N) * X3(IN(E,N))
            JAC(2,1) = JAC(2,1) + DNDETA(N) * X1(IN(E,N))
            JAC(2,2) = JAC(2,2) + DNDETA(N) * X2(IN(E,N))
            JAC(2,3) = JAC(2,3) + DNDETA(N) * X3(IN(E,N))
            JAC(3,1) = JAC(3,1) + DNDZETA(N) * X1(IN(E,N))
            JAC(3,2) = JAC(3,2) + DNDZETA(N) * X2(IN(E,N))
            JAC(3,3) = JAC(3,3) + DNDZETA(N) * X3(IN(E,N))
   60     CONTINUE
          DETJAC = JAC(1,1) * (JAC(2,2) * JAC(3,3) - JAC(3,2) *
     1             JAC(2,3)) - JAC(1,2) * (JAC(2,1) * JAC(3,3) -
     2             JAC(3,1) * JAC(2,3)) - JAC(1,3) * (JAC(2,1) *
     3             JAC(3,2) - JAC(3,1) * JAC(2,2))
          JACINV(1,1) = ( JAC(2,2) * JAC(3,3) - JAC(2,3) *
     1                   JAC(3,2)) / DETJAC
          JACINV(1,2) = (-JAC(2,1) * JAC(3,3) + JAC(2,3) *
     1                   JAC(3,1)) / DETJAC
          JACINV(1,3) = ( JAC(2,1) * JAC(3,2) - JAC(3,1) *
     1                   JAC(2,2)) / DETJAC
          JACINV(2,1) = (-JAC(1,2) * JAC(3,3) + JAC(1,3) *
     1                   JAC(3,2)) / DETJAC
          JACINV(2,2) = ( JAC(1,1) * JAC(3,3) - JAC(1,3) *
     1                   JAC(3,1)) / DETJAC
          JACINV(2,3) = (-JAC(1,1) * JAC(3,2) + JAC(1,2) *
     1                   JAC(3,1)) / DETJAC
          JACINV(3,1) = ( JAC(1,2) * JAC(2,3) - JAC(1,3) *
     1                   JAC(2,2)) / DETJAC
          JACINV(3,2) = (-JAC(1,1) * JAC(2,3) + JAC(1,3) *
     1                   JAC(2,1)) / DETJAC
          JACINV(3,3) = ( JAC(1,1) * JAC(2,2) - JAC(1,2) *
     1                   JAC(2,1)) / DETJAC
          DO 70 N = 1, 20
            DNDX(N) = JACINV(1,1) * DNDXI(N) + JACINV(1,2) *
     1                DNDETA(N) + JACINV(1,3) * DNDZETA(N)
            DNDY(N) = JACINV(2,1) * DNDXI(N) + JACINV(2,2) *
     1                DNDETA(N) + JACINV(2,3) * DNDZETA(N)
            DNDZ(N) = JACINV(3,1) * DNDXI(N) + JACINV(3,2) *
     1                DNDETA(N) + JACINV(3,3) * DNDZETA(N)
```

```
 70          CONTINUE
             DO 90 L = 1, 20
               DO 80 N = 1, 20
                 KE(L,N) = KE(L,N) + W(I) * W(J) * W(K) *
      1                   (KXE * DNDX(L) * DNDX(N) + KYE * DNDY(L) *
      2                   DNDY(N) + KZE * DNDZ(L) * DNDZ(N)) * DETJAC
 80            CONTINUE
 90          CONTINUE
100        CONTINUE
110      CONTINUE
120    CONTINUE
       RETURN
       END




       SUBROUTINE KPAR32(E,KE)
C*********************************************************************
C
C      PURPOSE:
C      TO COMPUTE THE ELEMENT CONDUCTANCE MATRIX FOR THREE-
C      DIMENSIONAL, CUBIC PARALLELEPIPED ELEMENT
C
C      DEFINITIONS OF VARIABLES:
C          DETJAC = DETERMINANT OF JACOBIAN MATRIX
C        DNDXI(I) = PARTIAL DERIVATIVE OF INTERPOLATION
C                   FUNCTION WITH RESPECT TO XI AT NODE I
C         DNDX(I) = PARTIAL DERIVATIVE OF INTERPOLATION
C                   FUNCTION WITH RESPECT TO X AT NODE I
C       DNDETA(I) = PARTIAL DERIVATIVE OF INTERPOLATION
C                   FUNCTION WITH RESPECT TO ETA AT NODE I
C         DNDY(I) = PARTIAL DERIVATIVE OF INTERPOLATION
C                   FUNCTION WITH RESPECT TO Y AT NODE I
C      DNDZETA(I) = PARTIAL DERIVATIVE OF INTERPOLATION
C                   FUNCTION WITH RESPECT TO ZETA AT NODE I
C         DNDZ(I) = PARTIAL DERIVATIVE OF INTERPOLATION
C                   FUNCTION WITH RESPECT TO Z AT NODE I
C               E = ELEMENT NUMBER
C          ETA(I) = GAUSS POINT IN ETA COORDINATE DIRECTION
C        JAC(I,J) = JACOBIAN MATRIX
C     JACINV(I,J) = INVERSE OF JACOBIAN MATRIX
C        KE(I,J) = ELEMENT CONDUCTANCE MATRIX
C             KXE = HYDRAULIC CONDUCTIVITY IN X
C                   COORDINATE DIRECTION
C             KYE = HYDRAULIC CONDUCTIVITY IN Y
C                   COORDINATE DIRECTION
C             KZE = HYDRAULIC CONDUCTIVITY IN Z
C                   COORDINATE DIRECTION
C            W(I) = WEIGHT FOR GAUSS POINT I
C           XI(I) = LOCATION OF GAUSS POINT IN XI
C                   COORDINATE DIRECTION
C         ZETA(I) = LOCATION OF GAUSS POINT IN ZETA
C                   COORDINATE DIRECTION
C
C      REFERENCES:
C      ISTOK,J.D. GROUNDWATER FLOW AND SOLUTE TRANSPORT
C      MODELING BY THE FINITE ELEMENT METHOD, CHAPTER 4,
C      FIGURE 4.15, EQUATION 4.63
C
C*********************************************************************
```

```
$INCLUDE: 'COMALL'
      REAL JAC(3,3),JACINV(3,3),KE(MAX3,MAX3),DNDXI(32),DNDX(32),
     1     DNDETA(32),DNDY(32),DNDZETA(32),DNDZ(32),W(4),XI(32),
     2     ETA(32),ZETA(32),SIGN1(32),SIGN2(32),SIGN3(32),KXE,KYE,KZE
      DATA SIGN1 /2*-1.,6*1.,5*-1.,2*1.,2*-1.,2*1.,3*-1.,6*1.,4*-1./
      DATA SIGN2 /5*-1.,6*1.,3*-1.,2*1.,2*-1.,2*1.,5*-1.,6*1.,-1./
      DATA SIGN3 /16*-1., 16*1./
C
      XI(1) = SQRT((3. - 2. * SQRT(6. / 5.)) / 7.)
      XI(2) = -XI(1)
      XI(3) = SQRT((3. + 2. * SQRT(6. / 5.)) / 7.)
      XI(4) = -XI(3)
      ETA(1) = XI(1)
      ETA(2) = XI(2)
      ETA(3) = XI(3)
      ETA(4) = XI(4)
      ZETA(1) = XI(1)
      ZETA(2) = XI(2)
      ZETA(3) = XI(3)
      ZETA(4) = XI(4)
      W(1) = 0.5 + 1. / (6. * SQRT(6. / 5.))
      W(2) = W(1)
      W(3) = 0.5 - 1. / (6. * SQRT(6. / 5.))
      W(4) = W(3)
      KXE = PROP(MATSET(E),1)
      KYE = PROP(MATSET(E),2)
      KZE = PROP(MATSET(E),3)
      DO 20 K = 1, 32
         DO 10 N = 1, 32
            KE(K,N) = 0.
 10      CONTINUE
 20   CONTINUE
      DO 120 I = 1, 4
         DO 110 J = 1, 4
            DO 100 K = 1, 4
               DO 40 L = 1, 3
                  DO 30 N = 1, 3
                     JAC(L,N) = 0.
 30               CONTINUE
 40            CONTINUE
               DO 50 N = 1, 32
                  IF ((N .EQ. 1) .OR. (N .EQ. 4) .OR. (N .EQ. 7) .OR.
     1                (N .EQ. 10) .OR. (N .EQ. 21) .OR. (N .EQ. 24) .OR.
     2                (N .EQ. 27) .OR. (N .EQ. 30)) THEN
                     DNDXI(N) = (9. / 64.) * (1. + SIGN2(N) * ETA(J)) *
     1                          (1. + SIGN3(N) * ZETA(K)) * (SIGN1(N) *
     2                          (-(19. / 9.) + 3. * XI(I)**2 +
     3                          ETA(J)**2 + ZETA(K)**2) + 2. * XI(I))
                     DNDETA(N) = (9. / 64.) * (1. + SIGN1(N) * XI(I)) *
     1                           (1. + SIGN3(N) * ZETA(K)) * (SIGN2(N) *
     2                           (-(19. / 9.) + XI(I)**2 + 3. *
     3                           ETA(J)**2 + ZETA(K)**2) + 2. * ETA(J))
                     DNDZETA(N) = (9. / 64.) * (1. + SIGN1(N) * XI(I)) *
     1                            (1. + SIGN2(N) * ETA(J)) * (SIGN3(N) *
     2                            (-(19. / 9.) + XI(I)**2 + ETA(J)**2 +
     3                            3. * ZETA(K)**2) + 2. * ZETA(K))
                  ELSEIF ((N .EQ. 2) .OR. (N .EQ. 3) .OR.
     1                    (N .EQ. 8) .OR. (N .EQ. 9) .OR.
     2                    (N .EQ. 22) .OR. (N .EQ. 23) .OR.
     3                    (N .EQ. 28) .OR. (N .EQ. 29)) THEN
                     DNDXI(N) = (81. / 64.) * (1. + SIGN2(N) * ETA(J))
```

```
      1                           * (1. + SIGN3(N) * ZETA(K))
      2                           * (SIGN1(N)/3. - 2./9. * XI(I) -
      3                             3.*SIGN1(N)/3. * XI(I)**2)
                 DNDETA(N) = (81. / 64.) * SIGN2(N) * (1. - XI(I)**2)
      1                           * (1. / 9. + SIGN1(N)/3.*XI(I))
      2                           * (1. + SIGN3(N) * ZETA(K))
                 DNDZETA(N) = (81. / 64.) * SIGN3(N) * (1. - XI(I)**2)
      1                           * (1. / 9. + SIGN1(N)/3.*XI(I))
      2                           * (1. + SIGN2(N) * ETA(J))
              ELSEIF ((N .EQ. 5) .OR. (N .EQ. 6) .OR.
      1               (N .EQ. 11) .OR. (N .EQ. 12) .OR.
      2               (N .EQ. 25) .OR. (N .EQ. 26) .OR.
      3               (N .EQ. 31) .OR. (N .EQ. 32)) THEN
                 DNDXI(N) = (81./64.) * SIGN1(N) * (1. - ETA(J)**2)
      1                           * (1./9. + SIGN2(N)/3. * ETA(J))
      2                           * (1. + SIGN3(N) * ZETA(K))
                 DNDETA(N) = (81./64.) * (1. + SIGN1(N)*XI(I))
      1                           * (1. + SIGN3(N)*ZETA(K)) * (SIGN2(N)/3. -
      2                             2./9.*ETA(J) - 3.*SIGN2(N)/3. * ETA(J)**2)
                 DNDZETA(N) = (81./64.) * SIGN3(N) * (1.- ETA(J)**2)
      1                           * ( 1./9. + SIGN2(N)/3. * ETA(J) )
      2                           * ( 1. + SIGN1(N) * XI(I) )
              ELSEIF ((N .GE. 13) .AND. (N .LE. 20)) THEN
                 DNDXI(N) = (81./64.) * SIGN1(N) * (1. - ZETA(K)**2)
      1                           * (1. / 9. + SIGN3(N)/3. * ZETA(K))
      2                           * (1. + SIGN2(N) * ETA(J))
                 DNDETA(N) = (81./64.) * SIGN2(N) * (1. - ZETA(K)**2)
      1                           * (1. / 9. + SIGN3(N)/3. * ZETA(K))
      2                           * (1. + SIGN1(N) * XI(I))
                 DNDZETA(N) = (81./64.) * (1. + SIGN1(N) * XI(I))
      1                           * (1. + SIGN2(N)*ETA(J)) * (SIGN3(N)/3. -
      2                             2./9.*ZETA(K) - 3.*SIGN3(N)/3.*ZETA(K)**2)
              ENDIF
   50    CONTINUE
         DO 60 N = 1, 32
            JAC(1,1) = JAC(1,1) + DNDXI(N)   * X1(IN(E,N))
            JAC(1,2) = JAC(1,2) + DNDXI(N)   * X2(IN(E,N))
            JAC(1,3) = JAC(1,3) + DNDXI(N)   * X3(IN(E,N))
            JAC(2,1) = JAC(2,1) + DNDETA(N)  * X1(IN(E,N))
            JAC(2,2) = JAC(2,2) + DNDETA(N)  * X2(IN(E,N))
            JAC(2,3) = JAC(2,3) + DNDETA(N)  * X3(IN(E,N))
            JAC(3,1) = JAC(3,1) + DNDZETA(N) * X1(IN(E,N))
            JAC(3,2) = JAC(3,2) + DNDZETA(N) * X2(IN(E,N))
            JAC(3,3) = JAC(3,3) + DNDZETA(N) * X3(IN(E,N))
   60    CONTINUE
         DETJAC = JAC(1,1) * (JAC(2,2) * JAC(3,3) - JAC(3,2) *
      1            JAC(2,3)) - JAC(1,2) * (JAC(2,1) * JAC(3,3) -
      2            JAC(3,1) * JAC(2,3)) - JAC(1,3) * (JAC(2,1) *
      3            JAC(3,2) - JAC(3,1) * JAC(2,2))
         JACINV(1,1) = ( JAC(2,2) * JAC(3,3) - JAC(2,3) *
      1                 JAC(3,2)) / DETJAC
         JACINV(1,2) = (-JAC(2,1) * JAC(3,3) + JAC(2,3) *
      1                 JAC(3,1)) / DETJAC
         JACINV(1,3) = ( JAC(2,1) * JAC(3,2) - JAC(3,1) *
      1                 JAC(2,2)) / DETJAC
         JACINV(2,1) = (-JAC(1,2) * JAC(3,3) + JAC(1,3) *
      1                 JAC(3,2)) / DETJAC
         JACINV(2,2) = ( JAC(1,1) * JAC(3,3) - JAC(1,3) *
      1                 JAC(3,1)) / DETJAC
         JACINV(2,3) = (-JAC(1,1) * JAC(3,2) + JAC(1,2) *
      1                 JAC(3,1)) / DETJAC
```

```
          JACINV(3,1) = ( JAC(1,2) * JAC(2,3) - JAC(1,3) *
     1                  JAC(2,2)) / DETJAC
          JACINV(3,2) = (-JAC(1,1) * JAC(2,3) + JAC(1,3) *
     1                  JAC(2,1)) / DETJAC
          JACINV(3,3) = ( JAC(1,1) * JAC(2,2) - JAC(1,2) *
     1                  JAC(2,1)) / DETJAC
          DO 70 N = 1, 32
            DNDX(N) = JACINV(1,1) * DNDXI(N) + JACINV(1,2) *
     1                DNDETA(N) + JACINV(1,3) * DNDZETA(N)
            DNDY(N) = JACINV(2,1) * DNDXI(N) + JACINV(2,2) *
     1                DNDETA(N) + JACINV(2,3) * DNDZETA(N)
            DNDZ(N) = JACINV(3,1) * DNDXI(N) + JACINV(3,2) *
     1                DNDETA(N) + JACINV(3,3) * DNDZETA(N)
70        CONTINUE
          DO 90 L = 1, 32
            DO 80 N = 1, 32
              KE(L,N) = KE(L,N) + W(I) * W(J) * W(K) *
     1                  (KXE * DNDX(L) * DNDX(N) + KYE * DNDY(L) *
     2                  DNDY(N) + KZE * DNDZ(L) * DNDZ(N)) * DETJAC
80          CONTINUE
90        CONTINUE
100     CONTINUE
110     CONTINUE
120   CONTINUE
      RETURN
      END
```

Chapter 13

SUBROUTINES DECOMP AND SOLVE

13.1 PURPOSE

Subroutines DECOMP and SOLVE solve a system of linear equations of the form

$$[M] \{X\} = \{B\} \qquad (13.1)$$

where [M] is a banded matrix of known coefficients (symmetric or not), {X} are the unknowns, and {B} is a vector of known values.

13.2 INPUT

None

13.3 OUTPUT

None

13.4 DEFINITIONS OF VARIABLES

B(I)   = Vector of known values.

M(IJ)   = Matrix of known cofficients in vector storage.

NDOF   = Number of unknown values in {X}.

SBW   = Semi-band width of [M].

SYMM   = Logical variable
        = 'True' if [M] is symmetric
        = 'False' if [M] is nonsymmetric.

X(I)   = Vector of unknown values to be compted.

13.5 USAGE

Subroutine DECOMP performs triangular decomposition on the matrix of known coefficients in vector matrix storage {M} (see Chapter 5). The resulting upper-, and lower-triangular matrices are stored in {M} (the original contents of {M} are overwritten during the decomposition process). Subroutine SOLVE solve for values of the unknowns by backward substitution. Once {M} has been decomposed SOLVE can be used to obtain values of {X} for any number of different vectors {B}.

For example, consider the system of linear equations

$$\begin{bmatrix} 2 & 1 & -1 & 1 \\ -1 & 4 & 1 & -1 \\ -2 & -1 & 4 & 1 \\ -1 & -2 & 1 & .2 \end{bmatrix} \begin{Bmatrix} x_1 \\ x_2 \\ x_3 \\ x_4 \end{Bmatrix} = \begin{Bmatrix} 1 \\ 0 \\ 0 \\ 0 \end{Bmatrix}$$

[M] is a nonsymmetric matrix (SYMM = 'False') with NDOF = SBW = 4. For use in DECOMP and SOLVE [M] must be in vector storage. Using the procedure in Section 5.1.3 we can write

$$\{M\} \underset{\substack{\text{Before} \\ \text{Decompostion}}}{=} \begin{Bmatrix} 2 \\ 1 \\ -1 \\ 1 \\ -1 \\ 4 \\ 1 \\ -1 \\ -2 \\ -1 \\ 4 \\ 1 \\ -1 \\ -2 \\ 1 \\ 2 \end{Bmatrix}$$

After executing subroutine DECOMP, {M} contains the upper and lower triangular matrices for {M} in vector storage

$$\{M\} \underset{\substack{\text{After} \\ \text{Decomposition}}}{=} \begin{Bmatrix} 2.0 \\ 0.5 \\ -0.5 \\ 0.5 \\ -1.0 \\ 4.5 \\ 0.11 \\ -0.11 \\ -2.00 \\ 0.0 \\ 3.0 \\ 0.67 \\ -1.0 \\ -1.5 \\ 0.67 \\ 1.89 \end{Bmatrix} = \begin{Bmatrix} l_{11} \\ u_{12} \\ u_{13} \\ u_{14} \\ l_{21} \\ u_{23} \\ u_{24} \\ l_{31} \\ l_{32} \\ l_{33} \\ u_{34} \\ l_{41} \\ l_{42} \\ l_{43} \\ l_{44} \end{Bmatrix}$$

where l_{ij} and u_{ij} are the entries in the ith row and j^{th} column of the lower and upper triangular matrices that would have been stored in [L] and [U] if the decomposition had been performed in full matrix storage.

After {M} has been decomposed, subroutine SOLVE can be used to find a solution {X} for any known right-hand-side vector {B}. For this example

$$\{B\} = \begin{Bmatrix} 1 \\ 0 \\ 0 \\ 0 \end{Bmatrix}$$

and after executing SOLVE (with NDOF = SBW = 4, and SYMM = 'False') we have

$$\{X\} = \begin{Bmatrix} 0.41 \\ 0.12 \\ 0.17 \\ 0.24 \end{Bmatrix}$$

SOLVE can be executed repeatedly to obtain a set of solutions $\{X_1\}$, $\{X_2\}$, ... for a set of known vectors $\{B_1\}$, $\{B_2\}$, ...

As another example consider the two system of equations

$$\begin{bmatrix} 3 & 2 & 0 & 0 \\ 2 & 4 & 2 & 0 \\ 0 & 2 & 4 & 2 \\ 0 & 0 & 2 & 3 \end{bmatrix} \begin{Bmatrix} x_1 \\ x_2 \\ x_3 \\ x_4 \end{Bmatrix} = \begin{Bmatrix} 1 \\ 0 \\ 0 \\ 0 \end{Bmatrix}$$

and

$$\begin{bmatrix} 3 & 2 & 0 & 0 \\ 2 & 4 & 2 & 0 \\ 0 & 2 & 4 & 2 \\ 0 & 0 & 2 & 3 \end{bmatrix} \begin{Bmatrix} x_1 \\ x_2 \\ x_3 \\ x_4 \end{Bmatrix} = \begin{Bmatrix} 0 \\ 1 \\ 0 \\ 0 \end{Bmatrix}$$

Writing the matrix of coefficients in vector storage gives

$$\{M\} = \begin{Bmatrix} 3 \\ 2 \\ 2 \\ 4 \\ 2 \\ 2 \\ 4 \\ 2 \\ 2 \\ 3 \end{Bmatrix}$$

{M} =
Before
Decompostion

After executing DECOMP with NDOF = 4, SBW = 2, and SYMM = 'True', we have

$$
\{M\} \underset{\substack{\text{After}\\\text{Decompostion}}}{=} \begin{Bmatrix} 1.732 \\ 1.155 \\ 1.633 \\ 1.225 \\ 1.581 \\ 1.265 \\ 1.183 \end{Bmatrix} = \begin{Bmatrix} u_{11} \\ u_{12} \\ u_{22} \\ u_{23} \\ u_{33} \\ u_{34} \\ u_{44} \end{Bmatrix}
$$

wher u_{ij} is an entry in the ith row and j^{th} column of the upper triangular matrix that would have been stored in [U] if the decomposition had been perfored in full matrix storage.

Subroutine SOLVE will be executed twice, once for each vector {B}. The first time SOLVE is executed, NDOF = 4, SBW = 2, SYMM = 'True',

$$
\{B\} = \begin{Bmatrix} 1 \\ 0 \\ 0 \\ 0 \end{Bmatrix}
$$

and the solution is

$$
\{X\} = \begin{Bmatrix} 0.71 \\ -0.57 \\ 0.43 \\ -0.29 \end{Bmatrix}
$$

The second time SOLVE is executed, NDOF = 4, SBW = 2, SYMM = 'True',

$$
\{B\} = \begin{Bmatrix} 0 \\ 1 \\ 0 \\ 0 \end{Bmatrix}
$$

and the solution is

$$
\{X\} = \begin{Bmatrix} -0.57 \\ 0.86 \\ -0.64 \\ 0.43 \end{Bmatrix}
$$

13.6 SOURCE CODE LISTING

```
          SUBROUTINE DECOMP(NDOF,SBW,SYMM,M)
C**********************************************************************
C
C   13.1   PURPOSE:
C             SUBROUTINES DECOMP AND SOLVE SOLVE A SYSTEM OF
C             LINEAR EQUATIONS OF THE FORM
C                       [M] {X} = {B}
C             WHERE [M] IS A BANDED MATRIX OF KNOWN COEFFICIENTS
C             (SYMMETRIC OR NONSYMMETRIC), {X} ARE THE UNKNOWNS,
C             AND {B} IS A VECTOR OF KNOWN VALUES
C
C   13.2   INPUT:
C             NONE
C
C   13.3   OUTPUT:
C             NONE
C
C   13.4   DEFINITIONS OF VARIABLES:
C             B(I) = VECTOR OF KNOWN VALUES
C             M(IJ) = MATRIX OF KNOWN COEFFICIENTS IN VECTOR STORAGE
C             NDOF = NUMBER OF UNKNOWN VALUES IN {X}
C              SBW = SEMI-BANDWIDTH OF [M]
C             SYMM = LOGICAL VARIABLE
C                  = 'TRUE' IF [M] IS SYMMETRIC
C                  = 'FALSE' IF [M] IS NONSYMMETRIC
C             X(I) = VECTOR OF UNKNOWN VALUES TO BE COMPUTED
C
C   13.5   USAGE:
C             SUBROUTINE DECOMP PERFORMS TRIANGULAR DECOMPOSITION
C             ON THE MATRIX OF KNOWN COEFFICIENTS IN VECTOR MATRIX
C             STORAGE, {M}.  THE RESULTING UPPER-, AND LOWER-
C             TRIANGULAR MATRICES ARE STORED IN {M} (THE ORIGINAL
C             CONTENTS OF {M} ARE OVERWRITTEN DURING THE
C             DECOMPOSITION PROCESS).  SUBROUTINE SOLVE SOLVES FOR
C             VALUES OF THE UNKNOWNS BY BACKWARD SUBSTITUTION.  ONCE
C             {M} HAS BEEN DECOMPOSED SOLVE CAN BE USED TO OBTAIN
C             VALUES OF {X} FOR ANY NUMBER OF DIFFERENT VECTORS {B}.
C
C          SUBROUTINES CALLED:
C            LOC
C
C          REFERENCES:
C            ISTOK,J.D. GROUNDWATER FLOW AND SOLUTE TRANSPORT
C            MODELING BY THE FINITE ELEMENT METHOD, CHAPTER 13.
C
C**********************************************************************
          INTEGER NDOF,SBW
          LOGICAL SYMM
          REAL M(1)
C
          IF (SYMM) THEN
C             M IS A SYMMETRIC MATRIX
              J2 = SBW
              IJ = 0
              DO 30 I = 1, NDOF
                 II = IJ + 1
```

```
         DO 20 J = I, J2
           IJ = IJ + 1
           IF (I .GT. 1) THEN
             K1 = J - SBW + 1
             IF (K1 .LT. I) THEN
               IF (K1 .LE. 0) K1 = 1
               DO 10 K = K1, (I-1)
                 CALL LOC(K,I,KI,NDOF,SBW,SYMM)
                 CALL LOC(K,J,KJ,NDOF,SBW,SYMM)
                 M(IJ) = M(IJ) - M(KI) * M(KJ)
10               CONTINUE
             ENDIF
           ENDIF
           IF (I .EQ. J) THEN
             M(IJ) = SQRT(M(IJ))
           ELSE
             M(IJ) = M(IJ) / M(II)
           ENDIF
20       CONTINUE
         IF (J2 .LT. NDOF) J2 = J2 + 1
30     CONTINUE
      ELSE
C     M IS A NONSYMMETRIC MATRIX
      J1 = 1
      J2 = SBW
      IJ = 0
      DO 60 I = 1, NDOF
        II = IJ + I - J1 + 1
        K1 = J1
        IKBEG = IJ + 1
        DO 50 J = J1, J2
          IJ = IJ + 1
          IF (J .GT. SBW .AND. I .LT. J) THEN
            K1 = K1 + 1
            IKBEG = IKBEG + 1
          ENDIF
          K2 = MIN(I,J) - 1
          IF (K2 .GE. K1) THEN
            IK = IKBEG
            DO 40 K = K1, K2
              CALL LOC(K,J,KJ,NDOF,SBW,SYMM)
              M(IJ) = M(IJ) - M(IK) * M(KJ)
              IK = IK + 1
40          CONTINUE
          ENDIF
          IF (I .LT. J) THEN
            M(IJ) = M(IJ) / M(II)
          ENDIF
50      CONTINUE
        IF (I .GE. SBW) J1 = J1 + 1
        IF (J2 .LT. NDOF) J2 = J2 + 1
60    CONTINUE
      ENDIF
      RETURN
      END
```

```
          SUBROUTINE SOLVE(NDOF,SBW,SYMM,M,B,X)
C**********************************************************************
C
C   13.1   PURPOSE:
C             SUBROUTINES DECOMP AND SOLVE SOLVE A SYSTEM OF
C             LINEAR EQUATIONS OF THE FORM
C                       [M] {X} = {B}
C             WHERE [M] IS A BANDED MATRIX OF KNOWN COEFFICIENTS
C             (SYMMETRIC OR NONSYMMETRIC), {X} ARE THE UNKNOWNS,
C             AND {B} IS A VECTOR OF KNOWN VALUES
C
C   13.2   INPUT:
C             NONE
C
C   13.3   OUTPUT:
C             NONE
C
C   13.4   DEFINITIONS OF VARIABLES:
C             B(I) = VECTOR OF KNOWN VALUES
C             M(IJ) = MATRIX OF KNOWN COEFFICIENTS IN VECTOR STORAGE
C             NDOF = NUMBER OF UNKNOWN VALUES {X}
C              SBW = SEMI-BANDWIDTH OF [M]
C             SYMM = LOGICAL VARIABLE
C                  = 'TRUE' IF [M] IS SYMMETRIC
C                  = 'FALSE' IF [M] IS NONSYMMETRIC
C             X(I) = VECTOR OF UNKNOWN VALUES TO BE COMPUTED
C
C   13.5   USAGE:
C             SUBROUTINE DECOMP PERFORMS TRIANGULAR DECOMPOSITION
C             ON THE MATRIX OF KNOWN COEFFICIENTS IN VECTOR MATRIX
C             STORAGE, {M}.  THE RESULTING UPPER-, AND LOWER-
C             TRIANGULAR MATRICES ARE STORED IN {M} (THE ORIGINAL
C             CONTENTS OF {M} ARE OVERWRITTEN DURING THE
C             DECOMPOSITION PROCESS). SUBROUTINE SOLVE SOLVES FOR
C             VALUES OF THE UNKNOWNS BY BACKWARD SUBSTITUTION.  ONCE
C             {M} HAS BEEN DECOMPOSED SOLVE CAN BE USED TO OBTAIN
C             VALUES OF {X} FOR ANY NUMBER OF DIFFERENT VECTORS {B}.
C
C          SUBROUTINES CALLED:
C            LOC
C
C          REFERENCES:
C            ISTOK,J.D. GROUNDWATER FLOW AND SOLUTE TRANSPORT
C            MODELING BY THE FINITE ELEMENT METHOD, CHAPTER 13.
C
C**********************************************************************
          INTEGER NDOF,SBW
          LOGICAL SYMM
          REAL M(1),B(1),X(1)
C
          DO 10 I = 1, NDOF
            X(I) = B(I)
   10     CONTINUE
          IF (SYMM) THEN
C         M IS A SYMMETRIC MATRIX
            K2 = SBW
            IK = 0
            DO 30 I = 1, NDOF
              DO 20 K = I, K2
```

```
          IK = IK + 1
          IF (K .EQ. I) THEN
             X(K) = X(K) / M(IK)
          ELSE
             X(K) = X(K) - M(IK) * X(I)
          ENDIF
20        CONTINUE
          IF (K2 .LT. NDOF) K2 = K2 + 1
30     CONTINUE
       K2 = 0
       DO 50 I = NDOF, 1, -1
          IF (K2 .GT. 0) THEN
             DO 40 K = (I + K2), (I + 1), -1
                X(I) = X(I) - M(IK) * X(K)
                IK = IK - 1
40           CONTINUE
          ENDIF
          X(I) = X(I) / M(IK)
          IK = IK - 1
          IF (K2 .LT. (SBW - 1)) K2 = K2 + 1
50     CONTINUE
     ELSE
C      M IS A NONSYMMETRIC MATRIX
       X(1) = X(1) / M(1)
       IF (NDOF .GT. 1) THEN
          K2 = 1
          DO 70 I = 2, NDOF
             IF (I .GT. SBW) K2 = K2 + 1
             CALL LOC(I,I,II,NDOF,SBW,SYMM)
             IF (I .GT. K2) THEN
                IK = II - 1
                DO 60 K = (I - 1), K2, -1
                   X(I) = X(I) - M(IK) * X(K)
                   IK = IK - 1
60              CONTINUE
             ENDIF
             X(I) = X(I) / M(II)
70        CONTINUE
       ENDIF
       J = NDOF - SBW + 1
       K2 = NDOF
       IF (NDOF .GT. 1) THEN
          DO 90 I = (NDOF - 1), 1, -1
             IF (I .LT. J) K2 = K2 - 1
             IF (I .LT. K2) THEN
                CALL LOC(I,I,II,NDOF,SBW,SYMM)
                IK = II + 1
                DO 80 K = (I + 1), K2
                   X(I) = X(I) - M(IK) * X(K)
                   IK = IK + 1
80              CONTINUE
             ENDIF
90        CONTINUE
       ENDIF
     ENDIF
   ENDIF
   RETURN
   END
```

Chapter 14

SUBROUTINE VELOCITY

14.1 PURPOSE

To compute the components of apparent groundwater velocity for each element in the mesh.

14.2 INPUT

None

14.3 OUTPUT

The components of apparent groundwater velocity are written to the user-defined file assigned to unit "OUTF".

14.4 DEFINITIONS OF VARIABLES

DIM	=	Type of coordinate system used in this problem (Figure 8.1).
	=	1, problem is one-dimensional.
	=	2, problem is two-dimensional.
	=	3, problem is three-dimensional.
	=	4, problem is two-dimensional (axisymmetric).
E	=	Element number.
ELEMTYP(I)	=	Element type for element I.
NUMELM	=	Number of elements in the mesh.
V1(I)	=	Apparent groundwater velocity in x coordinate direction (DIM =1,2, or 3).
	=	Apparent groundwater velocity in r coordinate direction (DIM = 4).
V2(I)	=	Unused (DIM = 1).
	=	Apparent groundwater velocity in y coordinate direction (DIM = 2 or 3).
	=	Apparent groundwater velocity in z coordinate direction (DIM = 4).
V3(I)	=	Unused (DIM = 1, 2, or 4).
	=	Apparent groundwater velocity in z coordinate direction (DIM = 3).

14.5 USAGE

The components of apparent groundwater velocity are computed in a set of subroutines, one subroutine for each element type (Table 14.1). Each subroutine name in this set begins with the letter "V" (for velocity) followed by three or four letters that identify the element type and the number of nodes in elements of that type. For example, subroutine VBAR2 computes the single component of apparent groundwater velocity $(v_x^{(e)})$ for one–dimensional, linear bar elements and subroutine VPAR32 computes the three components of apparent groundwater velocity $(v_x^{(e)}, v_y^{(e)}, v_z^{(e)})$ for three–dimensional, cubic parallelepiped elements.

Table 14.1 Subroutines used to compute components of apparent groundwater velocity in VELOCITY.

Element Type	Description	Subroutine Name	DIM
1	Linear bar	VBAR2	1
2	Quadratic bar	VBAR3	1
3	Cubic bar	VBAR4	1
4	Linear triangle	VTRI3	2
5	Linear rectangle	VREC4	2
6	Linear quadrilateral	VQUA4	2
7	Quadratic quadrilateral	VQUA8	2
8	Cubic quadrilateral	VQUA12	2
9	Linear parallelepiped	VPAR8	3
10	Quadratic parallelepiped	VPAR20	3
11	Cubic parallelepiped	VPAR32	3
12	Linear triangle (axisymmetric)	VTRI3A*	4
13	Linear rectangle (axisymmetric)	VREC4A*	4

* Source code listing not provided for these subroutines.

The source code listings for each of the element velocity subroutines gives the figure number that shows the interpolation functions for that element type and the equation used to compute the velocity components. For the linear rectangle and quadrilateral elements the components of apparent groundwater velocity are computed at the center of the element. A list of many of the FORTRAN variable names used in these subroutines and the corresponding textbook symbols are in Table 12.2. The variable names and symbols for the velocity components are in Table 14.2

The mesh in Figure 14.1 consists of three, linear bar elements (ELEMTYP(I) = 1, I = 1, 2, 3). Apparent groundwater velocities are computed using subroutine VBAR2. The output lists the computed value of apparent groundwater velocity in the x coordinate direction for each element.

The mesh in Figure 14.2 consists of twelve linear triangle elements (ELEMTYP(I) = 4, I = 1, ... , 12). Apparent groundwater velocities are computed using subroutine VTRI3. The output lists the computed values of apparent groundwater velocity in the x and y coordinate directions for each element.

Table 14.2   FORTRAN variable names and textbook symbols for components of apparent groundwater velocity in VELOCITY.

FORTRAN Variable Name	Definition	Symbol in Text
VXE	Velocity component in x coordinate direction	$v_x^{(e)}$
VYE	Velocity component in y coordinate direction	$v_y^{(e)}$
VZE	Velocity component in z coordinate direction	$v_z^{(e)}$
VRE	Velocity component in r coordinate direction	$v_r^{(e)}$

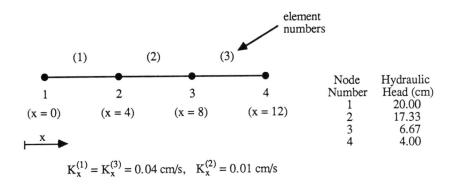

$$K_x^{(1)} = K_x^{(3)} = 0.04 \text{ cm/s}, \quad K_x^{(2)} = 0.01 \text{ cm/s}$$

Output:

COMPUTED VALUES OF APPARENT GROUNDWATER VELOCITY

ELEMENT	VX
1	2.666668E-02
2	2.666666E-02
3	2.666666E-02

Figure 14.1   Example output for subroutine VELOCITY.

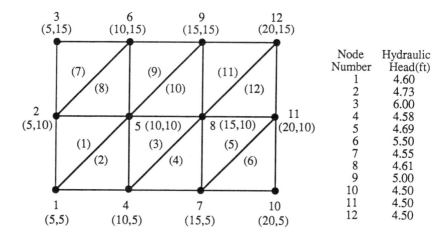

$$K_x^{(1)} = K_x^{(2)} = \ldots K_x^{(6)} = 0.01 \ \text{ft/s}$$

$$K_x^{(7)} = K_x^{(8)} = \ldots K_x^{(12)} = 0.02 \ \text{ft/s}$$

$$K_y^{(1)} = K_y^{(2)} = \ldots K_y^{(12)} = 0.001 \ \text{ft/s}$$

Output :

COMPUTED VALUES OF APPARENT GROUNDWATER VELOCITY

ELEMENT	VX	VY
1	7.999995E-05	-2.600000E-05
2	4.000018E-05	-2.200002E-05
3	1.599997E-04	-2.200002E-05
4	5.999957E-05	-1.200001E-05
5	2.200000E-04	-1.200001E-05
6	1.000001E-05	2.235174E-11
7	2.000000E-03	-2.540000E-04
8	1.600003E-04	-1.620000E-04
9	2.000000E-03	-1.620000E-04
10	3.199996E-04	-7.799997E-05
11	2.000000E-03	-7.799997E-05
12	4.400007E-04	2.235174E-11

Figure 14.2   **Example output for subroutine VELOCITY.**

14.6 SOURCE CODE LISTING

```
          SUBROUTINE VELOCITY
C**********************************************************************
C
C 14.1  PURPOSE:
C            TO COMPUTE THE COMPONENTS OF APPARENT GROUNDWATER
C            VELOCITY FOR EACH ELEMENT IN THE MESH
C
C 14.2  INPUT:
C          NONE
C
C 14.3  OUTPUT:
C            THE COMPONENTS OF APPARENT GROUNDWATER VELOCITY ARE
C            WRITTEN TO THE USER-DEFINED FILE ASSIGNED TO UNIT
C            "OUTF".
C
C 14.4  DEFINITIONS OF VARIABLES:
C                  DIM = COORDINATE SYSTEM TYPE
C                    E = ELEMENT NUMBER
C            ELEMTYP(I) = ELEMENT TYPE FOR ELEMENT I
C               NUMELM = NUMBER OF ELEMENTS IN THE MESH
C                V1(I) = APPARENT GROUNDWATER VELOCITY IN X
C                        COORDINATE DIRECTION (DIM=1, 2, OR 3)
C                      = APPARENT GROUNDWATER VELOCITY IN R
C                        COORDINATE DIRECTION (DIM=4)
C                V2(I) = UNUSED (DIM=1)
C                      = APPARENT GROUNDWATER VELOCITY IN Y
C                        COORDINATE DIRECTION (DIM=2 OR 3)
C                      = APPARENT GROUNDWATER VELOCITY IN Z
C                        COORDINATE DIRECTION (DIM=4)
C                V3(I) = UNUSED (DIM=1, 2, OR 4)
C                      = APPARENT GROUNDWATER VELOCITY IN Z
C                        COORDINATE DIRECTION (DIM=3)
C
C 14.5  USAGE:
C            THE COMPONENTS OF APPARENT GROUNDWATER VELOCITY ARE
C            COMPUTED IN A SET OF SUBROUTINES, ONE SUBROUTINE
C            FOR EACH ELEMENT TYPE.
C
C          SUBROUTINES CALLED:
C          VBAR2,VBAR3,VBAR4,VTRI3,VREC4,VQUA4,VQUA8,VQUA12,VPAR8,
C          VPAR20,VPAR32,VTRI3A,VREC4A
C
C          REFERENCES:
C          ISTOK,J.D. GROUNDWATER FLOW AND SOLUTE TRANSPORT
C          MODELING BY THE FINITE ELEMENT METHOD, CHAPTERS 6
C          AND 14.
C
C**********************************************************************
$INCLUDE:'COMALL'
C
          WRITE(OUTF,10)
   10     FORMAT(//70('*')//11X,'COMPUTED VALUES OF APPARENT ',
        1        'GROUNDWATER VELOCITY'/11X,48('-'))
          IF (DIM .EQ. 1) THEN
              WRITE(OUTF,20) '                   ','VX',' ',' '
          ELSEIF (DIM .EQ. 2) THEN
              WRITE(OUTF,20) '          ','VX','VY',' '
          ELSEIF (DIM .EQ. 3) THEN
              WRITE(OUTF,20) ' ','VX','VY','VZ'
```

```
          ELSEIF (DIM .EQ. 4) THEN
             WRITE(OUTF,20) '             ','VR','VZ',' '
          ENDIF
   20     FORMAT(/7X,A,'ELEMENT',10X,A,2(13X,A)/)
C         COMPUTE THE COMPONENTS OF APPARENT GROUNDWATER VELOCITY
C         FOR EACH ELEMENT
          DO 40 E = 1, NUMELM
             IF (ELEMTYP(E) .EQ. 1) THEN
C               ELEMENT IS A LINEAR BAR
                CALL VBAR2(E,V1(E))
             ELSEIF (ELEMTYP(E) .EQ. 2) THEN
C               ELEMENT IS A QUADRATIC BAR
                CALL VBAR3(E,V1(E))
             ELSEIF (ELEMTYP(E) .EQ. 3) THEN
C               ELEMENT IS A CUBIC BAR
                CALL VBAR4(E,V1(E))
             ELSEIF (ELEMTYP(E) .EQ. 4) THEN
C               ELEMENT IS A LINEAR TRIANGLE
                CALL VTRI3(E,V1(E),V2(E))
             ELSEIF (ELEMTYP(E) .EQ. 5) THEN
C               ELEMENT IS A LINEAR RECTANGLE
                CALL VREC4(E,V1(E),V2(E))
             ELSEIF (ELEMTYP(E) .EQ. 6) THEN
C               ELEMENT IS A LINEAR QUADRILATERAL
                CALL VQUA4(E,V1(E),V2(E))
             ELSEIF (ELEMTYP(E) .EQ. 7) THEN
C               ELEMENT IS A QUADRATIC QUADRILATERAL
                CALL VQUA8(E,V1(E),V2(E))
             ELSEIF (ELEMTYP(E) .EQ. 8) THEN
C               ELEMENT IS A CUBIC QUADRILATERAL
                CALL VQUA12(E,V1(E),V2(E))
             ELSEIF (ELEMTYP(E) .EQ. 9) THEN
C               ELEMENT IS A LINEAR PARALLELEPIPED
                CALL VPAR8(E,V1(E),V2(E),V3(E))
             ELSEIF (ELEMTYP(E) .EQ. 10) THEN
C               ELEMENT IS A QUADRATIC PARALLELEPIPED
                CALL VPAR20(E,V1(E),V2(E),V3(E))
             ELSEIF (ELEMTYP(E) .EQ. 11) THEN
C               ELEMENT IS A CUBIC PARALLELEPIPED
                CALL VPAR32(E,V1(E),V2(E),V3(E))
             ELSEIF (ELEMTYP(E) .EQ. 12) THEN
C               ELEMENT IS A LINEAR TRIANGLE (AXISYMMETRIC)
                CALL VTRI3A(E,V1(E),V2(E))
             ELSEIF (ELEMTYP(E) .EQ. 13) THEN
C               ELEMENT IS A LINEAR RECTANGLE (AXISYMMETRIC)
                CALL VREC4A(E,V1(E),V2(E))
             ENDIF
             IF (DIM .EQ. 1) THEN
                WRITE(OUTF,30) '                    ',E,V1(E)
             ELSEIF (DIM .EQ. 2) THEN
                WRITE(OUTF,30) '          ',E,V1(E),V2(E)
             ELSEIF (DIM .EQ. 3) THEN
                WRITE(OUTF,30) ' ',E,V1(E),V2(E),V3(E)
             ELSE
                WRITE(OUTF,30) '          ',E,V1(E),V2(E)
             ENDIF
   30        FORMAT(7X,A,I5,4X,1P3E15.6)
   40     CONTINUE
          RETURN
          END
```

```
      SUBROUTINE VBAR2(E,VXE)
C*************************************************************************
C
C        PURPOSE:
C           TO COMPUTE APPARENT GROUNDWATER VELOCITY FOR A
C           ONE-DIMENSIONAL, LINEAR BAR ELEMENT
C
C        DEFINITIONS OF VARIABLES:
C                  DHDX = PARTIAL DERIVATIVE OF HEAD WITH RESPECT TO X
C                     E = ELEMENT NUMBER
C                   KXE = HYDRAULIC CONDUCTIVITY IN X COORDINATE DIRECTION
C                    LE = ELEMENT LENGTH
C                   VXE = APPARENT GROUNDWATER VELOCITY IN
C                         X COORDINATE DIRECTION
C          X(IN(E,I)) = COMPUTED HEAD FOR NODE I, ELEMENT E
C          X1(IN(E,I)) = X COORDINATE FOR NODE I, ELEMENT E
C
C        REFERENCES:
C           ISTOK,J.D. GROUNDWATER FLOW AND SOLUTE TRANSPORT
C           MODELING BY THE FINITE ELEMENT METHOD, FIGURE 4.5,
C           EQUATIONS 6.10A, 6.12A.
C
C*************************************************************************
$INCLUDE:'COMALL'
      REAL KXE,LE
C
      KXE = PROP(MATSET(E),1)
      LE  = X1(IN(E,2)) - X1(IN(E,1))
      DHDX = (X(IN(E,2)) - X(IN(E,1))) / LE
      VXE = -KXE * DHDX
      RETURN
      END



      SUBROUTINE VBAR3(E,VXE)
C*************************************************************************
C
C        PURPOSE:
C           TO COMPUTE APPARENT GROUNDWATER VELOCITY FOR A
C           ONE-DIMENSIONAL, QUADRATIC BAR ELEMENT
C
C        DEFINITIONS OF VARIABLES:
C                  DHDX = PARTIAL DERIVATIVE OF HEAD WITH RESPECT TO X
C                DNDXI(I) = PARTIAL DERIVATIVE OF INTERPOLATION
C                         FUNCTION WITH RESPECT TO XI FOR NODE I
C                     E = ELEMENT NUMBER
C                   JAC = JACOBIAN MATRIX
C                JACINV = INVERSE OF JACOBIAN MATRIX
C                   KXE = HYDRAULIC CONDUCTIVITY IN X COORDINATE DIRECTION
C                   VXE = APPARENT GROUNDWATER VELOCITY IN
C                         X COORDINATE DIRECTION
C          X(IN(E,I)) = COMPUTED HEAD FOR NODE I, ELEMENT E
C          X1(IN(E,I)) = X COORDINATE FOR NODE I, ELEMENT E
C
C        REFERENCES:
C           ISTOK,J.D. GROUNDWATER FLOW AND SOLUTE TRANSPORT
C           MODELING BY THE FINITE ELEMENT METHOD, FIGURE 4.9B,
C           EQUATIONS 6.18 AND 6.26A.
C
C*************************************************************************
```

```
$INCLUDE: 'COMALL'
      REAL DNDXI(3),JAC,JACINV,KXE
C
      KXE = PROP(MATSET(E),1)
      DNDXI(1)= -0.5
      DNDXI(2)=  0.0
      DNDXI(3)=  0.5
      JAC = 0
      DO 10 I = 1, 3
        JAC = JAC + DNDXI(I) * X1(IN(E,I))
   10 CONTINUE
      JACINV = 1 / JAC
      DHDX = 0.
      DO 20 I = 1, 3
        DHDX = DHDX + JACINV * DNDXI(I) * X(IN(E,I))
   20 CONTINUE
      VXE = -KXE * DHDX
      RETURN
      END

      SUBROUTINE VBAR4(E,VXE)
C***********************************************************************
C
C        PURPOSE:
C          TO COMPUTE APPARENT GROUNDWATER VELOCITY FOR A
C          ONE-DIMENSIONAL, CUBIC BAR ELEMENT
C
C        DEFINITIONS OF VARIABLES:
C                 DHDX = PARTIAL DERIVATIVE OF HEAD WITH RESPECT
C                        TO X COORDINATE DIRECTION
C             DNDXI(I) = PARTIAL DERIVATIVE OF INTERPOLATION
C                        FUNCTION WITH RESPECT TO XI FOR NODE I
C                    E = ELEMENT NUMBER
C                  JAC = JACOBIAN MATRIX
C               JACINV = INVERSE OF JACOBIAN MATRIX
C                  KXE = HYDRAULIC CONDUCTIVITY IN X COORDINATE DIRECTION
C                  VXE = APPARENT GROUNDWATER VELOCITY IN
C                        X COORDINATE DIRECTION
C          X(IN(E,I)) = COMPUTED HEAD FOR NODE I, ELEMENT E
C         X1(IN(E,I)) = X COORDINATE OF NODE I, ELEMENT E
C
C        REFERENCES:
C          ISTOK,J.D. GROUNDWATER FLOW AND SOLUTE TRANSPORT
C          MODELING BY THE FINITE ELEMENT METHOD, FIGURE 4.9C,
C          EQUATIONS 6.18 AND 6.26A.
C
C***********************************************************************
$INCLUDE: 'COMALL'
      REAL DNDXI(4),JAC,JACINV,KXE
C
      KXE = PROP(MATSET(E),1)
      DNDXI(1) =  1. / 16.
      DNDXI(2) = -27. / 16.
      DNDXI(3) = -DNDXI(2)
      DNDXI(4) = -DNDXI(1)
      JAC = 0
      DO 20 I = 1, 4
        JAC = JAC + DNDXI(I) * X(IN(E,I))
   20 CONTINUE
      JACINV = 1 / JAC
```

```
      DHDX = 0.
      DO 30 I = 1, 4
         DHDX = DHDX + JACINV * DNDXI(I) * X(IN(E,I))
   30 CONTINUE
      VXE = -KXE * DHDX
      RETURN
      END

      SUBROUTINE VTRI3(E,VXE,VYE)
C***********************************************************************
C       PURPOSE:
C          TO COMPUTE COMPONENTS OF APPARENT GROUNDWATER
C          VELOCITY FOR A TWO-DIMENSIONAL, LINEAR TRIANGLE ELEMENT
C
C       DEFINITIONS OF VARIABLES:
C                  DHDX = PARTIAL DERIVATIVE OF HEAD WITH RESPECT TO X
C                  DHDY = PARTIAL DERIVATIVE OF HEAD WITH RESPECT TO Y
C              DNDX(I) = PARTIAL DERIVATE OF INTERPOLATION
C                         FUNCTION WITH RESPECT TO X FOR NODE I
C              DNDY(I) = PARTIAL DERIVATIVE OF INTERPOLATION
C                         FUNCTION WITH RESPECT TO Y FOR NODE I
C                     E = ELEMENT NUMBER
C                   KXE = HYDRAULIC CONDUCTIVITY IN X COORDINATE DIRECTION
C                   KYE = HYDRAULIC CONDUCTIVITY IN Y COORDINATE DIRECTION
C                   VXE = APPARENT GROUNDWATER VELOCITY IN
C                         X COORDINATE DIRECTION
C                   VYE = APPARENT GROUNDWATER VELOCITY IN
C                         Y COORDINATE DIRECTION
C            X(IN(E,I)) = COMPUTED HEAD FOR NODE I, ELEMENT E
C           X1(IN(E,I)) = X COORDINATE FOR NODE I, ELEMENT E
C           X2(IN(E,I)) = Y COORDINATE FOR NODE I, ELEMENT E
C
C       REFERENCES:
C          ISTOK,J.D. GROUNDWATER FLOW AND SOLUTE TRANSPORT
C          MODELING BY THE FINITE ELEMENT METHOD, FIGURE 4.6,
C          EQUATIONS 6.10A, 610B, 6.12A, AND 6.12B.
C***********************************************************************
$INCLUDE: 'COMALL'
      REAL DNDX(3),DNDY(3),KXE,KYE
C
      KXE = PROP(MATSET(E),1)
      KYE = PROP(MATSET(E),2)
      AE2 = X1(IN(E,2)) * X2(IN(E,3)) + X1(IN(E,1)) * X2(IN(E,2)) +
     1      X2(IN(E,1)) * X1(IN(E,3)) - X2(IN(E,3)) * X1(IN(E,1)) -
     2      X1(IN(E,3)) * X2(IN(E,2)) - X1(IN(E,2)) * X2(IN(E,1))
      DNDX(1) = (X2(IN(E,2)) - X2(IN(E,3))) / AE2
      DNDX(2) = (X2(IN(E,3)) - X2(IN(E,1))) / AE2
      DNDX(3) = (X2(IN(E,1)) - X2(IN(E,2))) / AE2
      DNDY(1) = (X1(IN(E,3)) - X1(IN(E,2))) / AE2
      DNDY(2) = (X1(IN(E,1)) - X1(IN(E,3))) / AE2
      DNDY(3) = (X1(IN(E,2)) - X1(IN(E,1))) / AE2
      DHDX = 0.
      DHDY = 0.
      DO 20 I = 1, 3
         DHDX = DHDX + DNDX(I) * X(IN(E,I))
         DHDY = DHDY + DNDY(I) * X(IN(E,I))
   20 CONTINUE
      VXE = -KXE * DHDX
      VYE = -KYE * DHDY
      RETURN
      END
```

```
      SUBROUTINE VREC4(E,VXE,VYE)
C******************************************************************************
C
C     PURPOSE:
C        TO COMPUTE COMPONENTS OF APPARENT GROUNDWATER
C        VELOCITY FOR A TWO-DIMENSIONAL, LINEAR RECTANGLE ELEMENT
C
C     DEFINITIONS OF VARIABLES:
C               DHDX = PARTIAL DERIVATIVE OF HEAD WITH RESPECT TO X
C               DHDY = PARTIAL DERIVATIVE OF HEAD WITH RESPECT TO Y
C           DNDX(I) = PARTIAL DERIVATE OF INTERPOLATION
C                     FUNCTION WITH RESPECT TO X FOR NODE I
C           DNDY(I) = PARTIAL DERIVATIVE OF INTERPOLATION
C                     FUNCTION WITH RESPECT TO Y FOR NODE I
C                  E = ELEMENT NUMBER
C                KXE = HYDRAULIC CONDUCTIVITY IN X COORDINATE DIRECTION
C                KYE = HYDRAULIC CONDUCTIVITY IN Y COORDINATE DIRECTION
C                VXE = APPARENT GROUNDWATER VELOCITY IN
C                      X COORDINATE DIRECTION
C                VYE = APPARENT GROUNDWATER VELOCITY IN
C                      Y COORDINATE DIRECTION
C        X(IN(E,I)) = COMPUTED HEAD FOR NODE I, ELEMENT E
C       X1(IN(E,I)) = X COORDINATE FOR NODE I, ELEMENT E
C       X2(IN(E,I)) = Y COORDINATE FOR NODE I, ELEMENT E
C
C     REFERENCES:
C        ISTOK,J.D. GROUNDWATER FLOW AND SOLUTE TRANSPORT
C        MODELING BY THE FINITE ELEMENT METHOD, FIGURE 4.7,
C        EQUATIONS 6.10A, 6.10B, 6.12A, AND 6.12B.
C
C******************************************************************************
$INCLUDE: 'COMALL'
      REAL DNDX(4),DNDY(4),KXE,KYE
C
      KXE = PROP(MATSET(E),1)
      KYE = PROP(MATSET(E),2)
      AE = ABS(X2(IN(E,1)) - X2(IN(E,3))) / 2.
      BE = ABS(X1(IN(E,1)) - X1(IN(E,3))) / 2.

      DNDX(1) = - 1. / (2.*BE)
      DNDX(2) = -DNDX(1)
      DNDX(3) =   0
      DNDX(4) =   0
      DNDY(1) = - 1. / (2.*AE)
      DNDY(2) =   0
      DNDY(3) =   0
      DNDY(4) = -DNDY(1)

      DHDX = 0.
      DHDY = 0.
      DO 10 I = 1, 4
        DHDX = DHDX + DNDX(I) * X(IN(E,I))
        DHDY = DHDY + DNDY(I) * X(IN(E,I))
   10 CONTINUE

      VXE = -KXE * DHDX
      VYE = -KYE * DHDY
      RETURN
      END
```

```
          SUBROUTINE VQUA4(E,VXE,VYE)
C************************************************************************
C
C         PURPOSE:
C           TO COMPUTE COMPONENTS OF APPARENT GROUNDWATER
C           VELOCITY FOR A TWO-DIMENSIONAL, LINEAR QUADRILATERAL
C           ELEMENT
C
C         DEFINITIONS OF VARIABLES:
C                 DETJAC = DETERMINANT OF JACOBIAN MATRIX
C                   DHDX = PARTIAL DERIVATIVE OF HEAD WITH RESPECT TO X
C                   DHDY = PARTIAL DERIVATIVE OF HEAD WITH RESPECT TO Y
C              DNDXI(I) = PARTIAL DERIVATIVE OF INTERPOLATION
C                         FUNCTION WITH RESPECT TO XI FOR NODE I
C               DNDX(I) = PARTIAL DERIVATIVE OF INTERPOLATION
C                         FUNCTION WITH RESPECT TO X FOR NODE I
C             DNDETA(I) = PARTIAL DERIVATIVE OF INTERPOLATION
C                         FUNCTION WITH RESPECT TO ETA FOR NODE I
C               DNDY(I) = PARTIAL DERIVATIVE OF INTERPOLATION
C                         FUNCTION WITH RESPECT TO Y FOR NODE I
C                     E = ELEMENT NUMBER
C               JAC(I,J) = JACOBIAN MATRIX
C            JACINV(I,J) = INVERSE OF JACOBIAN MATRIX
C                    KXE = HYDRAULIC CONDUCTIVITY IN X COORDINATE DIRECTION
C                    KYE = HYDRAULIC CONDUCTIVITY IN Y COORDINATE DIRECTION
C                    VXE = APPARENT GROUNDWATER VELOCITY IN
C                          X COORDINATE DIRECTION
C                    VYE = APPARENT GROUNDWATER VELOCITY IN
C                          Y COORDINATE DIRECTION
C             X(IN(E,I)) = COMPUTED HEAD FOR NODE I, ELEMENT E
C            X1(IN(E,I)) = X COORDINATE FOR NODE I, ELEMENT E
C            X2(IN(E,I)) = Y COORDINATE FOR NODE I, ELEMENT E
C
C         REFERENCES:
C           ISTOK,J.D. GROUNDWATER FLOW AND SOLUTE TRANSPORT
C           MODELING BY THE FINITE ELEMENT METHOD, FIGURE 4.10,
C           EQUATIONS 6.14A, 6.14B, 6.17, 6.22A, AND 6.22B.
C
C************************************************************************

$INCLUDE: 'COMALL'
          REAL JAC(2,2),JACINV(2,2),DNDXI(4),DNDX(4),
     1        DNDETA(4),DNDY(4),SIGN1(4),SIGN2(4),KXE,KYE
          DATA SIGN1/-1.,1.,1.,-1./
          DATA SIGN2/-1.,-1.,1.,1./
C
          KXE = PROP(MATSET(E),1)
          KYE = PROP(MATSET(E),2)

          DO 20 I = 1, 2
            DO 10 J = 1, 2
              JAC(I,J) = 0.
  10        CONTINUE
  20      CONTINUE

          DO 30 I = 1, 4
            DNDXI(I)  = 0.25 * SIGN1(I)
            DNDETA(I) = 0.25 * SIGN2(I)
```

```
30      CONTINUE

        DO 40 I = 1, 4
          JAC(1,1) = JAC(1,1) + DNDXI(I) * X1(IN(E,I))
          JAC(1,2) = JAC(1,2) + DNDXI(I) * X2(IN(E,I))
          JAC(2,1) = JAC(2,1) + DNDETA(I) * X1(IN(E,I))
          JAC(2,2) = JAC(2,2) + DNDETA(I) * X2(IN(E,I))
40      CONTINUE

        DETJAC = JAC(1,1) * JAC(2,2) - JAC(1,2) * JAC(2,1)
        JACINV(1,1) =  JAC(2,2) / DETJAC
        JACINV(1,2) = -JAC(1,2) / DETJAC
        JACINV(2,1) = -JAC(2,1) / DETJAC
        JACINV(2,2) =  JAC(1,1) / DETJAC

        DO 50 I = 1, 4
          DNDX(I) = JACINV(1,1) * DNDXI(I) + JACINV(1,2) * DNDETA(I)
          DNDY(I) = JACINV(2,1) * DNDXI(I) + JACINV(2,2) * DNDETA(I)
50      CONTINUE

        DHDX = 0.
        DHDY = 0.

        DO 60 I = 1, 4
          DHDX = DHDX + DNDX(I) * X(IN(E,I))
          DHDY = DHDY + DNDY(I) * X(IN(E,I))
60      CONTINUE

        VXE = -KXE * DHDX
        VYE = -KYE * DHDY
        RETURN
        END


        SUBROUTINE VQUA8(E,VXE,VYE)
C*********************************************************************
C
C       PURPOSE:
C       TO COMPUTE COMPONENTS OF APPARENT GROUNDWATER
C       VELOCITY FOR A TWO-DIMENSIONAL, QUADRATIC QUADRILATERAL
C       ELEMENT
C
C       DEFINITIONS OF VARIABLES:
C            DETJAC = DETERMINANT OF JACOBIAN MATRIX
C              DHDX = PARTIAL DERIVATIVE OF HEAD WITH RESPECT TO X
C              DHDY = PARTIAL DERIVATIVE OF HEAD WITH RESPECT TO Y
C          DNDXI(I) = PARTIAL DERIVATIVE OF INTERPOLATION
C                     FUNCTION WITH RESPECT TO XI FOR NODE I
C           DNDX(I) = PARTIAL DERIVATIVE OF INTERPOLATION
C                     FUNCTION WITH RESPECT TO X FOR NODE I
C         DNDETA(I) = PARTIAL DERIVATIVE OF INTERPOLATION
C                     FUNCTION WITH RESPECT TO ETA FOR NODE I
C           DNDY(I) = PARTIAL DERIVATIVE OF INTERPOLATION
C                     FUNCTION WITH RESPECT TO Y FOR NODE I
C                 E = ELEMENT NUMBER
C          JAC(I,J) = JACOBIAN MATRIX
C       JACINV(I,J) = INVERSE OF JACOBIAN MATRIX
C               KXE = HYDRAULIC CONDUCTIVITY IN X COORDINATE DIRECTION
C               KYE = HYDRAULIC CONDUCTIVITY IN Y COORDINATE DIRECTION
C               VXE = APPARENT GROUNDWATER VELOCITY IN X
C                     COORDINATE DIRECTION
C
```

```
C                   VYE = APPARENT GROUNDWATER VELOCITY IN Y
C                         COORDINATE DIRECTION
C          X(IN(E,I)) = COMPUTED HEAD FOR NODE I, ELEMENT E
C         X1(IN(E,I)) = X COORDINATE FOR NODE I, ELEMENT E
C         X2(IN(E,I)) = Y COORDINATE FOR NODE I, ELEMENT E
C
C      REFERENCES:
C        ISTOK,J.D. GROUNDWATER FLOW AND SOLUTE TRANSPORT
C        MODELING BY THE FINITE ELEMENT METHOD, FIGURE 4.11,
C        EQUATION 6.14A, 6.14B, 6.17, 6.22A, AND 6.22B.
C************************************************************************
$INCLUDE: 'COMALL'
      REAL JAC(2,2),JACINV(2,2),DNDXI(8),DNDX(8),DNDETA(8),
     1     DNDY(8),SIGN1(8),SIGN2(8),KXE,KYE
      DATA SIGN1/-1.,0.,1.,1.,1.,0.,-1.,-1./
      DATA SIGN2/-1.,-1.,-1.,0.,1.,1.,1.,0./
C
      KXE = PROP(MATSET(E),1)
      KYE = PROP(MATSET(E),2)
      DO 20 I = 1, 2
        DO 10 J = 1, 2
          JAC(I,J) = 0.
   10   CONTINUE
   20 CONTINUE
      DO 30 I = 1, 8
        IF ((I .EQ. 1) .OR. (I .EQ. 3) .OR.
     1      (I .EQ. 5) .OR. (I .EQ. 7)) THEN
          DNDXI(I) = 0.
          DNDETA(I) = 0.
        ELSEIF ((I .EQ. 2) .OR. (I .EQ. 6)) THEN
          DNDXI(I) = 0.
          DNDETA(I) = 0.5 * SIGN2(I)
        ELSEIF ((I .EQ. 4) .OR. (I .EQ. 8)) THEN
          DNDXI(I) = 0.5 * SIGN1(I)
          DNDETA(I) = 0.
        ENDIF
   30 CONTINUE
      DO 40 I = 1, 8
        JAC(1,1) = JAC(1,1) + DNDXI(I) * X1(IN(E,I))
        JAC(1,2) = JAC(1,2) + DNDXI(I) * X2(IN(E,I))
        JAC(2,1) = JAC(2,1) + DNDETA(I) * X1(IN(E,I))
        JAC(2,2) = JAC(2,2) + DNDETA(I) * X2(IN(E,I))
   40 CONTINUE
      DETJAC = JAC(1,1) * JAC(2,2) - JAC(1,2) * JAC(2,1)
      JACINV(1,1) =  JAC(2,2) / DETJAC
      JACINV(1,2) = -JAC(1,2) / DETJAC
      JACINV(2,1) = -JAC(2,1) / DETJAC
      JACINV(2,2) =  JAC(1,1) / DETJAC
      DO 50 I = 1, 8
        DNDX(I) = JACINV(1,1) * DNDXI(I) + JACINV(1,2) * DNDETA(I)
        DNDY(I) = JACINV(2,1) * DNDXI(I) + JACINV(2,2) * DNDETA(I)
   50 CONTINUE
      DHDX = 0.
      DHDY = 0.
      DO 60 I = 1, 8
        DHDX = DHDX + DNDX(I) * X(IN(E,I))
        DHDY = DHDY + DNDY(I) * X(IN(E,I))
   60 CONTINUE
      VXE = -KXE * DHDX
      VYE = -KYE * DHDY
      RETURN
      END
```

```
      SUBROUTINE VQUA12(E,VXE,VYE)
C*****************************************************************************
C
C        PURPOSE:
C           TO COMPUTE COMPONENTS OF APPARENT GROUNDWATER
C           VELOCITY FOR A TWO-DIMENSIONAL, CUBIC QUADRILATERAL
C           ELEMENT
C
C        DEFINITIONS OF VARIABLES:
C               DETJAC = DETERMINANT OF JACOBIAN MATRIX
C                 DHDX = PARTIAL DERIVATIVE OF HEAD WITH RESPECT TO X
C                 DHDY = PARTIAL DERIVATIVE OF HEAD WITH RESPECT TO Y
C            DNDXI(I) = PARTIAL DERIVATIVE OF INTERPOLATION
C                       FUNCTION WITH RESPECT TO XI FOR NODE I
C             DNDX(I) = PARTIAL DERIVATIVE OF INTERPOLATION
C                       FUNCTION WITH RESPECT TO X FOR NODE I
C           DNDETA(I) = PARTIAL DERIVATIVE OF INTERPOLATION
C                       FUNCTION WITH RESPECT TO ETA FOR NODE I
C             DNDY(I) = PARTIAL DERIVATIVE OF INTERPOLATION
C                       FUNCTION WITH RESPECT TO Y FOR NODE I
C                   E = ELEMENT NUMBER
C            JAC(I,J) = JACOBIAN MATRIX
C         JACINV(I,J) = INVERSE OF JACOBIAN MATRIX
C                 KXE = HYDRAULIC CONDUCTIVITY IN X COORDINATE DIRECTION
C                 KYE = HYDRAULIC CONDUCTIVITY IN Y COORDINATE DIRECTION
C                 VXE = APPARENT GROUNDWATER VELOCITY IN X
C                       COORDINATE DIRECTION
C                 VYE = APPARENT GROUNDWATER VELOCITY IN Y
C                       COORDINATE DIRECTION
C          X(IN(E,I)) = COMPUTED HEAD FOR NODE I, ELEMENT E
C         X1(IN(E,I)) = X COORDINATE FOR NODE I, ELEMENT E
C         X2(IN(E,I)) = Y COORDINATE FOR NODE I, ELEMENT E
C
C        REFERENCES:
C           ISTOK,J.D. GROUNDWATER FLOW AND SOLUTE TRANSPORT
C           MODELING BY THE FINITE ELEMENT METHOD, FIGURE 4.12,
C           EQUATIONS 6.14A, 6.14B, 6.17, 6.22A, AND 6.22B.
C
C*****************************************************************************

$INCLUDE: 'COMALL'
      REAL JAC(2,2),JACINV(2,2),DNDXI(12),DNDX(12),DNDETA(12),
     1     DNDY(12),SIGN1(12),SIGN2(12),KXE,KYE
      DATA SIGN1/-1.,-1.,1.,1.,1.,1.,1.,1.,-1.,-1.,-1.,-1./
      DATA SIGN2/-1.,-1.,-1.,-1.,-1.,1.,1.,1.,1.,1.,1.,-1./
C
      KXE = PROP(MATSET(E),1)
      KYE = PROP(MATSET(E),2)
      DO 20 I = 1, 2
        DO 10 J = 1, 2
          JAC(I,J) = 0.
   10   CONTINUE
   20 CONTINUE
      DO 30 I = 1, 12
        IF ((I .EQ. 1) .OR. (I .EQ. 4) .OR.
     1      (I .EQ. 7) .OR. (I .EQ. 10)) THEN
          DNDXI(I) = -(10. / 32.) * SIGN1(I)
          DNDETA(I) = -(10. / 32.) * SIGN2(I)
        ELSEIF ((I .EQ. 2) .OR. (I .EQ. 3) .OR.
     1          (I .EQ. 8) .OR. (I .EQ. 9)) THEN
          DNDXI(I) = (81. / 32.) * SIGN1(I)/3.
```

```
        DNDETA(I) = (9. / 32.) * SIGN2(I)
     ELSEIF ((I .EQ. 5) .OR. (I .EQ. 6) .OR.
   1         (I .EQ. 11) .OR. (I .EQ. 12)) THEN
        DNDXI(I) = (9. / 32.) * SIGN1(I)
        DNDETA(I) = (81. / 32.) * SIGN2(I)/3.
     ENDIF
30 CONTINUE
   DO 40 I = 1, 12
     JAC(1,1) = JAC(1,1) + DNDXI(I) * X1(IN(E,I))
     JAC(1,2) = JAC(1,2) + DNDXI(I) * X2(IN(E,I))
     JAC(2,1) = JAC(2,1) + DNDETA(I) * X1(IN(E,I))
     JAC(2,2) = JAC(2,2) + DNDETA(I) * X2(IN(E,I))
40 CONTINUE
   DETJAC = JAC(1,1) * JAC(2,2) - JAC(1,2) * JAC(2,1)
   JACINV(1,1) =  JAC(2,2) / DETJAC
   JACINV(1,2) = -JAC(1,2) / DETJAC
   JACINV(2,1) = -JAC(2,1) / DETJAC
   JACINV(2,2) =  JAC(1,1) / DETJAC
   DO 50 I = 1, 12
     DNDX(I) = JACINV(1,1) * DNDXI(I) + JACINV(1,2) * DNDETA(I)
     DNDY(I) = JACINV(2,1) * DNDXI(I) + JACINV(2,2) * DNDETA(I)
50 CONTINUE
   DHDX = 0.
   DHDY = 0.
   DO 60 I = 1, 12
     DHDX = DHDX + DNDX(I) * X(IN(E,I))
     DHDY = DHDY + DNDY(I) * X(IN(E,I))
60 CONTINUE
   VXE = -KXE * DHDX
   VYE = -KYE * DHDY
   RETURN
   END


      SUBROUTINE VPAR8(E,VXE,VYE,VZE)
C***********************************************************************
C
C      PURPOSE:
C         TO COMPUTE COMPONENTS OF APPARENT GROUNDWATER
C         VELOCITY FOR A THREE-DIMENSIONAL, LINEAR
C         PARALLELEPIPED ELEMENT
C
C      DEFINITIONS OF VARIABLES:
C            DETJAC = DETERMINANT OF JACOBIAN MATRIX
C              DHDX = PARTIAL DERIVATIVE OF HEAD WITH RESPECT TO X
C              DHDY = PARTIAL DERIVATIVE OF HEAD WITH RESPECT TO Y
C              DHDZ = PARTIAL DERIVATIVE OF HEAD WITH RESPECT TO Z
C          DNDXI(I) = PARTIAL DERIVATIVE OF INTERPOLATION
C                     FUNCTION WITH RESPECT TO XI FOR NODE I
C           DNDX(I) = PARTIAL DERIVATIVE OF INTERPOLATION
C                     FUNCTION WITH RESPECT TO X FOR NODE I
C         DNDETA(I) = PARTIAL DERIVATIVE OF INTERPOLATION
C                     FUNCTION WITH RESPECT TO ETA FOR NODE I
C           DNDY(I) = PARTIAL DERIVATIVE OF INTERPOLATION
C                     FUNCTION WITH RESPECT TO Y FOR NODE I
C        DNDZETA(I) = PARTIAL DERIVATIVE OF INTERPOLATION
C                     FUNCTION WITH RESPECT TO ZETA FOR NODE I
C           DNDZ(I) = PARTIAL DERIVATIVE OF INTERPOLATION
C                     FUNCTION WITH RESPECT TO Z FOR NODE I
C                 E = ELEMENT NUMBER
C          JAC(I,J) = JACOBIAN MATRIX
C       JACINV(I,J) = INVERSE OF JACOBIAN MATRIX
```

```
C               KXE = HYDRAULIC CONDUCTIVITY IN X COORDINATE DIRECTION
C               KYE = HYDRAULIC CONDUCTIVITY IN Y COORDINATE DIRECTION
C               KZE = HYDRAULIC CONDUCTIVITY IN Z COORDINATE DIRECTION
C               VXE = APPARENT GROUNDWATER VELOCITY IN X
C                     COORDINATE DIRECTION
C               VYE = APPARENT GROUNDWATER VELOCITY IN Y
C                     COORDINATE DIRECTION
C               VZE = APPARENT GROUNDWATER VELOCITY IN Z
C                     COORDINATE DIRECTION
C         X(IN(E,I)) = COMPUTED HEAD FOR NODE I, ELEMENT E
C        X1(IN(E,I)) = X COORDINATE FOR NODE I, ELEMENT E
C        X2(IN(E,I)) = Y COORDINATE FOR NODE I, ELEMENT E
C        X3(IN(E,I)) = Z COORDINATE FOR NODE I, ELEMENT E
C
C     REFERENCES:
C        ISTOK,J.D. GROUNDWATER FLOW AND SOLUTE TRANSPORT
C        MODELING BY THE FINITE ELEMENT METHOD, FIGURE 4.13,
C        EQUATIONS 6.14, 6.16, AND 6.22
C
C*********************************************************************
$INCLUDE: 'COMALL'
      REAL JAC(3,3),JACINV(3,3),DNDXI(8),DNDX(8),DNDETA(8),DNDY(8),
     1     DNDZETA(8),DNDZ(8),SIGN1(8),SIGN2(8),SIGN3(8),KXE,KYE,
     2     KZE
      DATA SIGN1/-1.,1.,1.,-1.,-1.,1.,1.,-1./
      DATA SIGN2/-1.,-1.,1.,1.,-1.,-1.,1.,1./
      DATA SIGN3/-1.,-1.,-1.,-1.,1.,1.,1.,1./
C
      KXE = PROP(MATSET(E),1)
      KYE = PROP(MATSET(E),2)
      KZE = PROP(MATSET(E),3)
      DO 20 I = 1, 3
        DO 10 J = 1, 3
          JAC(I,J) = 0.
  10    CONTINUE
  20 CONTINUE
      DO 30 I = 1, 8
        DNDXI(I) = 0.125 * SIGN1(I)
        DNDETA(I) = 0.125 * SIGN2(I)
        DNDZETA(I) = 0.125 * SIGN3(I)
  30 CONTINUE
      DO 40 I = 1, 8
        JAC(1,1) = JAC(1,1) + DNDXI(I) * X1(IN(E,I))
        JAC(1,2) = JAC(1,2) + DNDXI(I) * X2(IN(E,I))
        JAC(1,3) = JAC(1,3) + DNDXI(I) * X3(IN(E,I))
        JAC(2,1) = JAC(2,1) + DNDETA(I) * X1(IN(E,I))
        JAC(2,2) = JAC(2,2) + DNDETA(I) * X2(IN(E,I))
        JAC(2,3) = JAC(2,3) + DNDETA(I) * X3(IN(E,I))
        JAC(3,1) = JAC(3,1) + DNDZETA(I) * X1(IN(E,I))
        JAC(3,2) = JAC(3,2) + DNDZETA(I) * X2(IN(E,I))
        JAC(3,3) = JAC(3,3) + DNDZETA(I) * X3(IN(E,I))
  40 CONTINUE
      DETJAC = JAC(1,1) * (JAC(2,2) * JAC(3,3) - JAC(3,2) *
     1         JAC(2,3)) - JAC(1,2) * (JAC(2,1) * JAC(3,3) -
     2         JAC(3,1) * JAC(2,3)) - JAC(1,3) * (JAC(2,1) *
     3         JAC(3,2) - JAC(3,1) * JAC(2,2))
      IF ( DETJAC .EQ. 0 ) STOP ' DETERMINANT IS ZERO !!!!!!!'
      JACINV(1,1) = ( JAC(2,2) * JAC(3,3) - JAC(2,3) *
     1               JAC(3,2)) / DETJAC
      JACINV(1,2) = (-JAC(2,1) * JAC(3,3) + JAC(2,3) *
     1               JAC(3,1)) / DETJAC
```

```
      JACINV(1,3) = ( JAC(2,1) * JAC(3,2) - JAC(3,1) *
     1              JAC(2,2)) / DETJAC
      JACINV(2,1) = (-JAC(1,2) * JAC(3,3) + JAC(1,3) *
     1              JAC(3,2)) / DETJAC
      JACINV(2,2) = ( JAC(1,1) * JAC(3,3) - JAC(1,3) *
     1              JAC(3,1)) / DETJAC
      JACINV(2,3) = (-JAC(1,1) * JAC(3,2) + JAC(1,2) *
     1              JAC(3,1)) / DETJAC
      JACINV(3,1) = ( JAC(1,2) * JAC(2,3) - JAC(1,3) *
     1              JAC(2,2)) / DETJAC
      JACINV(3,2) = (-JAC(1,1) * JAC(2,3) + JAC(1,3) *
     1              JAC(2,1)) / DETJAC
      JACINV(3,3) = ( JAC(1,1) * JAC(2,2) - JAC(1,2) *
     1              JAC(2,1)) / DETJAC
      DO 50 I = 1, 8
        DNDX(I) = JACINV(1,1) * DNDXI(I) + JACINV(1,2) *
     1            DNDETA(I) + JACINV(1,3) * DNDZETA(I)
        DNDY(I) = JACINV(2,1) * DNDXI(I) + JACINV(2,2) *
     1            DNDETA(I) + JACINV(2,3) * DNDZETA(I)
        DNDZ(I) = JACINV(3,1) * DNDXI(I) + JACINV(3,2) *
     1            DNDETA(I) + JACINV(3,3) * DNDZETA(I)
   50 CONTINUE
      DHDX = 0.
      DHDY = 0.
      DHDZ = 0.
      DO 60 I = 1, 8
        DHDX = DHDX + DNDX(I) * X(IN(E,I))
        DHDY = DHDY + DNDY(I) * X(IN(E,I))
        DHDZ = DHDZ + DNDZ(I) * X(IN(E,I))
   60 CONTINUE
      VXE = -KXE * DHDX
      VYE = -KYE * DHDY
      VZE = -KZE * DHDZ
      RETURN
      END

      SUBROUTINE VPAR20(E,VXE,VYE,VZE)
C*********************************************************************
C
C     PURPOSE:
C       TO COMPUTE COMPONENTS OF APPARENT GROUNDWATER
C       VELOCITY FOR A THREE-DIMENSIONAL, QUADRATIC
C       PARALLELEPIPED ELEMENT
C
C     DEFINITIONS OF VARIABLES:
C            DETJAC = DETERMINANT OF JACOBIAN MATRIX
C              DHDX = PARTIAL DERIVATIVE OF HEAD WITH RESPECT TO X
C              DHDY = PARTIAL DERIVATIVE OF HEAD WITH RESPECT TO Y
C              DHDZ = PARTIAL DERIVATIVE OF HEAD WITH RESPECT TO Z
C         DNDXI(I) = PARTIAL DERIVATIVE OF INTERPOLATION
C                    FUNCTION WITH RESPECT TO XI FOR NODE I
C          DNDX(I) = PARTIAL DERIVATIVE OF INTERPOLATION
C                    FUNCTION WITH RESPECT TO X FOR NODE I
C        DNDETA(I) = PARTIAL DERIVATIVE OF INTERPOLATION
C                    FUNCTION WITH RESPECT TO ETA FOR NODE I
C          DNDY(I) = PARTIAL DERIVATIVE OF INTERPOLATION
C                    FUNCTION WITH RESPECT TO Y FOR NODE I
C       DNDZETA(I) = PARTIAL DERIVATIVE OF INTERPOLATION
C                    FUNCTION WITH RESPECT TO ZETA FOR NODE I
C          DNDZ(I) = PARTIAL DERIVATIVE OF INTERPOLATION
C                    FUNCTION WITH RESPECT TO Z FOR NODE I
```

```
C            JAC(I,J) = JACOBIAN MATRIX
C         JACINV(I,J) = INVERSE OF JACOBIAN MATRIX
C               KXE = HYDRAULIC CONDUCTIVITY IN X COORDINATE DIRECTION
C               KYE = HYDRAULIC CONDUCTIVITY IN Y COORDINATE DIRECTION
C               KZE = HYDRAULIC CONDUCTIVITY IN Z COORDINATE DIRECTION
C               VXE = APPARENT GROUNDWATER VELOCITY IN X
C                     COORDINATE DIRECTION
C               VYE = APPARENT GROUNDWATER VELOCITY IN Y
C                     COORDINATE DIRECTION
C               VZE = APPARENT GROUNDWATER VELOCITY IN Z
C                     COORDINATE DIRECTION
C        X(IN(E,I)) = COMPUTED HEAD FOR NODE I, ELEMENT E
C       X1(IN(E,I)) = X COORDINATE FOR NODE I, ELEMENT E
C       X2(IN(E,I)) = Y COORDINATE FOR NODE I, ELEMENT E
C       X3(IN(E,I)) = Z COORDINATE FOR NODE I, ELEMENT E
C
C     REFERENCES:
C       ISTOK,J.D. GROUNDWATER FLOW AND SOLUTE TRANSPORT
C       MODELING BY THE FINITE ELEMENT METHOD, FIGURE 4.14,
C       EQUATIONS 6.14, 6.16, AND 6.22
C
C**********************************************************************
$INCLUDE: 'COMALL'
      REAL JAC(3,3),JACINV(3,3),DNDXI(20),DNDX(20),DNDETA(20),
     1     DNDY(20),DNDZETA(20),DNDZ(20),SIGN1(20),SIGN2(20),
     3     SIGN3(20),KXE,KYE,KZE
      DATA SIGN1/-1.,0.,1.,1.,1.,0.,-1.,-1.,-1.,1.,1.,-1.,-1.,
     1           0.,1.,1.,1.,0.,-1.,-1./
      DATA SIGN2/-1.,-1.,-1.,0.,1.,1.,1.,0.,-1.,-1.,1.,1.,-1.,
     1           -1.,-1.,0.,1.,1.,1.,0./
      DATA SIGN3/-1.,-1.,-1.,-1.,-1.,-1.,-1.,-1.,0.,0.,0.,0.,1.,
     1           1.,1.,1.,1.,1.,1.,1./
C
      KXE = PROP(MATSET(E),1)
      KYE = PROP(MATSET(E),2)
      KZE = PROP(MATSET(E),3)
      DO 20 I = 1, 20
        DO 10 J = 1, 20
          JAC(I,J) = 0.
   10   CONTINUE
   20 CONTINUE
      DO 30 I = 1, 20
        IF ((I .EQ. 1) .OR. (I .EQ. 3) .OR. (I .EQ. 5) .OR.
     1      (I .EQ. 7) .OR. (I .EQ. 13) .OR. (I .EQ. 15) .OR.
     2      (I .EQ. 17) .OR. (I .EQ. 19)) THEN
            DNDXI(I) = -0.125 * SIGN1(I)
            DNDETA(I) = -0.125 * SIGN2(I)
            DNDZETA(I) = -0.125 * SIGN3(I)
        ELSEIF ((I .EQ. 2) .OR. (I .EQ. 6) .OR.
     1          (I .EQ. 14) .OR. (I .EQ. 18)) THEN
            DNDXI(I) = 0.
            DNDETA(I) = 0.25 * SIGN2(I)
            DNDZETA(I) = 0.25 * SIGN3(I)
        ELSEIF ((I .EQ. 4) .OR. (I .EQ. 8) .OR.
     1          (I .EQ. 16) .OR. (I .EQ. 20)) THEN
            DNDXI(I) = 0.25 * SIGN1(I)
            DNDETA(I) = 0.
            DNDZETA(I) = 0.25 * SIGN3(I)
        ELSEIF ((I .GE. 9) .AND. (I .LE. 12)) THEN
```

```
            DNDXI(I)  =  0.25 * SIGN1(I)
            DNDETA(I) =  0.25 * SIGN2(I)
           DNDZETA(I) =  0.
      ENDIF
   30 CONTINUE
      DO 40 I = 1, 20
         JAC(1,1) = JAC(1,1) + DNDXI(I) * X1(IN(E,I))
         JAC(1,2) = JAC(1,2) + DNDXI(I) * X2(IN(E,I))
         JAC(1,3) = JAC(1,3) + DNDXI(I) * X3(IN(E,I))
         JAC(2,1) = JAC(2,1) + DNDETA(I) * X1(IN(E,I))
         JAC(2,2) = JAC(2,2) + DNDETA(I) * X2(IN(E,I))
         JAC(2,3) = JAC(2,3) + DNDETA(I) * X3(IN(E,I))
         JAC(3,1) = JAC(3,1) + DNDZETA(I) * X1(IN(E,I))
         JAC(3,2) = JAC(3,2) + DNDZETA(I) * X2(IN(E,I))
         JAC(3,3) = JAC(3,3) + DNDZETA(I) * X3(IN(E,I))
   40 CONTINUE
      DETJAC = JAC(1,1) * (JAC(2,2) * JAC(3,3) - JAC(3,2) *
     1         JAC(2,3)) - JAC(1,2) * (JAC(2,1) * JAC(3,3) -
     2         JAC(3,1) * JAC(2,3)) - JAC(1,3) * (JAC(2,1) *
     3         JAC(3,2) - JAC(3,1) * JAC(2,2))
      JACINV(1,1) = ( JAC(2,2) * JAC(3,3) - JAC(2,3) *
     1              JAC(3,2)) / DETJAC
      JACINV(1,2) = (-JAC(2,1) * JAC(3,3) + JAC(2,3) *
     1              JAC(3,1)) / DETJAC
      JACINV(1,3) = ( JAC(2,1) * JAC(3,2) - JAC(3,1) *
     1              JAC(2,2)) / DETJAC
      JACINV(2,1) = (-JAC(1,2) * JAC(3,3) + JAC(1,3) *
     1              JAC(3,2)) / DETJAC
      JACINV(2,2) = ( JAC(1,1) * JAC(3,3) - JAC(1,3) *
     1              JAC(3,1)) / DETJAC
      JACINV(2,3) = (-JAC(1,1) * JAC(3,2) + JAC(1,2) *
     1              JAC(3,1)) / DETJAC
      JACINV(3,1) = ( JAC(1,2) * JAC(2,3) - JAC(1,3) *
     1              JAC(2,2)) / DETJAC
      JACINV(3,2) = (-JAC(1,1) * JAC(2,3) + JAC(1,3) *
     1              JAC(2,1)) / DETJAC
      JACINV(3,3) = ( JAC(1,1) * JAC(2,2) - JAC(1,2) *
     1              JAC(2,1)) / DETJAC
      DO 50 I = 1, 20
         DNDX(I) = JACINV(1,1) * DNDXI(I) + JACINV(1,2) *
     1             DNDETA(I) + JACINV(1,3) * DNDZETA(I)
         DNDY(I) = JACINV(2,1) * DNDXI(I) + JACINV(2,2) *
     1             DNDETA(I) + JACINV(2,3) * DNDZETA(I)
         DNDZ(I) = JACINV(3,1) * DNDXI(I) + JACINV(3,2) *
     1             DNDETA(I) + JACINV(3,3) * DNDZETA(I)
   50 CONTINUE
      DHDX = 0.
      DHDY = 0.
      DHDZ = 0.
      DO 60 I = 1, 20
         DHDX = DHDX + DNDX(I) * X(IN(E,I))
         DHDY = DHDY + DNDY(I) * X(IN(E,I))
         DHDZ = DHDZ + DNDZ(I) * X(IN(E,I))
   60 CONTINUE
      VXE = -KXE * DHDX
      VYE = -KYE * DHDY
      VZE = -KZE * DHDZ
      RETURN
      END
```

```
      SUBROUTINE VPAR32(E,VXE,VYE,VZE)
C************************************************************************
C
C      PURPOSE:
C      TO COMPUTE COMPONENTS OF APPARENT GROUNDWATER
C      VELOCITY FOR A THREE-DIMENSIONAL, CUBIC
C      PARALLELEPIPED ELEMENT
C
C      DEFINITIONS OF VARIABLES:
C          DETJAC = DETERMINANT OF JACOBIAN MATRIX
C            DHDX = PARTIAL DERIVATIVE OF HEAD WITH RESPECT TO X
C            DHDY = PARTIAL DERIVATIVE OF HEAD WITH RESPECT TO Y
C            DHDZ = PARTIAL DERIVATIVE OF HEAD WITH RESPECT TO Z
C        DNDXI(I) = PARTIAL DERIVATIVE OF INTERPOLATION
C                   FUNCTION WITH RESPECT TO XI FOR NODE I
C         DNDX(I) = PARTIAL DERIVATIVE OF INTERPOLATION
C                   FUNCTION WITH RESPECT TO X FOR NODE I
C       DNDETA(I) = PARTIAL DERIVATIVE OF INTERPOLATION
C                   FUNCTION WITH RESPECT TO ETA FOR NODE I
C         DNDY(I) = PARTIAL DERIVATIVE OF INTERPOLATION
C                   FUNCTION WITH RESPECT TO Y FOR NODE I
C      DNDZETA(I) = PARTIAL DERIVATIVE OF INTERPOLATION
C                   FUNCTION WITH RESPECT TO ZETA FOR NODE I
C         DNDZ(I) = PARTIAL DERIVATIVE OF INTERPOLATION
C                   FUNCTION WITH RESPECT TO Z FOR NODE I
C               E = ELEMENT NUMBER
C         JAC(I,J) = JACOBIAN MATRIX
C      JACINV(I,J) = INVERSE OF JACOBIAN MATRIX
C             KXE = HYDRAULIC CONDUCTIVITY IN X COORDINATE DIRECTION
C             KYE = HYDRAULIC CONDUCTIVITY IN Y COORDINATE DIRECTION
C             KZE = HYDRAULIC CONDUCTIVITY IN Z COORDINATE DIRECTION
C             VXE = APPARENT GROUNDWATER VELOCITY IN X
C                   COORDINATE DIRECTION
C             VYE = APPARENT GROUNDWATER VELOCITY IN Y
C                   COORDINATE DIRECTION
C             VZE = APPARENT GROUNDWATER VELOCITY IN Z
C                   COORDINATE DIRECTION
C      X(IN(E,I)) = COMPUTED HEAD FOR NODE I, ELEMENT E
C     X1(IN(E,I)) = X COORDINATE FOR NODE I, ELEMENT E
C     X2(IN(E,I)) = Y COORDINATE FOR NODE I, ELEMENT E
C     X3(IN(E,I)) = Z COORDINATE FOR NODE I, ELEMENT E
C
C      REFERENCES:
C      ISTOK,J.D. GROUNDWATER FLOW AND SOLUTE TRANSPORT
C      MODELING BY THE FINITE ELEMENT METHOD, FIGURE 4.15,
C      EQUATIONS 6.14, 6.16, AND 6.22.
C
C************************************************************************
$INCLUDE: 'COMALL'
      REAL JAC(3,3),JACINV(3,3),DNDXI(32),DNDX(32),DNDETA(32),
     1     DNDY(32),DNDZETA(32),DNDZ(32),SIGN1(32),SIGN2(32),
     3     SIGN3(32),KXE,KYE,KZE
      DATA SIGN1/ -1.,-1., 6*1., 5*-1.,
     1             1.,1., -1.,-1., 1.,1., 3*-1., 6*1., 4*-1. /
      DATA SIGN2/ 5*-1., 6*1., 3*-1.,
     1             1.,1., -1.,-1., 1.,1., 5*-1., 6*1., -1 /
      DATA SIGN3/ 16*-1., 16*1. /
C
      KXE = PROP(MATSET(E),1)
      KYE = PROP(MATSET(E),2)
      KZE = PROP(MATSET(E),3)
```

```
      DO 20 I = 1, 32
        DO 10 J = 1, 32
          JAC(I,J) = 0.
10    CONTINUE
20 CONTINUE
      DO 30 I = 1, 32
        IF ((I .EQ. 1) .OR. (I .EQ. 4) .OR. (I .EQ. 7) .OR.
   1        (I .EQ. 10) .OR. (I .EQ. 21) .OR. (I .EQ. 24) .OR.
   2        (I .EQ. 27) .OR. (I .EQ. 30)) THEN
            DNDXI(I) = (-19. / 64.) * SIGN1(I)
            DNDETA(I) = (-19. / 64.) * SIGN2(I)
            DNDZETA(I) = (-19. / 64.) * SIGN3(I)
        ELSEIF ((I .EQ. 2) .OR. (I .EQ. 3) .OR. (I .EQ. 8) .OR.
   1           (I .EQ. 9) .OR. (I .EQ. 22) .OR. (I.EQ.23) .OR.
   2           (I .EQ. 28) .OR. (I .EQ. 29)) THEN
            DNDXI(I) = (81. / 64.) * SIGN1(I) /3.
            DNDETA(I) = (9. / 64.) * SIGN2(I)
            DNDZETA(I) = (9. / 64.) * SIGN3(I)
        ELSEIF ((I .EQ. 5) .OR. (I .EQ. 6) .OR. (I .EQ. 11) .OR.
   1           (I .EQ. 12) .OR. (I .EQ. 25) .OR. (I .EQ. 26) .OR.
   2           (I .EQ. 31) .OR. (I .EQ. 32)) THEN
            DNDXI(I) = (9. / 64.) * SIGN1(I)
            DNDETA(I) = (81. / 64.) * SIGN2(I) /3.
            DNDZETA(I) = (9. / 64.) * SIGN3(I)
        ELSEIF ((I .GE. 13) .AND. (I .LE. 20)) THEN
            DNDXI(I) = (9. / 64.) * SIGN1(I)
            DNDETA(I) = (9. / 64.) * SIGN2(I)
            DNDZETA(I) = (81. / 64.) * SIGN3(I) /3.
        ENDIF
30 CONTINUE
      DO 40 I = 1, 32
        JAC(1,1) = JAC(1,1) + DNDXI(I) * X1(IN(E,I))
        JAC(1,2) = JAC(1,2) + DNDXI(I) * X2(IN(E,I))
        JAC(1,3) = JAC(1,3) + DNDXI(I) * X3(IN(E,I))
        JAC(2,1) = JAC(2,1) + DNDETA(I) * X1(IN(E,I))
        JAC(2,2) = JAC(2,2) + DNDETA(I) * X2(IN(E,I))
        JAC(2,3) = JAC(2,3) + DNDETA(I) * X3(IN(E,I))
        JAC(3,1) = JAC(3,1) + DNDZETA(I) * X1(IN(E,I))
        JAC(3,2) = JAC(3,2) + DNDZETA(I) * X2(IN(E,I))
        JAC(3,3) = JAC(3,3) + DNDZETA(I) * X3(IN(E,I))
40 CONTINUE
      DETJAC = JAC(1,1) * (JAC(2,2) * JAC(3,3) - JAC(3,2) *
   1         JAC(2,3)) - JAC(1,2) * (JAC(2,1) * JAC(3,3) -
   2         JAC(3,1) * JAC(2,3)) - JAC(1,3) * (JAC(2,1) *
   3         JAC(3,2) - JAC(3,1) * JAC(2,2))
      JACINV(1,1) = ( JAC(2,2) * JAC(3,3) - JAC(2,3) *
   1              JAC(3,2)) / DETJAC
      JACINV(1,2) = (-JAC(2,1) * JAC(3,3) + JAC(2,3) *
   1              JAC(3,1)) / DETJAC
      JACINV(1,3) = ( JAC(2,1) * JAC(3,2) - JAC(3,1) *
   1              JAC(2,2)) / DETJAC
      JACINV(2,1) = (-JAC(1,2) * JAC(3,3) + JAC(1,3) *
   1              JAC(3,2)) / DETJAC
      JACINV(2,2) = ( JAC(1,1) * JAC(3,3) - JAC(1,3) *
   1              JAC(3,1)) / DETJAC
      JACINV(2,3) = (-JAC(1,1) * JAC(3,2) + JAC(1,2) *
   1              JAC(3,1)) / DETJAC
      JACINV(3,1) = ( JAC(1,2) * JAC(2,3) - JAC(1,3) *
   1              JAC(2,2)) / DETJAC
      JACINV(3,2) = (-JAC(1,1) * JAC(2,3) + JAC(1,3) *
   1              JAC(2,1)) / DETJAC
```

```
      JACINV(3,3) = ( JAC(1,1) * JAC(2,2) - JAC(1,2) *
     1                   JAC(2,1)) / DETJAC
       DO 50 I = 1, 32
          DNDX(I) = JACINV(1,1) * DNDXI(I) + JACINV(1,2) *
     1              DNDETA(I) + JACINV(1,3) * DNDZETA(I)
          DNDY(I) = JACINV(2,1) * DNDXI(I) + JACINV(2,2) *
     1              DNDETA(I) + JACINV(2,3) * DNDZETA(I)
          DNDZ(I) = JACINV(3,1) * DNDXI(I) + JACINV(3,2) *
     1              DNDETA(I) + JACINV(3,3) * DNDZETA(I)
   50 CONTINUE
      DHDX = 0.
      DHDY = 0.
      DHDZ = 0.
      DO 60 I = 1, 32
        DHDX = DHDX + DNDX(I) * X(IN(E,I))
        DHDY = DHDY + DNDY(I) * X(IN(E,I))
        DHDZ = DHDZ + DNDZ(I) * X(IN(E,I))
   60 CONTINUE
      VXE = -KXE * DHDX
      VYE = -KYE * DHDY
      VZE = -KZE * DHDZ
      RETURN
      END
```

Chapter 15

SUBROUTINE DUMP

15.1 PURPOSE

Subroutine DUMP writes the contents of various variables and arrays to files for use in other computer programs (e.g. plotting packages). DUMP is also used to pass velocities computed in groundwater program GW1 (see Chapters 7 and 14) to the solute transport program ST1.

15.2 INPUT

Control information is read "free-format" from the user-supplied file assigned to unit "INF". The control information consists of a code (ICODE) that indicates which arrays are to be written to a user-defined file (FNAME), followed by the file name (two-lines of input for each choice of ICODE). This information can be repeated as often as desired. Input is terminated by placing a -1 in the first field of any line.

15.3 OUTPUT

The arrays are written to a set of user-defined files. DUMP opens the files using the file names read from INF. The contents of requested variables and arrays are written to the files "free-format" (i.e., without column headings or titles).

15.4 DEFINITIONS OF VARIABLES

FNAME = File name for a user-defined file (20 characters or less).

ICODE = Code indicating which variables and arrays are to be written to file FNAME.

       = 1, a list of node numbers and coordinates is written.

       = 2, a list of element numbers, types, and node numbers is written.

       = 3, a list of element numbers and material set numbers is written followed by the number of material properties and a list of properties for each material set.

       = 4, a list of node numbers and specified values of the field variable is written for Dirichlet nodes, followed by a list of node numbers and specified values of groundwater flow or solute flux for Neumann nodes.

= 5, the relaxation factor, ω, is written first. Then a list of time step intervals is written followed by a list of values of the time function $g(t)$, the total length of time for which calculations will be performed, and a list of node numbers and initial values of the field variable at each node.

= 6, a list of node numbers and computed values of the field variable at each node is written.

= 7, a list of element numbers and computed components of apparent groundwater velocity for each element is written.

15.5 USAGE

A list of the FORTRAN variable names written to file FNAME for each choice of ICODE is in Table 15.1. Additional information about these variables is in the chapters listed in Table 15.1. If the file "FNAME" does not yet exist, DUMP will create and open it for writing. If the file FNAME already exists, DUMP will overwrite it's contents. An exception is when DUMP is used to write out heads and velocities in programs GW2 and GW4, or solute concentrations in program ST1. In these cases DUMP appends the computed heads, velocities, and solute concentrations onto the bottom of the files after each time step.

An example input file for DUMP as used in program GW1 is shown below

```
1
NOD.LST
2
ELEM.LST
6
HEAD.LST
7
VELO.LST
-1
```

In this example DUMP would write a list of node numbers and coordinates to the file NOD.LST, a list of element numbers and element node numbers to file ELEM.LST, a list of node numbers and computed values of hydraulic head to file HEAD.LST, and a list of element numbers and computed components of apparent groundwater velocity for each element to file VELO.LST.

Table 15.1 Variable names written to user-defined file FNAME for each choice of ICODE.

ICODE	FORTRAN Variable Name(s)			Subroutine
1	1	X1(1)	X2(1)	X3(1)	NODES
	...	...	...	...	
	NUMNOD	X1(NUMNOD)	X2(NUMNOD)	X3(NUMNOD)	
2	1	ELEMTYP(1)	IN(1,1) ... IN(1,NODETBL(ELEMTYP(1)))		ELEMENT
	...	...	...		
	NUMELM	ELEMTYP(NUMELM)	IN(NUMELM,1) ... IN(NUMELM, NODETBL(ELEMTYP(NUMELM)))		
3	1	MATSET(1)			MATERL
	...	...			
	NUMELM	MATSET(NUMELM)			
	NUM PROP				
	1	PROP(1,1) ... PROP(1,NUMPROP)			
	...	...			
	NUMMAT	PROP(NUMMAT,1) ... PROP(NUMMAT, NUMPROP)			
4	1	X(1)			BOUND
	...	...	(For Dirichlet nodes only)		
	NUMNOD	X(NUMNOD)			

Table 15.1 Variable names written to user-defined file FNAME for each choice of ICODE. (continued)

ICODE	FORTRAN Variable Name(s)			Subroutine
5	1	FLUX(1)		INITIAL
		(For Neumann nodes only)		
	·	·		
	NUMNOD	FLUX(NUMNOD)		
	OMEGA			
	1	DTSTEP(1)	DELTAT(1)	
	·	·	·	
	MXSTEP	DTSTEP(MXSTEP)	DELTAT(MXSTEP)	
	TOTALT			
	TIME(1)		GT(1)	
	·		·	
	TIME(TOTALT)		GT(TOTALT)	
	1	X(1)		
	·	·		
	NUMNOD	X(NUMNOD)		
6	1	X(1)		SOLVE
	·	·		
	NUMNOD	X(NUMNOD)		
7	1	V1(1)	V2(1)	V3(1)	VELOCITY
	·	·	·	·
	NUMELM	V1(NUMELM)	V2(NUMELM)	V3(NUMELM)

15.6 SOURCE CODE LISTING

```
          SUBROUTINE DUMP(LOOP,HDF,VLF)
C*********************************************************************
C
C 15.1  PURPOSE:
C           TO WRITE CONTENTS OF ARRAYS TO USER-SUPPLIED DATA
C           FILES
C
C 15.2  INPUT:
C           CONTROL INFORMATION IS READ FROM USER-SUPPLIED FILE
C           ASSIGNED TO UNIT "INF".  FIRST LINE IS CODE INDICATING
C           WHICH ARRAYS ARE TO BE WRITTEN TO A OUTPUT FILE, THE
C           SECOND LINE IS THE NAME OF THE OUTPUT FILE.  THESE
C           TWO LINES CAN BE REPEATED AS OFTEN AS DESIRED.  INPUT
C           IS TERMINATED BY PLACING A -1 ON FILE "INF".
C
C 15.3  OUTPUT:
C           CONTENTS OF ARRAYS ARE WRITTEN TO A SET OF OUTPUT
C           FILES.
C
C 15.4  DEFINITIONS OF VARIABLES:
C           FNAME = OUTPUT FILE NAME
C           ICODE = ARRAYS TO BE WRITTEN TO FNAME:
C                 = 1, NODE NUMBERS AND COORDINATES
C                 = 2, ELEMENT NUMBERS, TYPES, AND NODE NUMBERS
C                 = 3, ELEMENT NUMBERS, MATERIAL SET NUMBERS,
C                      AND MATERIAL SET PROPERTIES
C                 = 4, NODE NUMBERS AND SPECIFIED VALUES OF HEAD
C                      OR SOLUTE CONCENTRATION (DIRICHLET BOUNDARY
C                      CONDITIONS) AND SPECIFIED RATES OF GROUNDWATER
C                      FLOW OR SOLUTE FLUX (DIRICHLET AND NEUMANN
C                      BOUNDARY CONDITIONS)
C                 = 5, RELAXATION FACTOR, TIME FUNCTION, AND INITIAL
C                      VALUES OF HEAD OR SOLUTE CONCENTRATION
C                 = 6, COMPUTED VALUES OF HEAD OR SOLUTE CONCENTRATION
C                 = 7, ELEMENT NUMBERS AND COMPONENTS OF APPARENT
C                      GROUNDWATER VELOCITY
C
C 15.5  USAGE:
C           THE CONTENTS OF THE ARRAYS ARE WRITTEN "FREE-FORMAT" TO
C           EACH DATA FILE.
C
C           SUBROUTINES CALLED:
C             NONE
C
C           REFERENCES:
C             ISTOK,J.D. GROUNDWATER FLOW AND SOLUTE TRANSPORT
C             MODELING BY THE FINITE ELEMENT METHOD, CHAPTER 15.
C
C*********************************************************************
$INCLUDE : 'COMALL'
          INTEGER DMPF,HDF,VLF
          LOGICAL LOOP,OPNED
          CHARACTER*20 FNAME
          DIMENSION XYZ(MAX1,3),V(MAX2,3),NODETBL(13)
          EQUIVALENCE (X1,XYZ(1,1)),(X2,XYZ(1,2)),(X3,XYZ(1,3)),
     1      (V1,V(1,1)),(V2,V(1,2)),(V3,V(1,3))
          DATA NODETBL/2,3,4,3,4,4,8,12,8,20,32,3,4/
C
          HDF = 0
          VLF = 0
```

```
10        READ(INF,*,END=140,ERR=140) ICODE
          IF (ICODE .LE. 0) GOTO 140
          READ(INF,20,END=140,ERR=10) FNAME
20        FORMAT(A)
          IF (ICODE .LE. 6) THEN
            DMPF = 1
          ELSE
            DMPF = 2
          ENDIF
          INQUIRE(UNIT=DMPF,OPENED=OPNED)
          IF (.NOT. OPNED)
     1      OPEN(DMPF,FILE=FNAME,STATUS='NEW',FORM='FORMATTED')
          IF (ICODE .EQ. 1) THEN
C
C         WRITE OUT NODE NUMBERS AND COORDINATES
C
            IF (DIM .LT. 4) THEN
              IDIM = DIM
            ELSE
              IDIM = 2
            ENDIF
            DO 30 I = 1, NUMNOD
              WRITE(DMPF,*) I,(XYZ(I,J),J=1,IDIM)
30          CONTINUE
          ELSEIF (ICODE .EQ. 2) THEN
C
C         WRITE OUT ELEMENT NUMBERS, TYPES, AND NODE NUMBERS
C
            DO 40 I=1, NUMELM
              WRITE(DMPF,*) I,ELEMTYP(I),(IN(I,J),J=1,NODETBL(ELEMTYP(I)))
40          CONTINUE
          ELSEIF (ICODE .EQ. 3) THEN
C
C         WRITE OUT ELEMENT AND MATERIAL SET NUMBERS AND MATERIAL
C         PROPERTIES
C
            DO 50 I = 1, NUMELM
              WRITE(DMPF,*) I,MATSET(I)
50          CONTINUE
            WRITE(DMPF,*) NUMPROP
            DO 60 I = 1, NUMMAT
              WRITE(DMPF,*) I,(PROP(I,J),J=1,NUMPROP)
60          CONTINUE
          ELSEIF (ICODE .EQ. 4) THEN
C
C         WRITE OUT SPECIFED VALUES OF FIELD VARIABLE AND GROUNDWATER
C         FLOW OR SOLUTE FLUX
C
            IF (NDN .GT. 0) THEN
              DO 70 I = 1, NUMNOD
                IF (ICH(I) .NE. 0) WRITE(DMPF,*) I,X(I)
70            CONTINUE
            ENDIF
            IF (NNN .GT. 0) THEN
              DO 80 I = 1, NUMNOD
                IF (FLUX(I) .NE. 0.) WRITE(DMPF,*) I,FLUX(I)
80            CONTINUE
            ENDIF
          ELSEIF (ICODE .EQ. 5) THEN
C
C         WRITE OUT RELAXATION FACTOR, TIME FUNCTION, AND INITIAL
```

```
C          VALUES OF FIELD VARIABLE
C
           IF (LOOP) THEN
             WRITE(DMPF,*) OMEGA
             ISTART = 1
             IT = 1
             TOTALT = 0.
   90        WRITE(DMPF,*) ISTART,DTSTEP(IT),DELTAT(IT)
             TOTALT = TOTALT + (DTSTEP(IT) - ISTART + 1) * DELTAT(IT)
             IF (DTSTEP(IT) .LT. MXSTEP) THEN
               ISTART = DTSTEP(IT) + 1
               IT = IT + 1
               GOTO 90
             ELSE
               WRITE(DMPF,*) TOTALT
               IT = 1
  100          WRITE(DMPF,*) TIME(IT),GT(IT)
               IF (TIME(IT) .GE. 0. .AND. TIME(IT) .LT. TOTALT) THEN
                 IT = IT + 1
                 GOTO 100
               ELSE
                 DO 110 I = 1, NUMNOD
                   IF (ICH(I) .EQ. 0) WRITE(DMPF,*) I,X(I)
  110            CONTINUE
               ENDIF
             ENDIF
           ENDIF
         ELSEIF (ICODE .EQ. 6) THEN
C
C          WRITE OUT COMPUTED VALUES OF FIELD VARIABLE
C
           IF (LOOP) THEN
             HDF = DMPF
           ELSE
             DO 120 I = 1, NUMNOD
               WRITE(DMPF,*) I,X(I)
  120        CONTINUE
           ENDIF
         ELSEIF (ICODE .EQ. 7) THEN
C
C          WRITE OUT ELEMENT NUMBERS AND COMPUTED COMPONENTS
C          OF APPARENT GROUNDWATER VELOCITY
C
           IF (LOOP) THEN
             VLF = DMPF
           ELSE
             IF (DIM .LT. 4) THEN
               IDIM = DIM
             ELSE
               IDIM = 2
             ENDIF
             DO 130 I = 1, NUMELM
               WRITE(DMPF,*) I,(V(I,J),J=1,IDIM)
  130        CONTINUE
           ENDIF
         ENDIF
         IF (ICODE .LE. 5 .OR. (ICODE .GT. 5 .AND. .NOT. LOOP))
     1     CLOSE(UNIT=DMPF)
         GOTO 10
  140    RETURN
         END
```

Chapter 16

SUBROUTINE INITIAL

16.1 PURPOSE

Subroutine INITIAL inputs control parameters and initial conditions needed to solve transient groundwater flow and solute transport problems. Subroutine INITIAL is also used to input control parameters and initial estimates for pressure head needed to solve, unsaturated flow problems.

16.2 INPUT

All data are read "free-format" from the user-supplied file assigned to unit "INF". INF is passed to INITIAL through a labeled common block contained in the file "COMALL" (Chapter 7). The relaxation factor used in the finite difference approximation of the time derivative, a list of time step intervals, and a list of values of the time function (see below) are also read. These are followed by a list of initial values of the field variable (hydraulic head, pressure head, or solute concentration) for each node in the mesh.

16.3 OUTPUT

The relaxation factor, time step intervals, values of the time function, and initial values of the field variable are written to the user-defined file assigned to unit "OUTF".

16.4 DEFINITIONS OF VARIABLES

DELTAT(I)	=	Size of time step I.
DTSTEP(I)	=	Last time step to take using a time step of size DELTAT(I).
GT(I)	=	Value of time function at time I.
ICH(I)	=	1 if the value of the field variable is specified at node I.
	=	0 otherwise.
LABEL1	=	Character variable used to label column headings for specified values of the field variable on file assigned to unit OUTF. LABEL1 = "HYDRAULIC HEAD", "PRESSURE HEAD", OR "SOLUTE FLUX".
MXSTEP	=	Number of different time step intervals.
NUMNOD	=	Number of nodes.
OMEGA	=	Relaxation factor, ω.

OMOMEGA   = $1 - \omega$.

TIME(I)   =   Starting time for time function value GT(I).

TOTALT   =   Total length of time for which calculations are performed.

X(I)   =   Value of the field variable (hydraulic head, pressure head, or solute concentration) at node I.

16.5 USAGE

The relaxation factor OMEGA is read first, followed by a list of time steps and time intervals. Each record in the list contains the last time step to take with a specified time step interval. Input of time steps and time step intervals is terminated by placing a -1 in both fields. This is followed by a list of times and values of the time function for each time. Input is terminated by placing a -1 in both fields. Finally, the initial values of the field variable are read for each node. The subroutine will "generate" initial values of the field variable for nodes "missing" from the input file. However, the specified values of the field variable for Dirichlet nodes read by subroutine BOUND will not be changed. Input is terminated by placing a -1 in both fields.

In the mesh in Figure 16.1, nodes 1, 2, 3, 12, 13, and 14 have specified values of pressure head. These *boundary* conditions would be read by subroutine BOUND. For this problem $\omega = 0.5$, and a solution is required for 6 time steps. The time step interval will be 1 minute for time steps 1, 2, and 3 and the time step interval will be 3 minutes for time steps 4, 5, and 6. The specified flow rate for the point sink at node 8, F(8), varies with time as shown below

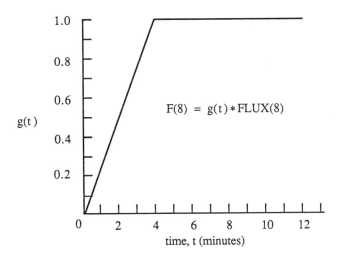

where $F(8) = g(t) * FLUX(8)$ and FLUX(8) is the specified flow rate for node 8 read in subroutine BOUND. For this problem we choose to use an initial value of pressure head of -15 for all nodes not on a constant head boundary. Note that the <u>same</u> time function is applied to every Neumann node.

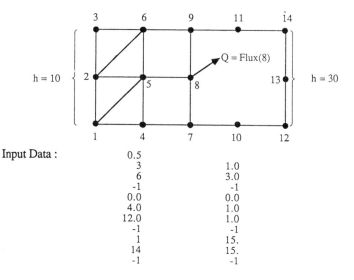

Input Data :
0.5	
3	1.0
6	3.0
-1	-1
0.0	0.0
4.0	1.0
12.0	1.0
-1	-1
1	15.
14	15.
-1	-1

Output :   OMEGA = 0.5000

START	END	DELTA T
1	3	1.0000
4	6	3.0000

TOTAL TIME = 12.0000

TIME T	G(T)
.0000	.0000
4.0000	1.0000
12.0000	1.0000

INITIAL VALUES OF HYDRAULIC HEAD

NODE NO.	HYDRAULIC HEAD
1	10.0000*
2	10.0000*
3	10.0000*
4	15.0000
5	15.0000
6	15.0000
7	15.0000
8	15.0000
9	15.0000
10	15.0000
11	15.0000
12	30.0000*
13	30.0000
14	30.0000*

* = SPECIFIED

Figure 16.1   **Example input data and output for subroutine INITIAL.**

16.6 SOURCE CODE LISTING

```
        SUBROUTINE INITIAL
C**********************************************************************
C
C 16.1  PURPOSE:
C           SUBROUTINE INITIAL INPUTS CONTROL PARAMETERS AND INITIAL
C           CONDITIONS NEEDED TO SOLVE TRANSIENT GROUNDWATER FLOW AND
C           SOLUTE TRANSPORT PROBLEMS.  SUBROUTINE INITIAL IS ALSO
C           USED TO INPUT CONTROL PARAMETERS AND TO SPECIFY INITIAL
C           ESTIMATES FOR PRESSURE HEAD NEEDED TO SOLVE STEADY-STATE,
C           UNSATURATED FLOW PROBLEMS.
C
C 16.2  INPUT:
C           ALL DATA ARE READ "FREE-FORMAT" FROM THE USER-SUPPLIED
C           FILE ASSIGNED TO UNIT "INF".  THE RELAXATION FACTOR
C           USED IN THE FINITE DIFFERENCE APPROXIMATION OF THE TIME
C           DERIVATIVE, A LIST OF TIME STEP INTERVALS, AND A LIST OF
C           VALUES OF THE TIME FUNCTION (SEE BELOW) ARE ALSO READ.
C           THESE ARE FOLLOWED BY A LIST OF INITIAL VALUES OF THE
C           FIELD VARIABLE FOR EACH NODE IN THE MESH.
C
C 16.3  OUTPUT:
C           THE RELAXATION FACTOR, TIME STEP INTERVALS, VALUES OF
C           THE TIME FUNCTION, AND INITIAL VALUES OF THE FIELD
C           VARIABLE ARE WRITTEN TO THE USER-DEFINED FILE ASSIGNED
C           TO UNIT "OUTF".
C
C 16.4  DEFINITIONS OF VARIABLES:
C           DELTAT(I) = SIZE OF TIME STEP I
C           DTSTEP(I) = NUMBER OF TIME STEPS TO TAKE USING A TIME
C                       STEP OF SIZE DELTAT(I)
C              GT(I) = VALUE OF TIME FUNCTION AT TIME I
C             ICH(I) = 1 IF THE VALUE OF THE FIELD VARIABLE IS
C                       SPECIFIED FOR NODE I,
C                     = 0 OTHERWISE
C             LABEL1 = CHARACTER VARIABLE USED TO LABEL COLUMN
C                       HEADINGS FOR SPECIFIED VALUES OF THE FIELD
C                       VARIABLE ON FILE ASSIGNED TO UNIT OUTF.
C                       LABEL1 = "HYDRAULIC HEAD", "PRESSURE HEAD",
C                       OR "SOLUTE FLUX"
C             MXSTEP = NUMBER OF DIFFERENT TIME STEP INTERVALS
C             NUMNOD = NUMBER OF NODES
C              OMEGA = RELAXATION FACTOR
C            OMOMEGA = 1. - OMEGA
C            TIME(I) = STARTING TIME FOR TIME FUNCTION VALUE GT(I)
C             TOTALT = TOTAL LENGTH OF TIME FOR WHICH CALCULATIONS
C                       ARE PERFORMED
C               X(I) = VALUE OF THE FIELD VARIABLE (HYDRAULIC
C                       HEAD, PRESSURE HEAD, OR SOLUTE CONCENTRATION)
C                       AT NODE I
C
C 16.5  USAGE:
C           THE RELAXATION FACTOR OMEGA IS READ FIRST.  THIS
C           IS FOLLOWED BY A LIST OF TIME STEPS AND TIME STEP
C           INTERVALS. EACH LINE OF INPUT CONTAINS THE NUMBER OF
C           TIME STEPS TO TAKE FOLLOWED BY A SPECIFIED TIME STEP
C           INTERVAL. INPUT OF TIME STEPS AND TIME STEP INTERVALS
C           IS TERMINATED BY PLACING A -1 IN BOTH FIELDS.
C           THIS IS FOLLOWED BY A LIST OF TIMES AND VALUES OF THE
C           TIME FUNCTION FOR EACH TIME.  INPUT IS TERMINATED BY
```

```
C          PLACING A -1 IN BOTH FIELDS.  FINALLY, THE INITIAL
C          VALUE OF THE FIELD VARIABLE IS READ FOR EACH NODE.
C          INPUT IS TERMINATED BY PLACING A -1 IN BOTH FIELDS.
C
C
C          SUBROUTINES CALLED:
C             NONE
C
C          REFERENCES:
C             ISTOK,J.D. GROUNDWATER FLOW AND SOLUTE TRANSPORT
C             MODELING BY THE FINITE ELEMENT METHOD, CHAPTER 16.
C
C********************************************************************
$INCLUDE:'COMALL'
C
C          INPUT OMEGA FROM INPUT FILE
           READ(INF,*) OMEGA
           WRITE(OUTF,10) OMEGA
   10      FORMAT(//2X,'OMEGA = ',F15.4)
           OMOMEGA = 1. - OMEGA
C          INPUT LIST OF TIME STEPS AND TIME STEP INTERVALS FROM INPUT FILE
           IT = 1
           MXSTEP = 0
   20      READ(INF,*) DTSTEP(IT),DELTAT(IT)
           IF (DTSTEP(IT) .LE. 0) GOTO 30
           IF (DTSTEP(IT) .GT. MXSTEP) MXSTEP = DTSTEP(IT)
           IT = IT + 1
           GOTO 20
   30      IT = IT - 1
           WRITE(OUTF,40)
   40      FORMAT(//2X,'START',8X,' END ',10X,'DELTA T'/
          1       2X,5('-'),8X,5('-'),8X,11('-'))
           ISTART = 1
           TOTALT = 0.
           DO 60 I = 1, IT
             WRITE(OUTF,50) ISTART,DTSTEP(I),DELTAT(I)
   50        FORMAT(2X,I4,9X,I4,3X,F15.4)
             TOTALT = TOTALT + (DTSTEP(I) - ISTART + 1) * DELTAT(I)
             ISTART = DTSTEP(I) + 1
   60      CONTINUE
           WRITE(OUTF,70) TOTALT
   70      FORMAT(/10X,'TOTAL TIME =',F15.4)
C          INPUT LIST OF TIME STEPS AND VALUES OF TIME FUNCTION
           IT = 1
   80      READ(INF,*) TIME(IT),GT(IT)
           IF (TIME(IT) .LT. 0.) GOTO 90
           IT = IT + 1
           GOTO 80
   90      IT = IT - 1
           IF (TIME(IT) .LT. TOTALT) TIME(IT) = TOTALT
           WRITE(OUTF,100)
  100      FORMAT(//8X,'TIME T',11X,'G(T)'/7X,8('-'),9X,6('-'))
           DO 120 I = 1, IT
             WRITE(OUTF,110) TIME(I),GT(I)
  110        FORMAT(2F15.4)
  120      CONTINUE
C          INPUT INITIAL VALUES OF FIELD VARIABLE FROM INPUT FILE
           ISTART = 1
  130      READ(INF,*) IT,HINIT
           IF (IT .LE. 0) GOTO 150
           IF (IT .GT. MAX1) IT = MAX1
```

```
        DO 140 I = ISTART, IT
          IF (ICH(I) .NE. 1) X(I) = HINIT
140     CONTINUE
        ISTART = IT + 1
        IF (ISTART .LE. MAX1) GOTO 130
150     WRITE(OUTF,160) LABEL1,LABEL1
160     FORMAT(//2X,'INITIAL VALUES OF ',A/2X,38('-')//
     1         2X,'NODE NO.',10X,A/2X,8('-'),10X,20('-'))
        DO 180 I = 1, NUMNOD
          IF (ICH(I) .EQ. 0) THEN
            WRITE(OUTF,170) I,X(I),' '
          ELSE
            WRITE(OUTF,170) I,X(I),'*'
          ENDIF
170       FORMAT(2X,I5,12X,F15.4,A)
180     CONTINUE
        WRITE(OUTF,190)
190     FORMAT(/23X,'* = SPECIFIED')
        RETURN
        END
```

Chapter 17

SUBROUTINE ASMBKC

17.1 PURPOSE

Subroutine ASMBKC assembles the combined global conductance and capacitance matrix (equation 5.30a) and the global specified flow matrix {F}. The global matrices are modified to account for specified values of the field variable (hydraulic head or pressure head) and for specified rates of groundwater flow. ASMBKC also computes the semi-bandwidth and number of degrees of freedom for the modified system of equations.

17.2 INPUT

None

17.3 OUTPUT

The semi-bandwidth and number of degrees of freedom for the modified system of equations are written to the user-defined file assigned to unit "OUTF".

17.4 DEFINITIONS OF VARIABLES

$B(I)$	=	Modified specified flow matrix.
$CE(I,J)$	=	Capacitance matrix for element e in full matrix storage.
E	=	Element number.
$ELEMTYP(E)$ =	=	Element type for element E (see Table 9.1 for a list of element types).
$FLUX(I)$	=	Specified rate of groundwater flow at node I.
$ICH(I)=$	=	1 if the value of the field variable is specified at node I,
	=	0 otherwise.
$IJSIZE$	=	Length of array [M] in vector storage ([M] is symmetric).
$KE(I,J)$	=	Conductance matrix for element E in full matrix storage.
$LCH(I)$	=	$\displaystyle\sum_{k=1}^{I} ICH(k)$. The arrays ICH and LCH are used to modify the global system of equations for specified values of the field variable.

M(IJ)   = Modified, combined global conductance and capacitance matrix in
                   vector storage.

NDOF   = Number of nodes where the value of the field variable is unknown.

NODETBL(I)   = Number of nodes in element type I.

NUMELM   = Number of elements in mesh.

SBW   = Semi-bandwidth of modified, combined global conductance and
                   capacitance matrix.

X(I)   = Value of the field variable (hydraulic head or pressure head) at node
                   I.

17.5 USAGE

Subroutine ASMBKC assembles the combined global conductance and capacitance
matrix [M]

$$[M] = [C] + \omega \Delta t\, [K] \tag{17.1}$$

and the specified groundwater flow matrix {F}. [M] and {F} are modified to account for
specified values of the field variable (hydraulic head or pressure head), during the assembly
process, using the procedures in Chapter 4. [M] is assembled and modified in vector
storage in the array M. The modified, global specified flow matrix is stored in the array B.
Further modifications to B are made in subroutine RHS (see Chapter 18).

The number of degrees of freedom (number of unknown values of the field variable)
NDOF is computed in subroutine BOUND as

$$NDOF = NUMNOD - NDN$$

where NUMNOD is the number of nodes in the mesh (Chapter 8) and NDN is the number
of nodes with specified values of the field variable (Chapter 11). The semi-bandwidth,
SBW for the modified system of equations is computed in ASMBKC using

$$SBW = R + 1 \tag{17.2}$$

where R is the maximum difference in node numbers for any two nodes within any element
in the mesh. However, if the value of the field variable is specified for a node that node is
not used in the calculation of R (because the row in [M] for that node will be eliminated
when [M] is modified for the specified value of head).

The element conductance and capacitance matrices are computed in two sets of
subroutines. The element conductance matrices are computed using the set of subroutines
in Table 12.1. The element capacitance matrices (consistent formulation) are computed
using the set of subroutines in Table 17.1. Each subroutine in this set begins with the letter
"C" (for the element capacitance matrix $[C^{(e)}]$) followed by three or four letters, that
identify the element type, and the number of nodes in elements of that type. For example,
subroutine CBAR2 computes the element capacitance matrix for one-dimensional, linear
bar elements and CPAR8 computes the element capacitance matrix for three-dimensional,
linear parallelepiped elements.

Table 17.1   Subroutines used to compute element conductance matrices in ASMBKC.

Element Type	Description	Subroutine Name	DIM
1	Linear bar	CBAR2	1
2	Quadratic bar	CBAR3*	1
3	Cubic bar	CBAR4*	1
4	Linear triangle	CTRI3	2
5	Linear rectangle	CREC4	2
6	Linear quadrilateral	CQUA4	2
7	Quadratic quadrilateral	CQUA8*	2
8	Cubic quadrilateral	CQUA12*	2
9	Linear parallelepiped	CPAR8	3
10	Quadratic parallelepiped	CPAR20*	3
11	Cubic parallelepiped	CPAR32*	3
12	Linear triangle (axisymmetric)	CTRI3A*	4
13	Linear rectangle (axisymmetric)	CREC4A*	4

*Source code listing not provided for these subroutines.

The source code listing for each element capacitance matrix gives the figure that shows the interpolation functions and the equation used to compute $[C^{(e)}]$ for that element type. A list of many of the important FORTRAN variables names and their symbols is in Table 12.2. Additional names and symbols for the subroutines in Tables 17.1 are shown below

FORTRAN Variable or Array Name	Definition	Symbols(s) in Text
N(I)	Interpolation function for node I	$N_i^{(e)}(\varepsilon)$ or $N_i^{(e)}(\varepsilon,\eta)$ or $N_i^{(e)}(\varepsilon,\eta,\zeta)$
SSE	Specific storage for element e	$S_s^{(e)}$

The operation of ASMBKC is most easily explained by considering a specific example. The mesh in Figure 17.1 contains four nodes (NUMNOD = 4) and three elements (NUMELM = 3).

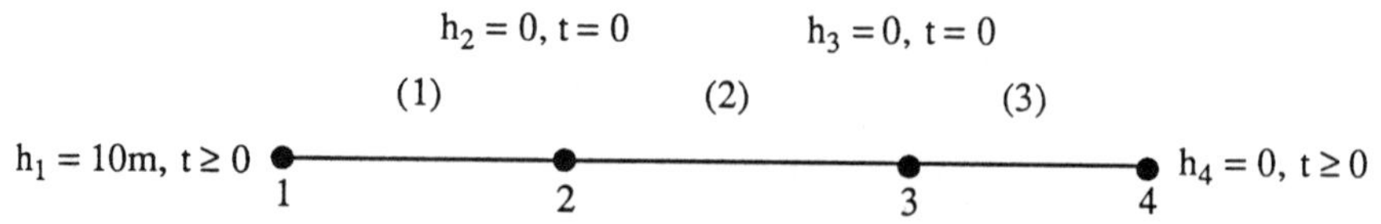

$$h_2 = 0, t = 0 \qquad\qquad h_3 = 0, t = 0$$

$$(1) \qquad\qquad (2) \qquad\qquad (3)$$

$$h_1 = 10m, t \geq 0 \quad\bullet\!\!-\!\!-\!\!-\!\!-\!\!-\!\!\bullet\!\!-\!\!-\!\!-\!\!-\!\!-\!\!\bullet\!\!-\!\!-\!\!-\!\!-\!\!-\!\!\bullet\quad h_4 = 0, t \geq 0$$
$$\quad 1 \qquad\qquad 2 \qquad\qquad 3 \quad 4$$

$$K_x^{(1)} = K_x^{(2)} = K_x^{(3)} = 0.02 \text{ m/day}$$

$$S_s^{(1)} = S_s^{(2)} = S_s^{(3)} = 0.10$$

$$L^{(1)} = 5, \ L^{(2)} = 6, \ L^{(3)} = 4\text{m}, \ \omega = 1.0, \ \Delta t = 1 \text{ day}$$

Figure 17.1   Example mesh for ASMBKC.

The value of hydraulic head is specified at node one and four (ICH(1) = ICH(4) = 1, NDN = 2) and will remain constant for all time steps. The initial values of head at nodes 2 and 3 are zero and we are to compute the head at these nodes for subsequent time steps. All elements are linear bar elements, ELEMTYP(1) = ELEMTYP(2) = ELEMTYP(3) = 1. This element type has two nodes (NODETBL(1) = 2). The element conductance matrices will be computed using subroutine KBAR2 (Table 12.1). The results are:

for element 1

$$E = 1 \qquad\qquad KE = \begin{bmatrix} 0.0040 & -0.0040 \\ -0.0040 & 0.0040 \end{bmatrix}$$

for element 2

$$E = 2 \qquad\qquad KE = \begin{bmatrix} 0.0033 & -0.0033 \\ -0.0033 & 0.0033 \end{bmatrix}$$

for element 3

$$E = 3 \qquad\qquad KE = \begin{bmatrix} 0.0050 & -0.0050 \\ -0.0050 & 0.0050 \end{bmatrix}$$

The element capacitance matrices will be computed using subroutine CBAR2 (Table 17.1). The results are:

for element 1

$$E = 1 \qquad\qquad CE = \begin{bmatrix} 0.1667 & 0.0833 \\ 0.0833 & 0.1667 \end{bmatrix}$$

for element 2

$$E = 2 \qquad\qquad CE = \begin{bmatrix} 0.2000 & 0.1000 \\ 0.1000 & 0.2000 \end{bmatrix}$$

for element 3

$$E = 3 \qquad\qquad CE = \begin{bmatrix} 0.1333 & 0.0667 \\ 0.0667 & 0.1333 \end{bmatrix}$$

The global system of equations is

$$([C] + \omega\Delta t\,[K])\{h\}_{t+\Delta t} = ([C] - (1-\omega)\,\Delta t\,[K]\{h\}_t + \Delta t\,((1-\omega)\{\overset{\{0\}}{F}\}_t + \omega\{\overset{\{0\}}{F}\}_{t+\Delta t})$$

$$\left(\begin{bmatrix} 0.1667 & 0.0833 & 0.0000 & 0.0000 \\ 0.0833 & 0.3667 & 0.1000 & 0.0000 \\ 0.0000 & 0.1000 & 0.3333 & 0.0667 \\ 0.0000 & 0.0000 & 0.0667 & 0.1333 \end{bmatrix} + \omega\Delta t \begin{bmatrix} 0.0040 & -0.0040 & 0.0000 & 0.0000 \\ -0.0040 & 0.0073 & -0.0033 & 0.0000 \\ 0.0000 & -0.0033 & 0.0083 & -0.0050 \\ 0.0000 & 0.0000 & -0.0050 & 0.0050 \end{bmatrix} \right) \begin{Bmatrix} h_1 \\ h_2 \\ h_3 \\ h_4 \end{Bmatrix}_{t+\Delta t} =$$

$$
\left(
\begin{bmatrix}
0.1667 & 0.0833 & 0.0000 & 0.0000 \\
0.0833 & 0.3667 & 0.1000 & 0.0000 \\
0.0000 & 0.1000 & 0.3333 & 0.0667 \\
0.0000 & 0.0000 & 0.0667 & 0.1333
\end{bmatrix}
- (1-\omega)\Delta t
\begin{bmatrix}
0.0040 & -0.0040 & 0.0000 & 0.0000 \\
-0.0040 & 0.0073 & -0.0033 & 0.0000 \\
0.0000 & -0.0033 & 0.0083 & -0.0050 \\
0.0000 & 0.0000 & -0.0050 & 0.0050
\end{bmatrix}
\right)
\begin{Bmatrix}
h_1 \\
h_2 \\
h_3 \\
h_4
\end{Bmatrix}_t
$$

There are no specified rates of groundwater flow in this problem and $\{F\} = 0$ for all time steps. With $\omega = 0.5$ and $\Delta t = 1$ day the global system of equations can be simplified to :

$$
\begin{bmatrix}
0.1687 & 0.0183 & 0.0000 & 0.0000 \\
0.0813 & 0.3704 & 0.0984 & 0.0000 \\
0.0000 & 0.0984 & 0.3375 & 0.0642 \\
0.0000 & 0.0000 & 0.0642 & 0.1358
\end{bmatrix}
\begin{Bmatrix}
h_1 \\
h_2 \\
h_3 \\
h_4
\end{Bmatrix}_{t+\Delta t}
$$

$$
=
\begin{bmatrix}
0.1647 & 0.0853 & 0.0000 & 0.0000 \\
0.0853 & 0.3631 & 0.1017 & 0.0000 \\
0.0000 & 0.1017 & 0.3292 & 0.0692 \\
0.0000 & 0.0000 & 0.0692 & 0.1308
\end{bmatrix}
\begin{Bmatrix}
h_1 \\
h_2 \\
h_3 \\
h_4
\end{Bmatrix}_t
$$

But $h_1 = 10$ and $h_4 = 0$ and for the first time step the modified global system of equations becomes

$$
\begin{bmatrix}
0.1687 & 0.0183 & 0.0000 & 0.0000 \\
0.0813 & 0.3704 & 0.0984 & 0.0000 \\
0.0000 & 0.0984 & 0.3375 & 0.0642 \\
0.0000 & 0.0000 & 0.0642 & 0.1358
\end{bmatrix}
\begin{Bmatrix}
10 \\
h_2 \\
h_3 \\
0
\end{Bmatrix}_{t=1\ \text{day}}
$$

$$
=
\begin{bmatrix}
0.1647 & 0.0853 & 0.0000 & 0.0000 \\
0.0853 & 0.3631 & 0.1017 & 0.0000 \\
0.0000 & 0.1017 & 0.3292 & 0.0692 \\
0.0000 & 0.0000 & 0.0692 & 0.1308
\end{bmatrix}
\begin{Bmatrix}
10 \\
0 \\
0 \\
0
\end{Bmatrix}_{t=0}
=
\begin{bmatrix}
1.647 \\
0.853 \\
0.000 \\
0.000
\end{bmatrix}
$$

or

$$
\begin{bmatrix}
0.3704 & 0.0984 \\
0.0484 & 0.3375
\end{bmatrix}
\begin{Bmatrix}
h_2 \\
h_3
\end{Bmatrix}_{t=1\ \text{dav}}
=
\begin{bmatrix}
0.853 \\
0.000
\end{bmatrix}
-
\begin{bmatrix}
(10)\,(0.0813) \\
(10)\,(0.0000)
\end{bmatrix}
=
\begin{bmatrix}
0.0400 \\
0.0000
\end{bmatrix}
$$

This system is stored in arrays M, X

$$
M =
\begin{Bmatrix}
0.3704 \\
0.0984 \\
0.3375
\end{Bmatrix}
\quad \text{(banded, symmetric matrix in vector storage)}
$$

$$
X =
\begin{Bmatrix}
h_2 \\
h_3
\end{Bmatrix}_{t=1\ \text{dav}}
\quad \text{(obtained using DECOMP and SOLVE)}
$$

The right-hand side terms are constructed in subroutine RHS (see Chapter 18).

17.6 SOURCE CODE LISTING

```
          SUBROUTINE ASMBKC
C*********************************************************************
C
C 17.1  PURPOSE:
C          TO ASSEMBLE THE COMBINED GLOBAL CONDUCTANCE AND
C          CAPACITANCE MATRIX (EQUATION 5.30A) AND THE  GLOBAL
C          SPECIFIED FLOW MATRIX FOR THE MESH AND TO MODIFY THE
C          SYSTEM OF EQUATIONS FOR SPECIFIED HEAD AND GROUNDWATER
C          FLOW BOUNDARY CONDITIONS
C
C 17.2  INPUT:
C          NONE
C
C 17.3  OUTPUT:
C          THE SEMI-BANDWIDTH AND NUMBER OF DEGREES OF FREEDOM
C          FOR THE MODIFIED, COMBINED GLOBAL CONDUCTANCE AND
C          CAPACITANCE MATRIX ARE WRITTEN TO THE USER-DEFINED
C          FILE ASSIGNED TO UNIT "OUTF"
C
C 17.4  DEFINITIONS OF VARIABLES:
C              B(I) = MODIFIED SPECIFIED FLOW MATRIX
C            E(I,J) = CAPACITANCE MATRIX FOR ELEMENT E IN
C                     FULL MATRIX STORAGE
C                 E = ELEMENT NUMBER
C        ELEMTYP(E) = ELEMENT TYPE FOR ELEMENT E (SEE TABLE
C                     9.1 FOR A LIST OF ELEMENT TYPES)
C           FLUX(I) = SPECIFIED RATE OF GROUNDWATER FLOW
C                     AT NODE I
C            ICH(I) = 1 IF THE VALUE OF HYDRAULIC HEAD OR
C                     PRESSURE HEAD IS SPECIFIED FOR NODE I,
C                   = 0 OTHERWISE
C            IJSIZE = LENGTH OF ARRAY [M] IN VECTOR STORAGE
C           KE(I,J) = CONDUCTANCE MATRIX FOR ELEMENT E IN
C                     FULL MATRIX STORAGE
C            LCH(I) = ICH(I) + ICH(I-1) + ICH(I-2) + ...
C                     THE ARRAYS ICH AND LCH ARE USED TO MODIFY
C                     THE GLOBAL SYSTEM OF EQUATIONS FOR SPECIFIED
C                     VALUES OF THE FIELD VARIABLE
C             M(IJ) = MODIFIED, COMBINED GLOBAL CONDUCTANCE
C                     AND CAPACITANCE MATRIX IN VECTOR STORAGE
C              NDOF = NUMBER OF NODES WHERE THE VALUE OF
C                     THE FIELD VARIABLE IS UNKNOWN
C        NODETBL(I) = NUMBER OF NODES IN ELEMENT TYPE I
C            NUMELM = NUMBER OF ELEMENTS IN MESH
C               SBW = SEMI-BANDWIDTH OF MODIFIED, COMBINED
C                     GLOBAL CONDUCTANCE AND CAPACITANCE MATRIX
C              X(I) = VALUE OF HYDRAULIC HEAD OR PRESSURE HEAD
C                     AT NODE I
C
C 17.5  USAGE:
C          THE SEMI-BANDWIDTH OF THE COMBINED GLOBAL CONDUCTANCE AND
C          CAPACITANCE MATRIX IS COMPUTED FIRST.  THEN THE ENTRIES
C          OF THE ELEMENT CONDUCTANCE AND CAPACITANCE MATRICES ARE
C          COMPUTED IN A SET OF SUBROUTINES, TWO SUBROUTINES FOR
C          EACH ELEMENT TYPE.  THE COMBINED GLOBAL CONDUCTANCE AND
C          CAPACITANCE MATRIX FOR THE MESH IS ASSEMBLED BY ADDING THE
C          CORRESPONDING ENTRIES OF THE ELEMENT MATRICES TO THE GLOBAL
C          MATRIX.  DURING THE ASSEMBLY PROCESS THE GLOBAL MATRIX IS
C          MODIFIED FOR SPECIFIED VALUES OF HEAD AND SPECIFIED VALUES
C          OF GROUNDWATER FLOW ARE ADDED TO THE GLOBAL FLOW MATRIX.
```

```
C
C         SUBROUTINES CALLED:
C            KBAR2,KBAR3,KBAR4,KTRI3,KREC4,KQUA4,KQUA8,KQUA12,KPAR8,
C            KPAR20,KBAR32,KTRI3A,KREC4A,LOC (LISTED WITH SUBROUTINE
C            ASMBK IN CHAPTER 12)
C            CBAR2,CBAR3,CBAR4,CTRI3,CREC4,CQUA4,CQUA8,CQUA12,CPAR8,
C            CPAR20,CBAR32,CTRI3A,CREC4A
C
C         REFERENCES:
C            ISTOK,J.D. GROUNDWATER FLOW AND SOLUTE TRANSPORT
C            MODELING BY THE FINITE ELEMENT METHOD, CHAPTER 17.
C
C**********************************************************************
$INCLUDE:'COMALL'
          REAL KE(MAX3,MAX3),CE(MAX3,MAX3)
          INTEGER NODETBL(13)
          DATA NODETBL/2,3,4,3,4,4,8,12,8,20,32,3,4/
C
C         COMPUTE THE SEMI-BANDWIDTH
C
          SBW = 1
          DO 30 E = 1, NUMELM
            DO 20 I = 1, NODETBL(ELEMTYP(E))
              KI = IN(E,I)
              IF (ICH(KI) .EQ. 0 .AND. I .LT. NODETBL(ELEMTYP(E))) THEN
                II = KI - LCH(KI)
                DO 10 J = I + 1, NODETBL(ELEMTYP(E))
                  KJ = IN(E,J)
                  IF (ICH(KJ) .EQ. 0) THEN
                    JJ = ABS(KJ - LCH(KJ) - II) + 1
                    IF (JJ .GT. SBW) SBW = JJ
                  ENDIF
10                CONTINUE
              ENDIF
20          CONTINUE
30        CONTINUE
          WRITE(OUTF,40) NDOF,SBW
40        FORMAT(//' NUMBER OF DEGREES OF FREEDOM IN MODIFIED,'/
     1           ' GLOBAL COMBINED CONDUCTANCE AND CAPACITANCE',
     2           ' MATRIX =',I5///' SEMI-BANDWIDTH OF MODIFIED,'/
     3           ' GLOBAL COMBINED CONDUCTANCE AND CAPACITANCE',
     4           ' MATRIX =',I5)
          IF (SBW .GT. MAX6) STOP'** EXCEEDS MAXIMUM SEMI-BAND WIDTH **'
C         INITIALIZE ENTRIES OF GLOBAL MATRIX TO ZERO
          IJSIZE = SBW * (NDOF - SBW + 1) + (SBW - 1) * SBW / 2
          DO 50 IJ = 1, IJSIZE
            M(IJ) = 0.0
            B1(IJ) = 0.0
50        CONTINUE

          DO 56 I = 1, MAX1
56          FC(I) = 0.

C         INITIALIZE ENTRIES OF THE GLOBAL GROUNDWATER MATRIX TO ZERO
          DO 60 I = NDOF
            B(I) = 0.0
60        CONTINUE
C         LOOP ON THE NUMBER OF ELEMENTS
          DO 90 E = 1, NUMELM
C           COMPUTE THE ELEMENT CONDUCTANCE AND CAPACITANCE MATRICES
C           FOR THIS ELEMENT TYPE
            IF (ELEMTYP(E) .EQ. 1) THEN
```

```
C             ELEMENT IS A ONE-DIMENSIONAL, LINEAR BAR
              CALL KBAR2(E,KE)
              CALL CBAR2(E,CE)
            ELSEIF (ELEMTYP(E) .EQ .2) THEN
C             ELEMENT IS A ONE-DIMENSIONAL, QUADRATIC BAR
              CALL KBAR3(E,KE)
              CALL CBAR3(E,CE)
            ELSEIF (ELEMTYP(E) .EQ. 3) THEN
C             ELEMENT IS A ONE-DIMENSIONAL, CUBIC BAR
              CALL KBAR4(E,KE)
              CALL CBAR4(E,CE)
            ELSEIF (ELEMTYP(E) .EQ. 4) THEN
C             ELEMENT IS A TWO-DIMENSIONAL, LINEAR TRIANGLE
              CALL KTRI3(E,KE)
              CALL CTRI3(E,CE)
            ELSEIF (ELEMTYP(E) .EQ. 5) THEN
C             ELEMENT IS A TWO-DIMENSIONAL, LINEAR RECTANGLE
              CALL KREC4(E,KE)
              CALL CREC4(E,CE)
            ELSEIF (ELEMTYP(E) .EQ. 6) THEN
C             ELEMENT IS A TWO-DIMENSIONAL, LINEAR QUADRILATERAL
              CALL KQUA4(E,KE)
              CALL CQUA4(E,CE)
            ELSEIF (ELEMTYP(E) .EQ. 7) THEN
C             ELEMENT IS A TWO-DIMENSIONAL, QUADRATIC QUADRILATERAL
              CALL KQUA8(E,KE)
              CALL CQUA8(E,CE)
            ELSEIF (ELEMTYP(E) .EQ. 8) THEN
C             ELEMENT IS A TWO-DIMENSIONAL, CUBIC QUADRILATERAL
              CALL KQUA12(E,KE)
              CALL CQUA12(E,CE)
            ELSEIF (ELEMTYP(E) .EQ. 9) THEN
C             ELEMENT IS A THREE-DIMENSIONAL, LINEAR PARALLELEPIPED
              CALL KPAR8(E,KE)
              CALL CPAR8(E,CE)
            ELSEIF (ELEMTYP(E) .EQ. 10) THEN
C             ELEMENT IS A THREE-DIMENSIONAL, QUADRATIC PARALLELEPIPED
              CALL KPAR20(E,KE)
              CALL CPAR20(E,CE)
            ELSEIF (ELEMTYP(E) .EQ. 11) THEN
C             ELEMENT IS A THREE-DIMENSIONAL, CUBIC PARALLELEPIPED
              CALL KPAR32(E,KE)
              CALL CPAR32(E,CE)
            ELSEIF (ELEMTYP(E) .EQ. 12) THEN
C             ELEMENT IS A TWO-DIMENSIONAL, LINEAR TRIANGLE (AXISYMMETRIC)
              CALL KTRI3A(E,KE)
              CALL CTRI3A(E,CE)
            ELSEIF (ELEMTYP(E) .EQ. 13) THEN
C             ELEMENT IS A TWO-DIMENSIONAL, LINEAR RECTANGLE (AXISYMMETRIC)
              CALL KREC4A(E,KE)
              CALL CREC4A(E,CE)
            ENDIF
C       ADD THE ELEMENT CONDUCTANCE AND CAPACITANCE MATRICES FOR
C       THIS ELEMENT TO THE GLOBAL MATRIX
C   KE(I,J),CE(I,J) -----------> M(IJ)   <=>         M(KI,KJ)
C (FULL MATRIX STORAGE)   (VECTOR MATRIX STORAGE)   (FULL MATRIX STORAGE)
            DO 80 I = 1, NODETBL(ELEMTYP(E))
              KI = IN(E,I)
              IF (ICH(KI) .EQ. 0) THEN
                II = KI - LCH(KI)
                DO 70 J = 1, NODETBL(ELEMTYP(E))
                  KJ = IN(E,J)
```

```
                      IF (ICH(KJ) .NE. 0) THEN
                         FC(II) = FC(II) - DELTAT(IDT) * KE(I,J) * X(KJ)
                      ELSEIF (J .GE. I) THEN
                         JJ = KJ - LCH(KJ)
                         CALL LOC(II,JJ,IJ,NDOF,SBW,SYMM)
                         M(IJ) = M(IJ) + CE(I,J) + OMEGA *
     1                           DELTAT(IDT) * KE(I,J)
                         B1(IJ) = B1(IJ) + CE(I,J) - OMOMEGA *
     1                           DELTAT(IDT) * KE(I,J)
                      ENDIF
 70                CONTINUE
                ENDIF
 80         CONTINUE
 90      CONTINUE
         DO 999 I = 1,IJSIZE
            WRITE(*,*) M(I),B(I)
 999     CONTINUE
         RETURN
         END
```

```
         SUBROUTINE CBAR2(E,CE)
C*********************************************************************
C
C        PURPOSE:
C           TO COMPUTE THE CONSISTENT FORM OF THE ELEMENT
C           CAPACITANCE MATRIX FOR A ONE-DIMENSIONAL, LINEAR
C           BAR ELEMENT
C
C        DEFINITIONS OF VARIABLES:
C                  E = ELEMENT NUMBER
C           CE(I,J) = ELEMENT CAPACITANCE MATRIX
C               SSE = ELEMENT SPECIFIC STORAGE
C                LE = ELEMENT LENGTH
C
C        REFERENCES:
C           ISTOK,J.D. GROUNDWATER FLOW AND SOLUTE TRANSPORT
C           MODELING BY THE FINITE ELEMENT METHOD, FIGURE 4.5,
C           EQUATION 4.16a.
C
C*********************************************************************
$INCLUDE:'COMALL'
         REAL CE(MAX3,MAX3),LE
C
         SSE = PROP(MATSET(E),2)
         LE = ABS(X1(IN(E,2)) - X1(IN(E,1)))
         CE(1,1) = SSE * LE / 3.
         CE(1,2) = SSE * LE / 6.
         CE(2,1) = CE(1,2)
         CE(2,2) = CE(1,1)
         RETURN
         END
```

```
      SUBROUTINE CTRI3 (E,CE)
C*********************************************************************
C
C       PURPOSE:
C        TO COMPUTE THE CONSISTENT FORM OF THE ELEMENT CAPACITANCE
C        MATRIX FOR TWO- DIMENSIONAL, LINEAR TRIANGLE ELEMENT
C
C       DEFINITIONS OF VARIABLES:
C            AE4 = FOUR TIMES ELEMENT AREA
C            E = ELEMENT NUMBER
C          CE(I,J) = ELEMENT CAPACITANCE MATRIX
C            SSE = ELEMENT SPECIFIC STORAGE
C
C       REFERENCES:
C        ISTOK,J.D. GROUNDWATER FLOW AND SOLUTE TRANSPORT
C        MODELING BY THE FINITE ELEMENT METHOD, FIGURE 4.7,
C        EQUATION 4.22a
C
C*********************************************************************
$INCLUDE: 'COMALL'
      REAL CE(MAX3,MAX3)
C
      SSE = PROP(MATSET(E),3)
      AE4 = 2 * (X1(IN(E,2)) * X2(IN(E,3)) + X1(IN(E,1)) *
     1      X2(IN(E,2)) +  X2(IN(E,1)) * X1(IN(E,3)) -
     2      X2(IN(E,3)) * X1(IN(E,1)) - X1(IN(E,3)) *
     3      X2(IN(E,2)) - X1(IN(E,2)) * X2(IN(E,1)))
      AE = AE4 / 4.
      CE(1,1) = SSE * AE / 6.
      CE(1,2) = CE(1,1) / 2.
      CE(1,3) = CE(1,2)
      CE(2,1) = CE(1,2)
      CE(2,2) = CE(1,1)
      CE(2,3) = CE(1,2)
      CE(3,1) = CE(1,2)
      CE(3,2) = CE(1,2)
      CE(3,3) = CE(1,1)
      RETURN
      END




      SUBROUTINE CREC4 (E,CE)
C*********************************************************************
C
C       PURPOSE:
C        TO COMPUTE THE CONSISTENT FORM OF THE ELEMENT CAPACITANCE
C        MATRIX FOR TWO- DIMENSIONAL, LINEAR RECTANGLE ELEMENT
C
C       DEFINITIONS OF VARIABLES:
C            E = ELEMENT NUMBER
C          CE(I,J) = ELEMENT CAPACITANCE MATRIX
C            SSE = ELEMENT SPECIFIC STORAGE
C
C       REFERENCES:
C        ISTOK,J.D. GROUNDWATER FLOW AND SOLUTE TRANSPORT
C        MODELING BY THE FINITE ELEMENT METHOD, FIGURE 4.6,
C        EQUATION 4.27a
C
C*********************************************************************
```

```
$INCLUDE: 'COMALL'
      REAL CE(MAX3,MAX3)
C
      SSE = PROP(MATSET(E),3)
      AE = ABS(X2(IN(E,1)) - X2(IN(E,3))) / 2.
      BE = ABS(X1(IN(E,1)) - X1(IN(E,3))) / 2.
      TEMP = (SSE * AE * BE) / 9
      CE(1,1) = 4. * TEMP
      CE(1,2) = 2. * TEMP
      CE(1,3) = TEMP
      CE(1,4) = CE(1,2)
      CE(2,1) = CE(1,2)
      CE(2,2) = CE(1,1)
      CE(2,3) = CE(1,2)
      CE(2,4) = CE(1,3)
      CE(3,1) = CE(1,3)
      CE(3,2) = CE(1,2)
      CE(3,3) = CE(1,1)
      CE(3,4) = CE(1,2)
      CE(4,1) = CE(1,2)
      CE(4,2) = CE(1,3)
      CE(4,3) = CE(1,2)
      CE(4,4) = CE(1,1)
      RETURN
      END

      SUBROUTINE CQUA4(E,CE)
C*********************************************************************
C
C        PURPOSE:
C           TO COMPUTE THE CONSISTENT FORM OF THE ELEMENT CAPACITANCE
C           MATRIX FOR A TWO-DIMENSIONAL, LINEAR QUADRILATERAL ELEMENT
C
C        DEFINITIONS OF VARIABLES:
C              CE(I,J) = ELEMENT CAPACITANCE MATRIX
C               DETJAC = DETERMINANT OF JACOBIAN MATRIX
C             DNDXI(I) = PARTIAL DERIVATIVE OF INTERPOLATION
C                        FUNCTION WITH RESPECT TO XI AT NODE I
C              DNDX(I) = PARTIAL DERIVATIVE OF INTERPOLATION
C                        FUNCTION WITH RESPECT TO X AT NODE I
C           DNDETA(I) = PARTIAL DERIVATIVE OF INTERPOLATION
C                        FUNCTION WITH RESPECT TO ETA AT NODE I
C              DNDY(I) = PARTIAL DERIVATIVE OF INTERPOLATION
C                        FUNCTION WITH RESPECT TO Y AT NODE I
C                XI(I) = LOCATION OF GAUSS POINT IN XI COORDINATE
C                        DIRECTION
C               ETA(I) = LOCATION OF GAUSS POINT IN ETA COORDINATE
C                        DIRECTION
C             JAC(I,J) = JACOBIAN MATRIX
C                    E = ELEMENT NUMBER
C                  SSE = ELEMENT SPECIFIC STORAGE
C                 N(I) = INTERPOLATION FUNCTION FOR NODE I
C                 W(I) = WEIGHT FOR GAUSS POINT I
C          X1(IN(E,I)) = X COORDINATE FOR NODE I, ELEMENT E
C          X2(IN(E,I)) = Y COORDINATE FOR NODE I, ELEMENT E
C
C        REFERENCES:
C           ISTOK,J.D. GROUNDWATER FLOW AND SOLUTE TRANSPORT
C           MODELING BY THE FINITE ELEMENT METHOD, FIGURE 4.10,
C           EQUATION 4.65
C
C*********************************************************************
```

```
$INCLUDE: 'COMALL'
      REAL JAC(2,2),JACINV(2,2),CE(MAX3,MAX3),N(4),DNDXI(4),
     1      DNDR(4),DNDETA(4),DNDZ(4),W(2),XI(2),ETA(2),SIGN1(4),
     2      SIGN2(4)

      DATA SIGN1/-1.,1.,1.,-1./
      DATA SIGN2/-1.,-1.,1.,1./
C
      XI(1) =   1. / SQRT(3.)
      XI(2) = -XI(1)
      ETA(1) = XI(1)
      ETA(2) = XI(2)
      W(1) = 1.
      W(2) = 1.
      SSE = PROP(MATSET(E),3)

      DO 30 I = 1, 4
        DO 20 J = 1, 4
          CE(I,J) = 0.
 20       CONTINUE
 30     CONTINUE

      DO 120 I = 1, 2
        DO 110 J = 1, 2

          DO 50 K = 1, 2
            DO 40 K1 = 1, 2
              JAC(K,K1) = 0.
 40         CONTINUE
 50       CONTINUE

          DO 60 K1 = 1, 4
                N(K1) = 0.25 * (1. + SIGN1(K1) * XI(I))
     1                     * (1. + SIGN2(K1) * ETA(J))
              DNDXI(K1) = 0.25 * SIGN1(K1) * (1. + SIGN2(K1) * ETA(J))
              DNDETA(K1) = 0.25 * SIGN2(K1) * (1. + SIGN1(K1) * XI(I))
 60       CONTINUE
          DO 70 K1 = 1, 4
              JAC(1,1) = JAC(1,1) + DNDXI(K1) * X1(IN(E,K1))
              JAC(1,2) = JAC(1,2) + DNDXI(K1) * X2(IN(E,K1))
              JAC(2,1) = JAC(2,1) + DNDETA(K1) * X1(IN(E,K1))
              JAC(2,2) = JAC(2,2) + DNDETA(K1) * X2(IN(E,K1))
 70       CONTINUE
          DETJAC = JAC(1,1) * JAC(2,2) - JAC(1,2) * JAC(2,1)
          DO 100 K = 1, 4
            DO 90 K1 = 1, 4
              CE(K,K1) = CE(K,K1) + W(I) * W(J) * SSE* N(K) *
     1                   N(K1) * DETJAC
 90         CONTINUE
 100      CONTINUE
 110    CONTINUE
 120  CONTINUE
      RETURN
      END
```

```
      SUBROUTINE CPAR8 (E,CE)
C*********************************************************************
C
C     PURPOSE:
C       TO COMPUTE THE CONSISTENT FORM OF THE ELEMENT
C       CAPACITANCE MATRIX FOR A THREE-DIMENSIONAL,
C       LINEAR QUADRILATERAL ELEMENT
C
C     DEFINITIONS OF VARIABLES:
C          CE(I,J) = ELEMENT CAPACITANCE MATRIX
C           DETJAC = DETERMINANT OF JACOBIAN MATRIX
C         DNDXI(I) = PARTIAL DERIVATIVE OF INTERPOLATION
C                    FUNCTION WITH RESPECT TO XI AT NODE I
C          DNDX(I) = PARTIAL DERIVATIVE OF INTERPOLATION
C                    FUNCTION WITH RESPECT TO X AT NODE I
C        DNDETA(I) = PARTIAL DERIVATIVE OF INTERPOLATION
C                    FUNCTION WITH RESPECT TO ETA AT NODE I
C          DNDY(I) = PARTIAL DERIVATIVE OF INTERPOLATION
C                    FUNCTION WITH RESPECT TO Y AT NODE I
C       DNDZETA(I) = PARTIAL DERIVATIVE OF INTERPOLATION
C                    FUNCTION WITH RESPECT TO ZETA AT NODE I
C          DNDZ(I) = PARTIAL DERIVATIVE OF INTERPOLATION
C                    FUNCTION WITH RESPECT TO Z AT NODE I
C            XI(I) = LOCATION OF GAUSS POINT IN XI COORDINATE
C                    DIRECTION
C           ETA(I) = LOCATION OF GAUSS POINT IN ETA COORDINATE
C                    DIRECTION
C          ZETA(I) = LOCATION OF GAUSS POINT IN ZETA COORDINATE
C                    DIRECTION
C          JAC(I,J) = JACOBIAN MATRIX
C                E = ELEMENT NUMBER
C              SSE = ELEMENT SPECIFIC STORAGE
C             N(I) = INTERPOLATION FUNCTION FOR NODE I
C             W(I) = WEIGHT FOR GAUSS POINT I
C      X1(IN(E,I) = X COORDINATE FOR NODE I, ELEMENT E
C      X2(IN(E,I) = Y COORDINATE FOR NODE I, ELEMENT E
C      X3(IN(E,I) = Z COORDINATE FOR NODE I, ELEMENT E
C
C     REFERENCES:
C       ISTOK,J.D. GROUNDWATER FLOW AND SOLUTE TRANSPORT
C       MODELING BY THE FINITE ELEMENT METHOD, FIGURE 4.10,
C       EQUATION 4.66
C
C*********************************************************************
$INCLUDE: 'COMALL'
      REAL JAC(3,3),CE(MAX3,MAX3),DNDX(8),DNDY(8),DNDZ(8),
     1     XI(8),ETA(8),ZETA(8),DNDXI(8),DNDETA(8),DNDZETA(8),W(2),
     2     N(8),SIGN1(8),SIGN2(8),SIGN3(8)
      DATA SIGN1/-1.,1.,1.,-1.,-1.,1.,1.,-1./
      DATA SIGN2/-1.,-1.,1.,1.,-1.,-1.,1.,1./
      DATA SIGN3/-1.,-1.,-1.,-1.,1.,1.,1.,1./
C
      XI(1) = 1. / SQRT(3.)
      XI(2) = -XI(1)
      ETA(1) = XI(1)
      ETA(2) = XI(2)
      ZETA(1) = XI(1)
      ZETA(2) = XI(2)
      W(1) = 1.
      W(2) = 1.
      SSE = PROP(MATSET(E),4)
```

```
      DO 20 K = 1, 8
        DO 10 N1 = 1, 8
          CE(K,N1) = 0.
10    CONTINUE
20 CONTINUE

      DO 120 I = 1, 2
        DO 110 J = 1, 2
          DO 100 K = 1, 2

            DO 40 L = 1, 3
              DO 30 N1 = 1, 3
                JAC(L,N1) = 0.
30          CONTINUE
40          CONTINUE

            DO 50 N1 = 1, 8
              N(N1) = 0.125 * (1.+SIGN1(N1)*XI(I)) * (1.+SIGN2(N1) *
     1                    ETA(J)) * (1. + SIGN3(N1) * ZETA(K))
              DNDXI(N1)    = 0.125 * SIGN1(N1) * (1. + SIGN2(N1) *
     1                    ETA(J)) * (1. + SIGN3(N1) * ZETA(K))
              DNDETA(N1)   = 0.125 * SIGN2(N1) * (1. + SIGN1(N1) *
     1                    XI(I)) * (1. + SIGN3(N1) * ZETA(K))
              DNDZETA(N1)  = 0.125 * SIGN3(N1) * (1. + SIGN1(N1) *
     1                    XI(I)) * (1. + SIGN2(N1) * ETA(J))
50          CONTINUE

            DO 60 M5 = 1, 8
              JAC(1,1) = JAC(1,1) + DNDXI(M5) * X1(IN(E,M5))
              JAC(1,2) = JAC(1,2) + DNDXI(M5) * X2(IN(E,M5))
              JAC(1,3) = JAC(1,3) + DNDXI(M5) * X3(IN(E,M5))
              JAC(2,1) = JAC(2,1) + DNDETA(M5) * X1(IN(E,M5))
              JAC(2,2) = JAC(2,2) + DNDETA(M5) * X2(IN(E,M5))
              JAC(2,3) = JAC(2,3) + DNDETA(M5) * X3(IN(E,M5))
              JAC(3,1) = JAC(3,1) + DNDZETA(M5) * X1(IN(E,M5))
              JAC(3,2) = JAC(3,2) + DNDZETA(M5) * X2(IN(E,M5))
              JAC(3,3) = JAC(3,3) + DNDZETA(M5) * X3(IN(E,M5))
60          CONTINUE

            DETJAC = JAC(1,1) * (JAC(2,2) * JAC(3,3) - JAC(3,2) *
     1              JAC(2,3)) - JAC(1,2) * (JAC(2,1) * JAC(3,3) -
     2              JAC(3,1) * JAC(2,3)) - JAC(1,3) * (JAC(2,1) *
     3              JAC(3,2) - JAC(3,1) * JAC(2,2))

            DO 90 L = 1, 8
              DO 80 M5 = 1, 8
              CE(L,M5) = CE(L,M5) + W(I) * W(J) * W(K) * SSE *
     1              N(L) * N(M5) * DETJAC

80          CONTINUE
90          CONTINUE

100     CONTINUE
110     CONTINUE
120 CONTINUE
      RETURN
      END
```

Chapter 18

SUBROUTINE RHS

18.1 PURPOSE

Subroutine RHS assembles the "right-hand-side" vector, for the transient, groundwater flow equation

$$([C] - (1 - \omega) \Delta t [K]) \{h\}_t + \Delta t ((1 - \omega) \{F\}_t + \omega \{F\}_{t+\Delta t}) \qquad (18.1)$$

and for the solute transport equation

$$([A] - (1 - \omega) \Delta t [D]) \{C\}_t + \Delta t ((1 - \omega) \{F\}_t + \omega \{F\}_{t+\Delta t}) \qquad (18.2)$$

where [C] is the gobal capacitance matrix, ω is the relaxation factor, Δt is the timestep interval, [K] is the global conductance matrix, $\{h\}_t$ are the heads at time t, $\{F\}_t$ and $\{F\}_{t+\Delta t}$ are specifies rates of groundwater flow (or solute flux) at times t and t+Δt, [A] is the global adsorption matrix, [D] is the gobal advection dispersion matrix, and $\{C\}_t$ and $\{C\}_{t+\Delta t}$ are the solute concentrations at times t and t+Δt. RHS performs the matrix multiplications and additions and modifys the resulting vector for specified values of head or solute concentration

18.2 INPUT

None

18.3 OUTPUT

None

18.4 DEFINITIONS OF VARIABLES

DELTAT(I) = Size of time step I

FLUX(I)   = Specified value of groundwater flow or solute flux at node I

GT(I)   = Value of time function at time t (see Chapter 16)

ICH(I)   = 1 if the value of the field variable is specified at node I
           = 0 otherwise

B1(I)   = Modified global matrix (equation 18.1 or 18.2) in vector storage

NDOF   = Number of nodes where the value of the field variable is unknown

NUMNOD   = Number of nodes

OMEGA   = Relaxation factor (ω)

OMOMEGA = 1 - ω

18.5 USAGE

Equation 18.1 or 18.2 is evaluated for each time step. The global matrices [C], [K], [A] and [D] are assembled in subroutines ASMBKC (Chapter 17) and ASMBAD (Chapter 19).

18.6 SOURCE CODE LISTING

```
          SUBROUTINE RHS
C**************************************************************************
C
C 18.1  PURPOSE:
C           SUBROUTINE RHS ASSEMBLES THE RIGHT-HAND-SIDE VECTOR
C           FOR TRANSIENT GROUNDWATER FLOW AND SOLUTE TRANSPORT
C           PROBLEMS.
C
C 18.2  INPUT:
C           NONE
C
C 18.3  OUTPUT:
C           NONE
C
C 18.4  DEFINITIONS OF VARIABLES:
C           DELTAT(I) = SIZE OF TIME STEP I
C             FLUX(I) = SPECIFIED VALUE OF GROUNDWATER FLOW OR
C                       SOLUTE FLUX AT NODE I
C               GT(I) = VALUE OF TIME FUNCITON AT TIME I
C              ICH(I) = 1 IF THE VALUE OF THE FIELD VARIABLE IS
C                       SPECIFIED AT NODE I
C                     = 0 OTHERWISE
C              B1(IJ) = MODIFIED GLOBAL MATRIX IN VECTOR STORAGE
C                NDOF = NUMBER OF NODES WHERE THE VALUE OF THE
C                       FIELD VARIABLE IS UNKNOWN (NAMED FOR
C                       NUMBER OF DEGREES OF FREEDOM)
C              NUMNOD = NUMBER OF NODES
C               OMEGA = RELAXATION FACTOR
C             OMOMEGA = 1 - OMEGA
C
C 18.5  USAGE:
C           FOR EACH TIME STEP, THE RIGHT-HAND-SIDE VECTOR IS
C           COMPUTED USING THE VALUES OF HEAD OR SOLUTE
C           CONCENTRATION FOR THE PREVIOUS TIME STEP, AND THE
C           MODIFIED COMBINED CONDUCTION AND CAPACITANCE MATRIX,
C           RELAXATION FACTOR, AND TIME STEP INTERVAL FOR THAT
C           TIME STEP
C
C           SUBROUTINES CALLED:
C             LOC
C
C           REFERENCES:
C             ISTOK,J.D. GROUNDWATER FLOW AND SOLUTE TRANSPORT
C             MODELING BY THE FINITE ELEMENT METHOD, CHAPTER 18.
C
C**************************************************************************
```

```
$INCLUDE:'COMALL'
C
        IF (T .GT. TIME(IGT)) IGT = IGT + 1
        T = T + DELTAT(IDT)
        IF (T .GT. TIME(IGTDT)) IGTDT = IGTDT + 1
        I = 0
        DO 10 J = 1, NUMNOD
          IF (ICH(J) .EQ. 0) THEN
            I = I + 1
            B(I) = FC(I) + DELTAT(IDT) * (OMOMEGA * GT(IGT) * FLUX(J)
     1             + OMEGA * GT(IGTDT) * FLUX(J))
          ENDIF
10      CONTINUE
        J1 = 1
        J2 = SBW
        DO 60 I = 1, NDOF
          J = 0
          DO 20 K = 1, NUMNOD
            IF (ICH(K) .EQ. 0) THEN
              J = J + 1
              IF (J .EQ. J1) GOTO 30
            ENDIF
20        CONTINUE
30        K = K - 1
          DO 50 J = J1, J2
40          K = K + 1
            IF (ICH(K) .NE. 0) GOTO 40
            CALL LOC(I,J,IJ,NDOF,SBW,SYMM)
            B(I) = B(I) + B1(IJ) * X(K)
50        CONTINUE
          IF (I .GE. SBW) J1 = J1 + 1
          IF (J2 .LT. NDOF) J2 = J2 + 1
60      CONTINUE
        RETURN
        END
```

Chapter 19

SUBROUTINE ASMBAD

19.1 PURPOSE

Subroutine ASMBAD assembles the combined global sorption and advection-dispersion matrix [M] (equation 5.48a) and the global specified flux matrix {F}. The global matrices are modified to account for specified values of solute concentration (at Dirichlet nodes) and for specified rates of solute flux (at Neumann nodes). [M] is assembled and modified in vector storage. ASMBAD also computes the semi-bandwidth and number of degrees of freedom for [M].

19.2 INPUT

None

19.3 OUTPUT

The semi-bandwidth and number of degrees of freedom for the modified, combined global conductance and capacitance matrix are written to the user-defined file assigned to unit "OUTF".

19.4 DEFINITIONS OF VARIABLES

$AE(I,J)$ =   Sorption matrix for element e in full matrix storage.

$DE(I,J)$ =   Advection-dispersion matrix for element e in full matrix storage.

$ELEMTYP(E)$ =   Element type for element E (see Table 9.1 for a list of element types).

$F(I)$ =   Global solute flux matrix.

$FLUX(I)$ =   Specified rate of solute flux at node I.

$X(I)$ =   Value of solute concentration at node I.

$ICH(I)$ =   1 if the value of solute concentration is specified at node I, 0 otherwise.

$IJSIZE$ =   Length of array [M] in vector storage.

$LCH(I)$ = $\sum_{k=1}^{I} ICH(k)$. The arrays ICH and LCH are used to modify the global system of equations for specified values of the field variable.

M(I,J)   =   Modified, combined global sorption and advection-dispersion matrix
                  in vector storage.

              = $([A] + \omega \, \Delta t[D])$

NDOF   =   Number of nodes where the value of solute concentration is
                  unknown.

NODETBL(I)   =   Number of nodes in element type I.

NUMELM   =   Number of elements in mesh.

SBW   =   Semi-bandwidth of modified, combined global sorption and
                  advection-dispersion matrix.

19.5 USAGE

Subroutine ASMBAD assembles the combined global sorption and advection-dispersion matrix [M]

$$[M] = ([A] + \omega \, \Delta t \, [D]) \qquad\qquad 19.1$$

and the specified solute flux matrix {F}. [M] and {F} are modified to account for specified values of solute concentration during the assembly process, using the procedures in Chapter 4. The modified, global specified flux matrix is stored in the array B. Further modifications to B are made in subroutine RHS (see Chapter 18).

Table 19.1 Subroutines used to compute element advection-dispersion matrices (DBAR2, DBAR3, etc.) and element sorption matrices (ABAR2, ABAR3, etc.) in ASMBAD.

Element Type	Description	Subroutine Names $[D^{(e)}]$	$[A^{(e)}]$	DIM
1	Linear bar	DBAR2	, ABAR2	1
2	Quadratic bar	DBAR3*	, ABAR3*	1
3	Cubic bar	DBAR4*	, ABAR4*	1
4	Linear triangle	DTRI3	, ATRI3	2
5	Linear rectangle	DREC4	, AREC4	2
6	Linear quadrilateral	DQUA4	, AQUA4	2
7	Quadratic quadrilateral	DQUA8*	, AQUA8*	2
8	Cubic quadrilateral	DQUA12*	, AQUA12*	2
9	Linear parallelepiped	DPAR8	, APAR8	3
10	Quadratic parallelepiped	DPAR20*	, APAR20*	3
11	Cubic parallelepiped	DPAR32*	, APAR32*	3
12	Linear triangle (axisymmetric)	DTRI3A*	, ATRI3A*	4
13	Linear rectangle (axisymmetric)	DREC4A*	, AREC4A*	4

*Source code listing not provided for these subroutines.

The number of degrees of freedom, NDOF, (number of unknown values of solute concentration) and semi-bandwith, SBW, are computed the same way as in ASMBKC (Chapter 17).

The element sorption and advection - dispersion matrices are computed in two sets of subroutines (Table 19.1). The first set of subroutines begins with the letter "A" (for the element sorption matrix $[A^{(e)}]$ and the second set begins with the letter "D" (for the element advection-dispersion matrix $[D^{(e)}]$). Additional letters and numbers in the subroutine names identify the element type and number of nodes in elements of that type. For example, subroutine ATRI3 computes the element sorption matrix for two-dimensional, linear triangle elements and subroutine DPAR8 computes the element advection-dispersion matrix for three-dimensional, linear parallelepiped elements.

The source code listing for each subroutine gives the figure that shows the interpolation functions and the equations used to compute [A] and [D] for that element type. Many of the important FORTRAN variable names and their symbols are in Table 12.2. Additional names and symbols for the subroutines in Table 19.1 are shown below

FORTRAN Variable	Definition	Symbol(s) in Text
ALE	Longitudinal dispersivity for element E	$\alpha_L^{(e)}$
ATE	Transverse dispersivity for element E	$\alpha_T^{(e)}$
KDE	Solute distribution coefficient for element E	$K_{d^{(e)}}$
LAMBAD	Solute decay coefficient	λ
NE	Porosity for element E	$n^{(e)}$
RHOBE	bulk density for element E	$\rho_b^{(e)}$
VXEP	Pore water velocity in x coordinate direction	$\dfrac{v_x^{(e)}}{n^{(e)}}$
VYEP	Pore water velocity in y coordinate direction	$\dfrac{v_y^{(e)}}{n^{(e)}}$

The operation of ASMBAD is very similar to ASMBKC and needs no special explanation. Remember that the global advection-dispersion matrix is *nonsymmetric* so the assembly and modification process is somewhat different (see Chapter 5).

19.6 SOURCE CODE LISTING

```
      SUBROUTINE ASMBAD
C**********************************************************************
C
C 19.1  PURPOSE:
C             TO ASSEMBLE THE COMBINED GLOBAL SORPTION AND
C             ADVECTION-DISPERSION MATRIX AND THE GLOBAL SPECIFIED
C             SOLUTE FLUX MATRIX FOR THE MESH AND TO MODIFY THE
C             SYSTEM OF EQUATIONS FOR SPECIFIED CONCENTRATION AND
C             SOLUTE FLUX BOUNDARY CONDITIONS
C
C 19.2  INPUT:
C             NONE
C
C 19.3  OUTPUT:
C             THE SEMI-BANDWIDTH AND NUMBER OF DEGREES OF FREEDOM
C             FOR THE MODIFIED, COMBINED GLOBAL SORPTION AND
C             ADVECTION-DISPERSION MATRIX ARE WRITTEN TO THE USER-
C             DEFINED FILE ASSIGNED TO UNIT OUTF
C
C 19.4  DEFINITIONS OF VARIABLES:
C                        AE(I,J) = SORPTION MATRIX FOR ELEMENT E IN FULL
C                                  MATRIX STORAGE
C                           B(I) = GLOBAL SPECIFIED SOLUTE FLUX MATRIX
C                        DE(I,J) = ADVECTION-DISPERSION MATRIX FOR ELEMENT
C                                  E IN FULL MATRIX STORAGE
C                              E = ELEMENT NUMBER
C                     ELEMTYP(E) = ELEMENT TYPE FOR ELEMENT E (SEE TABLE
C                                  9.1 FOR A LIST OF ELEMENT TYPES)
C                        FLUX(I) = SPECIFIED VALUE OF SOLUTE FLUX
C                                  AT NODE I
C                         ICH(I) = 1 IF THE VALUE OF SOLUTE CONCENTRATION
C                                  IS SPECIFIED FOR NODE I,
C                                = 0 OTHERWISE
C                         IJSIZE = LENGTH OF ARRAY ADGLOBAL
C                         LCH(I) = ICH(I) + ICH(I-1) + ICH(I-2) + ...
C                                  THE ARRAYS ICH AND LCH ARE USED TO
C                                  MODIFY THE GLOBAL MATRIX
C                          M(IJ) = MODIFIED, COMBINED GLOBAL SORPTION AND
C                                  ADVECTION-DISPERSION MATRIX IN VECTOR
C                                  STORAGE
C                           NDOF = NUMBER OF NODES WHERE THE VALUE OF
C                                  THE FIELD VARIABLE IS UNKNOWN
C          NODETBL(ELEMTYP(E)) = NUMBER OF NODES IN ELEMENT TYPE E
C                         NUMELM = NUMBER OF ELEMENTS IN MESH
C                            SBW = SEMI-BANDWIDTH OF MODIFIED, COMBINED
C                                  GLOBAL SORPTION AND ADVECTION-
C                                  DISPERSION MATRIX
C                           X(I) = VALUE OF SOLUTE CONCENTRATION
C                                  AT NODE I
C
C 19.5  USAGE:
C             THE SEMI-BANDWIDTH OF THE COMBINED GLOBAL SORPTION AND
C             ADVECTION-DISPERSION MATRIX IS COMPUTED FIRST.  THEN THE
C             ENTRIES OF THE ELEMENT SORPTION AND ADVECTION-DISPERSION
C             MATRICES ARE COMPUTED IN A SET OF SUBROUTINES, TWO
C             SUBROUTINES FOR EACH ELEMENT TYPE.  THE COMBINED GLOBAL
C             SORPTION AND ADVECTION-DISPERSION MATRIX FOR THE MESH IS
C             ASSEMBLED BY ADDING THE CORRESPONDING ENTRIES OF THE ELEMENT
C             SORPTION AND ADVECTION-DISPERSION MATRICES TO THE GLOBAL
```

```
C              MATRIX.  DURING THE ASSEMBLY PROCESS THE GLOBAL MATRIX IS
C              MODIFIED FOR SPECIFIED VALUES OF SOLUTE CONCENTRATION AND
C              SOLUTE FLUX ARE ADDED TO THE GLOBAL SOLUTE FLUX MATRIX.
C
C         SUBROUTINES CALLED:
C           ABAR2,ABAR3,ABAR4,ATRI3,AREC4,AQUA4,AQUA8,AQUA12,APAR8,
C           APAR20,ABAR32,ATRI3A,AREC4A
C           LOC (LISTED WITH SUBROUTINE ASMBK IN CHAPTER 12)
C           DBAR2,DBAR3,DBAR4,DTRI3,DREC4,DQUA4,DQUA8,DQUA12,DPAR8,
C           DPAR20,DBAR32,DTRI3A,DREC4A
C
C         REFERENCES:
C           ISTOK,J.D. GROUNDWATER FLOW AND SOLUTE TRANSPORT
C           MODELING BY THE FINITE ELEMENT METHOD, CHAPTER 19.
C
C*********************************************************************
$INCLUDE:'COMALL'
          REAL AE(MAX3,MAX3),DE(MAX3,MAX3)
          INTEGER NODETBL(13)
          DATA NODETBL/2,3,4,3,4,4,8,12,8,20,32,3,4/
C
C         COMPUTE THE SEMI-BANDWIDTH
          SBW = 1
          DO 30 E = 1, NUMELM
            DO 20 I = 1, NODETBL(ELEMTYP(E))
              KI = IN(E,I)
              IF (ICH(KI) .EQ. 0 .AND. I .LT. NODETBL(ELEMTYP(E))) THEN
                II = KI - LCH(KI)
                DO 10 J = I + 1, NODETBL(ELEMTYP(E))
                  KJ = IN(E,J)
                  IF (ICH(KJ) .EQ. 0) THEN
                    JJ = ABS(KJ - LCH(KJ) - II) + 1
                    IF (JJ .GT. SBW) SBW = JJ
                  ENDIF
10                CONTINUE
              ENDIF
20          CONTINUE
30        CONTINUE
          WRITE(OUTF,40) NDOF,SBW
40        FORMAT(//' NUMBER OF DEGREES OF FREEDOM IN MODIFIED,'/
     1           ' GLOBAL COMBINED SORPTION AND ADVECTION-DISPERSION',
     2           ' MATRIX =',I5///' SEMI-BANDWIDTH OF MODIFIED,'/
     3           ' GLOBAL COMBINED SORPTION AND ADVECTION-DISPERSION',
     4           ' MATRIX =',I5)
C         INITIALIZE ENTRIES OF GLOBAL CONDUCTANCE MATRIX TO ZERO
          IJSIZE = NDOF * NDOF - (NDOF - SBW) * (1 + NDOF - SBW)
          DO 50 I = 1, IJSIZE
            M(I) = 0.0
            B1(I) = 0.0
50        CONTINUE
C         INITIALIZE ENTRIES OF THE GLOBAL SOLUTE FLUX MATRIX TO ZERO
          DO 60 I = NDOF
            B(I) = 0.0
60        CONTINUE

          DO 56 I = 1, MAX1
56          FC(I) = 0.

C         LOOP ON THE NUMBER OF ELEMENTS
          DO 90 E = 1, NUMELM
C           COMPUTE THE ELEMENT SORPTION AND ADVECTION-DISPERSION MATRICES
```

```
C           FOR THIS ELEMENT TYPE
            IF (ELEMTYP(E) .EQ. 1) THEN
C               ELEMENT IS A ONE-DIMENSIONAL, LINEAR BAR
                CALL ABAR2(E,AE)
                CALL DBAR2(E,DE)
            ELSEIF (ELEMTYP(E) .EQ .2) THEN
C               ELEMENT IS A ONE-DIMENSIONAL, QUADRATIC BAR
                CALL ABAR3(E,AE)
                CALL DBAR3(E,DE)
            ELSEIF (ELEMTYP(E) .EQ. 3) THEN
C               ELEMENT IS A ONE-DIMENSIONAL, CUBIC BAR
                CALL ABAR4(E,AE)
                CALL DBAR4(E,DE)
            ELSEIF (ELEMTYP(E) .EQ. 4) THEN
C               ELEMENT IS A TWO-DIMENSIONAL, LINEAR TRIANGLE
                CALL ATRI3(E,AE)
                CALL DTRI3(E,DE)
            ELSEIF (ELEMTYP(E) .EQ. 5) THEN
C               ELEMENT IS A TWO-DIMENSIONAL, LINEAR RECTANGLE
                CALL AREC4(E,AE)
                CALL DREC4(E,DE)
            ELSEIF (ELEMTYP(E) .EQ. 6) THEN
C               ELEMENT IS A TWO-DIMENSIONAL, LINEAR QUADRILATERAL
                CALL AQUA4(E,AE)
                CALL DQUA4(E,DE)
            ELSEIF (ELEMTYP(E) .EQ. 7) THEN
C               ELEMENT IS A TWO-DIMENSIONAL, QUADRATIC QUADRILATERAL
                CALL AQUA8(E,AE)
                CALL DQUA8(E,DE)
            ELSEIF (ELEMTYP(E) .EQ. 8) THEN
C               ELEMENT IS A TWO-DIMENSIONAL, CUBIC QUADRILATERAL
                CALL AQUA12(E,AE)
                CALL DQUA12(E,DE)
            ELSEIF (ELEMTYP(E) .EQ. 9) THEN
C               ELEMENT IS A THREE-DIMENSIONAL, LINEAR PARALLELEPIPED
                CALL APAR8(E,AE)
                CALL DPAR8(E,DE)
            ELSEIF (ELEMTYP(E) .EQ. 10) THEN
C               ELEMENT IS A THREE-DIMENSIONAL, QUADRATIC PARALLELEPIPED
                CALL APAR20(E,AE)
                CALL DPAR20(E,DE)
            ELSEIF (ELEMTYP(E) .EQ. 11) THEN
C               ELEMENT IS A THREE-DIMENSIONAL, CUBIC PARALLELEPIPED
                CALL APAR32(E,AE)
                CALL DPAR32(E,DE)
            ELSEIF (ELEMTYP(E) .EQ. 12) THEN
C               ELEMENT IS A TWO-DIMENSIONAL, LINEAR TRIANGLE (AXISYMMETRIC)
                CALL ATRI3A(E,AE)
                CALL DTRI3A(E,DE)
            ELSEIF (ELEMTYP(E) .EQ. 13) THEN
C               ELEMENT IS A TWO-DIMENSIONAL, LINEAR RECTANGLE
   (AXISYMMETRIC)
                CALL AREC4A(E,AE)
                CALL DREC4A(E,DE)
            ENDIF
C           ADD THE ELEMENT SORPTION AND ADVECTION-DISPERSION MATRICES
C           FOR THIS ELEMENT TO THE GLOBAL MATRIX
C   AE(I,J),DE(I,J) ----------->    M(IJ)      <=>         M(KI,KJ)
C (FULL MATRIX STORAGE)      (VECTOR MATRIX STORAGE)     (FULL MATRIX
   STORAGE)
```

```
          DO 80 I = 1, NODETBL(ELEMTYP(E))
            KI = IN(E,I)
            IF (ICH(KI) .EQ. 0) THEN
              II = KI - LCH(KI)
              DO 70 J = 1, NODETBL(ELEMTYP(E))
                KJ = IN(E,J)
                  IF (ICH(KJ) .NE. 0) THEN
                    FC(II) = FC(II) - DELTAT(IDT) * DE(I,J) * X(KJ)
                  ELSE
                    JJ = KJ - LCH(KJ)
                    CALL LOC(II,JJ,IJ,NDOF,SBW,SYMM)
                    M(IJ) = M(IJ) + AE(I,J) + OMEGA *
     1                          DELTAT(IDT) * DE(I,J)
                    B1(IJ) = B1(IJ) + AE(I,J) - OMOMEGA *
     1                          DELTAT(IDT) * DE(I,J)
                  ENDIF
70            CONTINUE
            ENDIF
80        CONTINUE
90      CONTINUE
        RETURN
        END




        SUBROUTINE ABAR2(E,AE)
C***********************************************************************
C
C       PURPOSE:
C         TO COMPUTE THE CONSISTENT FORM OF THE ELEMENT
C         SORPTION MATRIX FOR A ONE-DIMENSIONAL, LINEAR
C         BAR ELEMENT
C
C       DEFINITIONS OF VARIABLES:
C         AE(I,J) = ELEMENT SORPTION MATRIX
C               E = ELEMENT NUMBER
C             KDE = ELEMENT DISTRIBUTION COEFFICIENT
C              LE = ELEMENT LENGTH
C           RHOBE = ELEMENT BULK DENSITY
C
C       REFERENCES:
C         ISTOK,J.D. GROUNDWATER FLOW AND SOLUTE TRANSPORT
C         MODELING BY THE FINITE ELEMENT METHOD, FIGURE 4.5,
C         EQUATION 4.19A.
C
C***********************************************************************
$INCLUDE:'COMALL'
        REAL AE(MAX3,MAX3),KDE,LE, NE
C
        RHOBE = PROP(MATSET(E),3)
        KDE = PROP(MATSET(E),4)
        NE  = PROP(MATSET(E),5)
        LE = ABS(X1(IN(E,2)) - X1(IN(E,1)))
        AE(1,1) = (1. + RHOBE*KDE/NE) * (LE / 6.) * 2.
        AE(1,2) = AE(1,1) / 2.
        AE(2,1) = AE(1,2)
        AE(2,2) = AE(1,1)
        RETURN
        END
```

```
      SUBROUTINE ATRI3(E,AE)
C**********************************************************************
C
C      PURPOSE:
C        TO COMPUTE THE CONSISTENT FORM OF THE ELEMENT SORPTION
C        MATRIX FOR A TWO-DIMENSIONAL, LINEAR TRIANGLE ELEMENT
C
C      DEFINITIONS OF VARIABLES:
C            AE4 = FOUR TIMES ELEMENT AREA
C        AE(I,J) = ELEMENT SORPTION MATRIX
C              E = ELEMENT NUMBER
C            KDE = ELEMENT DISTRIBUTION COEFFICIENT
C          RHOBE = ELEMENT BULK DENSITY
C
C      REFERENCES:
C        ISTOK,J.D. GROUNDWATER FLOW AND SOLUTE TRANSPORT
C        MODELING BY THE FINITE ELEMENT METHOD, FIGURE 4.6,
C        EQUATION 4.25A
C
C**********************************************************************
$INCLUDE: 'COMALL'
      REAL AE(MAX3,MAX3),KDE, NE
C
      RHOBE = PROP(MATSET(E),4)
      KDE = PROP(MATSET(E),5)
      NE  = PROP(MATSET(E),6)
      AE4 = 2. * (X1(IN(E,2)) * X2(IN(E,3)) + X1(IN(E,1)) *
     1      X2(IN(E,2)) +  X2(IN(E,1)) * X1(IN(E,3)) -
     2      X2(IN(E,3)) * X1(IN(E,1)) - X1(IN(E,3)) *
     3      X2(IN(E,2)) - X1(IN(E,2)) * X2(IN(E,1)))
      TEMP = AE4 / 12. / 4. * (1. + RHOBE*KDE/NE )
      AE(1,1) = 2. * TEMP
      AE(1,2) = TEMP
      AE(1,3) = TEMP
      AE(2,1) = TEMP
      AE(2,2) = AE(1,1)
      AE(2,3) = TEMP
      AE(3,1) = TEMP
      AE(3,2) = TEMP
      AE(3,3) = AE(1,1)
      RETURN
      END

      SUBROUTINE AREC4(E,AE)
C**********************************************************************
C
C      PURPOSE:
C        TO COMPUTE THE CONSISTENT FORM OF THE ELEMENT SORPTION
C        MATRIX FOR A TWO-DIMENSIONAL, LINEAR TRIANGLE ELEMENT
C
C      DEFINITIONS OF VARIABLES:
C        AE(I,J) = ELEMENT SORPTION MATRIX
C              E = ELEMENT NUMBER
C            KDE = ELEMENT DISTRIBUTION COEFFICIENT
C          RHOBE = ELEMENT BULK DENSITY
C
C      REFERENCES:
C        ISTOK,J.D. GROUNDWATER FLOW AND SOLUTE TRANSPORT
C        MODELING BY THE FINITE ELEMENT METHOD, FIGURE 4.7,
C        EQUATION 4.30A
C
C**********************************************************************
```

```
$INCLUDE: 'COMALL'
      REAL AE(MAX3,MAX3),KDE, NE
C
      RHOBE = PROP(MATSET(E),4)
      KDE = PROP(MATSET(E),5)
      NE  = PROP(MATSET(E),6)
      TEMP = (RHOBE*KDE/NE + 1.) * ABS( (X2(IN(E,1))-X2(IN(E,3)))/2.
     1        * (X1(IN(E,1)) - X1(IN(E,3)))/2. ) / 9.
      AE(1,1) = 4. * TEMP
      AE(1,2) = 2. * TEMP
      AE(1,3) = TEMP
      AE(1,4) = AE(1,2)
      AE(2,1) = AE(1,2)
      AE(2,2) = AE(1,1)
      AE(2,3) = AE(1,2)
      AE(2,4) = AE(1,3)
      AE(3,1) = AE(1,3)
      AE(3,2) = AE(1,2)
      AE(3,3) = AE(1,1)
      AE(3,4) = AE(1,2)
      AE(4,1) = AE(1,2)
      AE(4,2) = AE(1,3)
      AE(4,3) = AE(1,2)
      AE(4,4) = AE(1,1)
      RETURN
      END

      SUBROUTINE AQUA4(E,AE)
C***********************************************************************
C       PURPOSE:
C          TO COMPUTE THE CONSISTENT FORM OF THE ELEMENT SORPTION
C          MATRIX FOR A TWO-DIMENSIONAL, LINEAR QUADRILATERAL ELEMENT
C
C       DEFINITIONS OF VARIABLES:
C             AE(I,J) = ELEMENT CAPACITANCE MATRIX
C              DETJAC = DETERMINANT OF JACOBIAN MATRIX
C            DNDXI(I) = PARTIAL DERIVATIVE OF INTERPOLATION
C                       FUNCTION WITH RESPECT TO XI AT NODE I
C             DNDX(I) = PARTIAL DERIVATIVE OF INTERPOLATION
C                       FUNCTION WITH RESPECT TO X AT NODE I
C           DNDETA(I) = PARTIAL DERIVATIVE OF INTERPOLATION
C                       FUNCTION WITH RESPECT TO ETA AT NODE I
C             DNDY(I) = PARTIAL DERIVATIVE OF INTERPOLATION
C                       FUNCTION WITH RESPECT TO Y AT NODE I
C                   E = ELEMENT NUMBER
C               XI(I) = LOCATION OF GAUSS POINT IN XI COORDINATE
C                       DIRECTION
C              ETA(I) = LOCATION OF GAUSS POINT IN ETA COORDINATE
C                       DIRECTION
C             JAC(I,J) = JACOBIAN MATRIX
C                N(I) = INTERPOLATION FUNCTION FOR NODE I
C                W(I) = WEIGHT FOR GAUSS POINT I
C                 KDE = ELEMENT DISTRIBUTION COEFFICIENT
C               RHOBE = ELEMENT BULK DENSITY
C           X1(IN(E,I) = X COORDINATE FOR NODE I, ELEMENT E
C           X2(IN(E,I) = Y COORDINATE FOR NODE I, ELEMENT E
C
C       REFERENCES:
C          ISTOK,J.D. GROUNDWATER FLOW AND SOLUTE TRANSPORT
C          MODELING BY THE FINITE ELEMENT METHOD, FIGURE 4.10,
C          EQUATION 4.71
C***********************************************************************
```

```
$INCLUDE: 'COMALL'
      REAL JAC(2,2),JACINV(2,2),CE(MAX3,MAX3),N(4),DNDXI(4),
     1     DNDETA(4),W(2),XI(2),ETA(2),SIGN1(4),AE(MAX3,MAX3),
     2     SIGN2(4), KDE, NE
      DATA SIGN1/-1.,1.,1.,-1./
      DATA SIGN2/-1.,-1.,1.,1./
C
      XI(1) =  1. / SQRT(3.)
      XI(2) = -XI(1)
      ETA(1) = XI(1)
      ETA(2) = XI(2)
      W(1) = 1.
      W(2) = 1.
      RHOBE = PROP(MATSET(E),4)
      KDE = PROP(MATSET(E),5)
      NE  = PROP(MATSET(E),6)

      DO 30 I = 1, 4
        DO 20 J = 1, 4
          AE(I,J) = 0.
 20     CONTINUE
 30   CONTINUE

      DO 120 I = 1, 2
        DO 110 J = 1, 2

          DO 50 K = 1, 2
            DO 40 K1 = 1, 2
              JAC(K,K1) = 0.
 40         CONTINUE
 50       CONTINUE

          DO 60 K1 = 1, 4
              N(K1) = 0.25 * (1. + SIGN1(K1) * XI(I))
     1                     * (1. + SIGN2(K1) * ETA(J))
            DNDXI(K1) = 0.25 * SIGN1(K1) * (1. + SIGN2(K1) * ETA(J))
            DNDETA(K1) = 0.25 * SIGN2(K1) * (1. + SIGN1(K1) * XI(I))
 60       CONTINUE
          DO 70 K1 = 1, 4
            JAC(1,1) = JAC(1,1) + DNDXI(K1) * X1(IN(E,K1))
            JAC(1,2) = JAC(1,2) + DNDXI(K1) * X2(IN(E,K1))
            JAC(2,1) = JAC(2,1) + DNDETA(K1) * X1(IN(E,K1))
            JAC(2,2) = JAC(2,2) + DNDETA(K1) * X2(IN(E,K1))
 70       CONTINUE
          DETJAC = JAC(1,1) * JAC(2,2) - JAC(1,2) * JAC(2,1)
          DO 100 K = 1, 4
            DO 90 K1 = 1, 4
              AE(K,K1) = AE(K,K1) + W(I) * W(J) * (1. + RHOBE*KDE/NE)
     1                   * N(K) * N(K1) * DETJAC
 90         CONTINUE
 100      CONTINUE
 110    CONTINUE
 120  CONTINUE
      RETURN
      END
```

```
      SUBROUTINE APAR8 (E,AE)
C************************************************************************
C
C     PURPOSE:
C     TO COMPUTE THE CONSISTENT FORM OF THE ELEMENT
C     SORPTION MATRIX FOR A THREE-DIMENSIONAL,
C     LINEAR QUADRILATERAL ELEMENT
C
C     DEFINITIONS OF VARIABLES:
C            AE(I,J) = ELEMENT CAPACITANCE MATRIX
C             DETJAC = DETERMINANT OF JACOBIAN MATRIX
C           DNDXI(I) = PARTIAL DERIVATIVE OF INTERPOLATION
C                      FUNCTION WITH RESPECT TO XI AT NODE I
C            DNDX(I) = PARTIAL DERIVATIVE OF INTERPOLATION
C                      FUNCTION WITH RESPECT TO X AT NODE I
C          DNDETA(I) = PARTIAL DERIVATIVE OF INTERPOLATION
C                      FUNCTION WITH RESPECT TO ETA AT NODE I
C            DNDY(I) = PARTIAL DERIVATIVE OF INTERPOLATION
C                      FUNCTION WITH RESPECT TO Y AT NODE I
C         DNDZETA(I) = PARTIAL DERIVATIVE OF INTERPOLATION
C                      FUNCTION WITH RESPECT TO ZETA AT NODE I
C            DNDZ(I) = PARTIAL DERIVATIVE OF INTERPOLATION
C                      FUNCTION WITH RESPECT TO Z AT NODE I
C                  E = ELEMENT NUMBER
C              XI(I) = LOCATION OF GAUSS POINT IN XI COORDINATE
C                      DIRECTION
C             ETA(I) = LOCATION OF GAUSS POINT IN ETA COORDINATE
C                      DIRECTION
C            ZETA(I) = LOCATION OF GAUSS POINT IN ZETA COORDINATE
C                      DIRECTION
C           JAC(I,J) = JACOBIAN MATRIX
C               N(I) = INTERPOLATION FUNCTION FOR NODE I
C               W(I) = WEIGHT FOR GAUSS POINT I
C                KDE = ELEMENT DISTRIBUTION COEFFICIENT
C              RHOBE = ELEMENT BULK DENSITY
C      X1(IN(E,I) = X COORDINATE FOR NODE I, ELEMENT E
C      X2(IN(E,I) = Y COORDINATE FOR NODE I, ELEMENT E
C      X3(IN(E,I) = Z COORDINATE FOR NODE I, ELEMENT E
C
C     REFERENCES:
C     ISTOK,J.D. GROUNDWATER FLOW AND SOLUTE TRANSPORT
C     MODELING BY THE FINITE ELEMENT METHOD, FIGURE 4.10,
C     EQUATION 4.72
C
C************************************************************************
$INCLUDE: 'COMALL'
      REAL JAC(3,3),AE(MAX3,MAX3),DNDX(8),DNDY(8),DNDZ(8),
     1    XI(8),ETA(8),ZETA(8),DNDXI(8),DNDETA(8),DNDZETA(8),W(2),
     2    N(8),SIGN1(8),SIGN2(8),SIGN3(8),KDE, NE
      DATA SIGN1/-1.,1.,1.,-1.,-1.,1.,1.,-1./
      DATA SIGN2/-1.,-1.,1.,1.,-1.,-1.,1.,1./
      DATA SIGN3/-1.,-1.,-1.,-1.,1.,1.,1.,1./
C
      XI(1) = 1. / SQRT(3.)
      XI(2) = -XI(1)
      ETA(1) = XI(1)
      ETA(2) = XI(2)
      ZETA(1) = XI(1)
      ZETA(2) = XI(2)
      W(1) = 1.
      W(2) = 1.
```

```
      RHOBE  = PROP(MATSET(E),4)
      KDE    = PROP(MATSET(E),5)
      NE     = PROP(MATSET(E),6)

      DO 20 K = 1, 8
        DO 10 N1 = 1, 8
          AE(K,N1) = 0.
 10     CONTINUE
 20   CONTINUE

      DO 120 I = 1, 2
        DO 110 J = 1, 2
          DO 100 K = 1, 2

            DO 40 L = 1, 3
              DO 30 N1 = 1, 3
                JAC(L,N1) = 0.
 30         CONTINUE
 40         CONTINUE

            DO 50 N1 = 1, 8
              N(N1) = 0.125 * (1.+SIGN1(N1)*XI(I)) * (1.+SIGN2(N1) *
     1                   ETA(J)) * (1. + SIGN3(N1) * ZETA(K))
              DNDXI(N1)   = 0.125 * SIGN1(N1) * (1. + SIGN2(N1) *
     1                   ETA(J)) * (1. + SIGN3(N1) * ZETA(K))
              DNDETA(N1)  = 0.125 * SIGN2(N1) * (1. + SIGN1(N1) *
     1                   XI(I)) * (1. + SIGN3(N1) * ZETA(K))
              DNDZETA(N1) = 0.125 * SIGN3(N1) * (1. + SIGN1(N1) *
     1                   XI(I)) * (1. + SIGN2(N1) * ETA(J))
 50         CONTINUE

            DO 60 M5 = 1, 8
              JAC(1,1) = JAC(1,1) + DNDXI(M5)   * X1(IN(E,M5))
              JAC(1,2) = JAC(1,2) + DNDXI(M5)   * X2(IN(E,M5))
              JAC(1,3) = JAC(1,3) + DNDXI(M5)   * X3(IN(E,M5))
              JAC(2,1) = JAC(2,1) + DNDETA(M5)  * X1(IN(E,M5))
              JAC(2,2) = JAC(2,2) + DNDETA(M5)  * X2(IN(E,M5))
              JAC(2,3) = JAC(2,3) + DNDETA(M5)  * X3(IN(E,M5))
              JAC(3,1) = JAC(3,1) + DNDZETA(M5) * X1(IN(E,M5))
              JAC(3,2) = JAC(3,2) + DNDZETA(M5) * X2(IN(E,M5))
              JAC(3,3) = JAC(3,3) + DNDZETA(M5) * X3(IN(E,M5))
 60         CONTINUE

            DETJAC = JAC(1,1) * (JAC(2,2) * JAC(3,3) - JAC(3,2) *
     1                JAC(2,3)) - JAC(1,2) * (JAC(2,1) * JAC(3,3) -
     2                JAC(3,1) * JAC(2,3)) - JAC(1,3) * (JAC(2,1) *
     3                JAC(3,2) - JAC(3,1) * JAC(2,2))

            DO 90 L = 1, 8
              DO 80 M5 = 1, 8
                AE(L,M5) = AE(L,M5) + W(I) * W(J) * W(K) *
     1                   (1. + RHOBE*KDE/ NE) * N(L) * N(M5) * DETJAC
 80         CONTINUE
 90         CONTINUE

100       CONTINUE
110     CONTINUE
120   CONTINUE
      RETURN
      END
```

```
      SUBROUTINE DBAR2(E,DE)
C*********************************************************************
C
C       PURPOSE:
C         TO COMPUTE THE CONSISTENT FORM OF THE ELEMENT
C         ADVECTION-DISPERSION MATRIX FOR A ONE-DIMENSIONAL,
C         LINEAR BAR ELEMENT
C
C       DEFINITIONS OF VARIABLES:
C             ALE = LONGITUDINAL DISPERSIVITY FOR ELEMENT
C         DE(I,J) = ELEMENT ADVECTION-DISPERSION MATRIX
C             DXE = ELEMENT DISPERSION COEFFICIENT
C               E = ELEMENT NUMBER
C             KDE = ELEMENT DISTRIBUTION COEFFICIENT
C          LAMBDA = SOLUTE DECAY COEFFICIENT
C              LE = ELEMENT LENGTH
C              NE = ELEMENT POROSITY
C           RHOBE = ELEMENT BULK DENSITY
C             VXE = APPARENT GROUNDWATER VELOCITY IN
C                   X COORDINATE DIRECTION
C            VXEP = PORE WATER VELOCITY IN X COORDINATE DIRECTION
C
C       REFERENCES:
C         ISTOK,J.D. GROUNDWATER FLOW AND SOLUTE TRANSPORT
C         MODELING BY THE FINITE ELEMENT METHOD, FIGURE 4.5,
C         EQUATION 4.18A, EQUATION AIII.12
C
C*********************************************************************
$INCLUDE:'COMALL'
      REAL DE(MAX3,MAX3),KDE,LAMBDA,LE,NE
C
      ALE    = PROP(MATSET(E),1)
      LAMBDA = PROP(MATSET(E),2)
      RHOBE  = PROP(MATSET(E),3)
      KDE    = PROP(MATSET(E),4)
      NE     = PROP(MATSET(E),5)
      VXE    = V1(E)
      VXEP   = VXE / NE
      LE     = ABS(X1(IN(E,2)) - X1(IN(E,1)))
      DXE    = ALE * VXEP
      TEMP3  = LAMBDA * (1. + RHOBE * KDE/NE) * (LE / 6.)
      DE(1,1) =  DXE / LE - VXEP / 2. + 2. * TEMP3
      DE(1,2) = -DXE / LE + VXEP / 2. +      TEMP3
      DE(2,1) = -DXE / LE - VXEP / 2. +      TEMP3
      DE(2,2) =  DXE / LE + VXEP / 2. + 2. * TEMP3
      RETURN
      END


      SUBROUTINE DTRI3(E,DE)
C*********************************************************************
C
C       PURPOSE:
C         TO COMPUTE THE CONSISTENT FORM OF THE ELEMENT
C         ADVECTION- DISPERSION MATRIX FOR A TWO-DIMENSIONAL,
C         LINEAR TRIANGLE ELEMENT
C
C       DEFINITIONS OF VARIABLES:
C               AE4 = FOUR TIMES ELEMENT AREA
C               ALE = LONGITUDINAL DISPERSIVITY FOR ELEMENT
C               ATE = TRANSVERSE DISPERSIVITY FOR ELEMENT
```

```
C                DE(I,J) = ELEMENT ADVECTION-DISPERSION MATRIX
C            DXXE (ETC.) = ELEMENT DISPERSION COEFFCIENTS
C                      E = ELEMENT NUMBER
C                    KDE = ELEMENT DISTRIBUTION COEFFCIENT
C                 LAMBDA = SOLUTE DECAY COEFFICIENT
C                     NE = ELEMENT POROSITY
C                  RHOBE = ELEMENT BULK DENSITY
C                    VXE = APPARENT GROUNDWATER VELOCITY IN
C                          X COORDINATE DIRECTION
C                    VYE = APPARENT GROUNDWATER VELOCITY IN
C                          Y COORDINATE DIRECTION
C                   VXEP = PORE WATER VELOCITY IN X COORDINATE DIRECTION
C                   VYEP = PORE WATER VELOCITY IN Y COORDINATE DIRECTION
C
C        REFERENCES:
C        ISTOK,J.D. GROUNDWATER FLOW AND SOLUTE TRANSPORT
C        MODELING BY THE FINITE ELEMENT METHOD, FIGURE 4.6,
C        EQUATION 4.24A, EQUATION AIII.11
C
C************************************************************************
$INCLUDE: 'COMALL'
      REAL DE(MAX3,MAX3),LAMBDA,KDE,NE,BE(3),CE(3)
C
      ALE   = PROP(MATSET(E),1)
      ATE   = PROP(MATSET(E),2)
      LAMBDA = PROP(MATSET(E),3)
      RHOBE = PROP(MATSET(E),4)
      KDE   = PROP(MATSET(E),5)
      NE    = PROP(MATSET(E),6)
      VXE   = V1(E)
      VYE   = V2(E)
      VXEP  = VXE / NE
      VYEP  = VYE / NE
      DXXE  = (ATE * VYEP**2 + ALE * VXEP**2) / SQRT(VYEP**2+VXEP**2)
      DYYE  = (ATE * VXEP**2 + ALE * VYEP**2) / SQRT(VYEP**2+VXEP**2)
      DXYE  = ((ALE - ATE) * VXEP * VYEP) / SQRT(VYEP**2 + VXEP**2)
      DYXE  = DXYE
      BE(1) = X2(IN(E,2)) - X2(IN(E,3))
      BE(2) = X2(IN(E,3)) - X2(IN(E,1))
      BE(3) = X2(IN(E,1)) - X2(IN(E,2))
      CE(1) = X1(IN(E,3)) - X1(IN(E,2))
      CE(2) = X1(IN(E,1)) - X1(IN(E,3))
      CE(3) = X1(IN(E,2)) - X1(IN(E,1))
      AE4   = 2. * (X1(IN(E,2)) * X2(IN(E,3)) + X1(IN(E,1)) *
     1          X2(IN(E,2)) + X2(IN(E,1)) * X1(IN(E,3)) -
     2          X2(IN(E,3)) * X1(IN(E,1)) - X1(IN(E,3)) *
     3          X2(IN(E,2)) - X1(IN(E,2)) * X2(IN(E,1)))
      AE    = AE4 / 4.
      TEMP = AE / 12. * LAMBDA * (1. + RHOBE * KDE/NE )

      DO 20 I = 1, 3
        DO 10 J = 1, 3
          DE(I,J) = (DXXE*BE(I)*BE(J) + DYYE*CE(I)*CE(J) +
     1              DXYE*BE(I)*CE(J) + DYXE*CE(I)*BE(J) ) / AE4
     2              + VXEP/6.*BE(J) + VYEP/6.*CE(J) + TEMP
          IF (I .EQ. J) DE(I,J) = DE(I,J) + TEMP
10      CONTINUE
20    CONTINUE

      RETURN
      END
```

```
      SUBROUTINE DREC4(E,DE)
C************************************************************************
C
C       PURPOSE:
C          TO COMPUTE THE CONSISTENT FORM OF THE ELEMENT
C          ADVECTION-DISPERSION MATRIX FOR A TWO-DIMENSIONAL,
C          LINEAR TRIANGLE ELEMENT
C
C       DEFINITIONS OF VARIABLES:
C                   ALE = LONGITUDINAL DISPERSIVITY FOR ELEMENT
C                   ATE = TRANSVERSE DISPERSIVITY FOR ELEMENT
C               DE(I,J) = ELEMENT ADVECTION-DISPERSION MATRIX
C            DXXE (ETC.) = ELEMENT DISPERSION COEFFCIENTS
C                     E = ELEMENT NUMBER
C                   KDE = ELEMENT DISTRIBUTION COEFFCIENT
C                LAMBDA = SOLUTE DECAY COEFFICIENT
C                    NE = ELEMENT POROSITY
C                 RHOBE = ELEMENT BULK DENSITY
C                   VXE = APPARENT GROUNDWATER VELOCITY IN
C                         X COORDINATE DIRECTION
C                   VYE = APPARENT GROUNDWATER VELOCITY IN
C                         Y COORDINATE DIRECTION
C                  VXEP = PORE WATER VELOCITY IN X COORDINATE DIRECTION
C                  VYEP = PORE WATER VELOCITY IN Y COORDINATE DIRECTION
C
C       REFERENCES:
C          ISTOK,J.D. GROUNDWATER FLOW AND SOLUTE TRANSPORT
C          MODELING BY THE FINITE ELEMENT METHOD, FIGURE 4.7,
C          EQUATION 4.29a, EQUATION AIII.11
C
C************************************************************************
$INCLUDE: 'COMALL'
      REAL DE(MAX3,MAX3),LAMBDA,KDE,NE
C
      ALE    = PROP(MATSET(E),1)
      ATE    = PROP(MATSET(E),2)
      LAMBDA = PROP(MATSET(E),3)
      RHOBE  = PROP(MATSET(E),4)
      KDE    = PROP(MATSET(E),5)
      NE     = PROP(MATSET(E),6)
      VXE    = V1(E)
      VYE    = V2(E)
      VXEP   = VXE / NE
      VYEP   = VYE / NE
      DXXE = (ATE * VYEP**2 + ALE * VXEP**2) / SQRT(VYEP**2 + VXEP**2)
      DYYE = (ATE * VXEP**2 + ALE * VYEP**2) / SQRT(VYEP**2 + VXEP**2)
      DXYE = ((ALE - ATE) * VXEP * VYEP)     / SQRT(VYEP**2 + VXEP**2)
      DYXE = DXYE
      AE   = ABS(X2(IN(E,1)) - X2(IN(E,3))) / 2.
      BE   = ABS(X1(IN(E,1)) - X1(IN(E,3))) / 2.
      TEMP1 = (DXXE * AE) / (6. * BE)
      TEMP2 = (DYYE * BE) / (6. * AE)
      TEMP3 = DXYE / 4.
      TEMP4 = DYXE / 4.
      TEMP5 = VXEP * AE / 6.
      TEMP6 = VYEP * BE / 6.
      TEMP7 = LAMBDA * (1. + RHOBE * KDE/NE) * (AE * BE) / 9.
      DE(1,1)= 2.*TEMP1+2.*TEMP2+TEMP3+TEMP4-2.*TEMP5-2.*TEMP6+4.*TEMP7
      DE(1,2)=-2.*TEMP1+   TEMP2+TEMP3-TEMP4+2.*TEMP5-   TEMP6+2.*TEMP7
      DE(1,3) = - TEMP1-   TEMP2-TEMP3-TEMP4+   TEMP5+   TEMP6+   TEMP7
      DE(1,4) =   TEMP1-2.*TEMP2-TEMP3+TEMP4-   TEMP5+2.*TEMP6+2.*TEMP7
      DE(2,1)=-2.*TEMP1+   TEMP2-TEMP3+TEMP4-2.*TEMP5-   TEMP6+2.*TEMP7
```

```
      DE(2,2)= 2.*TEMP1+2.*TEMP2-TEMP3-TEMP4+2.*TEMP5-2.*TEMP6+4.*TEMP7
      DE(2,3) =    TEMP1-2.*TEMP2+TEMP3-TEMP4+   TEMP5+2.*TEMP6+2.*TEMP7
      DE(2,4) = -  TEMP1-   TEMP2+TEMP3+TEMP4-   TEMP5+   TEMP6+   TEMP7
      DE(3,1) = -  TEMP1-   TEMP2-TEMP3-TEMP4-   TEMP5-   TEMP6+   TEMP7
      DE(3,2) =    TEMP1-2.*TEMP2-TEMP3+TEMP4+   TEMP5-2.*TEMP6+2.*TEMP7
      DE(3,3) =2.*TEMP1+2.*TEMP2+TEMP3+TEMP4+2.*TEMP5+2.*TEMP6+4.*TEMP7
      DE(3,4) =-2.*TEMP1+   TEMP2+TEMP3-TEMP4-2.*TEMP5+   TEMP6+2.*TEMP7
      DE(4,1) =    TEMP1-2.*TEMP2+TEMP3-TEMP4-   TEMP5-2.*TEMP6+2.*TEMP7
      DE(4,2) = -  TEMP1-   TEMP2+TEMP3+TEMP4+   TEMP5-   TEMP6+   TEMP7
      DE(4,3) =-2.*TEMP1+   TEMP2-TEMP3+TEMP4+2.*TEMP5+   TEMP6+2.*TEMP7
      DE(4,4) =2.*TEMP1+2.*TEMP2-TEMP3-TEMP4-2.*TEMP5+2.*TEMP6+4.*TEMP7
      RETURN
      END



      SUBROUTINE DQUA4(E,DE)
C******************************************************************
C      PURPOSE:
C         TO COMPUTE THE CONSISTENT FORM OF THE ELEMENT
C         ADVECTION-DISPERSION MATRIX FOR A TWO-DIMENSIONAL,
C         LINEAR QUADRILATERAL ELEMENT
C
C      DEFINITIONS OF VARIABLES:
C           ALE = LONGITUDINAL DISPERSIVITY FOR ELEMENT
C           ATE = TRANSVERSE DISPERSIVITY FOR ELEMENT
C        DE(I,J) = ELEMENT ADVECTION-DISPERSION MATRIX
C         DETJAC = DETERMINANT OF JACOBIAN MATRIX
C       DNDXI(I) = PARTIAL DERIVATIVE OF INTERPOLATION
C                  FUNCTION WITH RESPECT TO XI AT NODE I
C        DNDX(I) = PARTIAL DERIVATIVE OF INTERPOLATION
C                  FUNCTION WITH RESPECT TO X AT NODE I
C      DNDETA(I) = PARTIAL DERIVATIVE OF INTERPOLATION
C                  FUNCTION WITH RESPECT TO ETA AT NODE I
C        DNDY(I) = PARTIAL DERIVATIVE OF INTERPOLATION
C                  FUNCTION WITH RESPECT TO Y AT NODE I
C              E = ELEMENT NUMBER
C          XI(I) = LOCATION OF GAUSS POINT IN XI COORDINATE
C                  DIRECTION
C         ETA(I) = LOCATION OF GAUSS POINT IN ETA COORDINATE
C                  DIRECTION
C        JAC(I,J) = JACOBIAN MATRIX
C      JACINV(I,J) = INVERSE OF JACOBIAN MATRIX
C           N(I) = INTERPOLATION FUNCTION FOR NODE I
C           W(I) = WEIGHT FOR GAUSS POINT I
C            KDE = ELEMENT DISTRIBUTION COEFFICIENT
C         LAMBDA = SOLUTE DECAY COEFFICIENT
C             NE = ELEMENT POROSITY
C          RHOBE = ELEMENT BULK DENSITY
C            VXE = APPARENT GROUNDWATER VELOCITY IN X
C                  COORDINATE DIRECTION
C            VYE = APPARENT GROUNDWATER VELOCITY IN Y
C                  COORDINATE DIRECTION
C           VXEP = PORE WATER VELOCITY IN X COORDINATE DIRECTION
C           VYEP = PORE WATER VELOCITY IN Y COORDINATE DIRECTION
C      X1(IN(E,I) = X COORDINATE FOR NODE I, ELEMENT E
C      X2(IN(E,I) = Y COORDINATE FOR NODE I, ELEMENT E
C
C      REFERENCES:
C         ISTOK,J.D. GROUNDWATER FLOW AND SOLUTE TRANSPORT
C         MODELING BY THE FINITE ELEMENT METHOD, FIGURE 4.10,
C         EQUATION 4.68
C******************************************************************
```

```
$INCLUDE: 'COMALL'
      REAL JAC(2,2),JACINV(2,2),DE(MAX3,MAX3),N(4),DNDXI(4),
     1      DNDX(4),DNDETA(4),DNDY(4),W(2),XI(2),ETA(2),SIGN1(4),
     2      SIGN2(4),NE,KDE,LAMBDA
      DATA SIGN1/-1.,1.,1.,-1./
      DATA SIGN2/-1.,-1.,1.,1./
C
      XI(1)  =  1. / SQRT(3.)
      XI(2)  = -XI(1)
      ETA(1) = XI(1)
      ETA(2) = XI(2)
      W(1)   = 1.
      W(2)   = 1.
      ALE    = PROP(MATSET(E),1)
      ATE    = PROP(MATSET(E),2)
      LAMBDA = PROP(MATSET(E),3)
      RHOBE  = PROP(MATSET(E),4)
      KDE    = PROP(MATSET(E),5)
      NE     = PROP(MATSET(E),6)
      VXE    = V1(E)
      VYE    = V2(E)
      VXEP   = VXE / NE
      VYEP   = VYE / NE
      DXXE = (ATE * VYEP**2 + ALE * VXEP**2) / SQRT(VYEP**2 + VXEP**2)
      DYYE = (ATE * VXEP**2 + ALE * VYEP**2) / SQRT(VYEP**2 + VXEP**2)
      DXYE = ((ALE - ATE) * VXEP * VYEP)     / SQRT(VYEP**2 + VXEP**2)
      DYXE = DXYE
      DO 30 I = 1, 4
        DO 20 J = 1, 4
          DE(I,J) = 0.
20      CONTINUE
30    CONTINUE
      DO 120 I = 1, 2
        DO 110 J = 1, 2

          DO 50 K = 1, 2
            DO 40 K1 = 1, 2
              JAC(K,K1) = 0.
40        CONTINUE
50        CONTINUE

          DO 60 K1 = 1, 4
                N(K1) = 0.25 * (1. + SIGN1(K1) * XI(I))
     1                       * (1. + SIGN2(K1) * ETA(J))
            DNDXI(K1) = 0.25 * SIGN1(K1) * (1. + SIGN2(K1) * ETA(J))
            DNDETA(K1) = 0.25 * SIGN2(K1) * (1. + SIGN1(K1) * XI(I))
60        CONTINUE
          DO 70 K1 = 1, 4
            JAC(1,1) = JAC(1,1) + DNDXI(K1) * X1(IN(E,K1))
            JAC(1,2) = JAC(1,2) + DNDXI(K1) * X2(IN(E,K1))
            JAC(2,1) = JAC(2,1) + DNDETA(K1) * X1(IN(E,K1))
            JAC(2,2) = JAC(2,2) + DNDETA(K1) * X2(IN(E,K1))
70        CONTINUE
          DETJAC = JAC(1,1) * JAC(2,2) - JAC(1,2) * JAC(2,1)
          JACINV(1,1) =  JAC(2,2) / DETJAC
          JACINV(1,2) = -JAC(1,2) / DETJAC
          JACINV(2,1) = -JAC(2,1) / DETJAC
          JACINV(2,2) =  JAC(1,1) / DETJAC
          DO 80 K1 = 1, 4
          DNDX(K1) = JACINV(1,1) * DNDXI(K1) + JACINV(1,2) * DNDETA(K1)
          DNDY(K1) = JACINV(2,1) * DNDXI(K1) + JACINV(2,2) * DNDETA(K1)
```

```
80        CONTINUE
          DO 100 K = 1, 4
            DO 90 K1 = 1, 4
              DE(K,K1) = DE(K,K1) + W(I) * W(J) *
     1                      (DXXE * DNDX(K) * DNDX(K1)
     2                      + DXYE * DNDX(K) * DNDY(K1)
     3                      + DYXE * DNDY(K) * DNDX(K1)
     4                      + DYYE * DNDY(K) * DNDY(K1)
     5                      + VXEP * N(K) * DNDX(K1)
     6                      + VYEP * N(K) * DNDY(K1)
     7                      + LAMBDA * (1. + RHOBE * KDE / NE )
     8                      * N(K) * N(K1)) * DETJAC
90        CONTINUE
100       CONTINUE
110     CONTINUE
120   CONTINUE
      RETURN
      END


      SUBROUTINE DPAR8 (E,DE)
C**********************************************************************
C       PURPOSE:
C         TO COMPUTE THE CONSISTENT FORM OF THE ELEMENT
C         ADVECTION-DISPERSION MATRIX FOR A THREE-DIMENSIONAL,
C         LINEAR QUADRILATERAL ELEMENT
C
C       DEFINITIONS OF VARIABLES:
C                     ALE = LONGITUDINAL DISPERSIVITY FOR ELEMENT
C                     ATE = TRANSVERSE DISPERSIVITY FOR ELEMENT
C                  DE(I,J) = ELEMENT ADVECTION-DISPERSION MATRIX
C                 DNDXI(I) = PARTIAL DERIVATIVE OF INTERPOLATION
C                            FUNCTION WITH RESPECT TO XI AT NODE I
C                  DNDX(I) = PARTIAL DERIVATIVE OF INTERPOLATION
C                            FUNCTION WITH RESPECT TO X AT NODE I
C                DNDETA(I) = PARTIAL DERIVATIVE OF INTERPOLATION
C                            FUNCTION WITH RESPECT TO ETA AT NODE I
C                  DNDY(I) = PARTIAL DERIVATIVE OF INTERPOLATION
C                            FUNCTION WITH RESPECT TO Y AT NODE I
C               DNDZETA(I) = PARTIAL DERIVATIVE OF INTERPOLATION
C                            FUNCTION WITH RESPECT TO ZETA AT NODE I
C                  DNDZ(I) = PARTIAL DERIVATIVE OF INTERPOLATION
C                            FUNCTION WITH RESPECT TO Z AT NODE I
C                        E = ELEMENT NUMBER
C                    XI(I) = LOCATION OF GAUSS POINT IN XI COORDINATE
C                            DIRECTION
C                   ETA(I) = LOCATION OF GAUSS POINT IN ETA COORDINATE
C                            DIRECTION
C                  ZETA(I) = LOCATION OF GAUSS POINT IN ZETA COORDINATE
C                            DIRECTION
C                  JAC(I,J) = JACOBIAN MATRIX
C                   DETJAC = DETERMINANT OF JACOBIAN MATRIX
C               JACINV(I,J) = INVERSE OF JACOBIAN MATRIX
C                     N(I) = INTERPOLATION FUNCTION FOR NODE I
C                     W(I) = WEIGHT FOR GAUSS POINT I
C                      KDE = ELEMENT DISTRIBUTION COEFFICIENT
C                   LAMBDA = SOLUTE DECAY COEFFICIENT
C                       NE = ELEMENT POROSITY
C                    RHOBE = ELEMENT BULK DENSITY
C                      VXE = APPARENT GROUNDWATER VELOCITY IN X
C                            COORDINATE DIRECTION
C                      VYE = APPARENT GROUNDWATER VELOCITY IN Y
C                            COORDINATE DIRECTION
```

```
C                     VXEP = PORE WATER VELOCITY IN X COORDINATE DIRECTION
C                     VYEP = PORE WATER VELOCITY IN Y COORDINATE DIRECTION
C            X1(IN(E,I) = X COORDINATE FOR NODE I, ELEMENT E
C            X2(IN(E,I) = Y COORDINATE FOR NODE I, ELEMENT E
C            X3(IN(E,I) = Z COORDINATE FOR NODE I, ELEMENT E
C
C     REFERENCES:
C        ISTOK,J.D. GROUNDWATER FLOW AND SOLUTE TRANSPORT
C        MODELING BY THE FINITE ELEMENT METHOD, FIGURE 4.10,
C        EQUATION 4.69
C********************************************************************
$INCLUDE: 'COMALL'
      REAL JAC(3,3),JACINV(3,3),DE(MAX3,MAX3),DNDX(8),DNDY(8),DNDZ(8),
     1     XI(2),ETA(2),ZETA(8),DNDXI(8),DNDETA(8),DNDZETA(8),W(2),
     2     N(8),SIGN1(8),SIGN2(8),SIGN3(8),NE,KDE,LAMBDA
      DATA SIGN1/-1.,1.,1.,-1.,-1.,1.,1.,-1./
      DATA SIGN2/-1.,-1.,1.,1.,-1.,-1.,1.,1./
      DATA SIGN3/-1.,-1.,-1.,-1.,1.,1.,1.,1./
C
      XI(1) = 1. / SQRT(3.)
      XI(2) = -XI(1)
      ETA(1) = XI(1)
      ETA(2) = XI(2)
      ZETA(1) = XI(1)
      ZETA(2) = XI(2)
      W(1) = 1.
      W(2) = 1.

      ALE    = PROP(MATSET(E),1)
      ATE    = PROP(MATSET(E),2)
      LAMBDA = PROP(MATSET(E),3)
      RHOBE  = PROP(MATSET(E),4)
      KDE    = PROP(MATSET(E),5)
      NE     = PROP(MATSET(E),6)

      VXE  = V1(E)
      VYE  = V2(E)
      VZE  = V3(E)
      VXEP = VXE / NE
      VYEP = VYE / NE
      VZEP = VZE / NE
      VXYZ = SQRT(VXEP**2 + VYEP**2 + VZEP**2)
      DXXE = (ATE * (VYEP**2 + VZEP**2) + ALE * VXEP**2 ) / VXYZ
      DXYE = ((ALE - ATE) * VXEP * VYEP) / VXYZ
      DYXE = DXYE
      DYYE = (ATE * (VXEP**2 + VZEP**2) + ALE * VYEP**2 ) / VXYZ
      DXZE = ((ALE - ATE) * VXEP * VZEP) / VXYZ
      DZXE = DXZE
      DZZE = (ATE * (VXEP**2 + VYEP**2) + ALE * VZEP**2 ) / VXYZ
      DYZE = ((ALE - ATE) * VYEP * VZEP) / VXYZ
      DZYE = DYZE

      DO 20 K = 1, 8
         DO 10 N1 = 1, 8
         DE(K,N1) = 0.
 10      CONTINUE
 20   CONTINUE

    DO 120 I = 1, 2
       DO 110 J = 1, 2
          DO 100 K = 1, 2
             DO 40 L = 1, 3
                DO 30 N1 = 1, 3
                JAC(L,N1) = 0.
```

```
 30              CONTINUE
 40              CONTINUE
                 DO 50 N1 = 1, 8
                   N(N1) = 0.125 * (1.+SIGN1(N1)*XI(I)) * (1.+SIGN2(N1) *
      1                   ETA(J)) * (1. + SIGN3(N1) * ZETA(K))
                   DNDXI(N1)   = 0.125 * SIGN1(N1) * (1. + SIGN2(N1) *
      1                   ETA(J)) * (1. + SIGN3(N1) * ZETA(K))
                   DNDETA(N1)  = 0.125 * SIGN2(N1) * (1. + SIGN1(N1) *
      1                   XI(I)) * (1. + SIGN3(N1) * ZETA(K))
                   DNDZETA(N1) = 0.125 * SIGN3(N1) * (1. + SIGN1(N1) *
      1                   XI(I)) * (1. + SIGN2(N1) * ETA(J))
 50              CONTINUE
                 DO 60 M5 = 1, 8
                   JAC(1,1) = JAC(1,1) + DNDXI(M5) * X1(IN(E,M5))
                   JAC(1,2) = JAC(1,2) + DNDXI(M5) * X2(IN(E,M5))
                   JAC(1,3) = JAC(1,3) + DNDXI(M5) * X3(IN(E,M5))
                   JAC(2,1) = JAC(2,1) + DNDETA(M5) * X1(IN(E,M5))
                   JAC(2,2) = JAC(2,2) + DNDETA(M5) * X2(IN(E,M5))
                   JAC(2,3) = JAC(2,3) + DNDETA(M5) * X3(IN(E,M5))
                   JAC(3,1) = JAC(3,1) + DNDZETA(M5) * X1(IN(E,M5))
                   JAC(3,2) = JAC(3,2) + DNDZETA(M5) * X2(IN(E,M5))
                   JAC(3,3) = JAC(3,3) + DNDZETA(M5) * X3(IN(E,M5))
 60              CONTINUE

                 DETJAC = JAC(1,1) * (JAC(2,2) * JAC(3,3) - JAC(3,2) *
      1                   JAC(2,3)) - JAC(1,2) * (JAC(2,1) * JAC(3,3) -
      2                   JAC(3,1) * JAC(2,3)) - JAC(1,3) * (JAC(2,1) *
      3                   JAC(3,2) - JAC(3,1) * JAC(2,2))
                 JACINV(1,1) = ( JAC(2,2)*JAC(3,3)-JAC(2,3)*JAC(3,2))/DETJAC
                 JACINV(1,2) = (-JAC(2,1)*JAC(3,3)+JAC(2,3)*JAC(3,1))/DETJAC
                 JACINV(1,3) = ( JAC(2,1)*JAC(3,2)-JAC(3,1)*JAC(2,2))/DETJAC
                 JACINV(2,1) = (-JAC(1,2)*JAC(3,3)+JAC(1,3)*JAC(3,2))/DETJAC
                 JACINV(2,2) = ( JAC(1,1)*JAC(3,3)-JAC(1,3)*JAC(3,1))/DETJAC
                 JACINV(2,3) = (-JAC(1,1)*JAC(3,2)+JAC(1,2)*JAC(3,1))/DETJAC
                 JACINV(3,1) = ( JAC(1,2)*JAC(2,3)-JAC(1,3)*JAC(2,2))/DETJAC
                 JACINV(3,2) = (-JAC(1,1)*JAC(2,3)+JAC(1,3)*JAC(2,1))/DETJAC
                 JACINV(3,3) = ( JAC(1,1)*JAC(2,2)-JAC(1,2)*JAC(2,1))/DETJAC

                 DO 70 M5 = 1, 8
                   DNDX(M5) = JACINV(1,1) * DNDXI(M5) + JACINV(1,2) *
      1                   DNDETA(M5) + JACINV(1,3) * DNDZETA(M5)
                   DNDY(M5) = JACINV(2,1) * DNDXI(M5) + JACINV(2,2) *
      1                   DNDETA(M5) + JACINV(2,3) * DNDZETA(M5)
                   DNDZ(M5) = JACINV(3,1) * DNDXI(M5) + JACINV(3,2) *
      1                   DNDETA(M5) + JACINV(3,3) * DNDZETA(M5)
 70              CONTINUE
                 DO 90 L = 1, 8
                   DO 80 M5 = 1, 8
                   DE(L,M5) = DE(L,M5) + W(I) * W(J) * W(K) * (
      1             DNDX(L)*(DXXE*DNDX(M5) + DXYE*DNDY(M5) + DXZE*DNDZ(M5)) +
      2             DNDY(L)*(DYXE*DNDX(M5) + DYYE*DNDY(M5) + DYZE*DNDZ(M5)) +
      3             DNDZ(L)*(DZXE*DNDX(M5) + DZYE*DNDY(M5) + DZZE*DNDZ(M5))
      4             + N(L) *(VXEP*DNDX(M5) + VYEP*DNDY(M5) + VZEP*DNDZ(M5))
      5                + LAMBDA * (1. + RHOBE*KDE/NE ) * N(L) * N(M5) )
      6                     * DETJAC

 80              CONTINUE
 90              CONTINUE
100            CONTINUE
110          CONTINUE
120        CONTINUE
           RETURN
           END
```

PART THREE

APPLICATIONS

Chapter 20

MODELING REGIONAL GROUNDWATER FLOW

20.1 PURPOSE OF GROUNDWATER FLOW MODELING

To "model regional groundwater" flow means to develop mathematical and numerical models of the aquifer system being studied and to use these models to predict the value of hydraulic head at points (and times) of interest. For example, the values of head may be needed to determine the impact of pumping on water table levels (e.g., to determine if a proposed well will cause excessive drawdown at an existing well) or to predict the direction and rate of groundwater flow (e.g., to compute groundwater travel times for site assessment or to predict the rate of movement of groundwater contaminants). The numerical procedures for solving the steady-state and transient groundwater flow equations by the finite element method were described in Part 1 and the implementation of these procedures in computer programs was described in Part 2. However, before these procedures and programs can be applied to an actual field problem the analyst must collect and analyze a variety of information about the study area:

1) to identify the type of model that should be used,

2) to identify the locations of aquifer boundaries,

3) to determine values for aquifer material properties,

4) to determine values and types of boundary and initial conditions, and

5) to calibrate and verify the model.

20.2 TYPES OF GROUNDWATER FLOW MODELS

Several types of models (e.g., one-, two-, and three-dimensional models; steady-state, saturated flow models; transient, unsaturated flow models; fracture flow models) can be used to study groundwater flow systems. The selection of the type of model to apply to a particular field problem can be difficult, particularly if field data are scarce or if the analyst has no previous experience in the study area. Ultimately, the choice is made by selecting a model 1) that represents the physical (and perhaps chemical and biological) processes that, in the opinion of the analyst, are most important in determining aquifer behavior, and 2) that is consistent with the available data. Particular attention should be paid to the assumptions used in the derivation of the differential equation(s) on which the model is based. The assumptions used to derive the four types of groundwater flow equations presented in this book (steady-state, saturated flow; steady-state, unsaturated flow; transient, saturated flow; and transient, unsaturated flow) are discussed in Appendices I and II. Care must be taken to avoid the application of these equations (and the procedures and computer programs in Parts 1 and 2 that are based on these equations) to field situations where the assumptions may not be valid.

For example, in these derivations, Darcy's Law is assumed to be valid and we can immediately conclude that problems involving flow through fractured rock, large cavities in Karst limestone, lava tubes, etc. can not be solved with models based on these equations because it is likely that groundwater velocities will be too large for Darcy's Law to be valid (Hillel, 1980, pp. 181-182). Darcy's Law may also not be valid when groundwater velocities are extremely small, e.g. in flow through compacted clay with small hydraulic gradients (Swarzendruber, 1962). Further, aquifer stress:strain behavior is assumed to be elastic, and the change in thickness of the aquifer in response to changes in head is assumed to be small; thus invalidating the use of these equations to solve problems of *consolidation* (the large and usually irreversible reduction in aquifer thickness that occurs beneath many foundations and earth structures as a result of surface loading, or that can occur in any area due to excessive groundwater withdrawls).

Recall also that groundwater density is assumed constant therefore invalidating the use of these equations to predict heads in problems where density variations are expected to be large e.g., near the fresh water:salty water interface that develops in coastal aquifers or in brine fields, or in problems involving multiple fluid phases such as the flow of non-aqueous phase liquids into groundwater at hazardous waste sites. Flow of groundwater above the water table as water vapor is also assumed to be negligibly small.

20.3 CONFINED VS UNCONFINED AQUIFERS

The steady-state and transient, saturated groundwater flow equations presented in this book can be applied to *confined* and *unconfined* aquifers (see e.g., Bear, 1979). In a confined aquifer, a true water table (the surface where water pressure equals atmospheric pressure) does not exist; the upper limit of the saturated zone is the base of a low-permeability layer called an *aquitard* (Figure 20.1). Hydraulic head is measured with *piezometers* and the height that water rises in the piezometers defines the *piezometric surface*. The procedures and computer programs in Parts 1 and 2 can be used to solve steady-state and transient, saturated groundwater flow problems for confined aquifers in one-, two-, and three-dimensions using as aquifer material properties the components of saturated hydraulic conductivity, K_x, K_y, and K_z, and specific storage, S_s. However, in two-dimensional (map view) problems it is common to use as aquifer properties *transmissivity*, T, and *storativty*, S. Storativity is the name given to specific storage in confined aquifers. The two components of transmissivity, T_x and T_y are defined as

$$T_x = bK_x \tag{20.1a}$$
$$T_y = bK_y \tag{20.1b}$$

where b is the saturated thickness of the aquifer (Figure 20.1). In this case the steady-state, saturated groundwater flow equation (e.g. equation 3.45) can be written

$$\frac{\partial}{\partial x}\left(T_x\frac{\partial h}{\partial x}\right) + \frac{\partial}{\partial y}\left(T_y\frac{\partial h}{\partial y}\right) = 0 \tag{20.2}$$

The method of weighted residuals can be applied to equation 20.2 and the results from Part 1 can be used to solve for unknown values of head, h(x, y). We simply substitute $T_x^{(e)}$ for $K_x^{(e)}$ and $T_y^{(e)}$ for $K_y^{(e)}$ in the equations for $[K_{(e)}]$ for any of the two-dimensional elements. Otherwise the solution procedure is identical and program GW1 can be used without modification (However, when specifying groundwater flow rates at Neumann nodes, we must be careful to use units for q that are consistent with T_x and T_y).

The transient, saturated groundwater flow equation (e.g. equation 3.79) can be written

$$\frac{\partial}{\partial x}\left(T_x\frac{\partial h}{\partial x}\right) + \frac{\partial}{\partial y}\left(T_y\frac{\partial h}{\partial y}\right) = S\frac{\partial h}{\partial t} \tag{20.3}$$

where S is the aquifer *storativity*

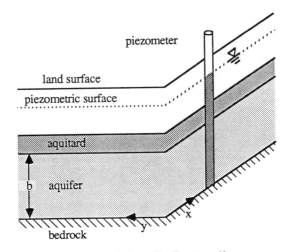

Figure 20.1   Confined aquifer.

The method of weighted residuals can also be applied to equation 20.3 and the results from Part 1 can be used to solve for unknown values of head, h(x, y, t). We simply substitute $T_x^{(e)}$ for $K_x^{(e)}$, $T_y^{(e)}$ for $K_y^{(e)}$, and $S^{(e)}$ for $S_s^{(e)}$ in the equations for $[K^{(e)}]$ and $[C^{(e)}]$ for any for the two-dimensional elements. Otherwise the solution procedure is identical and program GW3 can be used without modification (if the units for specified groundwater flow rates are consistent with T_x, T_y, and S).

In an *unconfined aquifer* the upper limit of the saturated zone is the water table (Figure 20.2). Hydraulic head is measured with *wells*. If the position of the water table is known (not common) the entire surface of the water table is treated as a Dirichlet boundary

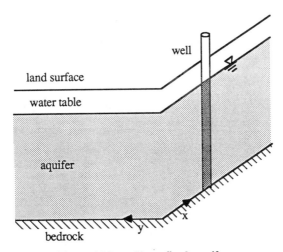

Figure 20.2   Unconfined aquifer.

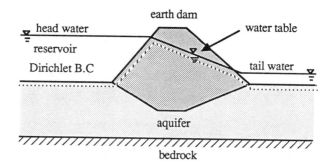

Figure 20.3   **Saturated flow through earth dam, position of water table is assumed known.**

condition. Programs GW1 or GW3 could be used to compute heads at points within the aquifer (Figure 20.3). However, in most situations the position of the water table is not known (except at a few locations) and we wish to compute it from the available data. Three approaches have been used to solve this type of problem.

In the first approach we assume that the slope of the water table is small so that 1) the saturated thickness of the aquifer is approximately constant and 2) groundwater flow is approximately horizontal (i.e., the Dupuit-Forchheimer assumption, see Freeze and Cherry, 1979). This approach is particularly useful in studies of regional groundwater flow where the lateral extent of the aquifer is much larger than the saturated thickness. With this approach the steady-state, saturated groundwater flow equation for two-dimensional flow in an unconfined aquifer can be written (see e.g. Bear, 1979)

$$\frac{\partial}{\partial x}\left(K_x h \frac{\partial h}{\partial x}\right) + \frac{\partial}{\partial y}\left(K_y h \frac{\partial h}{\partial y}\right) = 0 \tag{20.4}$$

But since

$$\frac{\partial^2 h^2}{\partial x^2} = 2h\frac{\partial h}{\partial x} \tag{20.5}$$

equation 20.4 can be written

$$\frac{\partial}{\partial x}\left(\frac{K_x}{2}\frac{\partial h^2}{\partial x}\right) + \frac{\partial}{\partial y}\left(\frac{K_y}{2}\frac{\partial h^2}{\partial y}\right) = 0 \tag{20.6}$$

The results from Part 1 can be used to solve equation 20.4 for unknown values of head, $h(x, y)$. To see this, define a new variable $u = h^2$ so that equation 20.6 becomes

$$\frac{\partial}{\partial x}\left(\frac{K_x}{2}\frac{\partial u}{\partial x}\right) + \frac{\partial}{\partial y}\left(\frac{K_y}{2}\frac{\partial u}{\partial y}\right) = 0 \tag{20.7}$$

Now we can substitue $\frac{K_x^{(e)}}{2}$ for $K_x^{(e)}$ and $\frac{K_y^{(e)}}{2}$ for $K_y^{(e)}$ in the equations for $[K^{(e)}]$ for any of the two-dimensional elements. Dirichlet boundary conditions are specified for u by squaring specified values of hydraulic head. Program GW1 can then be used without modification to solve form unknown values of u and values of head can be computed at each node, $h_i = \sqrt{u_i}$, for each node i. If velocities are required, program GW1 could easily be modified by the addition of the following FORTRAN statements just before (above) the statement "CALL VELOCITY"

```
        DO 99 I = 1, NUMNOD
          X(I) = SQRT(X(I))
     99 CONTINUE
```

The transient, saturated groundwater flow equation for two-dimensional flow in an unconfined aquifer can be written

$$\frac{\partial}{\partial x}\left(\frac{K_x}{2}\frac{\partial h^2}{\partial x}\right) + \frac{\partial}{\partial y}\left(\frac{K_y}{2}\frac{\partial h^2}{\partial y}\right) = S_y\frac{\partial h}{\partial t} \qquad (20.8)$$

where S_y is the aquifer specific yield. Defining $u = h^2$ with

$$\frac{\partial h}{\partial t} = \frac{\partial u^{1/2}}{\partial t} = \frac{1}{2\sqrt{u}}\frac{\partial u}{\partial t} \qquad (20.9)$$

equation 20.8 can be written

$$\frac{\partial}{\partial x}\left(\frac{K_x}{2}\frac{\partial u}{\partial x}\right) + \frac{\partial}{\partial y}\left(\frac{K_y}{2}\frac{\partial u}{\partial y}\right) = \frac{S_y}{2\sqrt{u}}\frac{\partial u}{\partial t} \qquad (20.10)$$

Equation 20.10 is a nonlinear differential equation (because of the term $1/\sqrt{u}$) and cannot be solved using program GW3 unless it is modified (e.g. by using Picard iteration as in program GW4). Program GW3 could be further modified to compute heads and velocities using the same FORTRAN statements given for program GW1 above.

The second approach that can be used to solve the transient and steady-state, saturated groundwater flow equations for an unconfined aquifer is based on the definition of the water table as a surface where water pressure is equal to atmospheric pressure (zero gage pressure). From the definition of hydraulic head

$$h = z + \overset{0 \quad \text{on water table}}{\psi} \qquad (20.11)$$

where z is the elevation head. The solution procedure is very simple. We guess the position of the water table and draw a finite element mesh. We then compute the value of head at each node in the mesh. For each node on the water table the computed value of head should equal the elevation of the node. If the values are not equal we set the coordinates of the nodes on the water table equal to the computed values of head. The process is repeated until a convergence criteria is satisfied (Neumann and Witherspoon, 1970). The *shape* of the mesh changes with each iteration (Figure 20.4). Programs GW1 and GW3 could be easily modified to use this method. This approach is useful for problems where the Dupuit-Forscheimer assumption is not valid (e.g. near a pumping well).

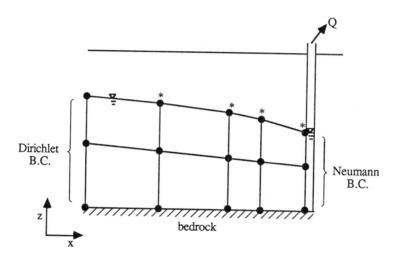

Figure 20.4   **Example problem for unconfined aquifer. z coordinate of nodes marked with asterisk change during solution procedure until $h_i = z_i$ for each node i.**

In the third approach we solve the steady-state or transient unsaturated flow equations and the position of the water table is indicated by nodes with computed pressure heads equal to zero. We use this approach when groundwater flow above the water table is considered to be significant (e.g., in a study of the response of a shallow water table to recharge during a rainstorm).

20.4 SENSITIVITY ANALYSIS

The development of a mathematical model for an aquifer system is a difficult task. Aquifer systems are complex and the interpretation of field and laboratory data for use in a regional groundwater flow model requires a considerable amount of professional judgement (which is why groundwater modeling is sometimes referred to as an "art"). Measured values of aquifer properties are usually scarce and well logs often give a rather incomplete description of the stratigraphy, structure, and lithology of subsurface materials. For example, the locations of aquifer boundaries are needed to specify the size and shape of the problem domain and to draw the finite element mesh. In many problems encountered in practice, there are insufficient data to precisely determine the position of aquifer boundaries e.g., in a valley-fill aquifer it may be difficult to determine the position of the contact between the alluvium and the underlying bedrock. In this case, the positions of aquifer boundaries must be inferred from the available data (e.g., the elevations of aquifer contacts recorded in well logs). This can be done quantitatively (e.g., using some form of interpolation) or qualitatively (e.g., using the judgement of persons knowledgable about the geology and geomorphology of the study area). In any case the effect of uncertainty in the positions of aquifer boundaries on model results should be investigated using a *sensitivity analysis*. In a sensitivity analysis, the values of model parameters (in this case the positions of aquifer boundaries) are varied across the range of likely values and the effect upon computed heads is noted. The most effort is expended to identify parameters that have the most effect on computed results (in most cases the positions of aquifer boundaries have relatively little effect relative to aquifer material properties and boundary and initial conditions).

Lack of data can make it particularly difficult to specify aquifer material properties. Real aquifers are rarely homogeneous and isotropic. The finite element method allows the analyst to specify a different set of material properties for each element in the mesh but a method is needed to obtain these properties from a (usually) limited data set (e.g., the results of pumping tests for a few wells). Although it may be possible to estimate aquifer properties using statistical methods (e.g., geostatistics) or by solving the inverse problem (see below) considerable uncertainty will remain and the effect of uncertainty in values of aquifer properties on model results should be investigated in a sensitivity analysis.

The most commonly occurring boundary conditions are the specified head (Dirichlet) and specified flow (Neumann) boundary conditions described in Part 1. Specified head boundary conditions are commonly used where a surface water body (lake, river, etc.) is in hydraulic connection with groundwater along a portion of the aquifer boundary. Specified flow boundary conditions are commonly used to represent groundwater withdrawal or recharge e.g., from wells, infiltration, and leakage between aquifer units. However, the interpretation of the available data (e.g., pumping rates, precipitation data, irrigation schedules, streamflow records, lake water surface elevations) to determine appropriate boundary head or flow values is rarely straightforward. Also, it can sometimes be very difficult to decide which type of boundary condition a particular feature represents (e.g., should a lake be represented as a constant head or specified flow boundary condition, or as some combination of the two). The effects of alternative types and values of boundary conditions on model results also should be investigated using a sensitivity analysis.

20.5 CALIBRATION, VERIFICATION AND PREDICTION

Calibration is the process of adjusting model parameters (material properties, boundary conditions, and initial conditions) until 1) the model is consistent with the analyst's understanding of the groundwater flow system and with all available data, and 2) computed values of head closely match measured values at selected points in the aquifer (locations of wells, springs, etc). The procedure is essentially an exercise in "trial and error" wherein a plausible set of model parameters are proposed, computed and measured values of head are compared, and model parameters are adjusted to improve the fit. Unfortunately there is no guarantee that the values of model parameters obtained by this procedure are unique. For this reason it is best to calibrate the model using only a portion of the available head data (or to make additional measurements after calibration). The fitted model is then used to predict these "reserved" head measurements. The results are used as a quasi-independent check on the model parameters arrived at by calibration. For example, it is sometimes possible to calibrate a model using measurements made at one time and to verify the model using measurements made at a different time (possibly using the same wells). This step is sometimes called model *verification*. Once the model is calibrated and verified it is ready for use in *prediction* (e.g., predicting water table response to pumping, predicting groundwater velocities for use in a solute transport model, etc.).

An alternative approach to calibration is to solve the *inverse problem*, i.e., to compute the values of model parameters directly from measured values of head. This approach is analogous to fitting a line to a data set using regression, except that the number of unknown parameters is much larger. An *objective function* is defined (e.g., the sum of the squares of the differences between measured and computed heads) and values of the parameters are sought that make the value of the function a minimum. There is a great deal of literature on this subject but the approach has not been widely used in practice (in part because of the theoretical and computational difficulties involved in succcessfully fitting a model with many plausible combinations of parameters, a common situation). Menke (1984) is an excellent introduction to techniques for solving the inverse problem. Reviews of different approaches for solving the inverse problem in groundwater hydrology are in Neuman and Yakowitz (1979), Neuman (1980), and Yeh et al. (1983).

20.6 MASS BALANCE CALCULATIONS

An additional check on model behavior that should always be performed is a *mass balance* for water. If the model is performing properly, the change in the amount of water stored in the aquifer should equal the inflow (e.g., through specified head boundaries or injection wells) minus the outflow (e.g., from pumping wells). For a steady-state flow problem, the change in storage will be zero. For a transient flow problem. the change in storage can be computed for each element in the mesh using the change in head for each node in the element and the value of storativity, specific yield, or storage coefficient for that element. Inflow and outflow at Neumann nodes will be known from the boundary conditions, and inflow and outflow across Dirichlet boundaries can be computed using the components of apparent groundwater velocity for each element on the boundary. If the results of the mass balance calculations are poor, it is probably an indication that the mesh is too coarse. Numerical errors in computed heads obtained using a coarse mesh will cause errors in the calculation of aquifer storage and apparent groundwater velocites. Errors in computed velocities will most impact mass balance calculations at Dirichlet boundaries, where water is entering or leaving the mesh. Refining the mesh will always improve the mass balance (unless there are gross errors, e.g., entering specified groundwater flows with the wrong sign or magnitude, etc.).

20.7 REPORTING MODEL RESULTS

Because of the variety of procedures that may be used to develop a groundwater flow model it is essential that the analyst document each step of the process used to obtain predictions in the project report. As a minimum such a report should contain the following information:

1. Assumptions about the groundwater flow processes considered:
   For example "two-dimensional, steady-state, saturated flow of groundwater with constant density through a rigid (nondeformable) aquifer". Always give the governing equation(s) used.

2. Description of Numerical Procedures Used:
   Show the finite element mesh. Label Dirichlet and Neumann nodes. Give a reference for the computer program used.

3. Data sources and procedures used to identify aquifer boundaries:
   For example "The lower boundary of the alluvial aquifer was assumed to vary linearly between alluvium-bedrock contacts reported in well logs". In this case the report should contain a map showing well locations and computed elevations of alluvium-bedrock contact and an appendix containing the well logs.

4. Data sources and procedures used to determine aquifer material properties.
   For example, "The aquifer was assumed to be homogeneous and isotropic. Aquifer hydraulic conductivity was set equal to the average value of hydraulic conductivity obtained for three wells using the Theis solution (see, e.g., Bear,1972) and the results of constant discharge pumping tests". In this case the report should contain a map showing pumping and observation well locations, drawdown curves for the pumping tests, and a summary of calculations.

5. Data sources and procedures used to determine boundary conditions.
   For example, "The portion of the aquifer boundary along the Red River was modeled as a specified head boundary. The value of head for this boundary was taken to be the average river stage for the months of October through December. Discharge from

several domestic water-supply wells in the study area were assumed to have negligible impact on model results and were neglected". In this case, the report should contain the stage and discharge records for the river and the location and estimated discharge rate for the wells.

6. Results of Model Calibration, Verification, and Mass Balance Calculations.
   For example, "The model was calibrated using ten of the available water level measurements (show well locations on a map). Values of model parameters were adjusted by trial and error until the difference between measured and predicted heads at each well was less than 0.5 m. The model was then verified using measured water levels in the five remaining wells. The maximum difference between measured and predicted head at these wells was 1.3 m. Results of a mass balance for the aquifer, performed after calibration, indicated that 95% of the water in the aquifer was conserved."

Chapter 21

MODELING SOLUTE TRANSPORT

21.1 PURPOSE OF SOLUTE TRANSPORT MODELING

To "model solute transport" means to develop mathematical and numerical models of the aquifer system being studied and to use these models to predict the concentration of a solute (radionuclide, hazardous waste, pesticide, plant nutrient, etc.) at points of interest for a set of specified times.. For example, it may be necessary to estimate the potential impacts on human health of a proposed waste disposal site, e.g., a municipal landfill. A solute transport model could be used to predict the likely concentration of contaminants leached from the site in the groundwater at nearby wells. This is an example of *site assessment*, the determination if a site is suitable for some purpose based on the likely impact of proposed activities on groundwater quality. Solute transport models are also used as a basis for the design of contaminant recovery and treatment systems at sites of existing contamination. For example, it may be necessary to install a set of capture wells at a hazardous waste site to prevent contaminant movement off-site. A solute transport model could be used to select the most effective combination of well locations and pumping rates. This is an example of *performance assessment*, the evaluation of how effective a proposed design is at meeting the project objectives.

The first step in developing a solute transport model is to calibrate and verify a groundwater flow model and the comments in Chapter 20 apply. During the calibration of the groundwater flow model the emphasis should be on producing a good fit between measured and predicted aquifer heads near solute sources and sinks (e.g., near a waste injection well). These heads will be used to compute groundwater velocities which are needed as input for the solute transport model (recall that apparent groundwater velocities are used to compute the rate of solute transport by advection and to compute dispersion coefficients, see Appendix III). The accuracy of predicted solute concentrations will to a large part be determined by the accuracy of predicted groundwater velocities near solute sources and sinks.

The development of a solute transport model will require additional information about the study area:

1) to identify the type of solute transport model that should be used,

2) to determine values for additional properties of the aquifer and the solute,

3) to determine values and types of boundary and initial conditions, and

4) to calibrate and verify the model.

21.2 TYPES OF SOLUTE TRANSPORT MODELS

Several types of models can be used to predict solute concentrations in groundwater flow systems. The models differ 1) in the type of groundwater flow equation used to obtain groundwater velocities (e.g., steady-state or transient flow, saturated or unsaturated flow), and 2) in the types of physical, chemical, and biological processes considered in the solute transport equation(s). In this book we have used a form of the solute transport equation that includes processes of advection, dispersion, diffusion, and decay. The assumptions used to derive this equation are discussed in Appendix III. Care must be taken to avoid the application of this equation to field situations where the assumptions may not be valid.

For example, in this derivation, the aquifer is assumed to be isotropic with respect to dispersion processes. This assumption is made primarily for convenience since field procedures for measuring all the coefficients of a general dispersion model are not available. Transport by advection is limited to Darcy-type flow through the pore space and the equation can not be used to predict rates of solute transport through fractured rock, large cavities in Karst limestone, lava tubes, etc. Recall also that the density of the solute-groundwater mixture is assumed constant therefore invalidating the use of this equation to predict solute concentrations in the presence of very high solute concentrations or multiple liquid phases (e.g.,cases with simultaneous flow of gasoline and groundwater phases). Transport in the gas phase was assumed to be small, thus invalidating the application of this equation to the transport of highly volatile compounds above the water table.

A very important assumption was that sorption processes can be described using an equilibrium distribution coefficient, K_d. Although this is a common assumption in practice, it should be considered a crude approximation because of the importance of other processes including competition among different solutes for exchange sites, reactions that require relatively long periods of time to reach equilibrium, and multiple-step sorption processes (e.g., involving diffusion through an immobile water layer before sorption can occur at the solid surface). Several alternative formulations for the sorption process are given in de Marsily (1986) and Bear (1979). Similarly, the assumption that solute decay can be described using a simple decay constant, λ, although appropriate for certain radionuclides, should be a considered a crude approximation for biological degradation (e.g., microbial metabolism).

21.3 SENSITIVITY ANALYSIS

Just as in the case of a groundwater flow model, the effect of uncertainty in the values of model parameters (boundary and initial conditions and the values of lateral and transverse dispersivity, distribution coefficient, and decay constant) on computed solute concentrations should be investigated using a sensitivity analysis. In most situations the greatest uncertainty involves the selection of dispersivities. Ideally these should be measured at the site using a tracer test but in most cases they must be estimated from tabulated values (e.g., Appendix V). However, for long times or large distances advection tends to be a much more important process than dispersion, and the effects of uncertainties in dispersivities tends to have less effect on computed solute concentrations, than for short times and small distances. Sometimes the effects of dispersion, sorption, and decay are neglected entirely and computed solute concentrations based only on advection are used to assess the greatest likely travel distances along a particular flow path (or the shortest likely travel times to a particular point), which is sometimes called a *worst case scenario*.

The most commonly occurring boundary conditions are the specified concentration (Dirichlet) and specified flux (Neumann) boundary conditions described in Part 1. Dirichlet boundary conditions are commonly used where a surface water body (waste storage lagoon, river, etc.) with a fixed solute concentration is in hydraulic connection with groundwater along a portion of the aquifer boundary. Specified flow boundary conditions are commonly used to represent solute leakage into the aquifer and solute withdrawal and injection by wells. However, the data needed to decide which type of boundary condition a particular feature represents are often unavailable, for example in the preliminary stages of an investigation at an uncontrolled waste site. The effects of alternative types and values of boundary conditions on model results also should be investigated using a sensitivity analysis.

21.4 CALIBRATION, VERIFICATION, AND PREDICTION

In the case of a solute transport model, calibration consists of proposing a plausible set of model parameters, comparing measured and predicted solute concentrations at a set of points, and adjusting model parameters to improve the fit. As in the case of the groundwater flow equation there is no guarantee that the values of model parameters obtained by this procedure are unique. For this reason it is best to calibrate the model using only a portion of the available data and to predict the remaining concentrations as a check on the model parameters arrived at by calibration. Once the model is calibrated and verified it is ready for use in prediction. It may also sometimes be possible to obtain values of certain model parameters by solving the inverse problem (e.g., dispersivities are often computed from measured concentrations in a tracer test).

21.5 MASS BALANCE CALCULATIONS

Just as in the case of groundwater flow a mass balance for the solute should be computed as a check on model behavior. If the model is performing properly, the change in the amount of solute stored in the aquifer should equal the inflow (e.g., through specified concentration boundaries or injection wells) minus the outflow (e.g., pumping wells). The change in storage can be computed for each element in the mesh using the change in concentration for each node in the element and the element's size, shape, and porosity. Inflow and outflow at Neumann nodes will be known from the boundary conditions, and inflow and outflow across specified concentration boundaries can be computed using the components of apparent groundwater velocity and the computed solute concentration at the nodes of each element on the boundary. If the results of the mass balance calculations are poor, it is probably an indication that the mesh is too coarse.

21.6 REPORTING MODEL RESULTS

Because of the variety of procedures that may be used to develop a solute transport model it is essential that the analyst document each step of the process used to obtain predictions in the project report. As a minimum such a report should contain the following information:

1. Assumptions about the solute transport processes considered:
   List the assumptions used to derive the governing equation(s) used. Comment on the applicability of these assumptions to the conditions at the site.

2. Description of Numerical Procedures Used:
   Show the finite element mesh. Label Dirichlet and Neumann nodes. Give a reference for the computer program used.

3. Data sources and procedures used to determine aquifer material properties.
   For example, "The aquifer was assumed to be homogeneous and isotropic with respect to dispersion. Lateral and transverse dispersivities were estimated using tabulated values (give reference). Decay and sorption were assumed to be negligible". Or, "Lateral and transverse dispersivities were estimated using a tracer test (give references and show data)".

4. Data sources and procedures used to determine boundary conditions.
   For example, "The portion of the aquifer boundary along the Red River was modeled as a specified concentration boundary. The value of concentration for this boundary was assumed to be zero based on water quality measurements taken upstream of point X."

5. <u>Results of Model Calibration, Verification, and Mass Balance Calculations.</u>
For example, "The model was calibrated using measured concentrations in five wells
(show well locations on a map). Values of model parameters were adjusted by trial and
error until the difference between measured and predicted concentrations at each well
was less than 25 ppm. The model was then verified using measured water levels in the
six remaining wells. The maximum difference between measured and predicted
concentration at these wells was 47 ppm. Results of a mass balance for the aquifer,
performed after calibration, indicated that 95% of the solute in the aquifer was
conserved.

Appendix I

DERIVATION OF EQUATIONS OF STEADY-STATE GROUNDWATER FLOW

Consider a unit volume of saturated porous media (Figure AI.1). In fluid mechanics, such a volume is called a *control volume*. The boundaries of the element are called *control surfaces*.

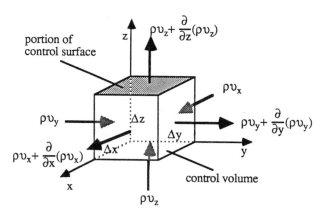

Figure AI.1   Control volume for groundwater flow through porous media.

The law of conservation of mass for steady-state flow requires that the rate at which fluid is entering the control volume is equal to the rate at which fluid is leaving the control volume or

$$\text{net rate of inflow} = \text{inflow} - \text{outflow} = 0 \tag{AI.1}$$

For purposes of analysis, consider the rate at which groundwater enters the control volume per unit surface area to consist of three components $\rho \upsilon_x$, $\rho \upsilon_y$, and $\rho \upsilon_z$ where ρ is the density of water and υ_x, υ_y and υ_z are the apparent velocities of groundwater flow entering the control volume through control surfaces perpendicular to the x, y, and z coordinate axes. The dimensions of $\rho \upsilon_x$, ρv_y, and ρv_z are $M/L^2 T$.

Using a Taylor Series approximation, the rate at which groundwater leaves the control volume in the x direction can be written

$$\rho \upsilon_x + \frac{\partial}{\partial x}(\rho \upsilon_x)\Delta x + \frac{\partial^2}{2!\partial x^2}(\rho \upsilon_x)\Delta x^2 + \frac{\partial^2}{3!\partial x^3}(\rho \upsilon_x)\Delta x^3 + \cdots \tag{AI.2}$$

If we make the size of the control volume small, we can neglect higher-order terms (i.e., those involving Δ^2, Δ^3 etc.) and , because we have chosen a unit control volume $(\Delta x = \Delta y = 1)$ the rate at which groundwater leaves the control volume is $\rho \upsilon_x + \frac{\partial}{\partial x}(\rho \upsilon_x)$. The net rate of inflow in the x direction is then

$$\begin{array}{ll} \text{net rate of inflow} & = \text{rate of inflow} - \text{rate of outflow} \\ \text{in x direction} & \quad\text{in x direction} \quad\; \text{in x directon} \end{array}$$

$$= \rho \upsilon_x - \left[\rho \upsilon_x + \frac{\partial}{\partial x}(\rho \upsilon_x) \right]$$

$$= -\frac{\partial}{\partial x}(\rho \upsilon_x) \tag{AI.3}$$

and the net rate of inflow in the y and z directions are $-\frac{\partial}{\partial y}(\rho \upsilon_y)$ and $-\frac{\partial}{\partial z}(\rho \upsilon_z)$, respectively. Because the net rate of inflow for the entire control volume must equal zero if the law of conservation of mass is to be satisfied, we can write

$$-\frac{\partial}{\partial x}(\rho \upsilon_x) - \frac{\partial}{\partial y}(\rho \upsilon_y) - \frac{\partial}{\partial z}(\rho \upsilon_z) = 0 \tag{AI.4}$$

If we assume that groundwater density, ρ is constant (i.e., the fluid is incompressible), we can use the product rule of calculus to evaluate a typical term in equation AI.4

$$-\frac{\partial}{\partial x}(\rho \upsilon_x) = -\left[\rho \frac{\partial \upsilon_x}{\partial x} + \upsilon_x \frac{\partial \rho}{\partial x}^{\;0} \right]$$

$$= -\rho \frac{\partial \upsilon_x}{\partial x} \tag{AI.5}$$

Similarly for the x and y directions. Because groundwater density appears outside the derivative it cancels from equation AI.4 and we have

$$-\frac{\partial \upsilon_x}{\partial x} - \frac{\partial \upsilon_y}{\partial y} - \frac{\partial \upsilon_z}{\partial z} = 0 \tag{AI.6}$$

Now the apparent groundwater velocities are given by Darcy's Law

$$\upsilon_x = -K_x \frac{\partial h}{\partial x} \tag{AI.7a}$$

$$\upsilon_y = -K_y \frac{\partial h}{\partial y} \tag{AI.7b}$$

$$\upsilon_z = -K_z \frac{\partial h}{\partial z} \tag{AI.7c}$$

where K_x, K_y and K_z are the hydraulic conductivities in the x, y, and z directions, respectively and h is the hydraulic head. Substituting equation AI.7 into equation AI.6. We arrive at the *steady-state, saturated flow equation.*

$$\frac{\partial}{\partial x}\left(K_x\frac{\partial h}{\partial x}\right) + \frac{\partial}{\partial y}\left(K_y\frac{\partial h}{\partial y}\right) + \frac{\partial}{\partial z}\left(K_z\frac{\partial h}{\partial z}\right) = 0$$

(AI.8)

If flow is two-dimensional, equation AI.8 simplifies to

$$\frac{\partial}{\partial x}\left(K_x\frac{\partial h}{\partial x}\right) + \frac{\partial}{\partial y}\left(K_y\frac{\partial h}{\partial y}\right) = 0$$

(AI.9)

and if the flow is one-dimensional, we have

$$\frac{\partial}{\partial x}\left(K_x\frac{\partial h}{\partial x}\right) = 0$$

(AI.10)

If a component of hydraulic conductivity is independent of position for a particular direction (i.e., is the same at all points along a line oriented in that direction), we can further simplify equation AI.8 using the product rule. For example, if K_x is independent of postion x

$$\frac{\partial}{\partial x}\left(K_x\frac{\partial h}{\partial x}\right) = K_x\frac{\partial^2 h}{\partial x^2} + \frac{\partial h}{\partial x}\overset{0}{\cancel{\frac{\partial K_x}{\partial x}}}$$

$$= K_x\frac{\partial^2 h}{\partial x^2}$$

(AI.11)

Similar terms can be obtained for K_y and K_z if $\frac{\partial K_y}{\partial y} = \frac{\partial K_z}{\partial z} = 0$. In this case we say the porous media is *homogenous* and equation AI.8 simplifies to

$$K_x\frac{\partial^2 h}{\partial x^2} + K_y\frac{\partial^2 h}{\partial y^2} + K_z\frac{\partial^2 h}{\partial z^2} = 0$$

(AI.12)

Finally, if $K_x = K_y = K_z = K$, a constant we say the porous media is homogeneous and *isotropic* and equation AI.8 simplifies to

$$\frac{\partial^2 h}{\partial x^2} + \frac{\partial^2 h}{\partial y^2} + \frac{\partial^2 h}{\partial z^2} = 0$$

(AI.13)

which is known to mathematicians as *La Place's equation.*

If the porous media is not saturated, the value of hydraulic conductivity at a point is a function of the pressure head of the water in the voids at that point

$$K = K(\Psi) \tag{AI.14}$$

where Ψ is the pressure head. Substituting equation AI.14 into equation AI.8 yields

$$\frac{\partial}{\partial x}\left(K_x(\psi)\frac{\partial h}{\partial x}\right) + \frac{\partial}{\partial y}\left(K_y(\psi)\frac{\partial h}{\partial y}\right) + \frac{\partial}{\partial z}\left(K_z(\psi)\frac{\partial h}{\partial z}\right) = 0 \tag{AI.15}$$

for the case where the unsaturated hydraulic conductivity function is different in the x, y, and z directions. Recalling the definition of hydraulic head

$$h = \Psi + z^* \tag{AI.16}$$

where z^* is the elevation head (i.e., the vertical distance from any point to an arbitrary datum). If the z coordinate axis is assumed to be vertical

$$\frac{\partial h}{\partial x} = \frac{\partial}{\partial x}(\psi + z^*) = \frac{\partial \psi}{\partial x} + \overset{0}{\cancel{\frac{\partial z^*}{\partial x}}}$$
$$= \frac{\partial \psi}{\partial x} \tag{AI.17}$$

similarly

$$\frac{\partial h}{\partial y} = \frac{\partial \psi}{\partial y} \tag{AI.18}$$

and

$$\frac{\partial h}{\partial z} = \frac{\partial}{\partial z}(\psi + z^*) = \frac{\partial \psi}{\partial z} + \overset{1}{\cancel{\frac{\partial z^*}{\partial z}}} = \frac{\partial \psi}{\partial z} + 1 \tag{AI.19}$$

Substituting equations AI.17, 18, and 19 into equation AI.15 gives

$$\boxed{\frac{\partial}{\partial x}\left(K_x(\psi)\frac{\partial \psi}{\partial x}\right) + \frac{\partial}{\partial y}\left(K_y(\psi)\frac{\partial \psi}{\partial y}\right) + \frac{\partial}{\partial z}\left(K_z(\psi)\left(\frac{\partial \psi}{\partial z} + 1\right)\right) = 0} \tag{AI.20}$$

which is the *steady-state, unsaturated flow equation.*

Problems

1. Appendix I has presented the derivation of the equations of steady-state groundwater flow for a *rectangular coordinate system* i.e., a coordinate system defined by the three orthogonal coordinate axes x, y, and z. In some situations, for example in the case of groundwater flow to a well, it is more convenient to work in a *cylindrical coordinate system* i.e., in a coordinate system defined by the two orthogonal coordinate axes r, θ, and z (Fig. AI.2).

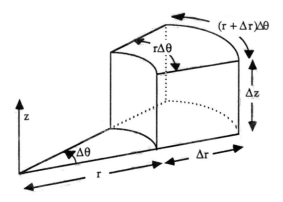

Figure AI.2   Control volume for groundwater flow through porous media in cylindrical coordinates.

a. Using the same approach presented in this chapter derive the *steady-state, saturated flow equation in cylindrical coordinates*

$$\frac{\partial}{\partial r}\left(K_r\frac{\partial h}{\partial r}\right) + \frac{K_r}{r}\frac{\partial h}{\partial r} + \frac{1}{r^2}\frac{\partial}{\partial \theta}\left(K_\theta\frac{\partial h}{\partial \theta}\right) + \frac{\partial}{\partial z}\left(K_z\frac{\partial h}{\partial z}\right) = 0 \qquad (AI.20)$$

b. Derive the *steady-state, unsaturated flow equation in cylindrical coordinates*

$$\frac{\partial}{\partial r}\left(K_r(\psi)\frac{\partial \psi}{\partial r}\right) + \frac{K_r}{r}\frac{\partial \psi}{\partial r} + \frac{1}{r^2}\frac{\partial}{\partial \theta}\left(K_\theta\frac{\partial \psi}{\partial \theta}\right) + \frac{\partial}{\partial z}\left(K_z(\psi)\left(\frac{\partial \psi}{\partial z} + 1\right)\right) = 0 \qquad (AI.21)$$

2. We can often use symmetry to reduce the dimensionality of a flow problem in cylindrical cordinates. In the case of groundwater flow to a well, it is common to consider the well to be an axis of symmetry. This is only true however if the aquifer geometry (i.e., the position of the soil surface and soil and rock layers), the components of hydraulic conductivity, and the specified boundary conditions are all independent of angular coordinate θ. In this case the derivatives of head with respect to θ vanish and we say the problem is *axisymmetric*. Show that the axisymmetric forms of the steady-state saturated and the steady-state unsaturated flow equations can be written

$$\frac{\partial}{\partial r}\left(K_r\frac{\partial h}{\partial r}\right) + \frac{K_r}{r}\frac{\partial h}{\partial r} + \frac{\partial}{\partial z}\left(K_z\frac{\partial h}{\partial z}\right) = 0 \qquad\qquad (AI.22a)$$

$$\frac{\partial}{\partial r}\left(K_r(\psi)\frac{\partial \psi}{\partial r}\right) + \frac{K_r(\psi)}{r}\frac{\partial \psi}{\partial r} + \frac{\partial}{\partial z}\left(K_z(\psi)\left(\frac{\partial \psi}{\partial z}+1\right)\right) = 0 \qquad (AI.22b)$$

Appendix II

DERIVATION OF EQUATIONS OF TRANSIENT GROUNDWATER FLOW

The law of conservation of mass for transient flow requires that the net rate at which fluid enters a control volume is equal to the time rate of change of fluid mass storage within the control volume.

$$\text{net rate of inflow} = \text{inflow} - \text{outflow} = \text{rate of change in storage} \qquad \text{(AII.1)}$$

From equations AI.1 and AI.4 we can write

$$\text{net rate of inflow} = -\frac{\partial}{\partial x}(\rho \upsilon_x) - \frac{\partial}{\partial y}(\rho \upsilon_y) - \frac{\partial}{\partial z}(\rho \upsilon_z) \qquad \text{(AII.2)}$$

In steady-state flow, the change in storage within the control volume is zero. In transient flow, the change in storage is not zero and equation AII.2 becomes

$$\underbrace{-\frac{\partial}{\partial x}(\rho \upsilon_x) - \frac{\partial}{\partial y}(\rho \upsilon_y) - \frac{\partial}{\partial z}(\rho \upsilon_z)}_{\text{net rate of inflow}} = \underbrace{\frac{\partial}{\partial t}(\rho n)}_{\substack{\text{rate of change} \\ \text{in storage}}} \qquad \text{(AII.3)}$$

where n is the porosity of the porous media. The dimensions of the term $\frac{\partial(\rho n)}{\partial t}$ are M/L^3T or the time rate of change of fluid mass per unit volume of the control volume. Now assume that the porous media is saturated. Then using the chain-rule we can expand the right-hand side of equation AII.3

$$\frac{\partial}{\partial t}(\rho n) = \frac{\partial}{\partial h}(\rho n)\frac{\partial h}{\partial t} \qquad \text{(AII.4)}$$

where we can see that, in transient, saturated flow, the rate of change in fluid storage in the control volume is related to the rate of change in hydraulic head. Using the product rule we can expand the first term on the right-hand side of equation AII.4

$$\frac{\partial}{\partial h}(\rho n) = \rho\frac{\partial n}{\partial h} + n\frac{\partial \rho}{\partial h} \qquad \text{(AII.5)}$$

The first term on the right-hand side of equation AII.5 is the mass of water produced by the expansion or compression of the porous media and the second term is the mass of water produce by the expansion or compression of the fluid. In the case of saturated flow, water can only enter the control volume if the porosity increases $\left(\frac{\partial n}{\partial h} > 0\right)$ or the fluid density increase $\left(\frac{\partial p}{\partial h} > 0\right)$.

To continue we must define two new terms: the *porous media compressibility* α and the *fluid compressibility* β. Compression or expansion of the porous media is caused by a change in effective stress σ_e. If the porous media is saturated

$$d\sigma_e = -\rho g d\psi \qquad (AII.6)$$

where ψ is the pressure head. But since $d\psi = d(h - z^*) = dh - dz^* \nearrow 0$ we can write

$$d\sigma_e = -\rho g dh \qquad (AII.7)$$

Now define the porous media compressibility α

$$\alpha = -\frac{dV_f}{V}\frac{1}{d\sigma_e} = \frac{dn}{d\sigma_e} \qquad (AII.8)$$

where V_f is the volume of fluid and V is the control volume. Combining equations AII.7 and AII.8 we have

$$\frac{dn}{dh} = \alpha\rho g \qquad (AII.9)$$

The fluid compressibility β is defined as

$$\beta = \frac{dV_f}{V_f}\frac{1}{dp} \qquad (AII.10)$$

where p is the fluid pressure. The change in pressure is given by

$$dp = \rho g d\psi = \rho g dh \qquad (AII.11)$$

and with $dV_f/V_f = d\rho/\rho$ equation AII.10 becomes

$$\beta = \frac{d\rho}{\rho}\frac{1}{\rho g dh} \qquad (AII.12)$$

or

$$\frac{d\rho}{dh} = \rho^2 g\beta \qquad (AII.13)$$

Substituting equations AII.9 and AII.13 into equation AII.4 gives

$$\frac{\partial}{\partial t}(\rho n) = \left(\rho\frac{\partial n}{\partial h} + n\frac{\partial \rho}{\partial h}\right)\frac{\partial h}{\partial t}$$

$$= (\rho^2 g\alpha + n\rho^2 g\beta)\frac{\partial h}{\partial t} \qquad (AII.14)$$

Now define the *specific storage* S_s as

$$S_s = \rho g(\alpha + n\beta) \qquad (AII.15)$$

The dimensions of S_s are L^{-1} representing the volume of water that a unit volume of aquifer releases from storage for a unit decline in hydraulic head. Substituting equation AII.15 into equation AII.14 gives

$$\frac{\partial}{\partial t}(\rho n) = \rho S_s \frac{\partial h}{\partial t} \qquad \text{(AII.16)}$$

and substituting equation AII.16 into equation AII.3 we have

$$-\frac{\partial}{\partial x}(\rho \upsilon_x) - \frac{\partial}{\partial y}(\rho \upsilon_y) - \frac{\partial}{\partial z}(\rho \upsilon_z) = \rho S_s \frac{\partial h}{\partial t} \qquad \text{(AII.17)}$$

If we assume that density is constant in the three coordinate directions equation AII.17 becomes

$$\rho \left(-\frac{\partial \upsilon_x}{\partial x} - \frac{\partial \upsilon_y}{\partial y} - \frac{\partial \upsilon_z}{\partial z} \right) = \rho S_s \frac{\partial h}{\partial t} \qquad \text{(AII.18)}$$

Cancelling ρ from both sides of equation AII.18 and using Darcy's Law we arrive at the *transient, saturated-flow equation.*

$$\boxed{\frac{\partial}{\partial x}\left(K_x \frac{\partial h}{\partial x} \right) + \frac{\partial}{\partial y}\left(K_y \frac{\partial h}{\partial y} \right) + \frac{\partial}{\partial z}\left(K_z \frac{\partial h}{\partial z} \right) = S_s \frac{\partial h}{\partial t}}$$
$$\text{(AII.19)}$$

If the porous media is homogeneous, K_x, K_y, and K_z are constant and equation AII.19 reduces to

$$K_x \frac{\partial^2 h}{\partial x^2} + K_y \frac{\partial^2 h}{\partial y^2} + K_z \frac{\partial^2 h}{\partial z^2} = S_s \frac{\partial h}{\partial t} \qquad \text{(AII.20)}$$

If the porous media is also isotropic, $K_x = K_y = K_z = K$, equation AII.20 is written

$$\frac{\partial^2 h}{\partial x^2} + \frac{\partial^2 h}{\partial y^2} + \frac{\partial^2 h}{\partial z^2} = \frac{S_s}{K} \frac{\partial h}{\partial t} \qquad \text{(AII.21)}$$

which is known to mathematicians as the *diffusion equation.* For the special case of horizontal, two-dimensional groundwater flow in a confined aquifer of constant thickness b equation AII.21 simplifies to

$$\frac{\partial^2 h}{\partial x^2} + \frac{\partial^2 h}{\partial y^2} = \frac{S}{T} \frac{\partial h}{\partial t} \qquad \text{(AII.22)}$$

where $S = S_s b$ and $T = Kb$.

In transient, unsaturated flow, the *degree of saturation* of the porous media within the control volume changes with time

$$\theta' = \theta'(t) \tag{AII.23}$$

where θ' is the degree of saturation. The mass of fluid, within the control volume is now $\rho n \theta'$ instead of ρn. Substituting this term into equation AII.3 gives

$$-\frac{\partial}{\partial x}(\rho \upsilon_x) - \frac{\partial}{\partial y}(\rho \upsilon_y) - \frac{\partial}{\partial z}(\rho \upsilon_z) = \frac{\partial}{\partial t}(\rho n \theta') \tag{AII.24}$$

Expanding the term on the right-hand side of equation AII.24 using the product rule gives

$$\frac{\partial}{\partial t}(\rho n \theta') = \rho n \frac{\partial \theta'}{\partial t} + \rho \theta' \frac{\partial n}{\partial t} + n \theta' \frac{\partial \rho}{\partial t} \tag{AII.25}$$

Now if we assume that $\frac{\partial \theta'}{\partial t} \gg \frac{\partial n}{\partial t}$ and $\frac{\partial \theta'}{\partial t} \gg \frac{\partial \rho}{\partial t}$, the last two terms on the right – hand side of equation AII.25 can be discarded. Taking ρ's outside the derivatives in the left-hand side of equation AII.24 and cancelling ρ from both sides of equation AII.24 gives

$$-\frac{\partial}{\partial x}(\upsilon_x) - \frac{\partial}{\partial y}(\upsilon_y) - \frac{\partial}{\partial z}(\upsilon_z) = n \frac{\partial \theta'}{\partial t} \tag{AII.26}$$

If we now substitute Darcy's law for unsaturated flow into equation AII.26 we have

$$\frac{\partial}{\partial x}\left(K_x(\psi)\frac{\partial h}{\partial x}\right) + \frac{\partial}{\partial y}\left(K_y(\psi)\frac{\partial h}{\partial y}\right) + \frac{\partial}{\partial z}\left(K_z(\psi)\frac{\partial h}{\partial z}\right) = n \frac{\partial \theta'}{\partial t} \tag{AII.27}$$

Recalling the definition of volumetric water content ($\theta = n\theta'$) we can write

$$n \frac{\partial \theta'}{\partial t} = \frac{\partial \theta}{\partial t} \tag{AII.28}$$

If we define the specific moisture capacity $C(\psi)$

$$C(\psi) = \frac{d\theta}{d\psi} \tag{AII.29}$$

where ψ is the pressure head and recall the definition of hydraulic head ($h = \psi + z^*$), we can rewrite equation AII.27 as

$$\frac{\partial}{\partial x}\left(K_x(\psi)\frac{\partial \psi}{\partial x}\right) + \frac{\partial}{\partial y}\left(K_y(\psi)\frac{\partial \psi}{\partial y}\right) + \frac{\partial}{\partial z}\left(K_z(\psi)\left(\frac{\partial \psi}{\partial z} + 1\right)\right) = C(\psi)\frac{\partial \psi}{\partial t} \tag{AII.30}$$

which is the equation for *transient unsaturated flow*. Equation AII.30 is also known as *Richards equation*.

Problems

1. Derive the transient, saturated flow equation in cylindrical coordinates

$$\frac{\partial}{\partial r}\left(K_r\frac{\partial h}{\partial x}\right) + \frac{K_r}{r}\frac{\partial h}{\partial r} + \frac{1}{r^2}\frac{\partial}{\partial \theta}\left(K_\theta\frac{\partial h}{\partial \theta}\right) + \frac{\partial}{\partial z}\left(K_z\frac{\partial h}{\partial z}\right) = S_s\frac{\partial h}{\partial t}$$

(AII.31)

2. Derive the transient, unsaturated flow equation in cylindrical coordinates

$$\frac{\partial}{\partial r}\left(K_r(\psi)\frac{\partial \psi}{\partial r}\right) + \frac{K_r}{r}\frac{\partial \psi}{\partial r} + \frac{1}{r^2}\frac{\partial}{\partial \theta}\left(K_\theta\frac{\partial \psi}{\partial \theta}\right) + \frac{\partial}{\partial z}\left(K_z(\psi)\left(\frac{\partial \psi}{\partial z} + 1\right)\right) = C(\psi)\frac{\partial \psi}{\partial t}$$

(AII.32)

3. Rewrite equations AII.31 and AII.32 for problems with axisymmetry.

Appendix III

DERIVATION OF EQUATIONS OF SOLUTE TRANSPORT

Consider a unit volume of porous media (Figure AIII.1). As in Appendix I, we refer to such a volume as a *control volume* with boundaries called *control surfaces*

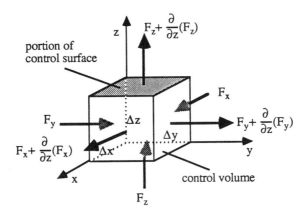

Figure AIII.1   **Control volume for solute transport through porous media.**

The law of conservation of mass for solute transport requires that the rate of change of solute mass within the control volume is equal to the net rate at which solute is entering the control volume through the control surfaces plus the net rate at which solute is produced within the control volume by various chemical and physical processes.

$$
\begin{array}{ccccc}
\text{rate of change} & = & \text{net rate of} & + & \text{net rate of} \\
\text{of solute mass} & & \text{solute inflow} & & \text{solute production}
\end{array}
\tag{AIII.1}
$$

For purposes of analysis, consider the rate at which solute enters the control volume to consist of three components F_x, F_y, and F_z that are parallel to the x, y, and z coordinate axes, respectively. The dimensions of F_x, F_y, and F_z are M/L^2T.

The rates at which solute leaves the control volume are

$$F_x + \frac{\partial}{\partial x}(F_x) \quad \text{in the x direction,}$$

$$F_y + \frac{\partial}{\partial y}(F_y) \quad \text{in the y direction, and}$$

$$F_z + \frac{\partial}{\partial z}(F_z) \quad \text{in the z direction,}$$

(which can be obtained from a Taylor's series approximation as in Appendix I). The net rate of solute inflow is the difference between the inflow and outflow for each component

$$\text{net rate of solute inflow} = F_x - \left(F_x + \frac{\partial}{\partial x}(F_x)\right) + F_y - \left(F_y + \frac{\partial}{\partial y}(F_y)\right) +$$
$$F_z - \left(F_z + \frac{\partial}{\partial z}(F_z)\right)$$
$$= -\frac{\partial}{\partial x}(F_x) - \frac{\partial}{\partial y}(F_y) - \frac{\partial}{\partial z}(F_z) \tag{AIII.2}$$

In porous media, solute transport occurs by three processes: advection, diffusion, and mechanical dispersion.

Advection

The process by which solutes are transported by the bulk motion of the flowing groundwater is called *advection*. The rate of solute transport that occurs by advection is given by the product of the solute concentration C and the components of the apparent groundwater velocity v_x, v_y, and v_z. In terms of the three components of solute transport in the x, y, and z directions, the rate of solute transport by advection is

$$F_x)_{\text{Advection}} = v_x C$$

$$F_y)_{\text{Advection}} = v_y C \tag{AIII.3}$$

$$F_z)_{\text{Advection}} = v_z C$$

Diffusion

The process by which solutes are transported by the random thermal motion of solute molecules is called *diffusion*. The rate of solute transport that occurs by diffusion is given by *Fick's* Law. In terms of the three components of solute transport in the x, y, and z directions, the rate of solute transport by diffusion is given by

$$F_x)_{\text{Diffusion}} = -D^* \frac{\partial C}{\partial x}$$

$$F_y)_{\text{Diffusion}} = -D^* \frac{\partial C}{\partial y} \tag{AIII.4}$$

$$F_z)_{\text{Diffusion}} = -D^* \frac{\partial C}{\partial z}$$

where D^* is the solute's *apparent diffusion coefficient*. The apparent diffusion coefficient for a solute in porous media is much smaller than the diffusion coefficient for the same solute in aqueous solution, D_0. An empirical relationship for D^* can be written

$$D^* = \omega(\theta)D_0 \tag{AIII.5}$$
$$\begin{pmatrix} \text{porous} \\ \text{media} \end{pmatrix} \quad \begin{pmatrix} \text{aqueous} \\ \text{solution} \end{pmatrix}$$

where $\omega(\theta)$ is an empirical *correction factor* that is a function of the volumetric water content. Values of ω typically range from 0.01 for very dry soils to 0.5 for saturated soils. Values of the apparent diffusion coefficients for the major, naturally-occurring constituents of groundwater (eg., Na^+, K^+, Mg^{2+}, Ca^{2+}, Cl^-, HCO_3^-, and SO_4^{2-}) are in the range 1×10^{-8} to 1×10^{-10} m^2/s at 25° C. Apparent diffusion coefficients are strongly temperature dependent (for example, values of the apparent diffusion coefficient are about 50% smaller at 5° C than at 25° C), but are only weakly dependent on the concentrations of other dissolved species.

The small size of apparent diffusion coefficient means that the rate of solute transport by diffusion is usually neglibly small relative to the rates of solute transport by advection and dispersion.

Mechanical Dispersion

Mechanical dispersion (or *hydraulic dispersion*) is a mixing or spreading process caused by small-scale fluctuations in groundwater velocity along the tortuous flow paths within individual pores. On a much larger scale mechanical dispersion can also be caused by the presence of heterogeneities (e.g. clay lenses or faults) within the aquifer. The rate of solute transport by mechanical dispersion is given by a generalized form of Fick's Law of diffusion. In terms of the three components of solute transport in the x, y, and z directions, the rate of solute transport by mechanical dispersion is given by

$$F_x)_{\text{Mechanical Dispersion}} = - D_{xx}\frac{\partial}{\partial x}(\theta C) - D_{xy}\frac{\partial}{\partial y}(\theta C) - D_{xz}\frac{\partial}{\partial z}(\theta C)$$

$$F_y)_{\text{Mechanical Dispersion}} = - D_{yx}\frac{\partial}{\partial x}(\theta C) - D_{yy}\frac{\partial}{\partial y}(\theta C) - D_{yz}\frac{\partial}{\partial z}(\theta C) \qquad \text{(AIII.6)}$$

$$F_z)_{\text{Mechanical Dispersion}} = - D_{zx}\frac{\partial}{\partial x}(\theta C) - D_{zy}\frac{\partial}{\partial y}(\theta C) - D_{zz}\frac{\partial}{\partial z}(\theta C)$$

where D_{xx}, D_{xy}, etc are the *coefficients* of *mechanical dispersion*. These coefficients can be computed from the expression

$$D_{ij} = a_{ij\,km}\frac{\bar{v}_m\,\bar{v}_n}{\sqrt{\bar{v}_m^2 + \bar{v}_n^2}} \qquad \text{(AIII.7)}$$

where the subscripts i and j refer to the three coordinate directions x, y, and z, $\bar{v}_m$ and $\bar{v}_n$ are the components of the *pore water velocity* (as opposed to the apparent groundwater velocity used in Darcy's Law), and the subscripts m and n refer to the directions of the principal components of pore water velocity. Components of the pore water velocity are computed from

$$\bar{v}_x = v_x/\theta$$

$$\bar{v}_y = v_y/\theta \qquad \text{(AIII.8)}$$

$$\bar{v}_z = v_z/\theta$$

where θ is the volumetric water content of the porous media.

The terms a_{ijkm} are the components of the aquifer's *dispersivity*. If the aquifer is assumed to be isotropic with respect to dispersion, all components of the aquifer's dispersivity are zero except for

$$a_{iiii} = a_L$$
$$a_{iijj} = a_T \qquad\qquad\qquad\qquad\qquad\qquad (AIII.9)$$
$$a_{ijij} = a_{ijji} = \frac{1}{2}(a_L - a_T), \ i \neq j$$

where a_L is the *longitudinal dispersivity* and a_T is the *transverse dispersivity* of the aquifer. "Longitudinal" refers to a direction along the flow path and "transverse" refers to a direction at right angles to the flow path. In this case, the coefficients of mechanical dispersion can be computed from the following expressions

$$D_{xx} = [a_T(\bar{v}_y^2 + \bar{v}_z^2) + a_L\bar{v}_x^2] \ / \ |\bar{v}|$$
$$D_{xy} = D_{yx} = [(a_L - a_T)\bar{v}_x\bar{v}_y] \ / \ |\bar{v}|$$
$$D_{xz} = D_{zx} = [(a_L - a_T)\bar{v}_x\bar{v}_z] \ / \ |\bar{v}| \qquad\qquad (AIII.10)$$
$$D_{yy} = [a_T(\bar{v}_x^2 + \bar{v}_z^2) + a_L\bar{v}_y^2] \ / \ |\bar{v}|$$
$$D_{yz} = D_{zy} = [(a_L - a_T)\bar{v}_y\bar{v}_z] \ / \ |\bar{v}|$$
$$D_{zz} = [a_T(\bar{v}_x^2 + \bar{v}_y^2) + a_L\bar{v}_z^2] \ / \ |\bar{v}|$$

where $|\bar{v}| = \sqrt{\bar{v}_x^2 + \bar{v}_y^2 + \bar{v}_z^2}$. In a two–dimensional problem equation AIII.10 becomes

$$D_{xx} = [a_T\bar{v}_y^2 + a_L\bar{v}_x^2] \ / \ |\bar{v}|$$
$$D_{yy} = [a_T\bar{v}_x^2 + a_L\bar{v}_y^2] \ / \ |\bar{v}|$$
$$D_{xy} = D_{yx} = [(a_L - a_T)\bar{v}_x\bar{v}_y] \ / \ |\bar{v}| \qquad\qquad (AIII.11)$$

where $|\bar{v}| = \sqrt{\bar{v}_x^2 + \bar{v}_y^2}$. In a one–dimensional problem, equation AIII.10 becomes

$$D_{xx} = D_x = a_L\bar{v}_x \qquad\qquad\qquad\qquad\qquad (AIII.12)$$

If we have *uniform* flow in the x-direction ($v_x \neq 0$, $v_y = v_z = 0$) in a three-dimensional aquifer, equation AIII.6 simplifies to

$$F_x)_{\text{Mechanical Dispersion}} = -D_x\frac{\partial}{\partial x}(\theta C)$$

$$F_y)_{\text{Mechanical Dispersion}} = -D_y\frac{\partial}{\partial y}(\theta C) \qquad\qquad (AIII.13)$$

$$F_z)_{\text{Mechanical Dispersion}} = -D_z\frac{\partial}{\partial z}(\theta C)$$

where $D_x = a_L v_x$, $D_y = D_z = a_T v_x$. If we substitute equations AIII.13 and AIII.3 into equation AIII.2 and neglect the contribution of diffusion we have

$$\text{net rate of solute inflow} = -\frac{\partial}{\partial x}\left(v_x C - D_x \frac{\partial}{\partial x}(\theta C)\right)$$

$$-\frac{\partial}{\partial y}\left(v_y C^{\,0} - D_y \frac{\partial}{\partial y}(\theta C)\right)$$

$$-\frac{\partial}{\partial z}\left(v_z C^{\,0} - D_z \frac{\partial}{\partial z}(\theta C)\right)$$

or

$$\text{net rate of solute inflow} = -\frac{\partial}{\partial x}(v_x C) + D_x \frac{\partial^2}{\partial x^2}(\theta C) + D_y \frac{\partial^2}{\partial y^2}(\theta C) + D_z \frac{\partial^2}{\partial z^2}(\theta C)$$

$$(\text{AIII.14})$$

Net Rate of Solute Production

Several processes can act as sources or sinks for solute within the control volume including sorption/desorption, chemical or biological reactions, and radioactive decay. Consider the case of transport involving a sorption/desorption reaction

$$A \Leftrightarrow \bar{A}$$

between a dissolved species A and a sorbed species $\bar{A}$. The net rate of reaction r, can be written

$$r = \theta \frac{\partial C}{\partial t} = -\rho_b \frac{\partial \bar{C}}{\partial t} \qquad (\text{AIII.15})$$

where θ and ρ_b, respectively are the porosity and bulk density of the porous media, C is the concentration of the dissolved species A (mass of solute / volume of groundwater), and $\bar{C}$ is the concentration of the sorbed species $\bar{A}$ (mass of solute / mass of dry porous media). Equation AIII.15 can also be written

$$r = -k_f C + k_r \bar{C} \qquad (\text{AIII.16})$$

where k_f is the constant for the *forward reaction* $(A \rightarrow \bar{A})$ and k_r is the rate constant for the *reverse reaction* $(\bar{A} \rightarrow A)$. A rate law of this mathematical form, for example could be used if the sorption process can be described by a first-order, reversible reaction or by a combination of linear diffusion and a linear equilibrium isotherm.

If we assume that the net rate of reaction is zero (i.e., the reaction is in equilibrium), equation AIII.16 can be solved directly for the concentration of the sorbed species $\bar{A}$

$$\bar{C} = \frac{k_f}{k_r} C = K_d C \qquad (\text{AIII.17})$$

where K_d is the *equilibrium distribution coefficient* (L^3/M). The net rate of solute production due to a sorption/desorption reaction between a solute and the porous media within the control volume can be obtained by combining equations AIII.15 and AIII.17 and introducing the volumetric water content of the porous media θ

$$\frac{\partial(\theta C)}{\partial t}\bigg)_{\text{sorption}} = -\rho_b K_d \frac{\partial C}{\partial t} \qquad\qquad\qquad (\text{AIII.18})$$

If the solute also undergoes radioactive decay or biological degradation, the net rate of solute production by this mechanism can be written

$$\frac{\partial(\theta C)}{\partial t}\bigg)_{\text{decay}} = -\lambda\,(\theta C + \rho_b K_d C) \qquad\qquad (\text{AIII.19})$$

where λ is the decay constant for the solute.

Integrating equation AIII.19 gives

$$(\theta C + \rho_b K_d C)_t = (\theta C + \rho_b K_d C)_{t_0} e^{-\lambda t} \qquad\qquad (\text{AIII.20})$$

where the left-hand side is the mass of solute (dissolved and sorbed) in the control volume at some future time t and the first term on the right-hand side is the initial mass of solute in the control volume. We can see that equation AIII.19 applies to processes that display exponential decay. The half-life T for such a process is defined by

$$\frac{(\theta C + \rho_b K_d C)_t}{(\theta C + \rho_b K_d C)_{t_0}} = \frac{1}{2} \qquad \text{at } t = T \qquad\qquad (\text{AIII.21})$$

which gives

$$e^{-\lambda T} = \frac{1}{2} \qquad \text{or} \qquad \lambda = \frac{\ln 2}{T} = \frac{0.693}{T} \qquad\qquad (\text{AIII.22})$$

Solute Transport Equation

If we substitute equations AIII.14, AIII.18, and AIII.19 into equation AIII.1 and write the rate of change of solute mass in the control volume as $\dfrac{\partial(\theta C)}{\partial t}$, we arrive at the *solute transport equation for uniform flow*

$$\boxed{\begin{aligned}
\frac{\partial(\theta C)}{\partial t} &= D_x \frac{\partial^2}{\partial x^2}(\theta C) + D_y \frac{\partial^2}{\partial y^2}(\theta C) + D_z \frac{\partial^2}{\partial z^2}(\theta C) \\
&\quad - \frac{\partial}{\partial x}(v_x C) - \frac{\partial}{\partial t}(\rho_b K_d C) - \lambda(\theta C + \rho_b K_d C)
\end{aligned}}$$

$$(\text{AIII.23})$$

If the porous media is *saturated,* $\theta = n$, and equation AIII.23 can be written

$$\boxed{\begin{aligned}
\frac{\partial C}{\partial t} &= D_x \frac{\partial^2 C}{\partial x^2} + D_y \frac{\partial^2 C}{\partial y^2} + D_z \frac{\partial^2 C}{\partial z^2} - \frac{\partial}{\partial x}\left(\frac{v_x C}{n}\right) \\
&\quad - \frac{\partial}{\partial t}\left(\frac{\rho_b K_d C}{n}\right) - \lambda\left(C + \frac{\rho_b K_d C}{n}\right)
\end{aligned}}$$

$$(\text{AIII.24})$$

Similar equations can be written for uniform groundwater flow in the y or z directions. If we define a *retardation factor*, R to be

$$R = 1 + \frac{\rho_b K_d}{n} \tag{AIII.25}$$

equation AIII.24 can be written

$$R\frac{\partial C}{\partial t} = D_x \frac{\partial^2 C}{\partial x^2} + D_y \frac{\partial^2 C}{\partial y^2} + D_z \frac{\partial^2 C}{\partial z^2} - \frac{\partial}{\partial x}\left(\frac{v_x C}{n}\right) - \lambda RC \tag{AIII.26}$$

If the groundwater flow is not uniform ($v_x \neq 0$, $v_y \neq 0$, $v_z \neq 0$) the rate of solute transport by mechanical dispersion is given by equation AIII.6. The net rate of solute inflow into the control volume becomes

$$\text{net rate of solute inflow} = -\frac{\partial}{\partial x}\left(v_x C - D_{xx}\frac{\partial(\theta C)}{\partial x} - D_{xy}\frac{\partial(\theta C)}{\partial y} - D_{xz}\frac{\partial(\theta C)}{\partial z}\right)$$

$$-\frac{\partial}{\partial y}\left(v_y C - D_{yx}\frac{\partial(\theta C)}{\partial x} - D_{yy}\frac{\partial(\theta C)}{\partial y} - D_{yz}\frac{\partial(\theta C)}{\partial z}\right)$$

$$-\frac{\partial}{\partial z}\left(v_z C - D_{zx}\frac{\partial(\theta C)}{\partial x} - D_{zy}\frac{\partial(\theta C)}{\partial y} - D_{zz}\frac{\partial(\theta C)}{\partial z}\right)$$

or

$$\text{net rate of solute inflow} = -\frac{\partial}{\partial x}(v_x C) - \frac{\partial}{\partial y}(v_y C) - \frac{\partial}{\partial z}(v_z C)$$

$$+ D_{xx}\frac{\partial^2(\theta C)}{\partial x^2} + D_{xy}\frac{\partial^2(\theta C)}{\partial x \partial y} + D_{xz}\frac{\partial^2(\theta C)}{\partial x \partial z}$$

$$+ D_{yx}\frac{\partial^2(\theta C)}{\partial y \partial x} + D_{yy}\frac{\partial^2(\theta C)}{\partial y^2} + D_{yz}\frac{\partial^2(\theta C)}{\partial y \partial z}$$

$$+ D_{zx}\frac{\partial^2(\theta C)}{\partial z \partial x} + D_{zy}\frac{\partial^2(\theta C)}{\partial z \partial y} + D_{zz}\frac{\partial^2(\theta C)}{\partial z^2}$$

$$\tag{AIII.27}$$

If we substitute equations AIII.18, AIII.19, and AIII.27 into equation AIII.1 and write the rate of change of solute mass in the control volume as $\frac{\partial(\theta C)}{\partial t}$, we arrive at the *solute transport equation for nonuniform flow*

$$
\begin{aligned}
\frac{\partial(\theta C)}{\partial t} = {}& D_{xx}\frac{\partial^2(\theta C)}{\partial x^2} + D_{xy}\frac{\partial^2(\theta C)}{\partial x \partial y} + D_{xz}\frac{\partial^2(\theta C)}{\partial x \partial z} \\
& + D_{yx}\frac{\partial^2(\theta C)}{\partial y \partial x} + D_{yy}\frac{\partial^2(\theta C)}{\partial y^2} + D_{yz}\frac{\partial^2(\theta C)}{\partial y \partial z} \\
& + D_{zx}\frac{\partial^2(\theta C)}{\partial z \partial x} + D_{zy}\frac{\partial^2(\theta C)}{\partial z \partial y} + D_{zz}\frac{\partial^2(\theta C)}{\partial z^2} \\
& - \frac{\partial}{\partial x}(v_x C) - \frac{\partial}{\partial y}(v_y C) - \frac{\partial}{\partial z}(v_z C) \\
& - \frac{\partial}{\partial t}(\rho_b K_d C) - \lambda(\theta C + \rho_b K_d C)
\end{aligned}
$$

$$\text{(AIII.28)}$$

If the porous media is *saturated*, $\theta = n$, and equation AIII.28 can be written

$$
\begin{aligned}
\frac{\partial C}{\partial t} = {}& D_{xx}\frac{\partial^2 C}{\partial x^2} + D_{xy}\frac{\partial^2 C}{\partial x \partial y} + D_{xz}\frac{\partial^2 C}{\partial x \partial z} + D_{yx}\frac{\partial^2 C}{\partial y \partial x} + D_{yy}\frac{\partial^2 C}{\partial y^2} + D_{yz}\frac{\partial^2 C}{\partial y \partial z} \\
& + D_{zx}\frac{\partial^2 C}{\partial z \partial x} + D_{zy}\frac{\partial^2 C}{\partial z \partial y} + D_{zz}\frac{\partial^2 C}{\partial z^2} - \frac{\partial}{\partial x}\left(\frac{v_x C}{n}\right) - \frac{\partial}{\partial y}\left(\frac{v_y C}{n}\right) - \frac{\partial}{\partial z}\left(\frac{v_z C}{n}\right) \\
& - \frac{\partial}{\partial t}\left(\frac{\rho_b K_d C}{n}\right) - \lambda\left(C + \frac{\rho_b K_d C}{n}\right)
\end{aligned}
$$

$$\text{(AIII.29)}$$

Equation AIII.29 can also be written using the retardation factor (equation AIII.25)

$$
\begin{aligned}
R\frac{\partial C}{\partial t} = {}& D_{xx}\frac{\partial^2 C}{\partial x^2} + D_{xy}\frac{\partial^2 C}{\partial x \partial y} + D_{xz}\frac{\partial^2 C}{\partial x \partial z} + D_{yx}\frac{\partial^2 C}{\partial y \partial x} + D_{yy}\frac{\partial^2 C}{\partial y^2} \\
& + D_{yz}\frac{\partial^2 C}{\partial y \partial z} + D_{zx}\frac{\partial^2 C}{\partial z \partial x} + D_{zy}\frac{\partial^2 C}{\partial z \partial y} + D_{zz}\frac{\partial^2 C}{\partial z^2} \\
& - \frac{\partial}{\partial x}\left(\frac{v_x C}{n}\right) - \frac{\partial}{\partial y}\left(\frac{v_y C}{n}\right) - \frac{\partial}{\partial z}\left(\frac{v_z C}{n}\right) - \lambda R C
\end{aligned}
$$

$$\text{(AIII.30)}$$

Problems

1. Derive the solute transport equation for problems with axisymmetry.

$$\frac{\partial(\theta C)}{\partial t} = \frac{1}{r}\frac{\partial}{\partial r}\left(D_r\, r\, \frac{\partial}{\partial r}(\theta C)\right) + D_z\frac{\partial^2}{\partial z^2}(\theta C) - \frac{1}{r}\frac{\partial}{\partial r}(r\, v_r C)$$

$$- \frac{\partial}{\partial z}(v_z C) - \frac{\partial}{\partial t}(\rho_b K_d C) - \lambda(\theta C + \rho_b K_d C) \qquad\text{(AIII.31)}$$

2. Rewrite equation AIII.26 if the porous media is saturated

$$\frac{\partial C}{\partial t} = \frac{1}{r}\frac{\partial}{\partial r}\left(D_r\, r\, \frac{\partial C}{\partial r}\right) + D_z\frac{\partial^2 C}{\partial z^2} - \frac{1}{r}\frac{\partial}{\partial r}\left(\frac{r\, v_r C}{n}\right)$$

$$- \frac{\partial}{\partial z}\left(\frac{v_z C}{n}\right) - \frac{\partial}{\partial t}\left(\frac{\rho_b K_d C}{n}\right) - \lambda\left(C + \frac{\rho_b K_d C}{n}\right) \qquad\text{(AIII.32)}$$

3. Using data from a field tracer test, the longitudinal and transverse dispersivity of an aquifer were determined to be 12 m and 1 m, respectively. Compute the coefficients for mechanical dispersion for each element in the mesh shown below

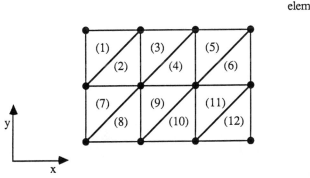

element	v_x	v_y
1	2	-2
2	3	-1
3	3	0
4	3	1
5	2	2
6	2	3
7	3	-1
8	3	0
9	3	0
10	2	1
11	2	1
12	2	2

Appendix IV

CONCEPTS FROM LINEAR ALGEBRA USED IN THE FINITE ELEMENT METHOD

The finite element method involves many operations on systems of equations and these are best handled using matrices. A typical system of linear algebraic equations has the form

$$
\begin{aligned}
a_{11}x_1 + a_{12}x_2 + \cdots + a_{1m}x_m &= f_1 \\
a_{21}x_1 + a_{22}x_2 + \cdots + a_{2m}x_m &= f_2 \\
\vdots \qquad \vdots \qquad\qquad \vdots \qquad \vdots \\
a_{n1}x_1 + a_{n2}x_2 + \cdots + a_{nm}x_m &= f_n
\end{aligned}
\tag{AIV.1}
$$

where each of the n equations contains m unknowns ($x_1, x_2, \ldots, x_m$) and m+1 known coefficients (the $a_{i1}, a_{i2}, \ldots, a_{im}, f_i$, where i is any equation). Equation AIV.1 can also be written in *matrix form* as

$$
\begin{bmatrix}
a_{11} & a_{12} & \cdots & a_{1m} \\
a_{21} & a_{22} & \cdots & a_{2m} \\
\vdots & \vdots & & \vdots \\
a_{n1} & a_{n2} & \cdots & a_{nm}
\end{bmatrix}
\begin{Bmatrix} x_1 \\ x_2 \\ \vdots \\ x_m \end{Bmatrix}
=
\begin{Bmatrix} f_1 \\ f_2 \\ \vdots \\ f_m \end{Bmatrix}
\tag{AIV.2}
$$

where each set of terms enclosed in brackets or braces is a *matrix* (plural matrices). A matrix is simply a rectangular array of numbers. If we use capital letters to denoted each matrix in equation AIV.2 we can rewrite that equation in the form

$$
[A] \{X\} = \{F\}
\tag{AIV.3}
$$

where

$$
[A] =
\begin{bmatrix}
a_{11} & a_{12} & \cdots & a_{1m} \\
a_{21} & a_{22} & \cdots & a_{2m} \\
\vdots & \vdots & & \vdots \\
a_{n1} & a_{n2} & \cdots & a_{nm}
\end{bmatrix}
\quad
\{X\} =
\begin{Bmatrix} x_1 \\ x_2 \\ \vdots \\ x_m \end{Bmatrix}
\quad
\{F\} =
\begin{Bmatrix} f_1 \\ f_2 \\ \vdots \\ f_m \end{Bmatrix}
\tag{AIV.4}
$$

A matrix consists of one or more *rows* of numbers and one or more *columns* of numbers. Thus the matrix A contains n rows and m columns, the matrices $\{X\}$ and $\{F\}$ contain m rows and 1 column. A matrix with 1 row is termed a *row matrix*. A matrix with 1 column is termed a *column matrix* or *vector*. Thus matrices $\{X\}$ and $\{F\}$ are vectors. Some other definitions are

1. The *size* of a matrix is the number of rows and columns the matrix contains. The size is written as two numbers separated by an "x" representing a cartesian product e.g. 3 x 2 where the first number is the number of rows and the second is the number of columns. Some examples are

$$\begin{bmatrix} 3 & 2 & 1 \\ 2 & 4 & 3 \\ 1 & 3 & 5 \end{bmatrix} \qquad \begin{bmatrix} 1 \\ 2 \\ 3 \end{bmatrix} \qquad \begin{bmatrix} 2 & 3 & 2 \\ 1 & 2 & 4 \end{bmatrix}$$

size: 3×3 3×1 2×3

2. A *square* matrix has an equal number of rows and columns ($n = m$). Some examples

$$[1] \qquad \begin{bmatrix} 2 & 1 \\ 1 & 5 \end{bmatrix} \qquad \begin{bmatrix} 3 & 2 & 1 \\ 2 & 4 & 3 \\ 1 & 3 & 5 \end{bmatrix}$$

1×1 2×2 3×3

3. The *main diagonal* of a matrix is the set of positions in the matrix where the row number and column numbers are equal. If we use a_{ij} to designate any number that is in row i and column j then the main diagonal is given by a_{ij} for all $i = j$. Some examples

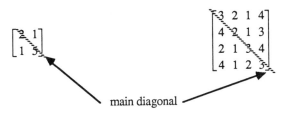

main diagonal

4. In a *symmetric matrix* the numbers in positions on opposites sides of the main diagonal are equal. That is $a_{ij} = a_{ji}$ for all i, j. Some examples

$a_{12} = a_{21}$ $a_{12} = a_{21}$

$a_{13} = a_{31}$

$a_{32} = a_{23}$

5. In a *diagonal matrix*, all positions in the matrix not on the main diagonal are zero. That is $a_{ij} = 0$ for all $i \neq j$. An example

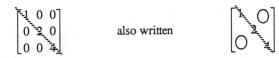

also written

6. The *identity matrix* is a diagonal matrix where $a_{ij} = 1$ for all $i = j$. An example

$$\begin{bmatrix} 1 & 0 & 0 \\ 0 & 1 & 0 \\ 0 & 0 & 1 \end{bmatrix} \qquad \text{also written} \qquad \begin{bmatrix} 1 & & O \\ & 1 & \\ O & & 1 \end{bmatrix}$$

An identity matrix is usually designated by the letter [I] regardless of the size of the matrix.

7. In *upper* and *lower triangular matrices* all positions below and above the main diagonal, respectively are occupied by zeros. Some examples

$$\begin{bmatrix} 1 & 0 & 0 \\ 4 & 2 & 0 \\ 5 & 4 & 3 \end{bmatrix} \qquad \text{also written} \qquad \begin{bmatrix} 1 & & O \\ 4 & 2 & \\ 5 & 4 & 3 \end{bmatrix}$$
lower triangular matrix

$$\begin{bmatrix} 1 & 4 & 5 \\ 0 & 2 & 4 \\ 0 & 0 & 3 \end{bmatrix} \qquad \text{also written} \qquad \begin{bmatrix} 1 & 4 & 5 \\ & 2 & 4 \\ O & & 3 \end{bmatrix}$$
upper triangular matrix

8. The *transpose of a matrix* is a matrix obtained by interchanging numbers using the rule

$$\begin{array}{ccc} a_{ij} & = & a_{ji} \\ \text{transpose} & & \text{original} \\ \text{matrix} & & \text{matrix} \end{array}$$

The superscript "T" is used to designate the transpose of a matrix. Some examples

$$[B] = \begin{bmatrix} 2 & 3 \\ 4 & 5 \end{bmatrix} \qquad\qquad [B]^T = \begin{bmatrix} 2 & 4 \\ 3 & 5 \end{bmatrix}$$

$$[C] = \begin{bmatrix} 1 & 2 & 3 \\ 4 & 5 & 6 \end{bmatrix} \qquad\qquad [C]^T = \begin{bmatrix} 1 & 4 \\ 2 & 5 \\ 3 & 6 \end{bmatrix}$$

9. *Matrix addition* involves the addition of entries in corresponding positions of two matrices to form a new matrix. If $[C] = [A] + [B]$ then $c_{ij} = a_{ij} + b_{ij}$ for all i and j. An example

$$[A] = \begin{bmatrix} 1 & 2 & 3 \\ 4 & 5 & 6 \\ 7 & 8 & 9 \end{bmatrix} \qquad\qquad [B] = \begin{bmatrix} 0 & 1 & 2 \\ 3 & 4 & 5 \\ 6 & 7 & 8 \end{bmatrix}$$

$$[C] = [A] + [B] = \begin{bmatrix} (1+0) & (2+1) & (3+2) \\ (4+3) & (5+4) & (6+5) \\ (7+6) & (8+7) & (9+8) \end{bmatrix} = \begin{bmatrix} 1 & 3 & 5 \\ 7 & 9 & 11 \\ 13 & 15 & 17 \end{bmatrix}$$

Matrix addition is *commutative* (i.e., $[A] + [B] = [B] + [A]$) and *associative* (i.e., $([A] + [B]) + [C] = [A] + ([B] + [C])$).

10. *Matrix subtraction* involves the subtraction of numbers in corresponding positions in two matrices to form a new matrix. If $[C] = [A] - [B]$ then $c_{ij} = a_{ij} - b_{ij}$ for all i and j. An example

$$[A] = \begin{bmatrix} 2 & 1 & 5 \\ 6 & 3 & 4 \end{bmatrix} \qquad [B] = \begin{bmatrix} 0 & 3 & 1 \\ 2 & 1 & 0 \end{bmatrix}$$

$$[C] = [A] - [B] = \begin{bmatrix} (2-0) & (1-3) & (5-1) \\ (6-2) & (3-1) & (4-0) \end{bmatrix} = \begin{bmatrix} 2 & -2 & 4 \\ 4 & 2 & 4 \end{bmatrix}$$

Matrix subtraction is commutative and associative.

11. *Matrix multiplication* of a pair of matrices [A] and [B] to form a new matrix [C] is only defined if the number of columns of [A] is equal to the number of rows of [B]. If the size of [A] is n x m and the the size of [B] is m x r, then multiplication of these two matrices is given by

$$\begin{array}{ccc} [C] & = & [A] \quad [B] \\ n \times r & & n \times m \quad m \times r \end{array}$$

where

$$c_{ij} = \sum_{k=1}^{m} a_{ik} b_{kj} \qquad i = 1 \text{ to } n, \quad j = 1 \text{ to } r$$

The number of rows of [C] is the same as the number of rows of [A] and the number of columns of [C] is the same as the number of columns of [B]. An example

$$[A] = \begin{bmatrix} 1 & 2 \\ 3 & 4 \end{bmatrix} \qquad [B] = \begin{bmatrix} 5 & 6 & 7 \\ 8 & 9 & 0 \end{bmatrix}$$

$$[C] = [A][B] = \begin{bmatrix} (1)(5)+(2)(8) & (1)(6)+(2)(9) & (1)(7)+(2)(0) \\ (3)(5)+(4)(8) & (3)(6)+(4)(9) & (3)(7)+(4)(0) \end{bmatrix}$$

$$2\times3 \quad 2\times2\,2\times3$$

$$= \begin{bmatrix} 21 & 24 & 7 \\ 47 & 54 & 21 \end{bmatrix}$$

Matrix multiplication is not communative (i.e., $[A][B] \neq [B][A]$) but it is associative (i.e., $([A][B])[C] = [A]([B][C])$).

12. The *determinant* of a matrix is a single number that is only defined for square matrices. The determinant has several uses, principly in matrix inversion (to be described next). Although it is possible to calculate the determinant for a square matrix of any size we only need to consider three cases:

a. Let $[A] = [a_{11}]$. Then the determinant of [A] written $|A| = a_{11}$
   1×1

b. Let $[A] = \begin{bmatrix} a_{11} & a_{12} \\ a_{21} & a_{22} \end{bmatrix}$

   2×2

   Then $|A| = a_{11}a_{22} - a_{12}a_{21}$

   Example

   $$[A] = \begin{bmatrix} 2 & 1 \\ 3 & 4 \end{bmatrix}$$

   $$|A| = (2)(4) - (1)(3) = 5$$

c. Let $[A] = \begin{bmatrix} a_{11} & a_{12} & a_{13} \\ a_{21} & a_{22} & a_{23} \\ a_{31} & a_{32} & a_{33} \end{bmatrix}$

   3×3

   Then $|A| = a_{11}(a_{22}a_{33} - a_{32}a_{23}) - a_{12}(a_{21}a_{33} - a_{31}a_{23}) + a_{13}(a_{21}a_{32} - a_{31}a_{22})$

   Example

   $$[A] = \begin{bmatrix} 1 & 2 & 3 \\ 0 & 1 & 0 \\ 3 & 2 & 1 \end{bmatrix}$$

   $$|A| = 1(1-0) - 2(0-0) + 3(0-3) = -8$$

13. The *inverse* of a matrix is a new matrix of the same size as the original matrix. The inverse operation is only defined for square matrices with nonzero determinants. Although it is possible to calculate the inverse of a square matrix of any size (although it is very difficult for large matrices), we only need to consider three cases:

   a. Let $[A] = [a_{11}]$. Then the inverse matrix for $[A]$ written $[A]^{-1}$ is $[A]^{-1} = \begin{bmatrix} \dfrac{1}{a_{11}} \end{bmatrix}$

   b. Let $[A] = \begin{bmatrix} a_{11} & a_{12} \\ a_{21} & a_{22} \end{bmatrix}$.   Then $[A]^{-1} = \dfrac{1}{|A|} \begin{bmatrix} a_{22} & -a_{12} \\ -a_{21} & a_{11} \end{bmatrix}$

   Example

   $$[A] = \begin{bmatrix} 2 & 5 \\ 1 & 3 \end{bmatrix}$$

   $$|A| = (2)(3)-(5)(1) = 1$$

   $$[A]^{-1} = \frac{1}{1}\begin{bmatrix} 3 & -5 \\ -1 & 2 \end{bmatrix} = \begin{bmatrix} 3 & -5 \\ -1 & 2 \end{bmatrix}$$

   c. Let $[A] = \begin{bmatrix} a_{11} & a_{12} & a_{13} \\ a_{21} & a_{22} & a_{23} \\ a_{31} & a_{32} & a_{33} \end{bmatrix}$.   Then $[A]^{-1} = \dfrac{1}{|A|}[B]^T$

The matrix $[B] = \begin{bmatrix} b_{11} & b_{12} & b_{13} \\ b_{21} & b_{22} & b_{23} \\ b_{31} & b_{32} & b_{33} \end{bmatrix}$ is called the *classical adjoint* matrix of the matrix [A].

The entries of [B] are given by

$$b_{11} = a_{22}a_{33} - a_{23}a_{32}$$
$$b_{12} = -a_{21}a_{33} + a_{23}a_{31}$$
$$b_{13} = a_{21}a_{32} - a_{31}a_{22}$$
$$b_{21} = -a_{12}a_{33} + a_{13}a_{32}$$
$$b_{22} = a_{11}a_{33} - a_{13}a_{31}$$
$$b_{23} = -a_{11}a_{32} + a_{12}a_{31}$$
$$b_{31} = a_{12}a_{23} - a_{13}a_{22}$$
$$b_{32} = -a_{11}a_{23} + a_{13}a_{21}$$
$$b_{33} = a_{11}a_{22} - a_{12}a_{21}$$

Example

Let $[A] = \begin{bmatrix} 2 & 3 & -4 \\ 0 & -4 & 2 \\ 1 & -1 & 5 \end{bmatrix}$

Then $|A| = -46$ and

$$b_{11} = (-4)(5) - (2)(-1) = -18$$
$$b_{12} = -(0)(5) + (2)(1) = 2$$
$$b_{13} = (0)(-1) - (-4)(1) = 4$$
$$b_{21} = -(3)(5) + (-4)(-1) = -11$$
$$b_{22} = (2)(5) - (-4)(1) = 14$$
$$b_{23} = -(2)(-1) + (3)(1) = 5$$
$$b_{31} = (3)(2) - (-4)(-4) = -10$$
$$b_{32} = -(2)(2) + (-4)(0) = -4$$
$$b_{33} = (2)(-4) - (0)(3) = -8$$

and $[A]^{-1} = \begin{bmatrix} 18/46 & 11/46 & 10/46 \\ -2/46 & -14/46 & 4/46 \\ -4/46 & -5/46 & 8/46 \end{bmatrix}$

Problems

1. Given the following matrices,

$$[A] = \begin{bmatrix} 2 & 3 \\ 6 & 4 \end{bmatrix} \qquad\qquad [B] = \begin{bmatrix} 3 & 1 \\ 1 & 3 \end{bmatrix}$$

find:

a. $[A]^T, [B]^T$ d. $[A][B], [B][A]$

b. $[A] + [B], [B] + [A]$ e. $|A|, |B|$

c. $[A] - [B], [B] - [A]$ f. $[A^{-1}], [B^{-1}]$

2. Given the following matrices,

$$[A] = \begin{bmatrix} 3 & 2 & 1 \\ 2 & 4 & 2 \\ 1 & 2 & 3 \end{bmatrix} \qquad\qquad [B] = \begin{bmatrix} 2 & 1 & 1 \\ 1 & 2 & 1 \\ 1 & 1 & 2 \end{bmatrix}$$

find:

a. $[A]^T, [B]^T$ d. $[A][B], [B][A]$

b. $[A] + [B], [B] + [A]$ e. $|A|, |B|$

c. $[A] - [B], [B] - [A]$ f. $[A^{-1}], [B^{-1}]$

3. Given the following matrices,

$$[B] = \begin{bmatrix} 3 & 2 & 1 & 3 \\ 1 & 2 & 1 & 2 \end{bmatrix} \qquad [K] = \begin{bmatrix} 2 & 0 \\ 0 & 3 \end{bmatrix} \qquad [J] = \begin{bmatrix} 2 & 1 \\ 1 & 2 \end{bmatrix}$$

find:

a. $[B]^T[K]$ d. $[B]^T[J^{-1}]^T [K][J^{-1}][B]$

b. $[K][B]$ e. Let $[C] = [B]^T [K][B]$, find $[C^{-1}]$ and $|C|$

c. $[J^{-1}][K][B]$

Appendix V

PROPERTIES OF SELECTED AQUIFER MATERIALS

Table AV.1   Physical properties of selected aquifer materials.

Material	Hydraulic Conductivity, K (m/s)	Specific Storage, S_s* (m^{-1})	Porosity, n	Bulk Density, ρ_b (kg/m^3)
Gravel	10^0 - 10^{-3}	0.1 - 0.3	0.20 - 0.40	1200 - 1800
Sand	10^{-2} - 10^{-6}	0.1 - 0.4	0.25 - 0.55	1300 - 1900
Silt	10^{-3} - 10^{-7}	0.2 - 0.4	0.35 - 0.60	1200 - 1800
Clay	10^{-7} - 10^{-10}	0.05 - 0.2	0.35 - 0.55	1000 - 1600
Sandstone	10^{-6} - 10^{-10}	0.01 - 0.2	0.25 - 0.50	2000 - 2400
Siltstone	10^{-8} - 10^{-12}	0.01 - 0.2	0.20 - 0.40	2000 - 2400
Shale	10^{-9} - 10^{-13}	0.01 - 0.08	0.01 - 0.10	2000 - 2400
Limestone				
(No solution cavities)	10^{-6} - 10^{-10}	0.01 - 0.05	0.01 - 0.20	2000 - 2500
(solution cavities)	10^{-2} - 10^{-6}	0.01 - 0.20	0 05 - 0.55	1800 - 2000
Igneous & Metamorphic				
(fractured)	10^{-4} - 10^{-8}	0.01 - 0.05	0.05 - 0.15	2000 - 2500
(unfractured)	10^{-10}- 10^{-14}	~ 0	0.01 - 0.05	2400 - 3000
Basalt				
(fractured)	10^{-2} - 10^{-7}	0.01 - 0.20	0.05 - 0.35	2000 - 2400
(unfractured)	10^{-10}- 10^{-14}	~ 0	0.01 - 0.10	2400 - 2800
Tuff/Breccia	10^{-5} - 10^{-9}	0.01 - 0.05	0.05 - 0.25	2000 - 2400

* These values are for unconfined aquifers (see Chapter 20). Values for confined aquifers will be 100 to 1000 times smaller.

Table AV.2   Aquifer dispersivities after Anderson(1979).

Material	Porosity, n	a_L (m)	a_T/a_L (m)
Alluvium	0.40	61	0.3
	0.40	61	0.01
	0.30	30.5	1.0
	-	30.5	1.0
	-	15	0.067
	0.20	12	0.33
	0.20	3.05	0.3
Glacial Deposits	0.35	21.3	0.2
Limestone	0.35	61	0.3
	0.25	6.7	0.1
Fractured Basalt	0.10	91	1.5
	0.10	91	1.0
	-	30.5	0.6

REFERENCES

Ames, W. F. 1977. Numerical methods for partial differential equations, 2nd Ed. Academic Press. New York.

Anderson, M. P. 1979. Using models to simulate the movement of contaminants through groundwater flow systems. Critical Reviews in Environmental Control, Vol. 9, No. 2, pp.97 - 156.

Ashcroft, G., D. D. Marsh, D. D. Evans, and L. Boersma. 1962. Numerical method for solving the diffusion equation : 1. Horizontal flow in semi-infinite media soil. Science Society of America Proceedings, Vol. 26, pp.522 - 525.

Bachmat, Y., B. Andrews, D. Holtz, S. Sebastian. 1978. Utilization of Numerical Groundwater Models for Water Resource Management. Report No. EPA -600/8-78-012, Robert S. Kerr Environmental Research Laboratory, Office of Research and Development, U.S. Environmental Protection Agency, Ada, OK 74820.

Bear, J. 1972. Dynamics of Fluids in Porous Media, Elsevier, New York.

Bear, J. 1979. Hydraulics of Groundwater, McGraw-Hill, New York.

Bear, J. and A. Verruijt. 1987. Modeling Groundwater Flow and Pollution, D. Reidel Publishing Company, Dordrecht.

Bennett, G. D. 1978. Introduction to Groundwater Hydraulics : A Programmed Text for Self-Instruction. Book 3, Techniques of Water-Resources Investigations of the United States Geological Survey, U.S. Government Printing Office.

Bruce, G. H., D. W. Peaceman, and H. H. Rachford, Jr. 1953. Calculation of unsteady-state gas flow through porous media. Petrol. Trans. AIME 198. pp.74 - 92.

Brutsaert, W. 1973. Numerical solution of multiphase well flow. Journal of the Hydraulic Division, American Society of Civil Engineers, Vol. 99, No. HY1, pp.1981 - 2001.

Concus, P. 1967. Numerical solution of the nonlinear magnetostatic field equation in two-dimensions. Journal of Computational Physics, Vol. 1, pp.330 - 342.

Cook, R.D. 1981. Concepts and applications of finite element analysis, 2nd ed. John Wiley & Sons, New York.

Davis, L. A. and S. P. Neumann. 1983. Documentation and user's guide UNSAT2 - Variably saturated flow model. Final Report NUREG/CR-3390, U.S. Nuclear Regulatory Commission, Division of Waste Management, Washington, D.C.

de Marsily, G. 1986. Quantitative Hydrogeology, Academic Press, New York.

de Wiest, R.J.M. 1969. Flow Through Porous Media, Academic Press, New York.

Dhatt, G. and G. Touzot 1984. The Finite Element Method Displayed, John Wiley & Sons, New York.

Ergatoudis, I., B. M. Irons, and O. C. Zienkiewicz. 1968. Curved, isoparametic "quadrilateral" elements for finite element analysis. International Journal Solids Structures, Vol. 4, pp.31 - 42.

Everstine, G. C. 1979. A comparison of three resequencing algorithms for the reduction of matrix, profile and wavefront. International Journal for Numerical Methods in Engineering, Vol. 14, No. 6, pp.837 - 853.

Fleck, W. B. and M .G. McDonald. 1978. Three-dimensional finite-difference model of ground-water system underlying the Muskegon County wastewater disposal system, Michigan. U.S. Geological Survey Journal of Research, Vol. 6, No. 3, pp.307 - 318.

Freeze, R. A. 1971. Three-dimensional, transient saturated-unsaturated flow in a groundwater basin. Water Resources Research, Vol. 7, No. 2, pp.347 - 366.

Freeze, R. A. and J. A. Cherry. 1979. Groundwater, Prentice-Hall, Englewood Cliffs, New Jersey.

Freeze, R. A. and P. A. Witherspoon. 1966. Theoretical analysis of regional groundwater flow; 1. Analytical and numerical solutions to the mathematical model. Water Resources Research, Vol. 2, pp.641 - 656.

Fried, I. 1979. Foundations of Solid Mechanics, Academic Press, New York.

Gambolati, G., P. Gatto, and R. A. Freeze. 1973. Mathematical simulation of the subsidence of Venice, 2, results. Water Resources Research, Vol. 9, pp.721 - 733.

Gardner, W. R. and M. S. Mayhugh. 1958. Solutions and tests of the diffusion equation for the movement of water in soil. Soil Science Society of America Proceedings, Vol. 22, pp.197 - 201.

Gupta, S. K. and K. K. Tanki. 1976. A three-dimensional Galerkin finite element solution of flow through multiaquifers in Sutter Basin, California. Water Resources Research, Vol. 12, No. 2, pp.155 - 162.

Gureghian, A. B., D. S. Ward, and R. W. Cleary. 1979. Simultaneous transport of water and reacting solutes through multilayered soils under transient unsaturated flow conditions. Journal of Hydrology, Vol. 41, pp.253 - 278.

Guymon, G. L., V. H. Scott, and L. R. Hermann. 1970. A general numerical solution of the two-dimensional diffusion - convection equation by finite element method. Water Resources Research, Vol. 6, No. 6, pp.1611 - 1617.

Hillel, D. 1980. Fundamentals of Soil Physics, Academic Press, New York.

Huyakorn, P. S. and G. F. Pinder. 1983. Computational Methods in Subsurface Flow, Academic Press, New York.

Irons, B. M. 1970. A frontal solution program for finite element analysis. International Journal for Numerical Methods in Engineering, Vol. 2, No. 1, pp.5 - 32.

Javandel, I. and P. A. Witherspoon. 1968. Application of the finite element method to transient flow in porous media. Society of Petroleum Engineers Journal, Vol. 8, No. 3, pp.241 - 252.

Javandel, I., C. Doughty, and C. F. Tsang. 1984. Groundwater Transport: Handbook of Mathematical Models, Water Resources Monograph 10, American Geophysical Union, Washington, D.C.

Jensen, H. G. and G. A. Parks. 1970. Efficient solutions for linear matrix equations. Journal of the structural Division, Proceedings of American Society of Civil Engineers, Vol. 96, No. ST1, pp.49 - 64.

Kirkner, D. J., T. L. Theis, and A. A. Jennings. 1984. Multicomponent solute transport with sorption and soluble complexation. Advances in Water Resources, Vol. 7, pp.120 - 125.

Klute, A., F. D. Whisler, and E. J. Scott. 1965. Numerical solution of the nonlinear diffusion equation for water flow in a horizontal soil column of finite length. Soil Science Society of America Proceedings, Vol. 29, pp.353 - 358.

Konikow, L. F. and J. D. Bredehoeft. 1978. Computer model of two-dimensional solute transport and dispersion in groundwater. U.S. Geological Investigation, Book 7.

Lapidus, L. and G. F. Pinder. 1982. Numerical Solution of Partial Differential Equations in Science and Engineering, John Wiley & Sons, New York.

Maadooliat, R. 1983. Element and Time Step Criteria for Solving Time - Dependent Field Problems Using the Finite Element Method, Ph.D Thesis, Michigan State Unversity.

Matanga, G. B. and E. D. Frind. 1981. An evaluation of mathematical models for mass transport in saturated - unsaturated porous media. Final Report, Department of Earth Science, University of Waterloo, Ontario, Canada.

Menke, W. 1984. Geophysical data analysis: discrete inverse theory, Academic Press, New York.

Meyer, C. 1973. Solution of linear equations - state of the art. Journal of the Structural Division, Proceedings of American Society of Civil Engineers, Vol. 99, No. ST7, pp.1507 - 1526.

Mualem, Y. 1976. A new model for predicting the hydraulic conductivity of unsaturated porous media. Water Resources Research, Vol. 12, No. 3, pp.513 - 522.

Myers, G. E. 1971. Analytical Methods in Conduction Heat Transfer, McGraw-Hill, New York.

Mitchell, A. R. and D .F. Griffiths. 1980. The finite difference method in partial differential equations, Wiley, New York.

Neuman, S. P. 1973. Saturated-unsaturated seepage by finite elements. Proceedings American Society of Civil Engineers, Vol. 99, No. HY12, pp.2233 - 2250.

Neuman, S. P. 1980. A statistical approach to the inverse problem in hydrology. 3. Improved solution method and added perspective. Water Resources Research, Vol. 16, No. 4, pp.845 - 860

Neuman, S. P. and S. Yakowitz. 1979. A statistical approach to the inverse problem in hydrology. 1. Theory. Water Resources Research, Vol. 15, No.4, pp.845 - 860.

Neumann, S. P. and P. A. Witherspoon. 1970. Finite-element method of analyzing steady seepage with a free surface. Water Resources Research, Vol. 6, No. 3, pp.889 - 897.

Ogata, A. 1970. Theory of dispersion in a granular medium. U.S. Geological Survey Professional Paper 411-I, pp.134.

Orlob, G. T. and P. C. Woods. 1967. Water-quality management in irrigation systems. Journal of the Irrigation and Drainage Division, American society of Civil Engineers, Vol. 93, pp.49 - 66.

Oster, C. A. 1982. Review of Ground-Water Flow and Transport Models in the Unsaturated Zone, Report No. NUREG/CR-2917, PNL-4427, Pacific Northwest Laboratory, P.O. Box 999, Richland, WA 99352.

Oster, C. A., J. C. Sonnichsen, and P. T. Jaske. 1970. Numerical solution of the convective diffusion equation. Water Resources Research, Vol. 6, pp.1746 - 1752.

Peaceman, D. W. and H. H. Rachford, Fr. 1962. Numerical calculation of multidimensional miscible displacement. Society of Petroleum Engineers Journal, Vol. 2, pp.327 - 339.

Philip, J. R. 1957. The theory of infiltration : 1. The infiltration equation and its solution. Soil Science, Vol. 83, pp.345 - 357.

Pickens, J. F. and R. W. Gillham. 1980. Finite element analysis of solute transport under hysteretic unsaturated flow conditions. Water Resources Research, Vol. 16, No. 6, pp.1071 - 1078.

Pickens, J. F., R. W. Gillham, and D. R. Cameron. 1979. Finite-element analysis of the transport of water and sloutes in tile-drained soils. Journal of Hydrology, Vol. 40, pp.243 - 264.

Pinder, G. F. 1973. A Galerkin finite element simulation of groundwater contamination in Long Island, New York. Water Resources Research, Vol. 9, No. 6, pp.1657 - 1669.

Pinder, G. F. and J. D. Bredehoeft. 1968. Application of the digital computer for aquifer evaluation. Water Resources Research, Vol. 4, pp.1069 - 1093.

Pinder, G. F. and E. O. Frind. 1972. Application of Galerkin's procedure to aquifer analysis. Water Resources Research, Vol. 8, No. 1, pp.108 - 120.

Pinder, G. F. and W. G. Gray. 1977. Finite Element Simulation in Surface and Subsurface Hydrology, Academic Press, New York.

Price, H. S., J. C. Cavendish, and R. A. Varga. 1968. Numerical methods of higher order accuracy for diffusion convection equations. Society of Petroleum Engineers Journal, pp.293 - 303.

Pricket, T. A. 1975. Modeling techniques for groundwater evaluation. In : Advances in Hydroscience, Vol. 10, Academic Press, New York, pp.1 - 143.

Raats, P, A. C. and W. R. Gardner, 1971, A comparison of some empirical relationships between pressure head and hydraulic conductivity, and some observations on radially symmetric flow. Water Resources Research, Vol. 7, pp.921 - 928.

Rao, S.S. 1982. The Finite Element Method in Engineering, Pergamon Press, New York.

Reeves, M. and J. O. Duguid. 1975. Water movement through saturated - unsaturated porous media : A finite-element Galerkin model. ORNL-4927, Oak Ridge National Laboratory, Oak Ridge, Tennessee 37830.

Remson, I., C. A. Appel, and R. A. Webster. 1965. Groundwater models solved by digital computer. Journal of the Hydraulics Division, American Society of Civil Engineers, Vol. 91, No. HY3, pp.133 - 147.

Remson, I., G. M. Hornberger, and F. J. Molz. 1971. Numerical Methods in Subsurface Hydrology with an Introduction to the Finite Element Method, John Wiley & Sons, New York.

Senger, R. K. and G. E.Fogg. 1987. Regional under pressuring in deep brine aquifers, Palo Dura Basin, Texas I, Effects of hydrostratigraphy and topography. Water Resources Research, Vol. 23, No. 8, pp.1481 - 1493.

Segerlind, L. J. 1984. Applied Finite Element Analysis, 2nd Ed., John Wiley & Sons, New York.

Segol, G., G. F. Pinder, and W. G. Gray. 1975. A Galerkin - finite element technique for calculating the transition position of the salt-water front. Water Resources Research, Vol. 11, No. 2, pp.343 - 347.

Stone, H. L. and P. T. L. Brian. 1963. Numerical solution of convective transport problems. AICHE J. Vol. 9, pp.681 - 688.

Swarzendruber, D. 1962. Non Darcy behavior in liquid saturated porous media. Journal of Geophysical Research, Vol. 67, pp.5205 - 5213.

Tanji, K. K., G. R. Dutt, J. L. Paul, and L. D. Doneen. 1967. A computer method for predicting salt concentrations in soils at variable moisture contents. Hilgardia, Vol. 38, pp.307 - 318.

Trescott, P. C. and S. P. Larson. 1977. Comparison of iterative methods for solving two-dimensional groundwater flow equations. Water Resources Research, Vol. 13, No. 1, pp.125 - 136.

Trescott, P. C., G. F. Pinder, and S. P. Larson. 1976. Finite difference model for aquifer simulation in two dimensions with results of numerical experiments. U.S. Geological Survey, Techniques of Water Resources Investigation, Book 7, Automated Data Processing and computations, pp.116.

Van Genuchten, M.Th., G. F. Pinder, and E. O. Frind. 1977. Simulation of two-dimensional contaminant transport with isoparametric - Hermitian finite elements. Water Resources Research, Vol. 13, No. 2, pp.451 - 458.

Wang, H. F. and M. P. Anderson. 1982. Introduction to Groundwater Modeling : Finite Difference and Finite Element Methods, W.H. Freeman and Company, San Francisco.

Wierenga, P. J. 1977. Solute distribution profiles computed with steady state and transient water movement models. Soil Science Society of America Journal, Vol. 41, pp.1050 - 1055.

Yeh, W. W., Y. S. Yoon, and K. S. Lee. 1983. Aquifer parameter identification with Kriging and optimum parameterization. Water Resources Research, Vol. 19, No. 1, pp.225 - 233.

Zienkiewicz, O. C. 1971. The finite element method in engineering science, McGraw-Hill, London.

Zienkiewicz, O. C., P. Meyer, and Y. K. Cheung. 1966. Solution of anisotropic seepage problems by finite elements. Proceedings American Society of Civil Engineers, Vol. 92 EMI, pp.111 - 120.

Zienkiewicz, O. C. and C. J. Parekh. 1970. Transient field problems : Two-dimensional and three-dimensional analysis by isoparametric finite elements. International Journal of Numerical Methods in Engineering, Vol. 2, pp.61 - 71.

INDEX